Unconventional Thin Film Photovoltaics

RSC Energy and Environment Series

Titles in the Series:

1: Thermochemical Conversion of Biomass to Liquid Fuels and Chemicals
2: Innovations in Fuel Cell Technologies
3: Energy Crops
4: Chemical and Biochemical Catalysis for Next Generation Biofuels
5: Molecular Solar Fuels
6: Catalysts for Alcohol-Fuelled Direct Oxidation Fuel Cells
7: Solid Oxide Fuel Cells: From Materials to System Modeling
8: Solar Energy Conversion: Dynamics of Interfacial Electron and Excitation Transfer
9: Photoelectrochemical Water Splitting: Materials, Processes and Architectures
10: Biological Conversion of Biomass for Fuels and Chemicals: Explorations from Natural Utilization Systems
11: Advanced Concepts in Photovoltaics
12: Materials Challenges: Inorganic Photovoltaic Solar Energy
13: Catalytic Hydrogenation for Biomass Valorization
14: Photocatalysis: Fundamentals and Perspectives
15: Photocatalysis: Applications
16: Unconventional Thin Film Photovoltaics

How to obtain future titles on publication:
A standing order plan is available for this series. A standing order will bring delivery of each new volume immediately on publication.

For further information please contact:
Book Sales Department, Royal Society of Chemistry, Thomas Graham House, Science Park, Milton Road, Cambridge, CB4 0WF, UK
Telephone: +44 (0)1223 420066, Fax: +44 (0)1223 420247
Email: booksales@rsc.org
Visit our website at www.rsc.org/books

Unconventional Thin Film Photovoltaics

Edited by

Enrico Da Como
University of Bath, UK
Email: e.da.como@bath.ac.uk

Filippo De Angelis
CNR Institute of Molecular Sciences and Technologies, Perugia, Italy
Email: filippo.d.angelis@gmail.com

Henry Snaith
University of Oxford, UK
Email: h.snaith1@physics.ox.ac.uk

Alison Walker
University of Bath, UK
Email: a.b.walker@bath.ac.uk

ROYAL SOCIETY OF CHEMISTRY

THE QUEEN'S AWARDS FOR ENTERPRISE: INTERNATIONAL TRADE 2013

RSC Energy and Environment Series No. 16

Print ISBN: 978-1-78262-293-2
PDF eISBN: 978-1-78262-406-6
EPUB eISBN: 978-1-78262-824-8
ISSN: 2044-0774

A catalogue record for this book is available from the British Library

Published by The Royal Society of Chemistry,
Thomas Graham House, Science Park, Milton Road,
Cambridge CB4 0WF, UK

Registered Charity Number 207890

For further information see our web site at www.rsc.org

Printed in the United Kingdom by CPI Group (UK) Ltd, Croydon, CR0 4YY, UK

Preface

By the end of 2014, almost 50% of the world's photovoltaic (PV) capacity of 178 GW_p was installed in Europe, with Germany alone counting for over 38 GW_p. The exponential growth of PV has continued in spite of the world-wide economic recession. China is now catching up rapidly and even the UK, which has traditionally taken a conservative view of PV, has seen an astonishing burst of activity, with the installed capacity exceeding 8GW_p by the second quarter of 2015. The vast majority of PV installations is based on highly developed first generation silicon technology. With efficiencies of over 20%, the best modern commercial silicon panels achieve around two thirds of the theoretical limit, setting the target that must be matched by any alternative PV technology. However, silicon has the drawback that its production is highly energy demanding, which also impacts on the carbon footprint of the technology. An important factor in this respect is the thickness of the silicon layer (and hence the amount of silicon per unit area of solar cell) needed to absorb sunlight efficiently. Because silicon is an indirect bandgap semiconductor, it absorbs red light rather weakly, and therefore, the typical thickness of the silicon absorber layer is of the order of 200 microns. The search for alternatives to silicon has generated second generation technologies based on thin film solar cells that utilise semiconductors such as cadmium telluride (CdTe) or copper indium gallium selenide (CIGS). These are direct band gap semiconductors that absorb light much more efficiently than silicon, so that the absorber layers only need to be a few microns thick. All of these technologies could be called 'conventional' in the sense that they have all progressed from lab bench to manufacture. By contrast, this book is concerned with newer types of photovoltaic cells that utilise what may loosely be called 'soft' materials, either organic compounds such as polymers and larger single molecules or the interesting class of hybrid perovskite materials such as

RSC Energy and Environment Series No. 16
Unconventional Thin Film Photovoltaics
Edited by Enrico Da Como, Filippo De Angelis, Henry Snaith and Alison Walker

Published by the Royal Society of Chemistry, www.rsc.org

methylammonium lead tri-iodide ($MAPbI_3$). As the title of this book indicates, these solar cells can be considered to be unconventional in so far as they are based on emerging types of materials that differ from conventional inorganic semiconductors. The field of unconventional PV is developing very rapidly, and the editors are to be congratulated for having managed to bring together some of the best known specialists in the fields of organic and hybrid perovskite photovoltaics to give an account of the current state of the art, both from an experimental and from a theoretical standpoint. These areas of research continue to attract some of our brightest minds, and the breadth of activity illustrated by this book underlines the huge importance of PV research in the context of the challenges that mankind faces as a consequence of its thirst for energy.

The chapters in this book are intended to present a balanced combination of in-depth explorations of the experimental and theoretical aspects of the subject, allowing the reader to gain a deeper understanding of the two main types of unconventional photovoltaic cells: hybrid organolead halide perovskite cells and organic photovoltaics (OPV). The list of authors may appear rather 'eurocentric', but this reflects in part the fact that recent research in this area has been particularly intense in Europe, although major contributions have also been made by groups in Asia and USA. In particular, in the case of hybrid perovskites, pioneering studies on their use as sensitizers in mesoporous solar cells have been carried out in Tokyo and Seoul (Tsutomu Miyasaka and Nam-Gyu Park, respectively). The subsequent development of solid state hybrid perovskite solar cells (both mesoporous and planar) then took place essentially simultaneously in labs in Lausanne (Michael Grätzel) and Oxford (Henry Snaith).

The first part of the book provides an excellent overview of current activity in the area of hybrid perovskites. Many aspects of the behaviour of these fascinating materials are still controversial, and this is reflected in the variety of different explanations suggested for the hysteretic behaviour and apparent ferroelectric properties, to give just two examples. The jury is still out on the validity of these explanations, but the ideas presented in this volume certainly represent a useful step forward. The importance of solution processing methods and nanostructuring for achieving high performance in hybrid perovskite cells will be evident from chapter 1 (Park) and chapter 2 (DoCampo *et al.*), whereas chapter 3 (Bisquert *et al.*) and chapter 4 (Petrozza *et al.*) focus on characterization methods and on the photophysical processes that determine cell performance. Chapters 6 (Yan *et al.*), 7 (Even *et al.*), 8 (De Angelis *et al.*) and 9 (Walker and Richardson) illustrate the wide range as well as the extraordinary sophistication of current theoretical work on hybrid perovskites. The theme of nanostructuring contacts, in this case for OPV devices, is addressed in chapter 5 (Schmidt-Mende and Dorman), and the OPV theme is the focus of the chapters in the latter part of the book, which deal with different kinds of theoretical approaches to understand the physics of OPV. Leo *et al.* give an authoritative review of small molecule OPV in chapter 10 and, in chapter 11, Beljonne *et al.* provide an excellent status assessment and identify the key issues facing OPV. In chapter 12, Zannoni and Roscioni describe molecular dynamics simulations and show how they can be applied to thin films, whereas Koster *et al.* address the problem of 3-D simulations of organic systems in chapter 13. The final chapter by Michels and Schaefer shows how modelling can be used to unravel the complex processes that take place during solution processing of OPV devices based on bulk heterojunctions, where the energetics and kinetics of phase separation are key issues.

I would like to thank the authors who have contributed to this book, and I am also grateful to the four editors who have guided the process from the outset. All have taken time from their busy schedules to put together a book that I hope will be an important landmark that will help propel unconventional PV from the laboratory to the market place.

Laurie Peter
Bath

Contents

RSC Energy and Environment Series No. 16
Unconventional Thin Film Photovoltaics
Edited by Enrico Da Como, Filippo De Angelis, Henry Snaith and Alison Walker

Published by the Royal Society of Chemistry, www.rsc.org

CHAPTER 1

High Efficiency Mesoscopic Organometal Halide Perovskite Solar Cells

NAM-GYU PARK

School of Chemical Engineering and Department of Energy Science, Sungkyunkwan University, Suwon 440-746, Republic of Korea
Email: naprk@skku.edu

1.1 Introduction

1.1.1 Emergence and Progress of Perovskite Solar Cells

Halide perovskites with ABX_3 formula (A = monovalent cation, B = divalent cation, X = halogen) form corner sharing (B–X) octahedra and cubo-octahedra (A–X). The recently studied methylammonium lead (or tin) halide perovskite materials for photovoltaics were originally discovered in 1978.[1,2] Colorless methylammonium lead chloride was observed to turn orange and black upon replacing the chloride anion with bromide and iodide, respectively, which is indicative of a decrease in the band gap energy. Despite the recent use of methylammonium lead iodide as light absorber in photovoltaic solar cells, little attention was initially paid to such a possibility, probably because the material has other interesting electrical properties (including superconductivity), which was discovered for layered organic–inorganic perovskite materials in 1994.[3] In 2009, Miyasaka *et al.* were the first to report that $CH_3NH_3PbX_3$ (X = Br and I) perovskite materials showed light harvesting properties when they were used as sensitizers in a liquid electrolyte

RSC Energy and Environment Series No. 16
Unconventional Thin Film Photovoltaics
Edited by Enrico Da Como, Filippo De Angelis, Henry Snaith and Alison Walker

Published by the Royal Society of Chemistry, www.rsc.org

based dye-sensitized solar cell.[4] In this cell, the bromide perovskite was deposited on nanocrystalline TiO_2 and displayed a power conversion efficiency (PCE) of 3.1%, with its iodide analogue showing a slightly higher PCE of 3.8%. However, this work was not cited for two years, probably due to problems in reproducing the recipe. In 2011, Park *et al.*[5] reported a higher PCE of 6.5% using $CH_3NH_3PbI_3$ and optimizing the coating solution concentration, TiO_2 film thickness and electrolyte formulation. As can be seen in Figure 1.1, $CH_3NH_3PbI_3$ nano-dots deposited sparsely on a nanocrystalline TiO_2 surface show a one order of magnitude higher absorption

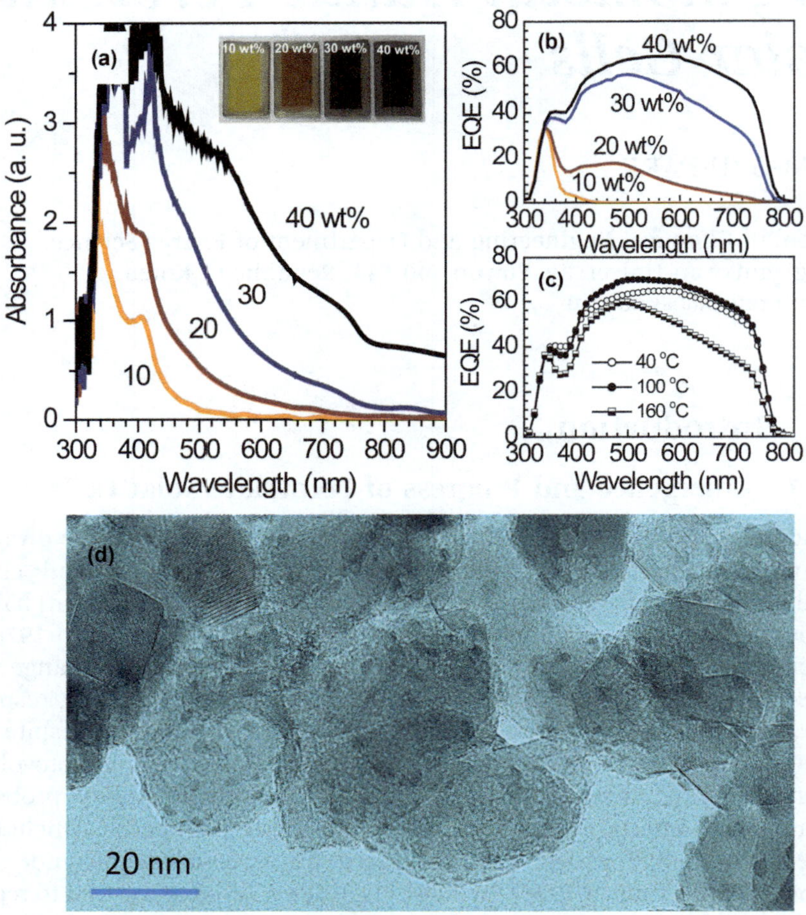

Figure 1.1 (a) Absorbance and (b) external quantum efficiency (EQE) spectra of $CH_3NH_3PbI_3$ depending on the coating solution concentration. (c) EQE spectra depending on the annealing temperature of the deposited $CH_3NH_3PbI_3$ on TiO_2 surface. (d) Scanning electron microscopy (SEM) image of the spin-coated and 100 °C-annealed $CH_3NH_3PbI_3$ showing nano-dot morphology.
Reproduced from ref. 5 with permission from the Royal Society of Chemistry.

coefficient than the ruthenium-based N719 dye fully covered nanocrystalline TiO_2 surface. Although this work demonstrated the potential of organic lead halide perovskites as photovoltaic materials, $CH_3NH_3PbX_3$ could not be considered as a promising sensitizer in liquid based dye-sensitized solar cells because of its rapid dissolution in polar electrolyte solutions (Figure 1.2).

In 2012, Park *et al.*[6] solved the instability problem of perovskites in polar liquid electrolytes by replacing the liquid with a solid hole-transporting material, leading to the first demonstration of a long-term stable all-solid-state mesoscopic perovskite solar cell with an efficiency of 9.7%. Figure 1.3a shows cross-sectional SEM images of a solid-state perovskite solar cell employing a nano-dot perovskite deposited on a TiO_2 surface. Since the perovskite nano-dots are isolated from each other, electron injection from the perovskite nano-dots to TiO_2 occurs even without encapsulation. This structure is quite stable up to 500 h, as can be seen in Figure 1.3b. This is due to the fact that hydrophobic spiro-OMeTAD protects the perovskite from moisture as it is coating completely the perovskite nano-dots.

Following these three important pioneering works,[4–6] there has been a huge surge of interest in perovskite solar cells. Based on Web of Science data, around 500 peer-reviewed research papers on perovskite solar cells were published in 2014, and the rate of publications continues to increase. As can be seen in Figure 1.4, the PCE of perovskite solar cells has increased rapidly as a consequence of improvements on the perovskite film quality and

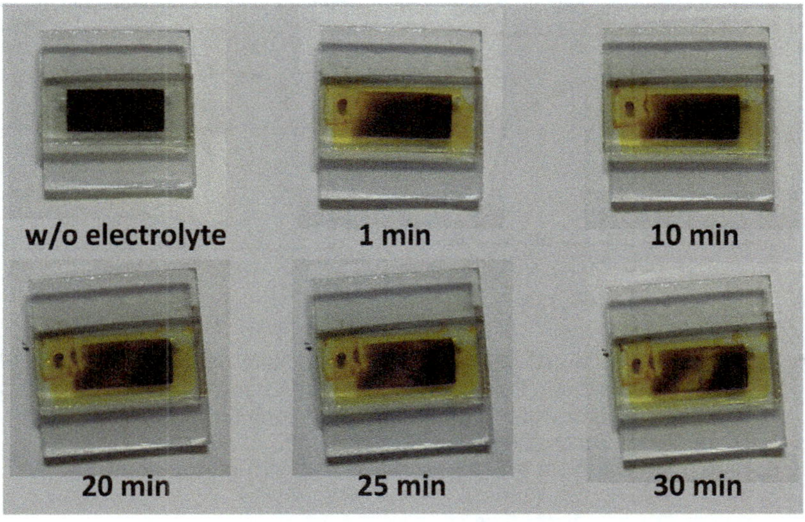

Figure 1.2 Instability of $CH_3NH_3PbI_3$ in a liquid electrolyte with time. $CH_3NH_3PbI_3$ was deposited on a nanocrystalline TiO_2 surface by spin-coating a 40 wt% gamma-butyrolactone (GBL) solution containing CH_3NH_3I and PbI_2. $CH_3NH_3PbI_3$ starts to dissolve within 20 min, and **dissolves** completely in 30 min.

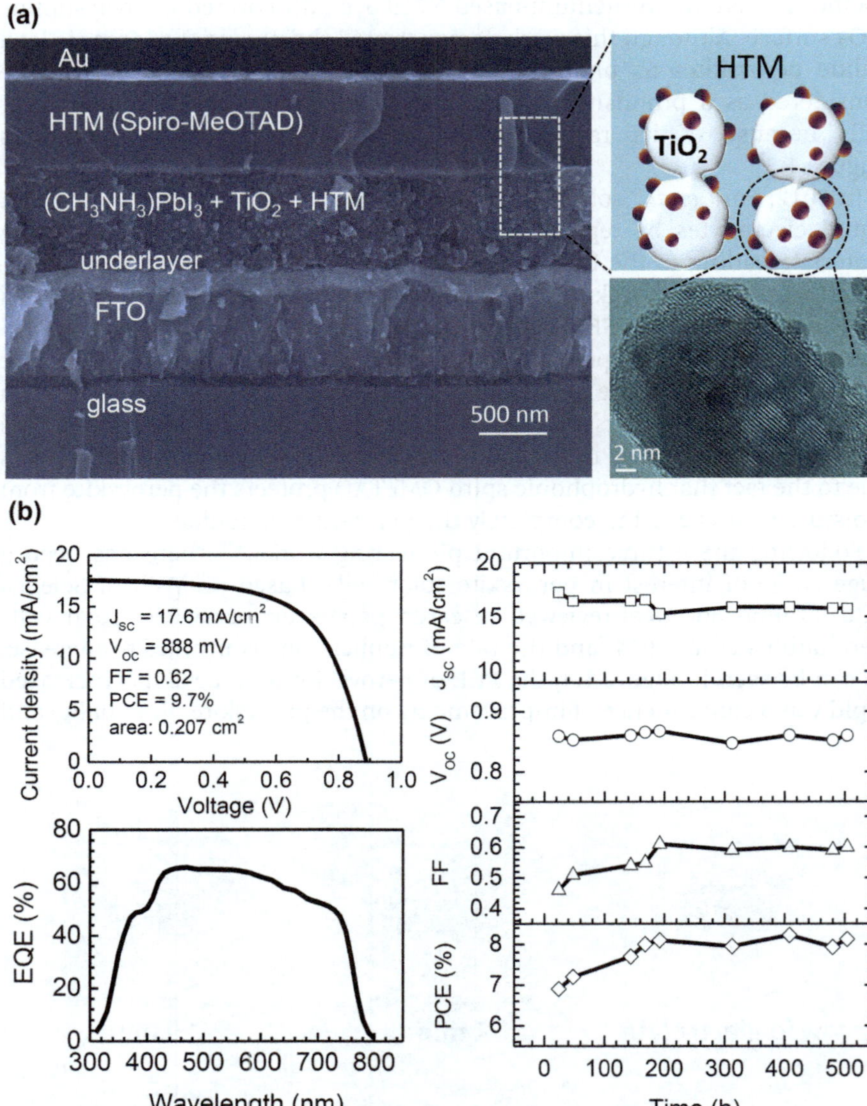

Figure 1.3 (a) Cross-sectional SEM images of solid-state perovskite ($CH_3NH_3PbI_3$)-sensitized solar cell employing perovskite nano dots deposited on TiO_2 surface. (b) Current–voltage, EQE and long-term stability data of the solid-state perovskite-sensitized solar cell.
Reproduced from *Sci. Rep.*, 2012, **2**, 591 with permission.[6]

other structural modifications. As a result, a certified PCE of 20.1% was reported in 2015.[7]

Efficiency improvements are related to the perovskite solar cell structure and perovskite film quality. Figure 1.5 shows the evolution of device

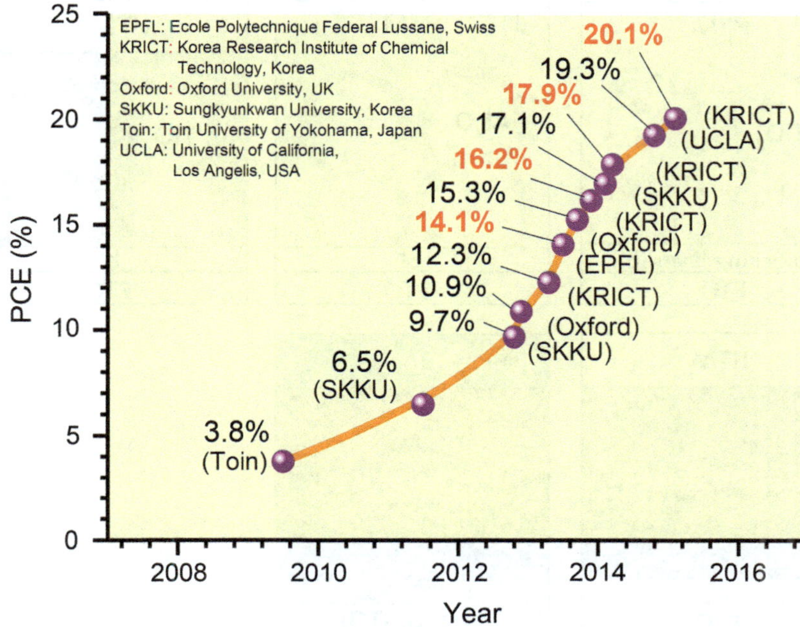

Figure 1.4 Progress of perovskite solar cell efficiency. A PCE of 3.8% in 2009 was improved to 20.1% in 2015. Values in red represent certified efficiencies.

structures. Diverse structures have been developed, ranging from the sensitized structure to the p-i-n junction structure. The first version of the perovskite solar cell was based on the perovskite-sensitized concept (Figure 1.5a), analogous to the solid-state dye-sensitized solar cell.[8] Since the perovskite is deposited on the TiO_2 surface without necking between the nano-dots, photocurrent generation can be explained by electron injection from the perovskite into TiO_2. However, it has been demonstrated that the perovskite also works without electron-accepting TiO_2, when Al_2O_3 is used as the scaffold instead, and a thin layer of perovskite is formed on the Al_2O_3 surface with perovskite networking.[9] The super-structured device configuration in Figure 1.5b underlines the importance of electron transport in the perovskite. Since the perovskite has the ability to transport electrons, it can be used as a bifunctional material acting as light harvester and n-type semiconductor. Based upon this bifunctionality, the perovskite can be used to completely fill the pores in the TiO_2 film, instead of the usual hole-transporting spiro-OMeTAD. This pore filling creates a capping layer comprising purely the perovskite, eventually leading to the mesoscopic perovskite solar cell structure shown in Figure 1.5c.[10] The mesoscopic structure has two different perovskite morphologies, with nanocrystals in the mesopores and a bulk thin film in the capping layer. Thus, two different pathways for electron transport can be expected. Since electron and hole diffusion lengths in the perovskite have been found to be almost

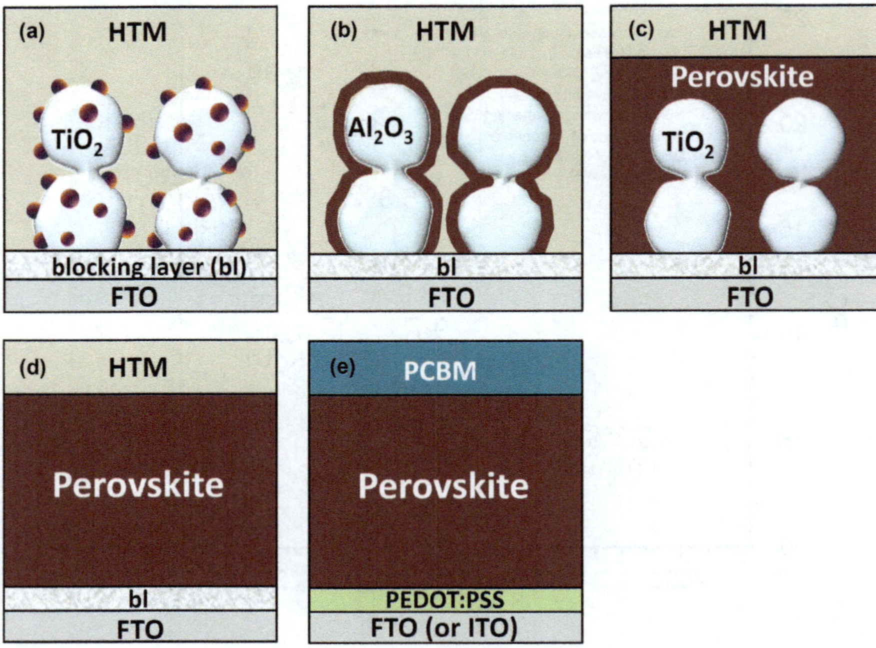

Figure 1.5 Diversity of perovskite solar cell structures. (a) sensitized configuration, (b) super-structured configuration, (c) mesoscopic structure with perovskite capping layer, (d) planar heterojunction type and (e) inverted layout analogous to that of organic photovoltaic devices.
Reproduced with permission from D.-Y. Son, J.-H. Im, H.-S. Kim and N.-G. Park, *J. Phys. Chem. C*, 2014, **118**, 16567–16573. Copyright (2014) American Chemical Society.[33]

identical,[11,12] the mesoporous oxide layer is not necessarily required to construct a perovskite solar cell. Figure 1.5d shows a planar heterojunction structure employing a compact thin TiO_2 electron-acceptor layer and a hole-accepting/transporting material with the perovskite layer in between.[13] Since the perovskite has unique opto-electronic properties, as mentioned previously, it can be applied to many types of solar cell structures. Figure 1.5e illustrates an inverted layout similar to the configuration of an organic photovoltaic device, with the perovskite layer being sandwiched between the transparent electrode with a hole-collecting PEDOT:PSS layer and the metal electrode with an electron-collecting PCBM.[14,15]

1.1.2 Role and Importance of the Organic Cation in Halide Perovskites: Phase Transitions, Ferroelectricity and Ion Migration

The three-dimensional organolead halide $CH_3NH_3PbI_3$ has been found to display a special feature involving the molecular motion of methylammonium ion.[16,17] The $CH_3NH_3^+$ ion can have two types of orientational

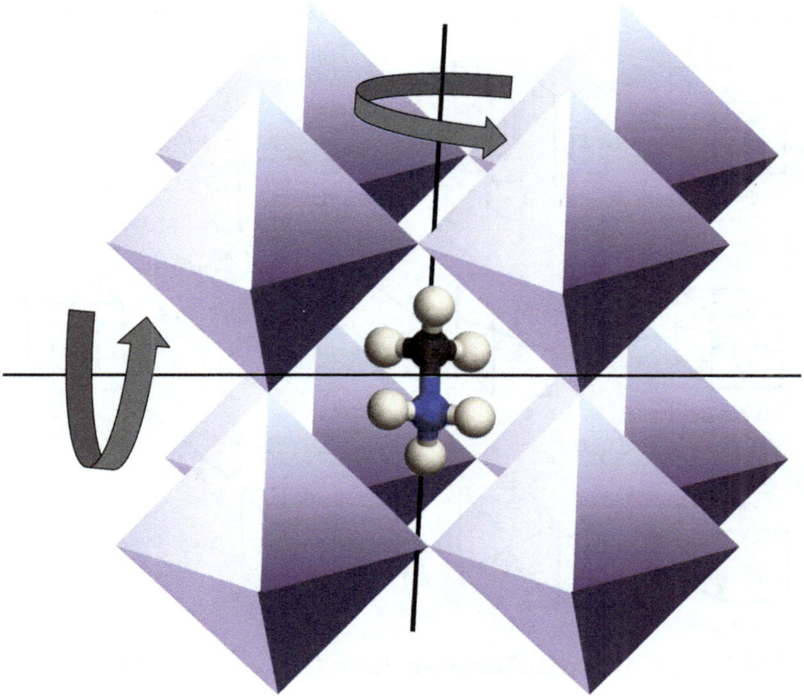

Figure 1.6 Molecular motion of $CH_3NH_3^+$ ion in the perovskite $CH_3NH_3PbI_3$ unit cell.
Reproduced from ref. 19 with permission from John Wiley and Sons. Copyright © 2014 WILEY-VCH Verlag GmbH & Co. KGaA, Weinheim.

disorder; one is the orientation of the C–N axis relative to the crystal axes, and the other is related to the rotation around the C–N axis, as can be seen in Figure 1.6.[18] It has been observed that $CH_3NH_3PbX_3$ undergoes a temperature-dependent phase transition for X = Cl, Br and I due to the molecular motion of $CH_3NH_3^+$.[17] For instance, the $CH_3NH_3PbI_3$ structure changes from orthorhombic to tetragonal at 162.2 K, and from tetragonal to cubic at 327.4 K.[17] According to temperature-dependent XRD analysis and complex permittivity results, dynamic disorder of methylammonium cation occurs in the high temperature phase,[17] resulting in a cubic phase. It is noteworthy that the phase transition is accompanied by a large volume change of the unit cell between the cubic (253.5×10^6 pm^3) and tetragonal (992.6×10^6 pm^3) structures. The molecular motion-induced phase transition is depicted in Figure 1.7.

The molecular motion of the $CH_3NH_3^+$ ion can induce or affect not only phase transitions, but also other physical properties such as ferroelectric polarization and ion migration. Ferroelectric polarization of $CH_3NH_3PbI_3$ has been observed directly by piezoresponse force microscopy (PFM).[20] Figure 1.8 shows PFM phases along with the topography observed in the dark and under illumination for 700 nm-sized $CH_3NH_3PbI_3$ cuboids. PFM

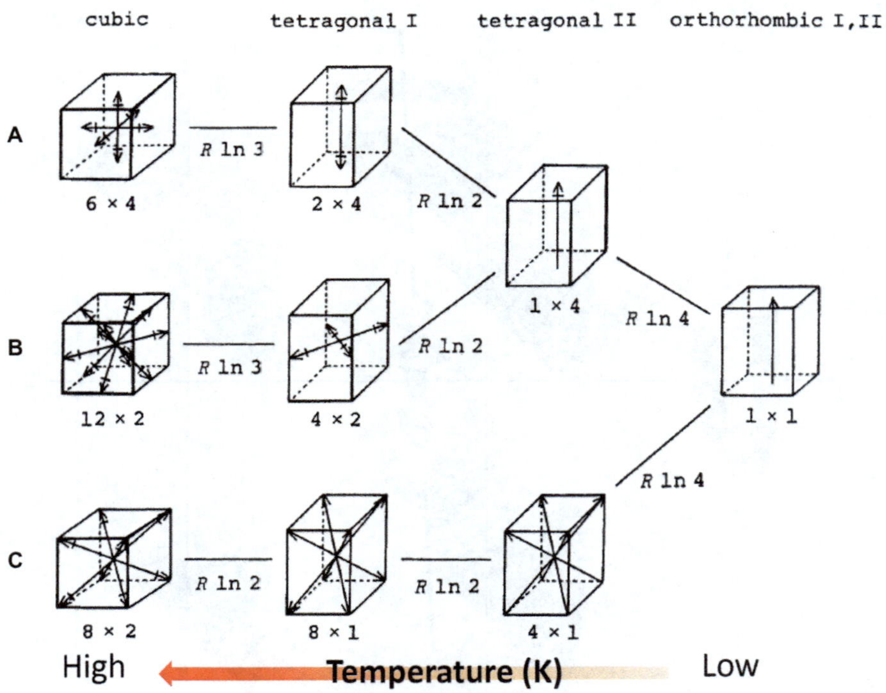

Figure 1.7 Possible orientations and ordering processes of the $CH_3NH_3^+$ ion. The arrows indicate the orientations of the C–N axis. ‡: disordered around the C–N axis. ↑: ordered around the C–N axis.
Reproduced from N. Onodo-Yamamuro *et al.*, Calorimetric and IR spectroscopic studies of phase transitions in methylammonium trihalogenplumbates (II), *J. Phys. Chem. Solids*, **51**, 1383–1395. Copyright (1990) with permission from Elsevier.[18]

phase images were obtained without bias voltage (unpoled) and with bias voltages of +3 V (positive poling) and −3 V (negative poling). In the dark, spontaneous polarization is observed even in the absence of an electric field (unpoled), which is due to the large freedom of rotation of the polar $CH_3NH_3^+$ ion.[21] The poling effect is evident for both positive and negative bias potentials, where dipoles do not rotate 180° completely upon changing from +3 V to −3 V. Such imperfect rotation of dipoles with respect to the applied electric field is probably due to their interaction with defects in the crystals. Light-induced polarization using the PFM measurement system equipped with a white light source (50 000 lux) has shown that the spontaneous (unpoled) polarization is partly attenuated by light irradiation, which indicates that the ferroelectric polarization is screened by photogenerated conduction electrons. However, photo-induced polarization is pronounced in the presence of an electric field for positive poling. Upon changing the poling polarity, a similar tendency is observed as in the dark.

The ferroelectric polarization remains unchanged even after removal of the external electric field, as can be seen in Figure 1.9. The PFM phase of the

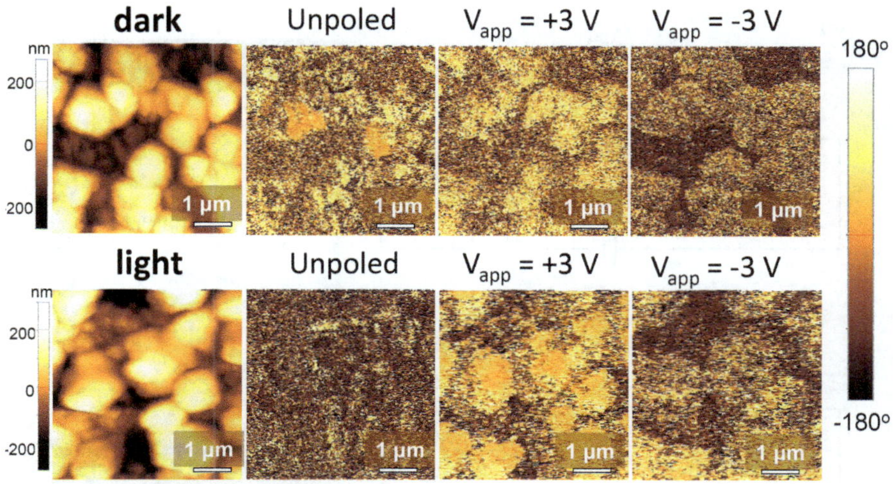

Figure 1.8 AFM topographic images and PFM phase images measured (top) in the dark and (bottom) under illumination for $CH_3NH_3PbI_3$ cuboids with size of 700 nm.
Reproduced with permission from H.-S. Kim *et al.*, *J. Phys. Chem. Lett.*, 2015, **6**, 1729–1735. Copyright (2015) American Chemical Society.[20]

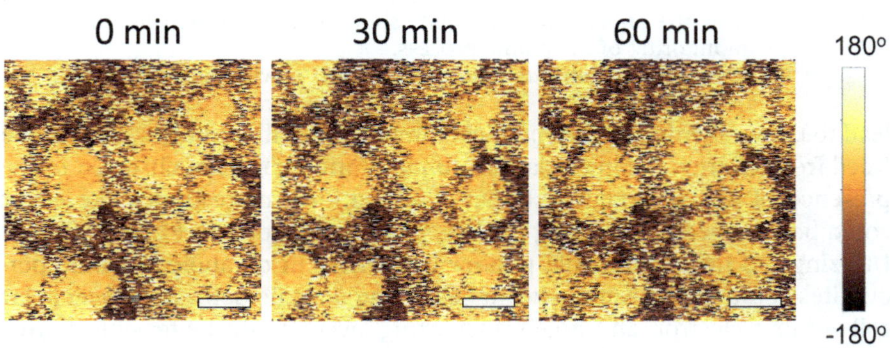

Figure 1.9 PFM phase images under illumination of 700 nm-sized $CH_3NH_3PbI_3$ cuboids, measured at +3 V positive poling (0 min) and at 30 min and 60 min after removal of the poling. Scale bars represent 1 µm.
Reproduced with permission from H.-S. Kim *et al.*, *J. Phys. Chem. Lett.*, 2015, **6**, 1729–1735. Copyright (2015) American Chemical Society.[20]

positively poled perovskite was measured at intervals of 30 min under illumination, and highly aligned dipoles were observed for 60 min. It was found that the retention of the light-induced polarization depends on the $CH_3NH_3PbI_3$ crystal size: larger crystals demonstrated a better retention behavior than smaller ones.[20] The observed size-dependent retention behavior is probably related to defects in the perovskite crystals.[22]

It is expected that iodide ions may be able to migrate in the perovskite ionic crystal due to molecular motion. Dynamic $CH_3NH_3^+$ motion is likely to

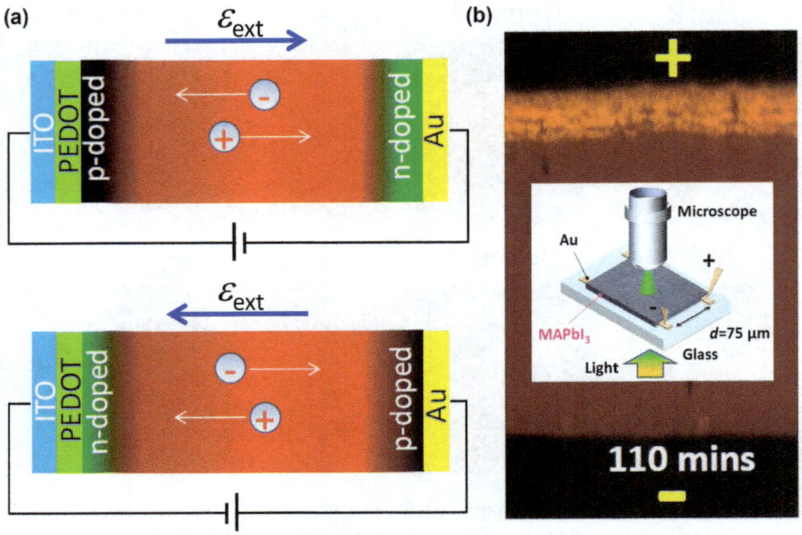

Figure 1.10 (a) Schematics of ion drift in perovskite during positive poling (top) and negative poling (bottom). The accumulated ions in the perovskite near the electrodes induce p- and n-doping. ε_{ext} represents an external electric field. (b) Snapshots of the *in situ* recorded video with the poled lateral structure device, showing the modified perovskite material near the anode side. The inset shows the setup used for the *in situ* monitoring of the poling process.
Reproduced from *Nat. Mater.*, 2015, **14**, 193 with permission.[23]

lead to the weakening of the chemical interactions with iodide, which can be freed from restrictions in the lattice. Ion migration has been observed in the presence of a bias potential under illumination. Interestingly, the effect could be switched by reversing the bias polarity as shown in Figure 1.10.[23,24] Utilizing the field switchable photovoltaic effect in organolead halide perovskites could offer several advantages. Using a single perovskite layer solar cell without electron and hole transporting layers might be feasible as free charges in the perovskite can be transported effectively by self-doping *via* poling. If one could fabricate a lateral structure with spacing between the metal electrodes, a TCO (transparent conductive oxide)-free solar cell with an extremely high photovoltage could be realized. The demonstration of electric-field-manipulated doping opens up new applications for perovskites in devices such as both electrically- and optically-readable memristors.

1.2 Mesoscopic Perovskite Solar Cells

1.2.1 Perovskite Dots and Extremely Thin Absorber Layers

In the first version of a solid-state perovskite solar cell shown previously in Figure 1.3, the perovskite deposited on the TiO_2 surface is in the form of nano-dots that are not interconnected. Therefore, perovskite nano-dots are expected to behave in the same way as a molecular sensitizer, where photoexcited electrons are injected to TiO_2 and holes are injected into spiro-OMeTAD. Femtosecond transient absorption spectra (TAS) of different perovskite assemblies were measured (Figure 1.11) to elucidate the mechanism of charge separation.[6] First, samples containing the perovskite material deposited on Al_2O_3 and TiO_2 without hole transporting layers were prepared to compare their electron transfer (Figure 1.11a and b). The negative signal at around 483 nm is ascribed to the bleaching of the absorption of the perovskite. The same bleaching is observed for both Al_2O_3 and TiO_2 samples. Since no electron transfer is expected between Al_2O_3 and the perovskite due to their large conduction band mismatch, the small difference in TA spectra observed between Al_2O_3 and TiO_2 indicates that no significant charge injection takes place from the excited state of the perovskite into TiO_2.

After introducing a spiro-OMeTAD hole transport medium (HTM) (Figure 1.11c and d), the bleaching peaks in the 480 nm-region are less pronounced than those for the perovskite without HTM, for both the Al_2O_3 and TiO_2 samples. The positive absorption signal in the 630–700 nm region is also attenuated for the TiO_2 sample and disappears for the Al_2O_3 one.

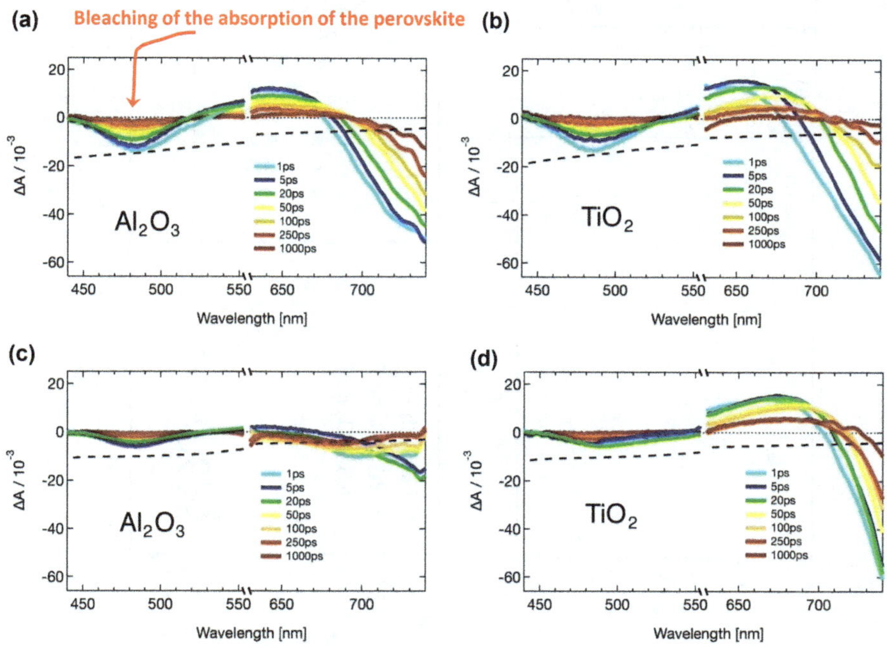

Figure 1.11 Femtosecond transient absorbance spectra obtained by excitation with a 580 nm laser and a white light continuum probe of (a) $CH_3NH_3PbI_3$/Al_2O_3, (b) $CH_3NH_3PbI_3$/TiO_2, (c) spiro-OMeTAD/$CH_3NH_3PbI_3$/Al_2O_3, and (d) spiro-OMeTAD/$CH_3NH_3PbI_3$/TiO_2, recorded at various time delays after excitation (color lines). Black dashed lines represent the absorbance spectrum of the sample (scaled by a factor of −0.01). Reproduced from *Sci. Rep.*, 2012, **2**, 591 with permission.[6]

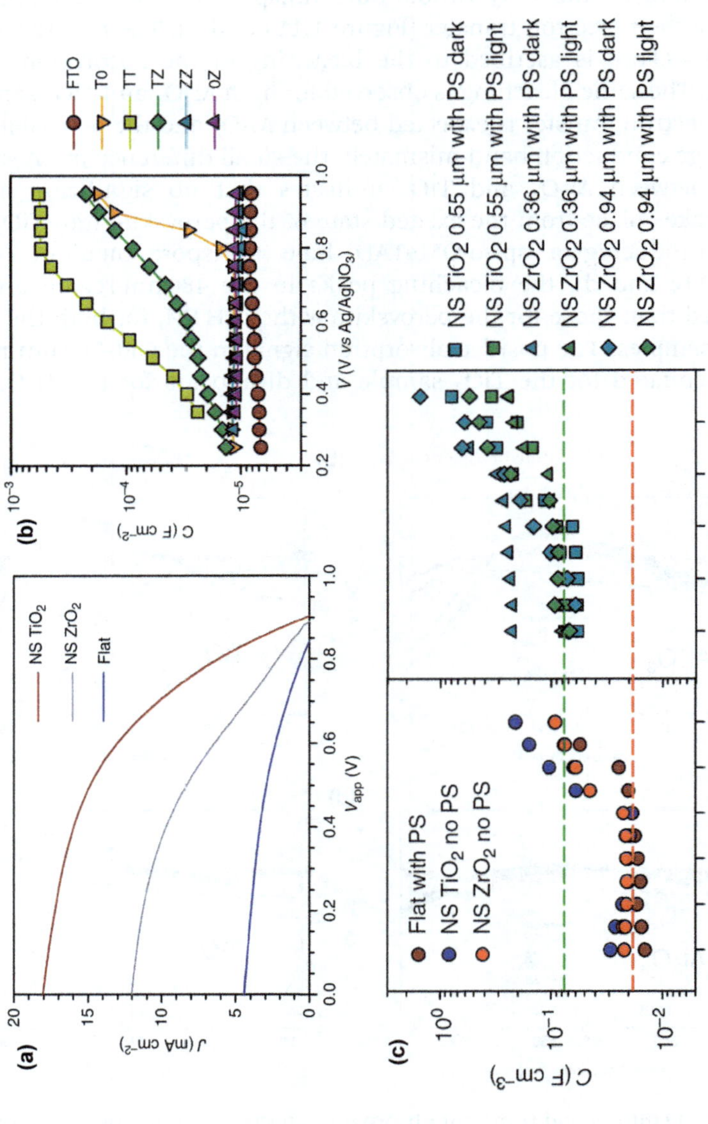

(a)
J (mA cm−2)
Vapp (V)
NS TiO2
NS ZrO2
Flat
(b)
C (F cm−2)
V (V vs Ag/AgNO3)
FTO
T0
TT
TZ
ZZ
0Z
(c)
C (F cm−3)
Vapp (V)
Flat with PS
NS TiO2 no PS
NS ZrO2 no PS
NS TiO2 0.55 μm with PS dark
NS TiO2 0.55 μm with PS light
NS ZrO2 0.36 μm with PS dark
NS ZrO2 0.36 μm with PS light
NS ZrO2 0.94 μm with PS dark
NS ZrO2 0.94 μm with PS light

These results suggest a rapid reductive quenching of the excited state of the perovskite by spiro-OMeTAD. From the TAS study, hole separation by the HTM is obvious, however, electron injection is not clear. Nevertheless, the photocurrent observed from the sensitization-type system provides strong evidence of electron injection under short-circuit conditions. Otherwise, charge collection at the FTO substrate could not be explained. Then, the absence of evidence of electron injection to TiO_2 in the TAS results is still questionable. As shown in Figure 1.3, the TiO_2 surface is not fully covered with perovskite, and the uncovered surfaces contacting directly with spiro-OMeTAD are likely to act as recombination sites. However, the solid-state sensitized perovskite solar cells deliver a remarkably high open circuit voltage of 888 mV, despite the bare TiO_2 surfaces, which indicates that photo-excited electrons may accumulate in perovskites without injection under open-circuit conditions.

The charge accumulation properties of $CH_3NH_3PbI_3$ have been studied using impedance spectroscopy. Figure 1.12 shows the *J–V* curves for $CH_3NH_3PbI_3$ perovskite solar cells with three different types of electrodes: FTO/Compact TiO_2 (Flat), FTO/Compact TiO_2/Nanostructured TiO_2 (NS · TiO_2) and FTO/Compact TiO_2/Nanostructured ZrO_2 (NS · ZrO_2), along with the capacitance as a function of the applied voltage.[25] All three samples demonstrate photovoltaic behavior with different performances. In Figure 1.12b, the capacitances of TiO_2 and ZrO_2 electrodes in a liquid redox electrolyte based on I^-/I_3^- are compared as a function of the applied bias potential to investigate electrochemical charging effects. This experiment proves that the NS ZrO_2 layer cannot be charged, whereas the NS TiO_2 one is charged with the applied bias (V_{app}).

Figure 1.12c shows the capacitance for TiO_2 (Flat and NS) and ZrO_2 electrodes with and without perovskite. For the flat sample with perovskite at low V_{app}, the capacitance is almost invariant while, at high V_{app}, an increase in the capacitance slope is observed. This exponential enhancement

Figure 1.12 (a) Current density (*J*)–voltage (*V*) curves of $CH_3NH_3PbI_3$ perovskite solar cells with three different electrodes: FTO/Compact TiO_2 (Flat), FTO/Compact TiO_2/Nanoestructured TiO_2 (NS TiO_2) and FTO/Compact TiO_2/Nanoestructured ZrO_2 (NS ZrO_2). (b) Capacitance extracted from impedance spectroscopic measurements in a liquid electrolyte with the I^-/I_3^- redox couple. The capacitance of the bare conductive Sn_2O:F (FTO) electrode is compared to that of different electrodes with notation XY, where X is the compact layer and Y the porous layer, and T and Z mean TiO_2 and ZrO_2 layer, respectively. 0 means no layer *i.e.* T0: compact TiO_2 layer without nanostructured layer. (c) Capacitance analysis of flat samples and NS samples with and without perovskite. Left: capacitance of flat with perovskite (PS) and blank samples without PS; NS TiO_2 and ZrO_2 samples represent NS layers of 0.35 μm TiO_2 and 0.39 μm ZrO_2, respectively. Right: capacitance of perovskite coated NS TiO_2 and ZrO_2 samples with different NS layer thickness in the dark and under 1 sun illumination.
Reproduced from *Nat. Commun.*, 2013, **4**, 2242 with permission.[25]

of the capacitance can be attributed to the chemical capacitance of the flat TiO_2 compact layer, since the blank NS samples without perovskite show a similar tendency in capacitance change with V_{app}. In the case of complete NS devices with the perovskite, higher capacitances are clearly observed for the samples with perovskite than for the blank samples. It can be seen that the capacitances display a similar behavior regardless of the electrode material (TiO_2 or ZrO_2), the NS layer thickness and the illumination conditions. This indicates that the capacitance is originated in the perovskite deposited in the NS layer, showing direct evidence of charge accumulation in the perovskite material.

From the meso-superstructure concept with a scaffold oxide layer whose surface is coated with an extremely thin perovskite layer as reported elsewhere,[26] electron transport in $CH_3NH_3PbI_3$ is proved because Al_2O_3 cannot be charged by injection from the perovskite due to a conduction band mismatch, as can be seen in Figure 1.13. To deposit the perovskite on an Al_2O_3 mesoporous film, one mole of $PbCl_2$ and 3 moles of CH_3NH_3I are mixed in a polar aprotic solvent, instead of using an equimolar mixture of PbI_2 and CH_3NH_3I. From the stoichiometric point of view, the reaction of one mole of $PbCl_2$ with three moles of CH_3NH_3I yields $CH_3NH_3PbI_3$ + $2CH_3NH_3Cl$. Since $CH_3NH_3^+$ is a strong acid (because CH_3NH_2 is a weak base), $CH_3NH_3^+$ tends to dissociate into CH_3NH_2 and H^+. The strong acid HCl gives rise to the weak conjugate base Cl^-. Thus, two moles of CH_3NH_3Cl are expected to form CH_3NH_2, H^+ and Cl^-, which are volatile when drying. Therefore, $CH_3NH_3PbI_3$ is eventually expected as the final product, along with the possibility of the presence of Cl as an impurity. Both the sensitized structure and meso-superstructure are beneficial for long-term stability because the moisture-sensitive perovskite is fully wrapped with the hydrophobic HTM.

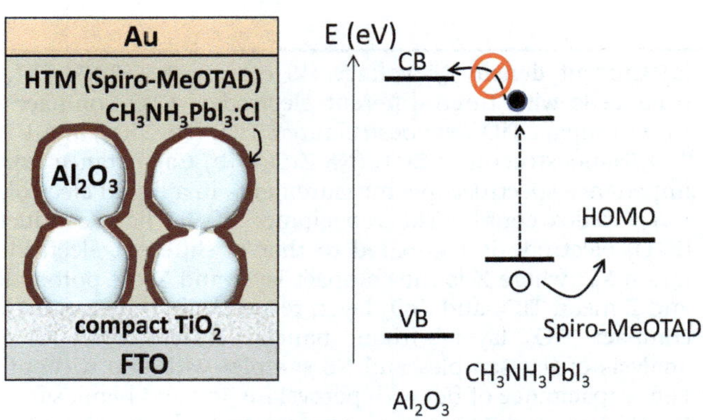

Figure 1.13 Meso-superstructured perovskite solar cell employing an Al_2O_3 scaffold layer and an extremely thin perovskite layer. The diagram on the right illustrates the band edge positions, showing that no electron injection from the perovskite to Al_2O_3 can occur.

1.2.2 Perovskite Hybrids with Mesoporous and Nanostructured TiO_2

1.2.2.1 3D $CH_3NH_3PbI_3$ Perovskite Cuboids

In Figure 1.5, we saw diverse structures for perovskite solar cells. Among them, the mesoscopic and planar heterojunction structures have been widely and intensively studied because they show an abnormally high photovoltaic performance. The mesoporous TiO_2 layer is included in the mesoscopic structure by depositing it on a compact thin TiO_2-blocking layer, while the compact TiO_2 (or other oxide) layer alone is used as an electron-accepting layer in the planar heterojunction configuration. Spin-coating of the perovskite solution usually leads to formation of a capping layer on the mesoporous TiO_2 film. Either one-step or two-step sequential coating is available for the preparation of the perovskite layer. It has been found that the $CH_3NH_3PbI_3$ crystal size is easily controlled by the concentration of CH_3NH_3I in the two-step spin coating procedure.[27] The size-controllable two-step procedure is presented schematically in Figure 1.14. First, a solution of PbI_2 in *N*,*N*′-dimethyl formamide (DMF) is spin-coated on the mesoporous

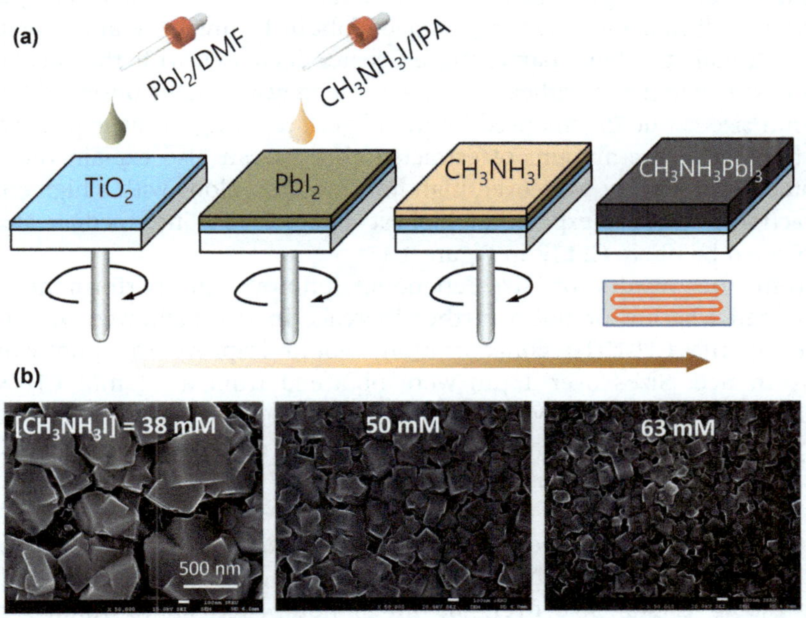

Figure 1.14 (a) Two-step spin-coating procedure for $CH_3NH_3PbI_3$ cuboids. A PbI_2 solution is spin-coated on the mesoporous TiO_2 film, dried at a temperature lower than 100 °C for a few minutes, and then a CH_3NH_3I solution is spin-coated. Finally, the film is heated at 100 °C to form a black perovskite film. (b) SEM images showing how the perovskite film structure depends on the concentration of the CH_3NH_3I solution in isopropanol.
Reproduced from *Nat. Nanotechnol.*, 2014, **9**, 927 with permission.[27]

TiO_2 film, and this is followed by spin-coating of a CH_3NH_3I solution in isopropanol (IPA). Finally, a dark black perovskite film is formed after mild heating. In the second step, the crystal size can be controlled by varying the CH_3NH_3I concentration. As can be seen in the SEM images in Figure 1.14b, a relatively low concentration is found to increase the perovskite crystal size, but smaller sizes are obtained at high concentrations. Perovskite crystals about 700–800 nm in size are obtained from 38 mM [CH_3NH_3I], while the use of a 63 mM solution leads to *ca.* 100 nm-sized crystals. It is noteworthy that the size distribution becomes broader when bigger crystals grow and also, large crystals are not closely packed on the substrate.

Interestingly, the photovoltaic performance is influenced significantly by the concentration of CH_3NH_3I, which indicates that the perovskite crystal size plays an important role in determining the photovoltaic parameters. Figure 1.15a–d show that the photocurrent density increases as the crystal size increases. The fill factor exceeds 70% when the size is larger than 200 nm, and the open-circuit voltage is maximized for the intermediate size. As a result, an average PCE exceeding 16% is achieved for crystal sizes larger than 200 nm, whereas less than 14% is observed for smaller sizes of around 100 nm. The high photocurrent achieved with large crystal films can be explained by the light harvesting efficiency and photo-CELIV (charge extraction with linearly increasing voltage) data in Figure 1.15e and f, respectively. Overall, the light harvesting efficiency increases with the increasing crystal size, where the highest efficiency at long wavelengths observed for the 800 nm size is due to enhanced internal light scattering in the gaps between crystals. The large amount of extracted charges can also explain the high photocurrent density. Moreover, high hole mobility, along with a high charge extraction ability, can explain the highest voltage for the intermediate size, as confirmed by photo-CELIV in Figure 1.15f.

From the results of size-dependent photovoltaic performance, we wondered whether or not a further increase in size (achieved by further lowering the CH_3NH_3I concentration) would improve the photovoltaic performance. Sizes over 1 μm were obtained from a 32 mM CH_3NH_3I solution, as confirmed by the SEM images in Figure 1.16. However, contrary to our expectation, the corresponding PCE decreased to about 11.3%, due to a relatively low photocurrent density of 17 mA cm^{-2} and open circuit voltage of 0.92 V. This is ~10% lower than the voltages exceeding 1.1 V observed for the intermediate size of ~200 nm. Taking into account the fact that $V_{oc} = 1.06$ V for the 800 nm size crystals, one can conclude that an increase in crystal size over 200 nm is not beneficial to voltage gain. According to a micro photoluminescence (PL) study,[28] strong PL was observed from intermediate sized crystals (*e.g.* 200 nm), whereas large crystals of 800 nm and over 1 μm did not exhibit bright PL. The weak PL from large crystals suggests that non-radiative recombination is important, which is probably due to the presence of many defects and/or grain boundaries in large crystals. Such non-radiative recombination seems to be the responsible for low open-circuit voltages.

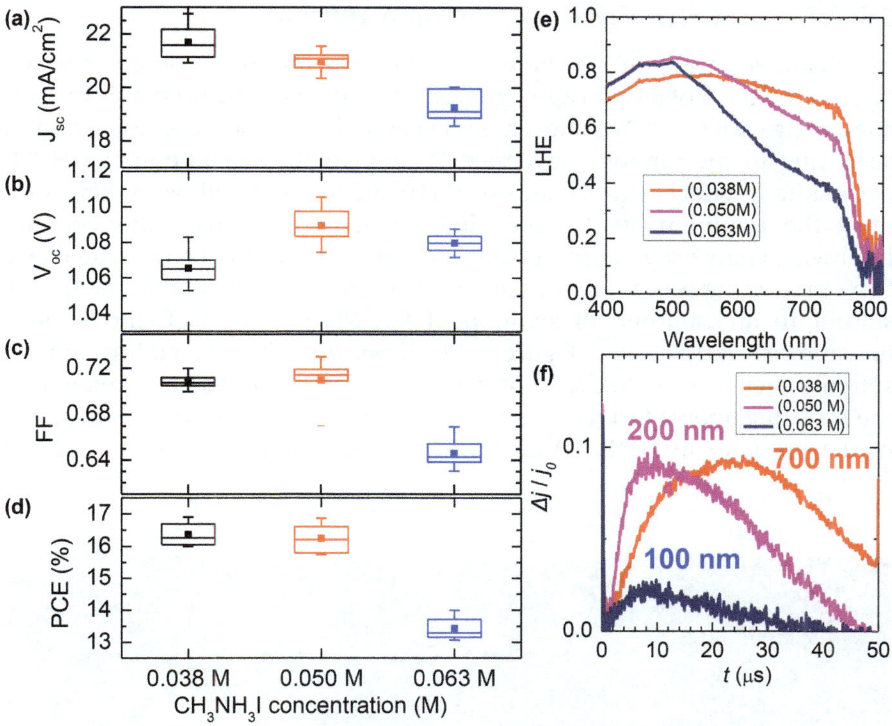

Figure 1.15 Effect of $CH_3NH_3PbI_3$ cuboid size on (a–d) photovoltaic parameters of short-circuit current density J_{sc}, open-circuit voltage V_{oc}, fill factor FF and power conversion efficiency PCE. (e) Light harvesting efficiency (LHE) and (f) photo-CELIV transients. Photo-CELIV transients were recorded for the devices comprising FTO/bl-Al_2O_3/mp-TiO_2/$MAPbI_3$/spiro-OMeTAD/Au at different CH_3NH_3I concentrations.
Reproduced from *Nat. Nanotechnol.*, 2014, **9**, 927 with permission.[27]

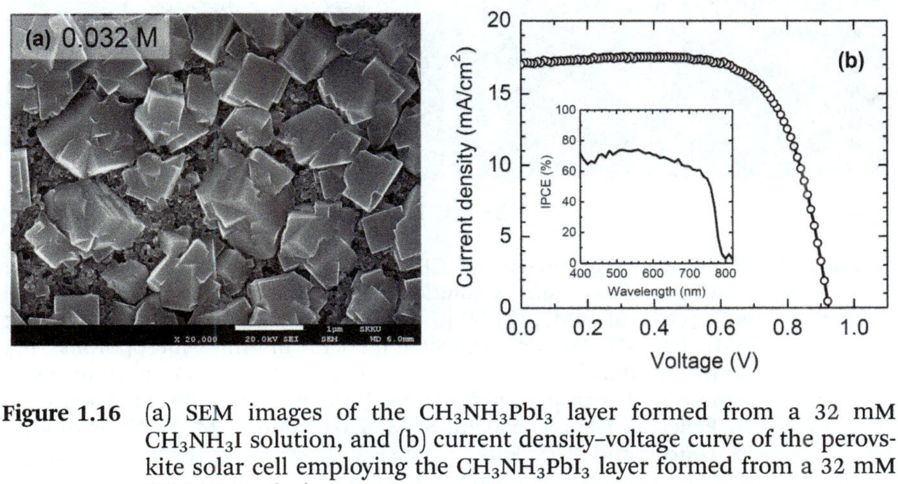

Figure 1.16 (a) SEM images of the $CH_3NH_3PbI_3$ layer formed from a 32 mM CH_3NH_3I solution, and (b) current density–voltage curve of the perovskite solar cell employing the $CH_3NH_3PbI_3$ layer formed from a 32 mM CH_3NH_3I solution.
Reproduced from *Nat. Nanotechnol.*, 2014, **9**, 927 with permission.[27]

1.2.2.2 1D $CH_3NH_3PbI_3$ Perovskite Nanowires

As illustrated in Figure 1.5, the mesoscopic structure has a pure perovskite layer in the form of a thin capping layer (around 100–200 nm) that is directly contacting with the HTM. Thus, holes generated in the perovskite layer travel according to the random walk model[29] and finally separate at the HTM/perovskite interface. One-dimensional (1D) nanostructured perovskites may meet the requirements for reducing the length of hole transport and improving charge separation. 1D nanowire perovskites can be fabricated by a two-step coating procedure, where the addition of a small amount of aprotic solvent to an isopropanol solution of CH_3NH_3I (MAI) is found to lead to nanowire formation.[30] Figure 1.17 shows the effect of adding aprotic solvents such as DMF to the isopropanol solution of CH_3NH_3I. Compared to the 3D nanocuboids formed in the absence of DMF, 1D nanowires are grown in the presence of DMF. 50 μL DMF in 5 mL isopropanol is the optimal

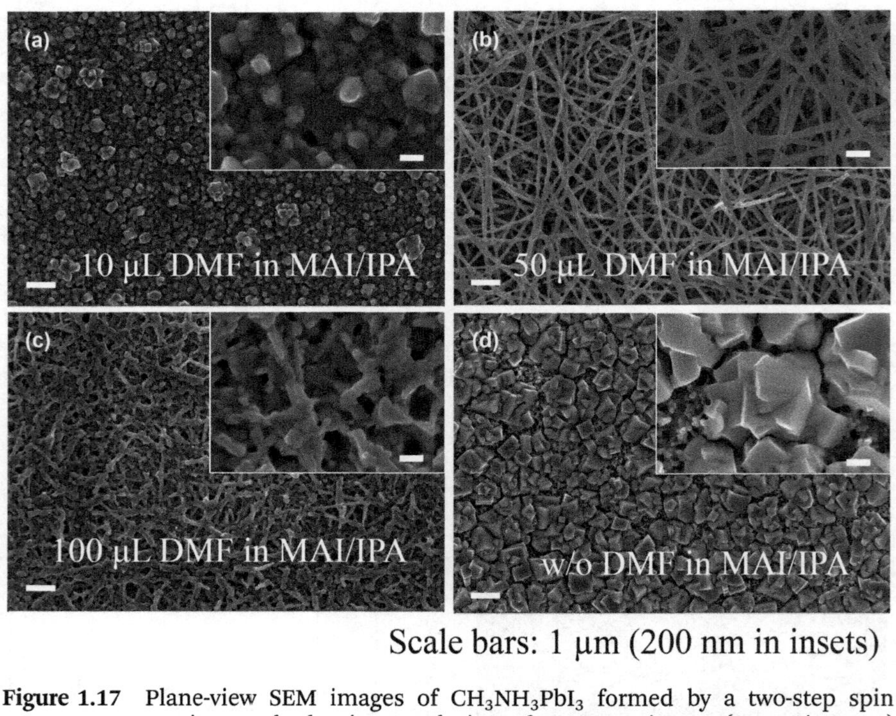

Figure 1.17 Plane-view SEM images of $CH_3NH_3PbI_3$ formed by a two-step spin coating method using a solution of CH_3NH_3I in IPA (35 mg in 5 mL) with (a) 10 μL DMF, (b) 50 μL DMF, (c) 100 μL DMF and (d) without DMF. $CH_3NH_3PbI_3$ was deposited on a 100 nm-thick mesoporous TiO_2 (diameter of about 40 nm) layer. Scale bars: 1 μm (200 nm in insets). Reproduced with permission from J.-H. Im, J. Luo, M. Franckevicius, N. Pellet, P. Gao, T. Moehl, S. M. Zakeeruddin, M. K. Nazeeruddin, M. Grätzel and N.-G. Park, *Nano Lett.*, 2015, **15**, 2120–2126. Copyright (2015) American Chemical Society.[30]

formulation for the growth of long perovskite nanowires with diameters ranging from ~30 to ~200 nm. In terms of the mechanism involved in the anisotropic growth of $MAPbI_3$, the locally dissolved PbI_2, caused by the small amount of DMF present during the second spinning step, serves as the preferential site for reaction with MAI and growth of 1D structures, like in a liquid catalyst cluster mode.[30] This underlying mechanism suggests that the solubility of PbI_2 in the aprotic solvent can affect the nanowire formation. The effect of adding different aprotic solvents, such as dimethyl sulfoxide (DMSO) and gamma butyrolactone (GBL) to the MAI solution has been investigated. DMSO is able to produce nanowires but not perfectly, whereas GBL cannot induce nanowire formation due to the low solubility of PbI_2 in it. Nanowire growth correlates well with the order of PbI_2 solubility, DMF > DMSO ≫ GBL, and thus it can be said that the liquid catalyst cluster model is likely to be involved in nanowire growth.

As expected, the nanowire morphology was found to be better for hole separation at the HTM/perovskite interface, as can be seen in the time-resolved PL spectra in Figure 1.18. Approximately 79% of the holes are injected into the HTM from the perovskite nanowires with a lifetime of about 700 ps, whereas hole extraction in 3D bulk perovskites is not so efficient (*ca.* 68% of holes are injected into the HTM with a lifetime of 2.9 ns). This more efficient hole extraction by the nanowire perovskite is related to the increased surface area, which ensures a better contact with the HTM. Electron extraction of the nanowire perovskite deposited on a TiO_2 layer is not as efficient as for the bulk 3D perovskite, which is probably due to poor contact with the TiO_2 layer.

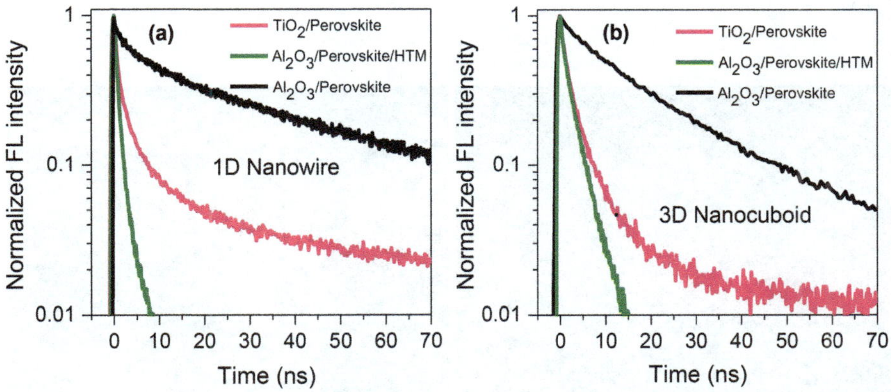

Figure 1.18 Normalized fluorescence (FL) decay kinetics of perovskite films of (a) 1D nanowire $CH_3NH_3PbI_3$ and (b) 3D nanocuboid $CH_3NH_3PbI_3$ following excitation at 406 nm.
Reproduced with permission from J.-H. Im, J. Luo, M. Franckevicius, N. Pellet, P. Gao, T. Moehl, S. M. Zakeeruddin, M. K. Nazeeruddin, M. Grätzel and N.-G. Park, *Nano Lett.*, 2015, **15**, 2120–2126. Copyright (2015) American Chemical Society.[30]

1.2.2.3 $CH_3NH_3PbI_3$ Perovskite with Nanorod ZnO

Nanorod oxides can be used alternatively to induce pseudo-one-dimensional perovskites. As nanorod oxides, TiO_2 and ZnO are the best candidates when considering matching the band positions with that of the perovskite. Since ZnO is better than TiO_2 in terms of electron transport kinetics[31,32] and fabrication process, ZnO nanorods are expected to be good candidates from the viewpoint of electron transport and modulation of the perovskite morphology. Figure 1.19 shows solution-processed ZnO nanorods on an FTO substrate and the corresponding perovskite solar cell. To grow ZnO nanorods, the FTO should be well covered with a ZnO seed layer. An ethanol solution of zinc acetate dihydrate ($Zn(CH_3COO)_2 \cdot 2H_2O$) is used to prepare the seed layer. In this case, the coverage and thickness of the seed layer are important because the final morphology is influenced by these factors. The aqueous solution for growing ZnO nanorods is composed of equimolar zinc nitrate hexahydrate ($Zn(NO_3)_2 \cdot 6H_2O$) and hexamethylenetetramine. The diameter of the ZnO nanorod can be controlled by varying the solution concentration. The length of the ZnO nanorods was found to depend on reaction time.

The photovoltaic performance of the ZnO nanorod perovskite solar cell is presented in Figure 1.20. The current density of 20.08 mA cm^{-2} observed under solar simulator illumination is fully consistent with the integrated value of 20.03 mA cm^{-2} calculated from external quantum efficiency (EQE) data. The relatively low PCE of 11.13% is mainly due to a relatively low fill factor of 0.56 and voltage of 0.99 V. It is interesting to see that the EQE spectral shape is almost squared, as shown in Figure 1.20b, which is related to the perovskite being embedded in the ZnO nanorod film with a negligible capping layer.

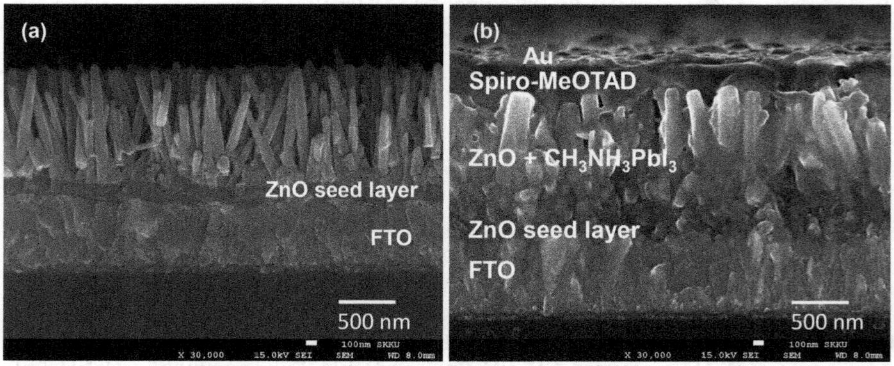

Figure 1.19 SEM images of (a) bare ZnO nanorods grown on a FTO substrate, and (b) full cell with $CH_3NH_3PbI_3$-deposited on ZnO nanorods.
Reproduced with permission from D.-Y. Son, J.-H. Im, H.-S. Kim and N.-G. Park, *J. Phys. Chem. C*, 2014, **118**, 16567–16573. Copyright (2014) American Chemical Society.[33]

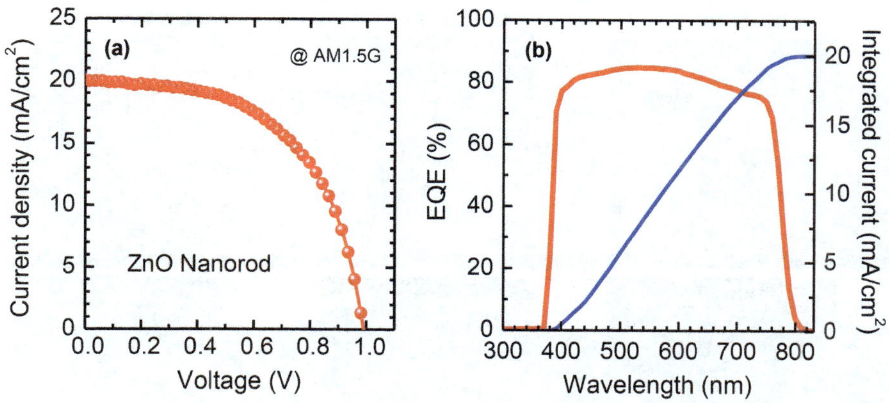

Figure 1.20 (a) Current density–voltage curve, and (b) EQE spectra together with EQE data-based integrated current density for the perovskite solar cell based on a ZnO nanorod (82 nm wide and 1 μm long) film. Reproduced with permission from D.-Y. Son, J.-H. Im, H.-S. Kim and N.-G. Park, *J. Phys. Chem. C*, 2014, **118**, 16567–16573. Copyright (2014) American Chemical Society.[33]

As mentioned previously, the coverage of the seed layer on the FTO surface is important for the subsequent growth of ZnO nanorods. The nature of the seed layer can be influenced by the coating solution. The effect of the seed layer on the growth of ZnO nanorods and the resulting photovoltaic performance has been investigated.[34] Three different seed layers were prepared using a clear zinc acetate solution (designated as 'solution' in Figure 1.21), a ZnO colloidal solution (designated as 'colloidal') and ZnO powder dispersed in ethanol (designated as 'powder'). Coverage of the seed layer is much better when using the colloidal solution compared to the coverage obtained from the other two solutions, as shown in Figure 1.21a–c. The difference in coverage is found to influence significantly the ZnO nanorod morphology. The seed layer formed from the colloidal precursor results in vertically grown ZnO nanorods, whereas some or most of the nanorods are tilted when grown on a seed layer formed from the solution or powder precursor (Figure 1.21d–f). Focused ion beam (FIB) assisted cross-sectional SEM images for the full cell structure reveal that the ZnO nanorods are well attached, while some of the ZnO nanorods are detached from the substrate for both the solution- and powder-based seed layers, resulting in some of the FTO surfaces being exposed directly to the perovskite layer (Figure 1.21g–i). The tilted nanorods are expected to be easily detached due to the reduction in anchoring surface. It is also noted that some of the detached ZnO nanorods are directly contacted with the HTM, acting potentially as sites for recombination.

The colloidal-based ZnO nanorod cell shows a higher voltage and fill factor compared to the solution- and powder-based ones (Figure 1.22a). The values of series resistance (R_S) and shunt resistance (R_{SH}) estimated from the current–voltage curves explain the differences in the fill factor. It is

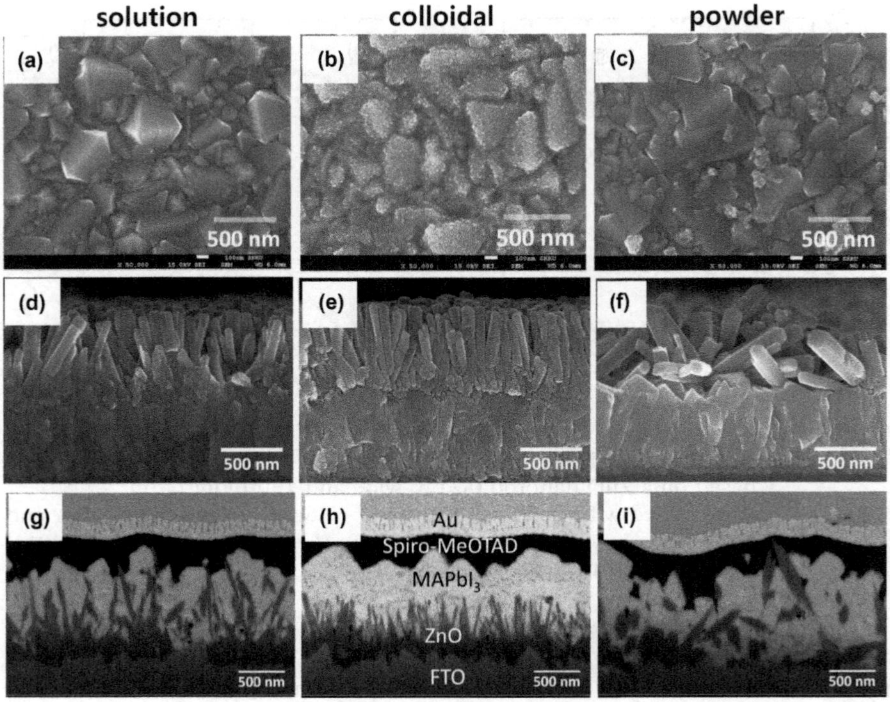

Figure 1.21 Plane-view SEM images of the seed layers and cross-sectional SEM images of ZnO nanorods formed by (a,d) an ethanolic solution of zinc acetate (solution), (b,e) a colloidal solution of zinc acetate in the presence of KOH (colloidal) and (c,f) ZnO nanopowder dispersed in ethanol (powder). FIB-assisted cross-sectional SEM images of the perovskite solar cells based on ZnO nanorods grown on the three different seed layers formed from (g) solution, (h) colloidal and (i) powder precursor solutions.
Reproduced with permission from D.-Y. Son, K.-H. Bae, H. S. Kim and N.-G. Park, *J. Phys. Chem. C*, 2015, **119**, 10321–10328. Copyright (2015) American Chemical Society.[34]

interesting to note that a similar short-circuit current density and EQE (also known as incident photon to current efficiency, IPCE) are observed (Figure 1.22b), which indicates that the photocurrent is hardly altered by the ZnO nanorod morphology. The significant difference in voltage is related to the recombination resistance (R_{rec}), associated with the interface and morphology of the ZnO nanorods. At low bias potential, R_{rec} (which is related to the interface or bulk of the perovskite) increases in the order of colloidal > solution > powder (Figure 1.22c). A higher recombination resistance leads to a higher V_{oc}. At high bias potential, R_{rec} is probably related to the ZnO nanorod/perovskite interface or perovskite/HTM interface. It decreases in the order of colloidal > solution > powder, which suggests that suppression of recombination at the ZnO nanorod/perovskite interface is greatest for the colloidal-based seed layer. The resistance of the selective contacts can be

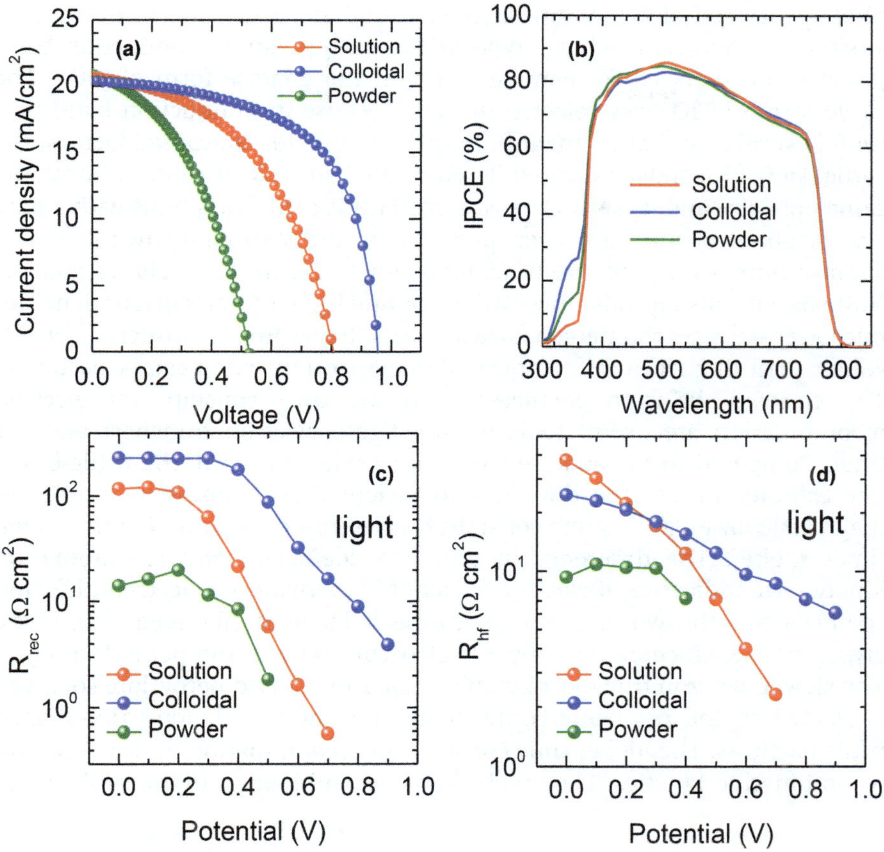

Figure 1.22 (a) Photo current–voltage curves, (b) IPCE spectra, (c) recombination resistance under light, and (d) high frequency resistance under light for the perovskite solar cells based on ZnO nanorods grown on three different seed layers.
Reproduced with permission from D.-Y. Son, K.-H. Bae, H. S. Kim and N.-G. Park, *J. Phys. Chem. C*, 2015, **119**, 10321–10328. Copyright (2015) American Chemical Society.[34]

obtained from the high frequency resistance (R_{hf}) in Figure 1.22d. In the bias potential range between 0.2 and 0.7 V, the higher R_{hf} observed for the colloidal-based ZnO nanorod cell compared to those of the solution- and powder-based cells is due to the full coverage on FTO, which is also responsible for the high voltage.

1.2.2.4 *Comparison of Anatase and Rutile TiO_2 in $CH_3NH_3PbI_3$ Perovskite Solar Cells*

In the case of mesoscopic perovskite solar cells comprising mesoporous TiO_2 films filled with perovskite, a contentious issue is whether or not photo-generated electrons are injected from the perovskite into TiO_2. To answer this question, other n-type oxides having similar conduction band position are required for comparison with the anatase form of TiO_2. The rutile form of TiO_2 was selected initially because its conduction band position is similar to that of anatase. In order to exclude unwanted factors, the rutile TiO_2 film must be carefully controlled in order to have almost the same pore size and layer thickness. Mesoporous rutile TiO_2 films with a pore size of about 40 nm have been prepared in order to compare them with anatase films with pore size of about 43 nm.[35] Figure 1.23a shows that the rutile-based cells reproducibly exhibit slightly higher photocurrent densities but lower voltages than anatase-based cells. Regarding the different photovoltaic performance, a different extent of electron injection can be assumed. The electron diffusion coefficient and the time constant for electron recombination are useful tools to investigate electron transport and life time. Compared to the anatase-based perovskite solar cell, the rutile-based one exhibits an electron diffusion coefficient that is almost one order of magnitude lower, and a time constant for electron recombination that is ten times higher. The difference in diffusion coefficient and recombination kinetics are indicative of electron injection, otherwise the electron diffusion coefficients in the two cases would be expected to be similar regardless of the crystal phase, since electron transport occurs only in the perovskite layer. The slower electron transport kinetic in the rutile-based perovskite solar cell is related to the poor interconnectivity associated with long rice-shaped rutile particles. The higher time constant for recombination indicates slower recombination kinetics. The much slower recombination in the rutile-based

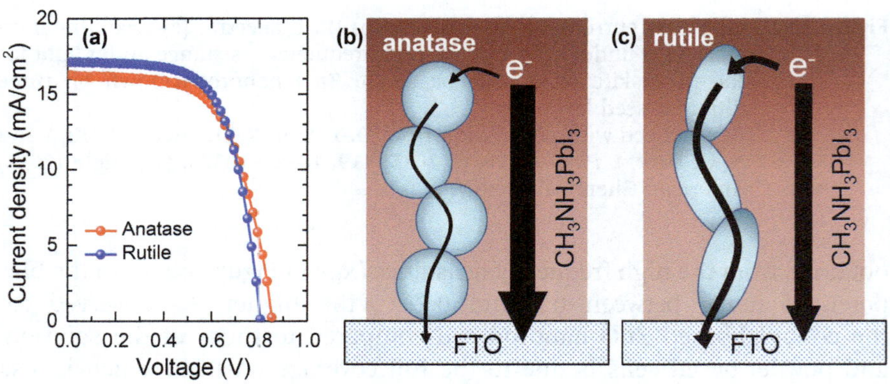

Figure 1.23 (a) Current–voltage curves of perovskite solar cells with mesoporous anatase or rutile TiO_2. (b,c) Comparison of the extent of electron injection and dual pathway for electron transport between anatase and rutile TiO_2. A larger number of electrons injected to rutile TiO_2 lower the overall Fermi energy level at equilibrium, leading to a lower voltage.
Reproduced from ref. 35 with permission from the Royal Society of Chemistry.

perovskite solar cell is a consequence of more injected electrons in rutile TiO_2, as recombination is accelerated in the case of ineffective charge separation at the oxide/perovskite interface. Moreover, the lower V_{oc} for the rutile-based device is also related to the extent of electron injection. When considering that the conduction band edge of TiO_2 is lower than that of the perovskite, an increase in the amount of the injected electrons will lower the Fermi level of the perovskite, and thereby decrease the voltage. According to transient photocurrent–voltage measurements along with the observed current–voltage characteristics, electron injection from the perovskite to the TiO_2 layer occurs partially and, evidently, the rutile layer accepts relatively more photo-excited electrons than the anatase layer. One can conclude that a dual pathway for electron transport seems to be present in the mesoscopic structure. The extent of electron injection is schematically compared between anatse and rutile in Figure 1.23.

1.2.2.5 $HC(NH_2)_2PbI_3$ Perovskite with Mesoporous and Nano-helical TiO_2

$CH_3NH_3PbI_3$ is a key component in current perovskite solar cells. However, the material needs to be engineered further to achieve a higher absorbance and lower band gap. The valence band of $CH_3NH_3PbI_3$ is related to the filled p orbitals of the X^- ions, and the conduction band is associated with the empty p orbitals of Pb^{2+} ions.[36] It is also mentioned in ref. 35 that the organic cation cannot influence the band position directly, but it can indirectly affect the band gap. Thus, it is expected that band gap changes without significant tuning of the band position will occur upon replacing methylammonium with other organic cations. Replacement of methylammonium by formamidinium ($HC(NH_2)_2$) results in a decrease in the band gap from 1.55 eV to 1.5 eV.[37] The corresponding red shift of the onset wavelength by around 30–40 nm is observed in the UV-Vis absorbance spectra going from the $CH_3NH_3PbI_3$ perovskite to the $HC(NH_2)_2PbI_3$ one (Figure 1.24), without altering the absorption coefficient.

A perovskite solar cell based on two-step deposited $HC(NH_2)_2PbI_3$ in a mesoscopic structure demonstrated a PCE of 16% in 2014.[37] The mesoporous TiO_2 layer has been found to play an important role in the photovoltaic performance of formamidinium perovskite cells, and the influence of the mesoporous TiO_2 film on charge collection kinetics has been investigated.[37] Figure 1.25a displays cross-sectional SEM images showing the mesoporous TiO_2 layer and the perovskite capping layer, showing that the pores are completely filled with perovskite. The layer structures are illustrated schematically in Figure 1.25b. The kinetics of charge collection is influenced significantly by the presence or absence of the mesoporous TiO_2 film, as well as by the thickness of the TiO_2 film (Figure 1.25c). The time-dependent photocurrent signal shown in Figure 1.25c confirms that the signal amplitude increases with the increasing TiO_2 film thickness, along

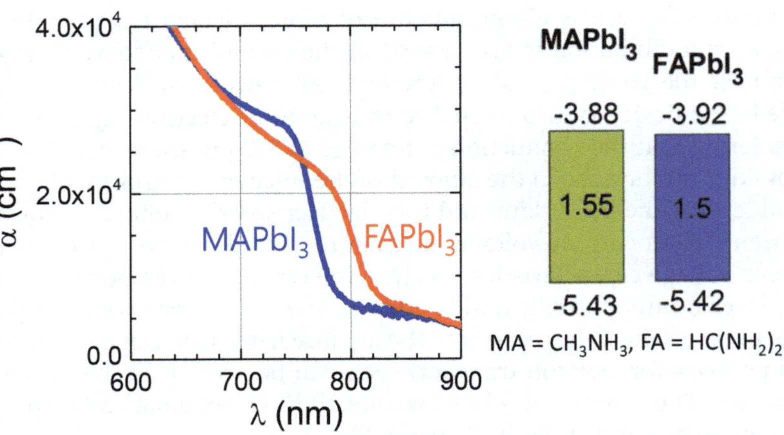

Figure 1.24 Absorption coefficient as a function of the wavelength for $FAPbI_3$ ($FA = CH(NH_2)_2$) and $MAPbI_3$ ($MA = CH_3NH_3$) (left) along with the band position and band gap (right).
Reproduced from ref. 37 with permission from John Wiley and Sons. Copyright © 2014 WILEY-VCH Verlag GmbH & Co. KGaA, Weinheim.

with fast collection kinetics. This underlines the fact that the mesoporous TiO_2 film plays an important role in facilitating photocurrent collection. Moreover, the fill factor is improved after the introduction of mesoporous TiO_2 and the increase of its thickness. The inclusion of mesoporous TiO_2 results in a decrease in the series resistance and an increase in the shunt resistance, which is responsible for the improved fill factor.

In order to investigate the electron injection behavior at the TiO_2/$HC(NH_2)_2PbI_3$ interface, a nano-helical TiO_2 layer was introduced. Helical TiO_2 arrays with different pitch (p) and radius (r), helix-1 ($p/2 = 118$ nm, $r = 42$ nm), helix-2 ($p/2 = 353$ nm, $r = 88$ nm) and helix-3 ($p/2 = 468$ nm, $r = 122$ nm), were grown on FTO glass by oblique-angle electron beam evaporation, as confirmed by the SEM images in Figure 1.26. It is essential to rotate the substrate at controlled time intervals and high substrate-tilt angles. $HC(NH_2)_2PbI_3$ perovskite was deposited in the helical TiO_2 films by the two-step technique. The dependence of the photovoltaic performance and electron transport behavior on the helix morphology was investigated.[38]

The cell fabricated using helix-1 showed a higher short-circuit current density, while the cell using helix-3 exhibited a slightly higher open-circuit voltage (Figure 1.27). The higher photocurrent from the helix-1 cell is related to the enhanced light scattering efficiency and higher absorbed photon-to-current conversion efficiency. The time-dependent photocurrent response also depends on the helical TiO_2 nanostructure used. At 4 Hz and 10 Hz, the helix-1 cell shows the fastest photocurrent response, while the helix-2 cell shows the slowest response, which implies that charge collection is less efficient for the helix-2 cell compared to the others. At an increased chopping rate of 50 Hz, the photocurrent response is less intense, and a fast

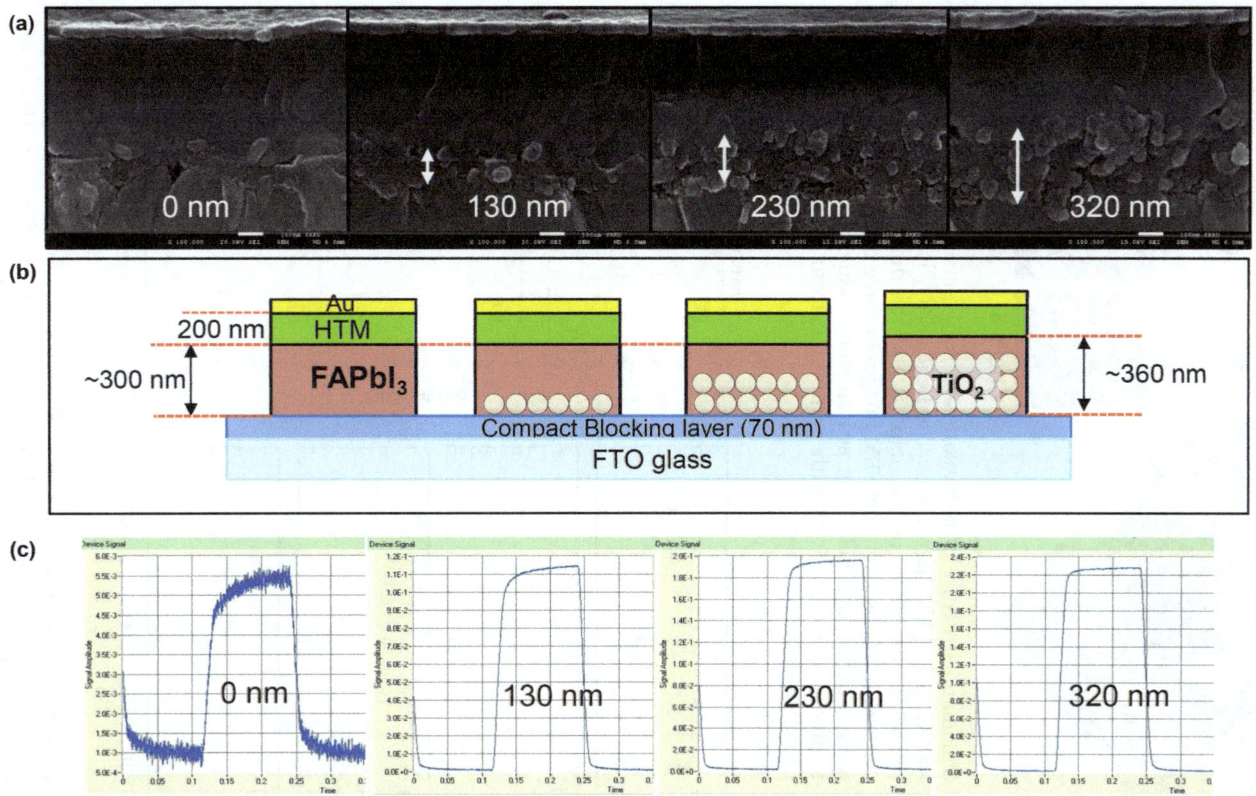

Figure 1.25 (a) Cross-sectional SEM images, and (b) schematic structure of $HC(NH_2)_2PbI_3$ layers deposited on 0 nm (compact blocking TiO_2 layer only), 130 nm, 230 nm and 320 nm thick mesoporous TiO_2 films. (c) Time-dependent photocurrent response as a function of the mesoporous TiO_2 film thickness.
Reproduced from ref. 37 with permission from John Wiley and Sons. Copyright © 2014 WILEY-VCH Verlag GmbH & Co. KGaA, Weinheim.

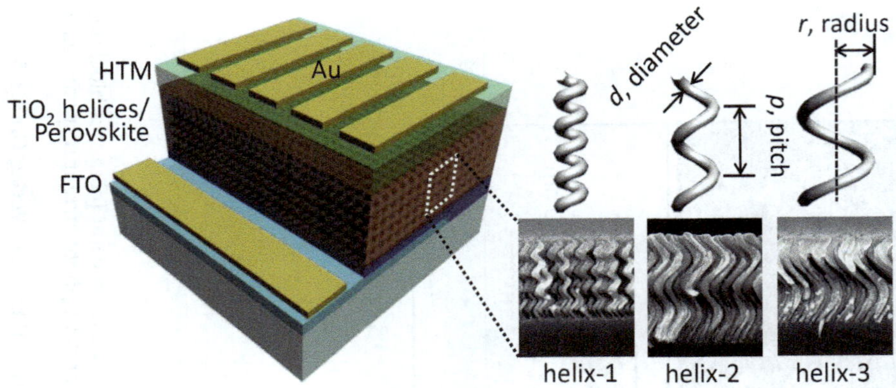

Figure 1.26 Schematic structure of perovskite solar cell consisting of TiO_2 helices and $HC(NH_2)_2PbI_3$ perovskite. The helical TiO_2 is defined by the helix pitch (p), the radius of the helix (r) and the diameter of the helical cylinder (d). Cross sectional SEM images of three different TiO_2 helices, helix-1, helix-2 and helix-3, grown on silicon wafers.
Reproduced from ref. 38 with permission from the Royal Society of Chemistry.

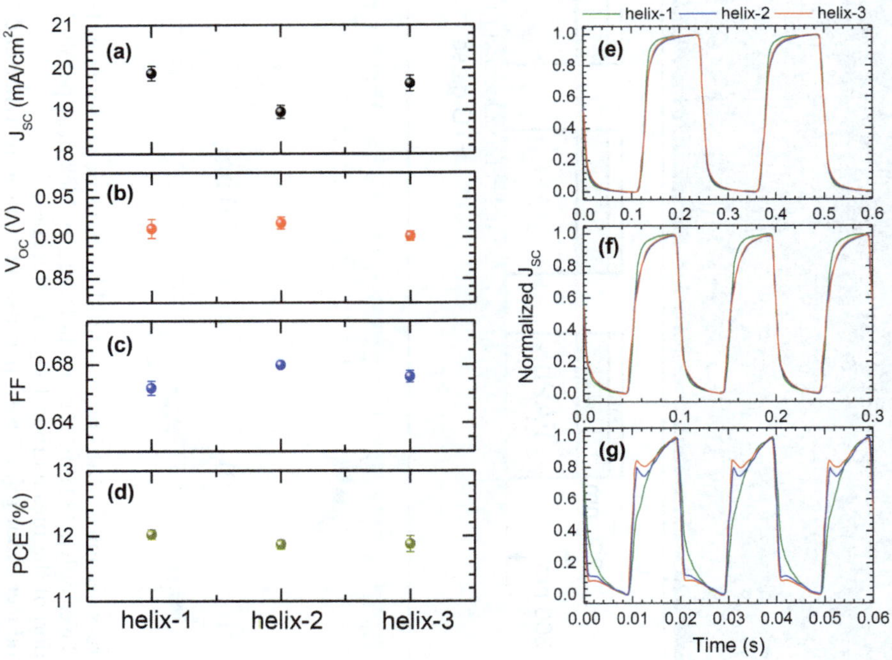

Figure 1.27 (a) Short-circuit current density (J_{SC}), (b) open-circuit voltage (V_{OC}), (c) fill factor (FF), and (d) power conversion efficiency (PCE) of $HC(NH_2)_2PbI_3$ perovskite solar cells based on TiO_2 nanohelices. Time-limited photocurrent response at different chopping frequencies of (e) 4 Hz, (f) 10 Hz and (g) 50 Hz. 530 nm monochromatic beam generated from a 75 W Xenon lamp was used as the light source.
Reproduced from ref. 38 with permission from the Royal Society of Chemistry.

component can be identified. This fast component is probably due to charge transport through the perovskite layer itself, given the three orders of magnitude higher diffusion coefficient of electrons in the perovskite compared to that in TiO_2. The slower component seems to be related to charge transport in helical TiO_2. The amplitude of the fast component increases in order helix-2 < helix-3 < helix-1, which is in reverse order of the surface area. With a larger surface area, a larger fraction of electrons generated in the perovskite appear to be injected into the TiO_2 network, so that the fraction of the electrons transported through the perovskite layer decreases. This result again proves partial electron injection from the $HC(NH_2)_2PbI_3$ perovskite to TiO_2, and a strong probability of injection in the vicinity of the interface.

1.3 Summary

The emergence and progress of perovskite solar cells have been described briefly. Thanks to the important pioneering works on perovskite-sensitized solar cells reported in 2009, 2011 and 2012, a surge of reports on perovskite solar cells has followed. There is no doubt that these pioneering works should be appreciated. Perovskite materials can be applied to any kind of junction structure, although mesoscopic solar cells containing mesoporous or nanostructured oxide layers have been the most intensively reviewed in this chapter. The basic fundamentals of $CH_3NH_3PbI_3$ have also been considered, and it is clear that careful attention should be paid to the molecular motion of the organic ion because of the underlying correlation between molecular motion and photovoltaic properties. Mesoporous TiO_2 was found to play an important role in electron transport and the resistance components in the devices. Electron injection was evident at the junction between nanocrystalline TiO_2 and the perovskite, which could be isolated by dual function fit of the data obtained from transient photocurrent measurements. In order to achieve even higher efficiencies with mesoscopic perovskite solar cells, defect-free high quality perovskite grains without non-radiative recombination are essential.

Acknowledgements

This work was supported by National Research Foundation of Korea (NRF) grants funded by the Ministry of Science, ICT & Future Planning (MSIP) of Korea under contracts No. NRF-2012M3A6A7054861 (Nano Material Technology Development Program) and NRF-2012M3A6A7054861 (Global Frontier R&D Program on Center for Multiscale Energy System).

References

1. D. Weber, *Z. Naturforsch.*, 1978, **33b**, 1443–1445.
2. D. Weber, *Z. Naturforsch.*, 1978, **33b**, 862–865.

3. D. B. Mitzi, C. A. Feild, W. T. A. Harrison and A. M. Guloy, *Nature*, 1994, **369**, 467–469.
4. A. Kojima, K. Teshima, Y. Shirai and T. Miyasaka, *J. Am. Chem. Soc.*, 2009, **131**, 6050–6051.
5. J.-H. Im, C.-R. Lee, J.-W. Lee, S.-W. Park and N.-G. Park, *Nanoscale*, 2011, **3**, 4088–4093.
6. H.-S. Kim, C.-R. Lee, J.-H. Im, K.-B. Lee, T. Moehl, A. Marchioro, S.-J. Moon, R. Humphry-Baker, J.-H. Yum, J. E. Moser, M. Grätzel and N.-G. Park, *Sci. Rep.*, 2012, **2**, 591.
7. W. S. Yang, J. H. Noh, N. J. Jeon, Y. C. Kim, S. Ryu, J. Seo and S. I. Seok, *Science*, 2015, **348**, 1234–1237.
8. U. Bach, D. Lupo, P. Comte, J. E. Moser, F. Weissörtel, J. Salbeck, H. Spreitzer and M. Grätzel, *Nature*, 1998, **395**, 583–585.
9. M. M. Lee, J. Teuscher, T. Miyasaka, T. N. Murakami and H. J. Snaith, *Science*, 2012, **338**, 643–647.
10. J. H. Heo, S. H. Im, J. H. Noh, T. N. Mandal, C.-S. Lim, J. A. Chang, Y. H. Lee, H.-j. Kim, A. Sarkar, Md. K. Nazeeruddin, M. Grätzel and S. I. Seok, *Nat. Nanophotonics*, 2013, **7**, 486–491.
11. S. D. Stranks, G. E. Eperon, G. Grancini, C. Menelaou, M. J. P. Alcocer, T. Leijtens, L. M. Herz, A. Petrozza and H. J. Snaith, *Science*, 2013, **342**, 341–344.
12. G. Xing, N. Mathews, S. Sun, S. S. Lim, Y. M. Lam, M. Grätzel, S. Mhaisalkar and T. C. Sum, *Science*, 2013, **342**, 344–347.
13. M. Liu, M. B. Johnston and H. J. Snaith, *Nature*, 2013, **501**, 395–398.
14. J.-Y. Jeng, Y.-F. Chiang, M.-H. Lee, S.-R. Peng, T.-F. Guo, P. Chen and T.-C. Wen, *Adv. Mater.*, 2013, **25**, 3727–3732.
15. S. Sun, T. Salim, N. Mathews, M. Duchamp, C. Boothroyd, G. Xing, T. C. Sum and Y. M. Lam, *Energy Environ. Sci.*, 2014, **7**, 399–407.
16. R. E. Wasylishen, O. Knop and J. B. Macdonald, *Solid State Commun.*, 1985, **56**, 581–582.
17. A. Poglitsch and D. Weber, *J. Chem. Phys.*, 1987, **87**, 6373–6378.
18. N. Onoda-Yamamuro, T. Matsuo and H. Suga, *J. Phys. Chem. Solids*, 1990, **51**, 1383–1395.
19. H. S. Jung and N.-G. Park, *Small*, 2015, **11**, 10–25.
20. H.-S. Kim, S. K. Kim, B. J. Kim, K.-S. Shin, M. K. Gupta, H. S. Jung, S.-W. Kim and N.-G. Park, *J. Phys. Chem. Lett.*, 2015, **6**, 1729–1735.
21. J. M. Frost, K. T. Butler, F. Brivio, C. H. Hendon, M. Van Schilfgaarde and A. Walsh, *Nano Lett.*, 2014, **14**, 2584–2590.
22. J. H. Lee, R. H. Shin and W. Jo, *Phys. Rev. B: Condens. Matter Mater. Phys.*, 2011, **84**, 094112.
23. Z. Xiao, Y. Yuan, Y. Shao, Q. Wang, Q. Dong, C. Bi, P. Sharma, A. Gruverman and J. Huang, *Nat. Mater.*, 2015, **14**, 193–198.
24. N.-G. Park, *Nat. Mater.*, 2015, **14**, 140–141.
25. H.-S. Kim, I. Mora-Sero, V. Gonzalez-Pedro, F. Fabregat-Santiago, E. J. Juarez-Perez, N.-G. Park and J. Bisquert, *Nat. Commun.*, 2013, **4**, 2242.

26. M. M. Lee, J. Teuscher, T. Miyasaka, T. N. Murakami and H. J. Snaith, *Science*, 2012, **338**, 643–647.
27. J.-H. Im, I.-H. Jang, N. Pellet, M. Grätzel and N.-G. Park, *Nat. Nanotechnol.*, 2014, **9**, 927–932.
28. S. Mastroianni, F. D. Heinz, J.-H. Im, W. Veurman, M. Padilla, M. C. Schubert, U. Würfel, M. Grätzel, N.-G. Park and A. Hinsch, *Nanoscale*, 2015, **7**, 19653–19662.
29. J. Nelson, *Phys. Rev. B: Condens. Matter Mater. Phys.*, 1999, **59**, 15374–15380.
30. J.-H. Im, J. Luo, M. Franckevicius, N. Pellet, P. Gao, T. Moehl, S. M. Zakeeruddin, M. K. Nazeeruddin, M. Grätzel and N.-G. Park, *Nano Lett.*, 2015, **15**, 2120–2126.
31. J. Hu, T. W. Odom and C. Lieber, *Acc. Chem. Res.*, 1999, **32**, 435–445.
32. D. Bi, G. Boschloo, S. Schwarzmuller, L. Yang, E. M. J. Johanssona and A. Hagfeldt, *Nanoscale*, 2013, **5**, 11686–11691.
33. D.-Y. Son, J.-H. Im, H.-S. Kim and N.-G. Park, *J. Phys. Chem. C*, 2014, **118**, 16567–16573.
34. D.-Y. Son, K.-H. Bae, H. S. Kim and N.-G. Park, *J. Phys. Chem. C*, 2015, **119**, 10321–10328.
35. J.-W. Lee, T.-Y. Lee, P. J. Yoo, M. Grätzel, S. Mhaisalkar and N.-G. Park, *J. Mater. Chem. A*, 2014, **2**, 9251–9259.
36. K. T. Butler, J. M. Frost and A. Walsh, *Mater. Horiz.*, 2015, **2**, 228–231.
37. J.-W. Lee, D.-J. Seol, A.-N. Cho and N.-G. Park, *Adv. Mater.*, 2014, **26**, 4991–4998.
38. J.-W. Lee, S. H. Lee, H.-S. Ko, J. Kwon, J. H. Park, S. M. Kang, N. Ahn, M. Choi, J. K. Kim and N.-G. Park, *J. Mater. Chem. A*, 2015, **3**, 9179–9186.

CHAPTER 2

Towards Optimum Solution-processed Planar Heterojunction Perovskite Solar Cells

FABIAN HANUSCH, MICHIEL PETRUS AND PABLO DOCAMPO*

Centre for Nanoscience (CeNS), LMU Munich, Munich, Germany
*Email: pablo.docampo@cup.lmu.de

2.1 Introduction

Over the last few years, metal halide perovskite materials have proven to be very well suited for their application in solid-state solar cells. In only 3 years, efficiencies have rocketed from an initial value of 3.8% to over 20% in recent reports.[1–3] While it was initially crucial to develop synthetic routes for phase-pure materials,[4,5] most of the high efficiency optimization has targeted both the optimum surface coverage and overall crystal morphology, as well as optimizing the energy level alignment with the charge selective contacts.

Currently, two main perovskite solar cell (PSC) architectures are being investigated: mesoporous and planar. In the first case, mesostructured semiconducting metal oxides, such as titanium dioxide, are used as scaffolds, while the perovskite absorber is infiltrated within the pores.[5–8] In this configuration, the TiO_2 layer covers the whole substrate, and the size of the perovskite crystals within the scaffold is limited by the pore size, which is typically tens of nanometers. There is wide debate in the community about the exact effect of the scaffold on perovskite solar cell operation.[9–12] Regardless of their influence on the solar cell, metal oxide scaffolds require

RSC Energy and Environment Series No. 16
Unconventional Thin Film Photovoltaics
Edited by Enrico Da Como, Filippo De Angelis, Henry Snaith and Alison Walker

Published by the Royal Society of Chemistry, www.rsc.org

high sintering temperatures in order to maximize transport properties.[9,13] Planar heterojunction perovskite solar cells, in contrast, employ a thin film of the perovskite absorber sandwiched between the charge selective contacts.[14,15] In this configuration, the high temperature sintered mesoporous scaffold can be replaced either by metal oxides prepared at lower temperatures or by organic contact materials.[16] To function efficiently, planar heterojunction perovskite solar cells require a homogeneous and dense layer of a light-absorbing material without gaps between the individual perovskite crystals.[17,18]

For PSCs to compete in the current photovoltaic sector, a compelling market advantage is required. Perovskite solar cell fabrication costs are low, and the materials are cheap and readily available. However, beating current competing technologies in price alone is unlikely to occur, as the price of the module in mainstream photovoltaics is only a small fraction of the total cost, with other manufacturing costs such as installation, module assembly, *etc.* taking the biggest shares. A simple solution to bring down the costs of established photovoltaic technologies using PSCs – and thus introducing them into the market – is by fabricating tandem architectures with a wide bandgap PSC and a narrow bandgap inorganic cell.[19] Low-temperature processing steps will be required for this type of application and thus, tandem PSCs based on mesoporous scaffolds are unlikely to be implemented. For this reason, this chapter focusses on the development of planar heterojunction PSCs.

The first part of this chapter deals with the impact of complete surface coverage and appropriate crystal morphology on device efficiency and considers which synthetic strategies can be applied to achieve these objectives. The second part of the chapter focusses on the contact materials that are used in different device architectures and on the issue of fabricating truly "low-cost" perovskite solar cells.

2.2 Towards Optimum Solution-processed Cells

2.2.1 Basic Cell Structure

Perovskite solar cells are generally based on an n-i-p junction, where the intrinsic perovskite material sits between hole and electron extraction layers, as depicted schematically in Figure 2.1. As a first approach, perovskite films are simply deposited *via* a 1-step spin-coating method with a subsequent annealing step, where the precursor materials are dissolved in an appropriate solvent (*e.g.* dimethylformamide or γ-butyrolactone).

Here, it soon became apparent that unfavorable substrate–solution interactions can lead to de-wetting of the precursor solution from the substrate.[15,17,21] This leads ultimately to a perovskite thin film that does not cover the whole surface and shows voids between island-like perovskite clusters. This, in turn, leads to shunting *via* direct contact between the n- and p-type layers and consequently reduce the device performance.

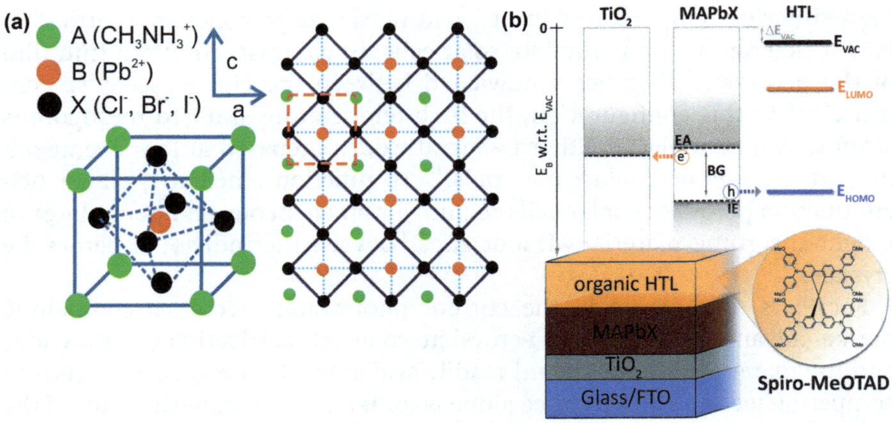

Figure 2.1 General crystal structure of ABX_3 perovskites (a)[20] and solar cell layout and energy alignment for planar heterojunction solar cells (b).[12] Figure 2.1a is reproduced from ref. 20 with permission from John Wiley and Sons © 2015 Wiley-VCH GmbH & Co. KGaA, Weinheim. Figure 2.1b is reproduced from ref. 12 with permission from the Royal Society of Chemistry.

These early results clearly demonstrated that a high substrate surface area coverage is essential in order to maximize the performance of planar heterojunction perovskite solar cells.

2.2.2 Towards 100% Surface Coverage

Every aspect of the preparation process, such as the thermal annealing step,[22–25] the addition of certain chemicals that aid film formation,[26–34] the solvent used,[35–37] the precursor composition,[18,26,29,32,38,39] the precursor concentration,[40] *etc.* has a critical influence on the perovskite crystallization rate and therefore, on the morphology of the resulting perovskite layer. Several strategies have been devised to improve the film formation properties of the perovskite materials. These include employing intermediate lead-based compounds which then slowly convert into the perovskite phase,[18,35,41] intercalation of organic compounds into lead-halide crystal structures,[6,16,17,42] the inclusion of additives into the precursor mixture[27–29,32–34] and solvent-induced immediate crystallization of the perovskite phase.[43–45] All these strategies can lead to the formation of films exhibiting 100% surface area coverage, and they are described in more detail below.

2.2.2.1 *The Influence of the Lead Precursor*

The conventional methylammonium lead iodide ($MAPbI_3$) perovskite is normally prepared by spin-coating a 1 : 1 molar mixture of PbI_2 and MAI in a suitable solvent like dimethylformamide (DMF), dimethylsulfoxide (DMSO) or γ-butyrolactone (γ-GBL).[5] The resulting film is then heat-treated to remove the excess solvent and induce the perovskite formation.[24] When using GBL as the solvent, the crystallization of the perovskite phase starts as soon as a critical precursor concentration is reached upon drying. For DMF and DMSO, however, an intermediate phase is formed, where layers of lead iodide (PbI_2) are separated by a mixture of the solvent and MAI.[35,44] This precursor phase has high ion mobility due to the remaining solvent. Upon further heat treatment, the excess solvent is removed and the perovskite crystallizes. The morphology of the resulting perovskite film is strongly dependent on the temperature and time of the heating step and can range from full surface coverage to sponge-like disrupted films.[24]

Halide mixtures, such as those formed by addition of chloride to the iodide-based perovskites, can have beneficial effects on the resulting perovskite solar cells.[17] Although the general spin-coating procedure is the same for $MAPbI_3$ and the $MAPbI_{3-x}Cl_x$ perovskite, it is necessary to use a 3 : 1 molar ratio of MAI to $PbCl_2$ for the mixed halide perovskite in order to achieve the correct final stoichiometry.[4] Detailed analysis of the crystallization process has shown that $MAPbI_{3-x}Cl_x$ forms an intermediate organolead mixed halide phase, which is later transformed into the pure perovskite phase through the evaporation of MACl.[46,47] Little or no remaining chloride is found in samples derived from $PbCl_2$ and MAI.[47,48] The chloride ions and the excess of organic cations compared to lead cations in the solution slow down the crystallization rate, leading to the growth of large, continuous crystal domains within the film and a high degree of surface coverage (Figure 2.2a).[48] An additional mechanism, proposed by Williams *et al.*, involves the formation of an intermediate short-lived phase of pure $MAPbCl_3$ that acts as a topotactic template for the crystallization of the $MAPbI_{3-x}Cl_x$ perovskite and therefore aids the formation of smooth films.[46]

Even more homogeneous and smoother layers of $MAPbI_3$ perovskite have been obtained by starting from lead acetate ($Pb(OAc)_2$) as the precursor salt (Figure 2.2e).[18,49] All three processing routes result in the formation of $MAPbI_3$ with very few impurities. The films processed from PbI_2 and $PbCl_2$ are non-continuous with large gaps in between the perovskite clusters and with inhomogeneous film thickness (Figure 2.2a–d). However, films prepared from $Pb(OAc)_2$ achieve nearly 100% surface coverage and are very smooth and highly consistent in thickness over the whole substrate (Figure 2.2e,f).

The smoothness and high surface coverage of the resulting films is attributed to the rapid crystallization of perovskite crystals with a high nucleation rate. MAOAc evaporates at temperatures much lower than the sublimation temperature of MACl and MAI, namely 130 K and 150 K, respectively. With this in mind, it is clear that the use of an acetate based lead precursor has a strong influence on the evaporation and therefore the crystallization dynamics, resulting in the observed large difference in morphology for the different preparation routes. The obtained efficiencies for optimized solar cells for each lead precursor follow the same trend as the morphology and surface coverage of the thin films. For solar cells prepared

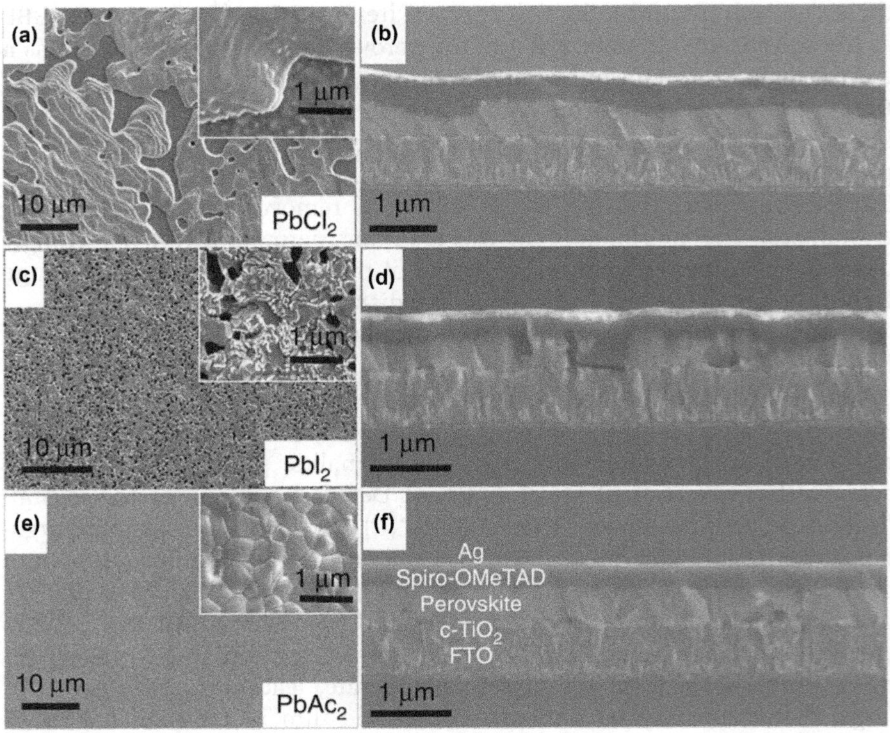

Figure 2.2 Comparative SEM micrographs in top and cross-section view using $PbCl_2$ (a,b) PbI_2 (c,d) or $Pb(OAc)_2$ (e,f) as the lead source.
Adapted by permission from Macmillan Publishers Ltd: *Nat. Commun.*[18] Copyright 2015.

Table 2.1 Summary of the performance and morphology quality of the assembled films starting from different lead sources.[18]

Lead precursor	Reported PCE_{max}	Grain size (approx.)	Surface coverage	RMS roughness
PbI_2	9.3	100–500 nm	90%	52.2 nm
$PbCl_2$	12.0	10–50 µm	90%	62.4 nm
$Pb(OAc)_2$	14.0	0.5–1 µm	100%	12.3 nm

with comparable architectures, *i.e.* employing undoped TiO_2 and spiro-OMeTAD as the charge extraction layers, the lead acetate route gave the highest efficiency, followed by $PbCl_2$ and PbI_2 in this order (see Table 2.1).

2.2.2.2 Additive-controlled Crystallization

A different approach to ensure the formation of homogeneous films is the use of additives in the spin-coating solution that affect the crystallization and formation of the perovskite layer. The most frequently used additives are ammonium or phosphonium halides,[29,30,34] dihalide alkanes[27,33] or protic acids.[28,31] When incorporating small amounts of ammonium chloride (NH_4Cl) or methylammonium chloride (MACl) into the spin-coating solution for the $MAPbI_3$ perovskite, the film homogeneity is greatly improved (Figure 2.3a and b). A similar effect has been reported for the phosphonium

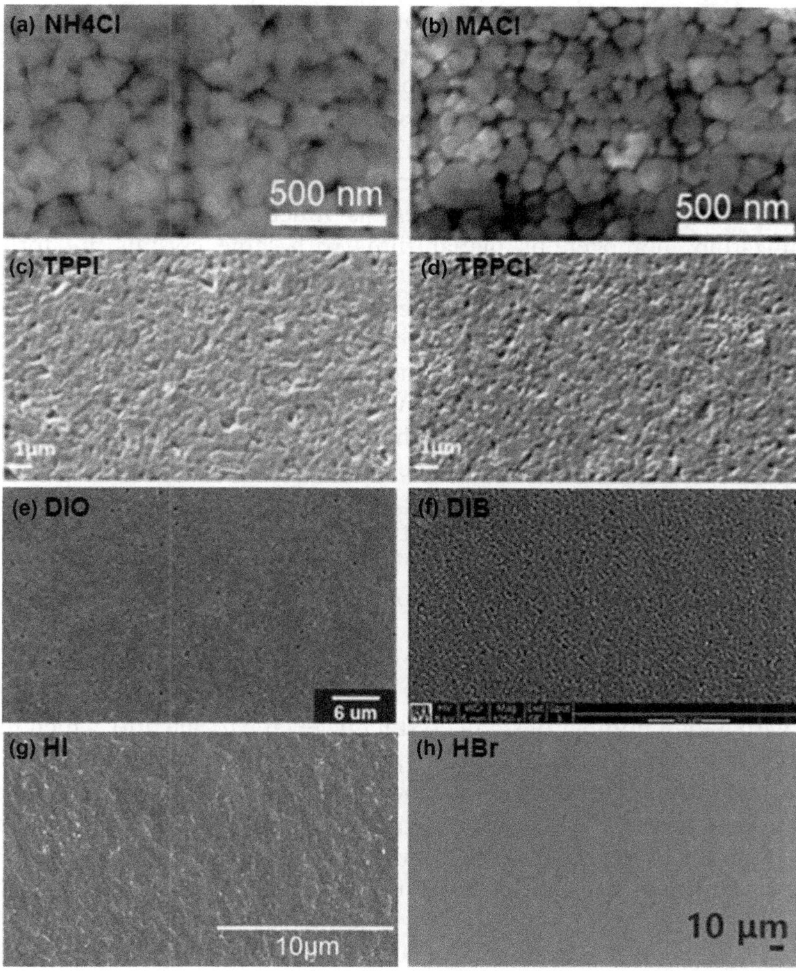

Figure 2.3 Perovskite film morphologies obtained using various additives (a) ammonium chloride (NH_4Cl), (b) methylammonium chloride (MACl), (c) tetraphenylphosphonium iodide (TPPI), (d) tetraphenylphosphonium chloride (TPPCl), (e) diiodooctane (DIO), (f) diiodobutane (DIB), (g) hydriodic acid (HI) and (h) hydrobromic acid (HBr).
Figure 2.3a, b, f and g are reproduced from ref. 27, 31, and 34 with permission from the Royal Society of Chemistry. Figure 2.3c, d, e and h are reproduced from ref. 28, 29, and 33 with permission from John Wiley and Sons © 2014/2015 Wiley-VCH Verlag GmbH& Co. KGaA, Weinheim.

halides tetraphenylphosphonium iodide and chloride (TPPI and TPPCl).[29] Since the TPP^+ cation is stable on its own, there is an excess of iodide and chloride anions in the solution. These free ions speed up the growth kinetics, leading to a larger number of crystal nuclei,[29] smoother films and therefore more efficient solar cells (Figure 2.3c and d).

The addition of dihalide alkanes such as 1,8-diiodooctane (DIO) or the smaller 1,4-diiodobutane (DIB) has a similarly profound effect on the morphology of the perovskite thin films.[27,33] Not only do these additives increase the solubility of the lead precursors in DMF,[33] but they also slow down the crystallization rate of the perovskite during thermal annealing, which leads to films consisting of fewer, but larger crystals. These effects are attributed to chelation of the Pb^{2+} ions in the solution by DIO or DIB.[27,33] Before the perovskite can form crystals, the chelating agents have to be thermally removed. Since this process is much slower than the actual crystallization, homogenous planar films with large crystals are formed (Figure 2.3e and f).

A comparable slow-down of the crystallization rate can be achieved by adding solvents to the reaction mixture that increase the precursor solubility. This approach has been demonstrated for the formation of $FAPbI_3$ and $MAPbBr_3$. The solvents used were HI for $FAPbI_3$ and HBr for $MAPbBr_3$.[28,31] In each of these cases, the addition of only 10% of the second solvent was enough to increase the concentration of the precursors in the solution to a level where, upon drying, a highly super-saturated solution is formed. Further heating of this solution leads to the rapid formation of many crystal nuclei. This, in turn, leads to a homogeneous crystallite size distribution and therefore to homogeneous thin films (Figure 2.3g and h).[28,31]

In general, the addition of easily removable species, like ammonium or phosphonium halides, speeds up the crystallization reaction and leads to films with a high number of small crystal nuclei (100 to 500 nm), which form a polycrystalline dense layer. However, addition of less volatile species like dihalide alkanes or aqueous acids slows down the reaction, leading to the formation of fewer crystal nuclei, which can grow to sizes above 1 μm. The crystallite sizes and the approximate surface coverage for the different additives are summarized in Table 2.2. When looking at the

Table 2.2 Influence of different additives on the morphological quality of perovskite thin films and the efficiency of the resulting solar cells.

Additive	Perovskite	Reported PCE_{max}	Grain size (approx.)	Surface coverage
NH_4Cl[30,34]	$MAPbI_3$	9.93	250 nm	~99%
TPPI/TPPCl[29]	$MAPbI_3$	11.7	<1 μm	~99%
DIO[33]	$MAPbI_3$	11.8	1 μm	95%
DIB[27]	$MAPbI_3$	13.1		95%
HI[31]	$FAPbI_3$	14.2	1 μm	100%
HBr[28]	$MAPbBr_3$	10.4		100%

performance of the resulting solar cells, it appears to be correlated with the size of the crystalline domains, where larger crystals result in more efficient solar cells.

2.2.2.3 Solvent-induced Crystallization

Solvent-induced crystallization is generally based on the deposition of an initial wet film of the perovskite precursor mixture on a substrate. This is followed by the quick addition of or the immersion into a second solvent in which the precursors are insoluble, so as to force rapid crystallization of the perovskite layer.

For the solvent engineering approach, Seok *et al.* dissolved the perovskite precursors in a mixture of GBL and DMSO which was spin-coated on a 200 nm thick mesoporous TiO_2 layer.[44] In a consecutive spin-coating step, the still wet film was treated with a toluene drop to force the precursors to precipitate out of the GBL–DMSO solution and to crystallize in the PbI_2–MAI–DMSO intermediate phase described in the previous section. After heating for 10 minutes at 100 °C, the DMSO was removed, and an extremely smooth film was obtained with crystallite size of a few 100 nm (Figure 2.4b).

A similar approach for the preparation of extremely smooth $MAPbI_3$ layers has been reported by Spiccia and co-workers.[43] 4–5 seconds after the precursor film has been deposited from DMF, the rapid crystallization of the perovskite phase is induced by dropping chlorobenzene (CB) on the still spinning film (the solubility of the precursors in the DMF phase is strongly decreased by the addition of CB). In contrast to the standard spin-coating procedure, where the crystallization is slowly induced by evaporation of the solvent and the crystals grow in a more or less undefined manner (Figure 2.4a), the high number of simultaneously created nuclei in the fast deposition conversion (FDC) procedure results in an extremely smooth film with micrometer sized grains (Figure 2.4c).

Another "anti-solvent" approach has been reported by Zhou *et al.*[45] In this case, the precursors are dissolved in a high boiling solvent ($T_B = 200$ °C). The high boiling temperature of the solvent ensures that the film still contains some residual solvent after the spin-coating process. The still wet film is then immersed in a bath of diethylether (DEE) at room temperature, in which neither the precursors nor the perovskite are soluble. The ether extracts the solvent out of the wet film, leading to an abrupt supersaturation of the liquid film and the formation of a high number of perovskite nuclei. This process is called solvent–solvent extraction (SSE), and it can be used to prepare extremely smooth perovskite thin films with grain sizes of about 100 nm (Figure 2.4d). All these procedures enable the formation of films with extremely high surface coverage and comparably large crystal grains (see Table 2.3). All solar cells fabricated with these methods exhibit high efficiencies above 13%.

Figure 2.4 SEM micrographs of perovskite thin films obtained *via* (a) conventional spin-coating, (b) solvent engineering, (c) fast deposition–crystallization and (d) the solvent–solvent extraction approach.
Figure 2.4a and c are reproduced from ref. 43 with permission from John Wiley and Sons © 2014 Wiley-VCH Verlag GmbH & Co. KGaA, Weinheim. Figure 2.4b is reproduced by permission from Macmillan Publishers Ltd: *Nat. Mater.*[44] Copyright 2014. Figure 2.4d is reproduced from ref. 45 with permission from the Royal Society of Chemistry.

Table 2.3 Film quality parameters and solar cell efficiencies for perovskite films prepared *via* different solvent-induced crystallization approaches.

Process	Solvent/antisolvent	Reported PCE_{max}	Grain size (approx.)	Surface coverage
Solvent engineering[44]	GBL + DMSO/toluene	16.7	500 nm	100%
Fast deposition conversion[43]	DMF/chlorobenzene	13.9	0.5–1 μm	100%
Solvent–solvent extraction[45]	NMP/ether	15.2	100–500 nm	100%

2.2.2.4 *Perovskite Conversion* via *Cation Infiltration*

The two-step deposition approach is an alternative solution-processing route for the preparation of perovskite thin films first reported by Liang *et al.* in the late 90s, later expanded by Burschka *et al.* for mesoporous substrates, and finally introduced by Liu *et al.* in planar heterojunction solar cells.[6,16,51] It consists of two main steps (Figure 2.5a). First, a layer of the lead precursor is deposited on the substrate *via* spin-coating from DMF or a similar solvent. To obtain a smooth precursor film, it may be necessary to preheat the solution and the substrate to induce rapid crystallization of the lead precursor.[50] After obtaining a smooth lead precursor film, the sample is immersed in a solution of the ammonium halide salt (MAI, MACl, FAI, *etc.*) in isopropanol (IPA). The lead precursor is insoluble in IPA, but the organic and halide ions can diffuse from the immersion bath through the lead precursor layer and induce the crystallization of the perovskite phase.[51] Since the diffusion speed of the ions and the crystallization rate of the

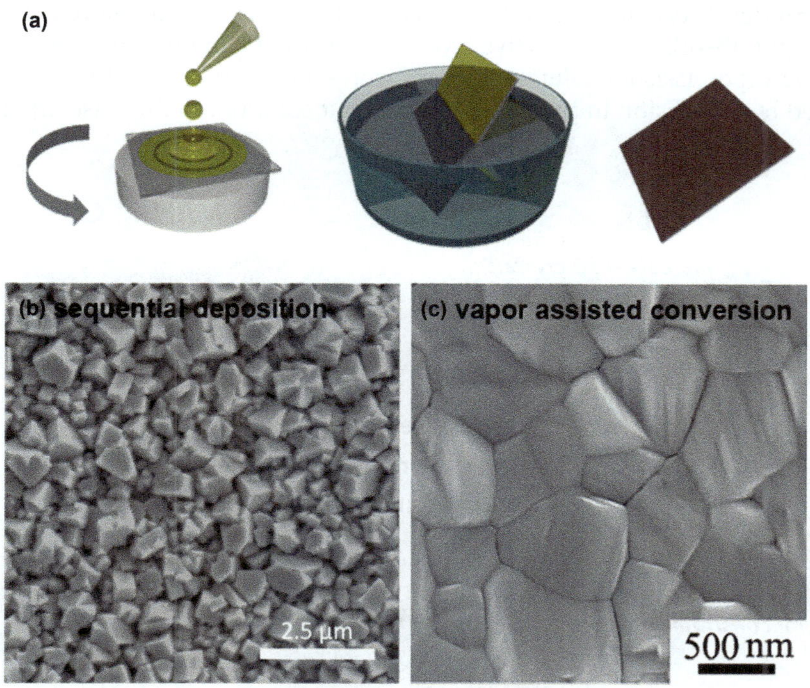

Figure 2.5 General procedure for the sequential deposition of perovskite thin films (a) and SEM micrographs of films obtained from sequential deposition (b) and vapor assisted conversion (c).
Figure 2.5b is reproduced from ref. 50 with permission from John Wiley and Sons © 2014 Wiley-VCH Verlag GmbH & Co. KGaA, Weinheim. Figure 2.5c is reprinted with permission from ref. 42. Copyright (2014) American Chemical Society.

perovskite depend on the temperature and on the concentration of the solution, fine-tuning of these parameters is necessary to obtain full surface coverage and complete conversion.[52]

Recent studies performed with 2D X-ray scattering have provided a deeper insight into the actual crystallization process for the $MAPbI_3$ perovskite (Figure 2.6).[53] Starting from a film of PbI_2, which consists of crystalline sheets on top of each other, MAI diffuses in between these sheets and starts the crystallization. As the PbI_2 layers expand upon inclusion of MAI, growth is constrained in the lateral direction. This leads to a predominant growth direction perpendicular to the substrate. The crystallites closer to the substrate experience a stronger confinement and start to crack, leaving smaller grains of perovskite, while the crystals closer to the surface can expand in a unrestricted way and form large cuboid crystals.[53]

By analogy with the 1-step protocols, addition of small amounts of MACl to the MAI immersion solution (up to 5%) improves the solar cell performance.[50] These films consist of $MAPbI_3$ cuboids, which cover nearly 100% of the surface (Figure 2.5b). In addition, the rough surface created by the randomly oriented cuboids can act as a light trap and therefore enhance the light absorbed in the active layer without modifying its thickness.[54]

The vapor-assisted solution process works in a similar way to those prepared by immersion in the organic cation salt solution. In this case, the PbI_2

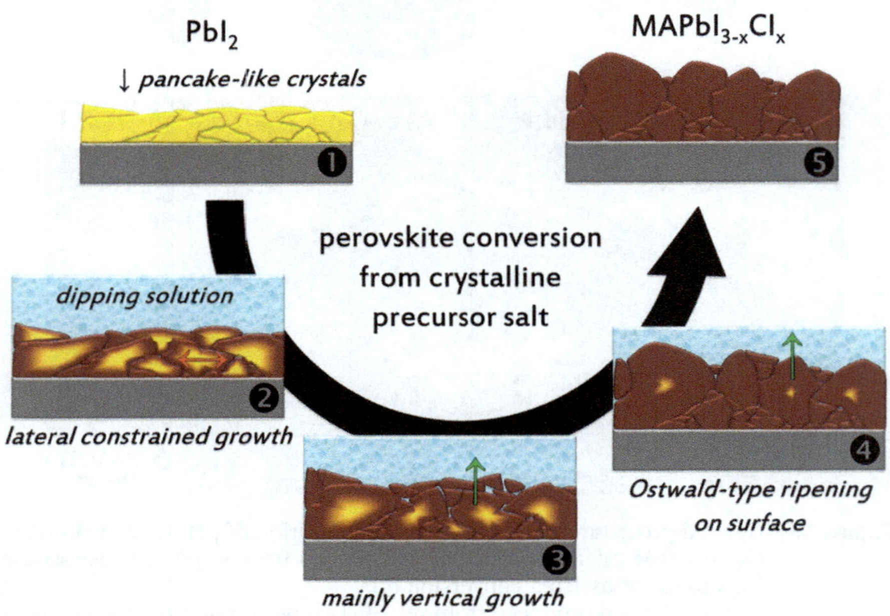

Figure 2.6 Schematic representation of the crystallization procedure of perovskite films prepared *via* two-step deposition.
Reprinted with permission from ref. 53. Copyright (2015) American Chemical Society.

Table 2.4 Film quality parameters and solar cell efficiencies for perovskite films prepared *via* sequential deposition or vapor assisted conversion.

Synthetic procedure	Reported PCE_{max}	Grain size (approx.)	Surface coverage
MAI[16]	15.7	0.1–1 μm	100%
MAI + MACl[50]	14.8	100–600 nm	100%
Vapor-assisted conversion[42]	12.1	>1 μm	100%

layer is exposed to vapor of the ammonium halide salt.[42] As MAI evaporates, it diffuses into the PbI_2 layer and starts forming the perovskite phase. As there is equilibrium between the MAI atmosphere surrounding the PbI_2 film and the perovskite phase formed, crystallization takes place very slowly, forming large crystal plates (Figure 2.5c). The resulting films are much smoother than those obtained by solution-induced crystallization, with crystallites reaching through the whole thickness of the film.[42]

The crystal size, surface coverage and the obtained efficiencies for the different 2-step approaches are shown in Table 2.4.

2.2.3 Contact Materials

There are two configurations in which the solar cells can be built. The conventional structure, in which electrons are collected at the substrate and holes at the top contact (Figure 2.7a); and the inverted structure in which the holes are collected at the substrate and the electrons are collected at the top contact (Figure 2.7b). We note that perovskite solar cell conventional and inverted architectures have the opposite polarity compared to the terminology used in the field of organic photovoltaics. Generally, hole selective and electron selective layers are used between the perovskite layer and the electrodes, not only to extract the charges selectively, but also to maximize the open circuit voltage of the solar cells.[55,56] In general, the highest occupied molecular orbital (HOMO) energy level of the hole transporting material should be well aligned with the valence band of the perovskite, while the conduction band of the perovskite should be aligned with the lowest unoccupied molecular orbital (LUMO) energy level of the electron transporting material, in order to minimize energy losses.

2.2.3.1 Electron Transporting Layer

Initial results were obtained with a TiO_2 electron extraction layer and spiro-OMeTAD as the hole extraction layer.[4,5] While efficient solar cells can be prepared in this way, the energy level alignment of the system could be improved. The first reports on alternative metal oxides included the solution-processable ZnO nanoparticle layers prepared by Liu *et al.*[16] These showed higher conductivities than the usual sol–gel derived TiO_2 layers, leading to higher fill factors and better performing solar cells. As the

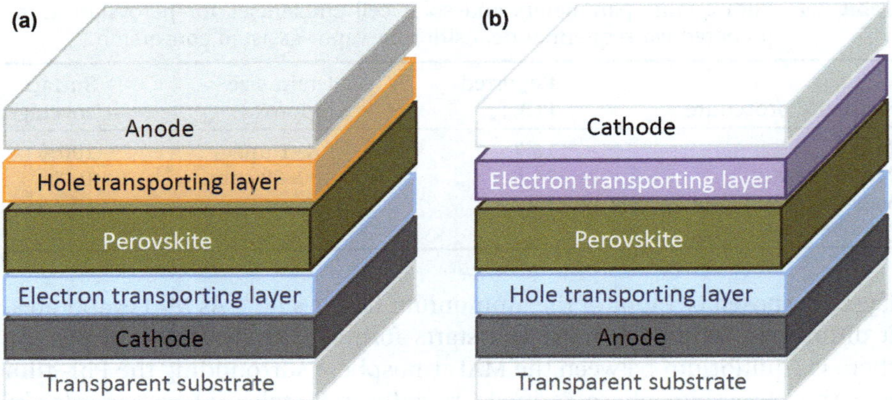

Figure 2.7 Architecture of different device configurations: (a) conventional device architecture, (b) inverted device architecture.

particles can be deposited at low temperatures, the authors were also able to demonstrate flexible perovskite solar cells with performances exceeding 10%. Other metal oxides that are more conductive than TiO_2, such as SnO_2, have also been used successfully.[32] A further step in this direction was achieved by the group of Yang,[3] who modified the TiO_2 layer with yttrium-doped TiO_2. By additionally incorporating an extra polymeric layer on the ITO surface, they brought the work function of the electrode closer to vacuum, resulting in an overall better alignment with the perovskite conduction band, resulting in a record-breaking PCE of 19.3%. Despite their good performance and low cost, metal oxides suffer from a large degree of electronic disorder, resulting in a significant number of sub-bandgap states below the conduction band edge which may limit the maximum voltage attainable by the system and have an impact on the hysteresis.[13,57,58]

Molecular electron transporters, such as fullerene derivatives, have well-defined energy levels and are thus very well suited to act as electron acceptors in the system. Fullerenes have already proven to be excellent organic electron transporting materials in organic photovoltaics, and they are often used in inverted perovskite devices.[59] The most commonly-used fullerene derivatives are ([6,6]-phenyl-C_{61}-butyric acid methyl ester ([60]PCBM) and ([6,6]-phenyl-C_{71}-butyric acid methyl ester ([70]PCBM), as they are solution processable from orthogonal solvents and possess a sufficiently high electron mobility and energy alignment with the perovskite. Chiang *et al.* demonstrated an inverted device using [70]PCBM as the electron-transporting layer with a PCE exceeding 16%,[60] while Heo *et al.* reported inverted devices based on [60]PCBM with efficiencies of up to 18.1%.[61] It is worth noting that inverted planar heterojunction solar cells based on organic contacts have been reported to not show the hysteresis that is commonly observed with other contacts.[62]

Buffer layers have been used in order to obtain a good alignment between the LUMO of the electron accepting material (PCBM) and the Fermi level of the metal electrode. Interface materials like LiF,[63] Ca[60] and bathocuproine (BCP)[60] between PCBM and the metal electrode can improve the device performance, mainly by increasing the fill factor. Docampo *et al.* used a TiO_x layer as the interface material between PCBM and Al, which they could solution process.[15] Other examples of solution-processable interface materials are PEIE,[3] bis-C_{60} surfactants,[38] amino-functionalized polymers (PFN and PN_4N)[64] and perylene-diimide (PDINO).[29]

2.2.3.2 Hole Transporting Materials

In order to extract the holes selectively from the perovskite, a hole transporting layer is deposited on top of the perovskite. Organic hole transporting layers can be processed from orthogonal solvents at room temperature, and their energy levels can be fine-tuned in order to match the energy levels of the perovskite. 2,2′,7,7′-Tetrakis[*N*,*N*-di-*p*-methoxyphenylamine]-9,9′-spirobifluorene (spiro-OMeTAD, Figure 2.8), is currently the most used hole

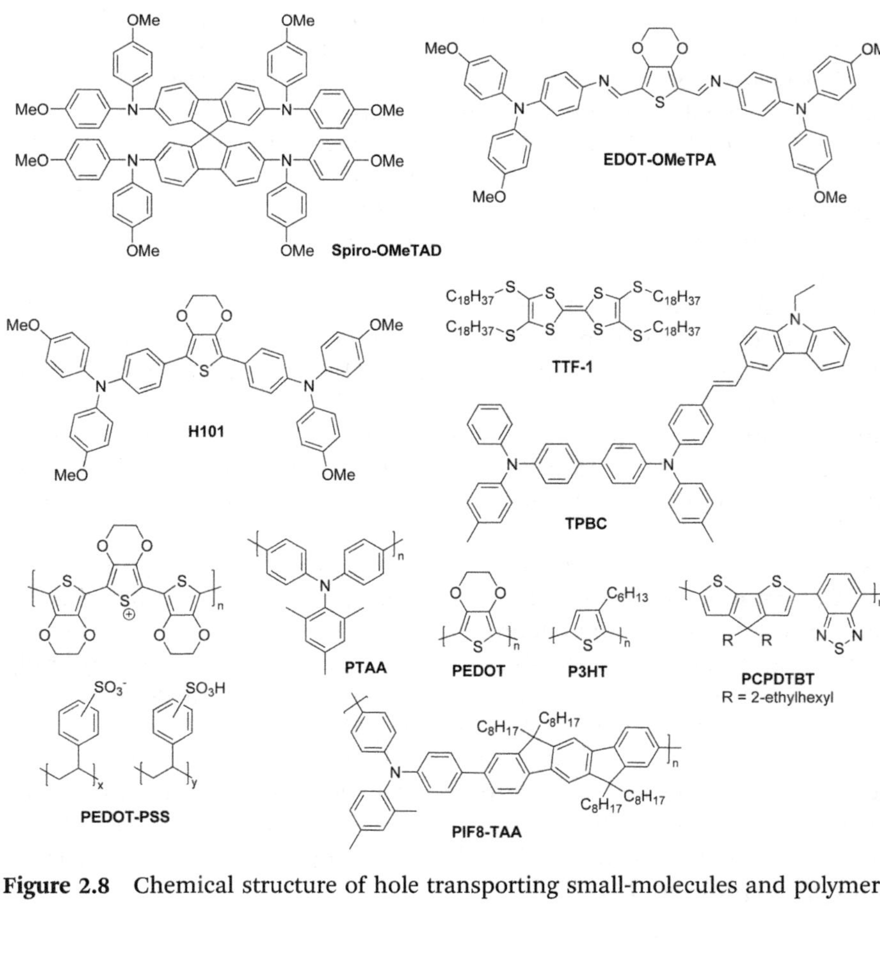

Figure 2.8 Chemical structure of hole transporting small-molecules and polymers.

transporter. The material was first introduced in perovskite-based solar cells in 2012 by Lee *et al.* to overcome the stability problems that occurred when liquid electrolytes were used.[4,5] However, since spiro-OMeTAD possesses a relatively low charge carrier mobility (like most conjugated organic materials), doping of the organic hole transporter is necessary to enhance the conductivity. Oxygen doping using bis(trifluoromethane)sulfonimide lithium salt (LiTFSI) or chemical doping with cobalt complexes like tris(2-(1*H*-pyrazol-1-yl)-4-tertbutylpyridine)cobalt(III) bis(trifluoromethylsulphonyl) imide (FK209) are the most commonly used ways to increase the performance. A PCE of 19.3% was published by the group of Yang[3] using spiro-OMeTAD as the hole transporter, which was doped with a combination of LiTFSI and FK209.

Seok *et al.*[65] synthesized spiro-OMeTAD derivatives in which the position of the methoxy group on the aromatic ring was varied. This resulted in a relative increase of 10% in the PCE for the derivative with *ortho*-methoxy substituents, which was mainly ascribed to an increase in the fill factor to 78% as a result of lower series and shunt resistances.

Despite the promising properties of spiro-OMeTAD and its derivatives, the demanding multi-step synthesis and extensive purification result in high costs and limit industrial application. In recent years, a plethora of polymeric, (such as P3HT,[66,67] PCDTBT,[56] PCPDTBT,[56] PEDOT,[68] diketopyrrolopyrrole-based polymers[69] and triarylamine-based polymers[2,5]) and small-molecule hole transporting materials (such as H101,[70] PST1,[71] TTF-1,[72] and EDOT-OMeTPA[73]) have been reported as alternatives to spiro-OMeTAD. However, cells fabricated using the majority of these materials gave power conversion efficiencies that are lower than those of cells made using spiro-OMeTAD, and only a few alternatives have been able to outperform this state-of-the-art material.

It is worth noting here that several published HTMs are based on triphenylamines (TPA), which are often substituted with methoxy groups. The three-dimensional propeller-like structure of TPA increases the solubility of the material, which also has a good hole mobility and a relatively deep ionization potential. Heo *et al.* showed that the triphenylamine-based polymer PTAA is able to outperform spiro-OMeTAD on a like-to-like comparison[56] and, more recently, this group has claimed efficiencies exceeding 17%.[62]

Ryu *et al.* studied the effect of optimizing the energy alignment for methylammonium lead bromide cells by using a series of TPA-based polymers with different HOMO energy levels.[55] Their results show that higher open circuit voltages could be obtained by lowering the HOMO energy level, and open circuit voltages exceeding 1.4 V were reported for devices using polymer PIF8-TAA, which has a relatively deep HOMO energy level of 5.51 eV. Since dopants are generally considered to reduce the stability of the photovoltaic device, some groups have developed HTMs with good performance without doping. For example, TTF-1 and TPBC have given – without the use of dopants – PCEs that are comparable to those achieved with doped spiro-OMeTAD.[16,74]

Inorganic hole transporting materials can also be used in regular architecture perovskite solar cells, as demonstrated by Christians *et al.*[75] who achieved an efficiency of 8% using CuI. Shortly after, the Grätzel group reported devices with PCEs up to 12.4% by using copper thiocyanate (CuSCN) and recently, devices with PCEs of 14.9% were reported using NiO as the hole selective material.[8] In general, inorganic hole transporters display higher conductivities than molecular and polymeric hole transporters, but their performance still lags behind their organic counterparts.

The most widely used hole transporting material for inverted devices is poly(3,4-ethylenedioxythiophene) polystyrene sulfonate (PEDOT-PSS, Figure 2.8). Although it was originally developed as an antistatic agent, its high transparency, high conductivity and appropriate energy levels make it a suitable transparent electrode for various photovoltaic applications.[76] PEDOT-PSS is generally spin-coated from water and is insoluble in most organic solvents, making it possible to deposit the perovskite precursors on top without redissolving the PEDOT-PSS layer. The most efficient inverted devices reported to date use PEDOT-PSS as the hole transporting layer and show efficiencies exceeding 18%.[61] Despite the good PCEs, PEDOT-PSS also has some disadvantages as it is hygroscopic and the interface with ITO is not very stable, especially in the presence of oxygen.[77] Inorganic anode buffer materials based on metal oxides are generally more stable in ambient environments and offer an alternative to PEDOT-PSS. For example, nickel oxide has been studied as a hole transporting material. Zhu *et al.*[95] reported a sol–gel processed NiO nanocrystal layer as the hole transporting layer in inverted devices with efficiencies exceeding 9%. More recently, Jen *et al.*[78] improved the device performance by doping the NiO_x layer with copper, resulting in efficiencies exceeding 15%, demonstrating the potential of inorganic materials as hole transporting layers in inverted architectures.

2.2.3.3 Material Cost

Most of the organic HTMs discussed previously are synthesized in expensive cross-coupling reactions that require stringent reaction conditions and extensive product purification. Some of these materials are commercially available, but most of them are very expensive with costs above $200 per gram (Table 2.5). In order to commercialize perovskite solar cells successfully, the hole transporting materials need to become significantly cheaper, as stressed recently by Petrus *et al.* and Liu *et al.*[68,73] In order to compete with the current silicon solar cell technology, the target module cost needs to be well below $0.50 per peak Watt ($W_p^{-1}$).[79] For present-day devices, the material cost contribution of a layer of spiro-OMeTAD (assuming a 300 nm thick layer and a PCE of 20%) would be around $0.30 W_p^{-1}, making it very difficult to reach the target module cost. It is therefore crucial to reduce the cost of the hole transporter by an order of magnitude in order to make this technology industrially interesting.

Table 2.5 Cost of various organic hole transporting materials.[83–87]

HTM	Cost ($ g^{-1})
Spiro-OMeTAD	200
TTF-1	2000
H101	250
EDOT-OMeTPA[a]	10
PTAA	2000
PIF8-TAA	4500
PCPDTBT	3000
P3HT	500
PEDOT-PSS[b]	10

[a]Estimated materials cost.[73]
[b]Calculated for solid content.

Recently, efficient and low-cost hole transporting azomethine-based materials have been reported for organic photovoltaics.[80,81] These materials can be prepared by simple condensation chemistry under ambient conditions, with water as the only byproduct as shown in Figure 2.9, making the purification very straightforward.[82] Petrus *et al.* introduced the azomethine-based small-molecule EDOT-OMeTPA as the HTM for perovskite solar cells and showed efficiencies comparable to those of spiro-OMeTAD.[73] In a cost analysis, the authors reported an estimated material cost of only $10 per gram, which results in a negligible material cost contribution of ~$0.004 W_p^{-1}. Liu *et al.* used the inexpensive highly conductive polymer PEDOT (not to be confused with PEDOT-PSS) as the hole transporting material on top of the perovskite, with reported efficiencies comparable to those of spiro-OMeTAD.[68]

2.3 Outlook and Conclusions

The perovskite films described above generally follow the trend that the larger the crystallites, the higher the resulting performance of the solar cell, as long as the surface area coverage approaches 100%. This result is rather intuitive, as larger grains are expected to result in fewer grain boundaries and thus fewer trap states and recombination centers.[89] A more direct insight into these phenomena has been provided by the work of Xiao *et al.*[88,90] Here, the introduction of an additional solvent annealing step after perovskite deposition led to micrometer sized perovskite grains spanning the whole thickness of the sample.[88,90] This resulted in a significant performance improvement over thermally annealed films from 8.2% to 14.8%, as shown in Figure 2.10c. The superior photovoltaic suitability of solvent annealing *vs.* thermal annealing for device fabrication is also indicated by a longer open circuit voltage decay, which implies slower charge recombination rates (Figure 2.10d). The concept was recently extended by Nie *et al.*[89] to achieve millimeter-sized grains. Here, the precursor solution was spin-coated on a hot substrate. The presence of excess high-boiling point solvent

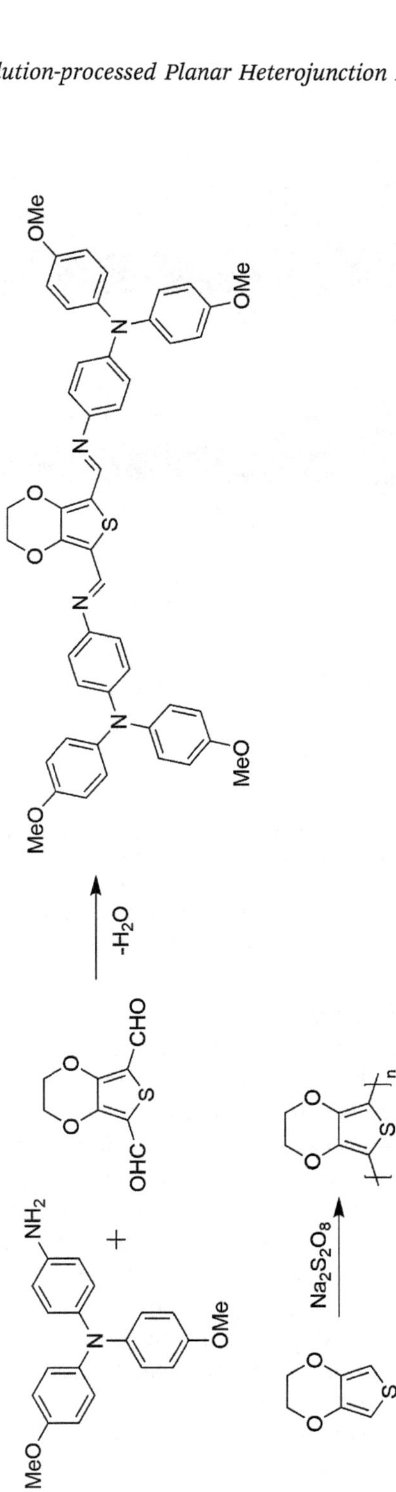

Figure 2.9 Synthetic schemes of low-cost hole transporting materials EDOT-OMeTPA (top) and PEDOT (bottom).

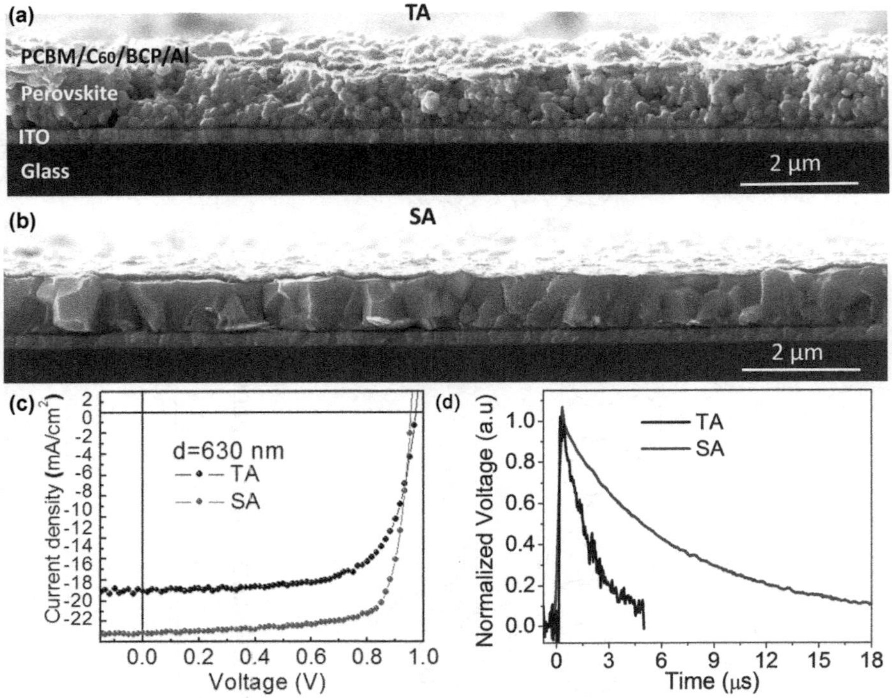

Figure 2.10 Morphology of perovskite layers prepared by thermal annealing (TA, a) and solvent annealing (SA, b). The influence of the crystallite size can be seen in the J–V measurements (c) and the open circuit voltage decay (d). Adapted from ref. 88 with permission from John Wiley and Sons © 2014 Wiley-VCH Verlag GmbH & Co. KGaA, Weinheim.

combined with the high temperature above the super-saturation point prolongs the growth period of the perovskite crystals.

Not only the grain size but also the grain orientation has an influence on the device performance. For instance, Docampo *et al.*[52] have shown for perovskite converted samples that a higher degree of orientation of the (110) plane perpendicular to the substrate leads, on average, to higher short circuit currents. A similar effect was observed by Saliba *et al.*,[24] who compared flash-annealing to slow annealing both in devices and with grazing incidence wide angle scattering (GIWAXS). They observed that for the fast and hot annealing step, highly uniform micron sized crystallites were obtained with a higher degree of orientation in the 2D-X-ray scattering experiments, which resulted in higher short circuit currents. These experimental results have been substantiated recently by theoretical studies which point towards anisotropy in terms of charge transfer for different perovskite surfaces.[91] Overall, films exhibiting a preferential orientation on the (002) plane parallel to the substrate are expected to exhibit optimum performance.

In addition to optimizing the perovskite layer, tuning of the contact materials can lead to large improvements in the device performance. In particular, good energy level alignment is essential for maximizing the open circuit voltage. Recent results from the Yang group[3] have shown that a combination of the reduction of the work function of ITO and doped titania layers can achieve open circuit voltages of over 1.1 V for methylammonium lead iodide (MAPI). This is higher than the voltage of approximately 1 V that can be achieved with current state-of-the-art fullerenes.[89,92] We note that, as the architecture is inverted in this case, spiro-OMeTAD as the hole transporter is usually replaced with PEDOT-PSS, which may also play a role in determining the open circuit voltage that can be obtained.

The HTM HOMO energy level alignment is a much better studied property, with dozens of different molecules and polymers having been already investigated.[71,93,94] Currently, the strongest competitors in this area are spiro-OMeTAD and PTAA, both of which achieve similar open circuit voltages of 1.1–1.15 V on MAPI, with similar HOMO energy levels of 5.1–5.2 eV. Modifications to the PTAA system to match the valence band of bromide-based perovskites have allowed open circuit voltages of up to 1.48 V in this system, although the overall efficiency is limited by the bandgap of the absorber.

The best HTMs developed so far require expensive cross-coupling reactions, stringent reaction conditions and extensive product purification and therefore cannot be produced at a low cost. For perovskite solar cells to be able to enter the photovoltaic market, along with the stability, their cost needs to be competitive with current established technologies. With the current state-of-the-art hole transporters, this is difficult, as they are one of the largest cost fractions, with an approximate cost contribution of \$0.30 W_p^{-1}, which is comparable to readily available commercial silicon module costs. Developing new chemistries that result in simple compound purification, such as those based on azomethine bonds, will be an essential part of the future molecular design for perovskite solar cells.

References

1. A. Kojima, K. Teshima, Y. Shirai and T. Miyasaka, *J. Am. Chem. Soc.*, 2009, **131**, 6050–6051.
2. N. J. Jeon, J. H. Noh, W. S. Yang, Y. C. Kim, S. Ryu, J. Seo and S. I. Seok, *Nature*, 2015, **517**, 476–480.
3. H. Zhou, Q. Chen, G. Li, S. Luo, T.-b. Song, H.-S. Duan, Z. Hong, J. You, Y. Liu and Y. Yang, *Science*, 2014, **345**, 542–546.
4. M. M. Lee, J. Teuscher, T. Miyasaka, T. N. Murakami and H. J. Snaith, *Science*, 2012, **338**, 643–647.
5. H.-S. Kim, C.-R. Lee, J.-H. Im, K.-B. Lee, T. Moehl, A. Marchioro, S.-J. Moon, R. Humphry-Baker, J.-H. Yum, J. E. Moser, M. Gratzel and N.-G. Park, *Sci. Rep.*, 2012, **2**, 591.

6. J. Burschka, N. Pellet, S.-J. Moon, R. Humphry-Baker, P. Gao, M. K. Nazeeruddin and M. Gratzel, *Nature*, 2013, **499**, 316–319.
7. J.-H. Im, I.-H. Jang, N. Pellet, M. Grätzel and N.-G. Park, *Nat. Nanotechnol.*, 2014, **9**, 927–932.
8. P. Qin, S. Tanaka, S. Ito, N. Tetreault, K. Manabe, H. Nishino, M. K. Nazeeruddin and M. Gratzel, *Nat. Commun.*, 2014, **5**, 3834.
9. J. M. Ball, M. M. Lee, A. Hey and H. J. Snaith, *Energy Environ. Sci.*, 2013, **6**, 1739–1743.
10. V. Roiati, E. Mosconi, A. Listorti, S. Colella, G. Gigli and F. De Angelis, *Nano Lett.*, 2014, **14**, 2168–2174.
11. J.-H. Im, C.-R. Lee, J.-W. Lee, S.-W. Park and N.-G. Park, *Nanoscale*, 2011, **3**, 4088–4093.
12. P. Schulz, E. Edri, S. Kirmayer, G. Hodes, D. Cahen and A. Kahn, *Energy Environ. Sci.*, 2014, **7**, 1377–1381.
13. P. Docampo, S. Guldin, U. Steiner and H. J. Snaith, *J. Phys. Chem. Lett.*, 2013, **4**, 698–703.
14. M. Liu, M. B. Johnston and H. J. Snaith, *Nature*, 2013, **501**, 395–398.
15. P. Docampo, J. M. Ball, M. Darwich, G. E. Eperon and H. J. Snaith, *Nat. Commun.*, 2013, **4**, 2761.
16. D. Liu and T. L. Kelly, *Nat. Photonics*, 2014, **8**, 133–138.
17. G. E. Eperon, V. M. Burlakov, P. Docampo, A. Goriely and H. J. Snaith, *Adv. Funct. Mater.*, 2014, **24**, 151–157.
18. W. Zhang, M. Saliba, D. T. Moore, S. K. Pathak, M. T. Hörantner, T. Stergiopoulos, S. D. Stranks, G. E. Eperon, J. A. Alexander-Webber, A. Abate, A. Sadhanala, S. Yao, Y. Chen, R. H. Friend, L. A. Estroff, U. Wiesner and H. J. Snaith, *Nat. Commun.*, 2015, **6**, 6142.
19. M. A. Green, A. Ho-Baillie and H. J. Snaith, *Nat. Photonics*, 2014, **8**, 506–514.
20. S. D. Stranks, P. K. Nayak, W. Zhang, T. Stergiopoulos and H. J. Snaith, *Angew. Chem., Int. Ed.*, 2015, **54**, 3240–3248.
21. V. M. Burlakov, G. E. Eperon, H. J. Snaith, S. J. Chapman and A. Goriely, *Appl. Phys. Lett.*, 2014, **104**, 091602.
22. H.-L. Hsu, C.-P. Chen, J.-Y. Chang, Y.-Y. Yu and Y.-K. Shen, *Nanoscale*, 2014, **6**, 10281–10288.
23. J. Troughton, C. Charbonneau, M. J. Carnie, M. L. Davies, D. A. Worsley and T. M. Watson, *J. Mater. Chem. A*, 2015, **3**, 9123–9127.
24. M. Saliba, K. W. Tan, H. Sai, D. T. Moore, T. Scott, W. Zhang, L. A. Estroff, U. Wiesner and H. J. Snaith, *J. Phys. Chem. C*, 2014, **118**, 17171–17177.
25. A. Dualeh, N. Tétreault, T. Moehl, P. Gao, M. K. Nazeeruddin and M. Grätzel, *Adv. Funct. Mater.*, 2014, **24**, 3250–3258.
26. C.-Y. Chang, C.-Y. Chu, Y.-C. Huang, C.-W. Huang, S.-Y. Chang, C.-A. Chen, C.-Y. Chao and W.-F. Su, *ACS Appl. Mater. Interfaces*, 2015, **7**, 4955–4961.

27. C.-C. Chueh, C.-Y. Liao, F. Zuo, S. T. Williams, P.-W. Liang and A. K. Y. Jen, *J. Mater. Chem. A*, 2015, **3**, 9058–9062.
28. J. H. Heo, D. H. Song and S. H. Im, *Adv. Mater.*, 2014, **26**, 8179–8183.
29. C. Sun, Q. Xue, Z. Hu, Z. Chen, F. Huang, H.-L. Yip and Y. Cao, *Small*, 2015, **11**, 3344–3350.
30. Y. Chen, Y. Zhao and Z. Liang, *Chem. Mater.*, 2015, **27**, 1448–1451.
31. G. E. Eperon, S. D. Stranks, C. Menelaou, M. B. Johnston, L. M. Herz and H. J. Snaith, *Energy Environ. Sci.*, 2014, **7**, 982–988.
32. X. Song, W. Wang, P. Sun, W. Ma and Z.-K. Chen, *Appl. Phys. Lett.*, 2015, **106**, 033901.
33. P.-W. Liang, C.-Y. Liao, C.-C. Chueh, F. Zuo, S. T. Williams, X.-K. Xin, J. Lin and A. K. Y. Jen, *Adv. Mater.*, 2014, **26**, 3748–3754.
34. C. Zuo and L. Ding, *Nanoscale*, 2014, **6**, 9935–9938.
35. D. Shen, X. Yu, X. Cai, M. Peng, Y. Ma, X. Su, L. Xiao and D. Zou, *J. Mater. Chem. A*, 2014, **2**, 20454–20461.
36. H.-B. Kim, H. Choi, J. Jeong, S. Kim, B. Walker, S. Song and J. Y. Kim, *Nanoscale*, 2014, **6**, 6679–6683.
37. S. Paek, N. Cho, H. Choi, H. Jeong, J. S. Lim, J.-Y. Hwang, J. K. Lee and J. Ko, *J. Phys. Chem. C*, 2014, **118**, 25899–25905.
38. S. Pang, H. Hu, J. Zhang, S. Lv, Y. Yu, F. Wei, T. Qin, H. Xu, Z. Liu and G. Cui, *Chem. Mater.*, 2014, **26**, 1485–1491.
39. F. Wang, H. Yu, H. Xu and N. Zhao, *Adv. Funct. Mater.*, 2015, **25**, 1120–1126.
40. T. Leijtens, B. Lauber, G. E. Eperon, S. D. Stranks and H. J. Snaith, *J. Phys. Chem. Lett.*, 2014, **5**, 1096–1102.
41. Y. Wu, A. Islam, X. Yang, C. Qin, J. Liu, K. Zhang, W. Peng and L. Han, *Energy Environ. Sci.*, 2014, **7**, 2934–2938.
42. Q. Chen, H. Zhou, Z. Hong, S. Luo, H.-S. Duan, H.-H. Wang, Y. Liu, G. Li and Y. Yang, *J. Am. Chem. Soc.*, 2014, **136**, 622–625.
43. M. Xiao, F. Huang, W. Huang, Y. Dkhissi, Y. Zhu, J. Etheridge, A. Gray-Weale, U. Bach, Y.-B. Cheng and L. Spiccia, *Angew. Chem.*, 2014, **53**, 9898–9903.
44. N. J. Jeon, J. H. Noh, Y. C. Kim, W. S. Yang, S. Ryu and S. I. Seok, *Nat. Mater.*, 2014, **13**, 897–903.
45. Y. Zhou, M. Yang, W. Wu, A. L. Vasiliev, K. Zhu and N. P. Padture, *J. Mater. Chem. A*, 2015, **3**, 8178–8184.
46. S. T. Williams, F. Zuo, C.-C. Chueh, C.-Y. Liao, P.-W. Liang and A. K. Y. Jen, *ACS Nano*, 2014, **8**, 10640–10654.
47. E. L. Unger, A. R. Bowring, C. J. Tassone, V. L. Pool, A. Gold-Parker, R. Cheacharoen, K. H. Stone, E. T. Hoke, M. F. Toney and M. D. McGehee, *Chem. Mater.*, 2014, **26**, 7158–7165.
48. H. Yu, F. Wang, F. Xie, W. Li, J. Chen and N. Zhao, *Adv. Funct. Mater.*, 2014, **24**, 7102–7108.
49. A. Buin, P. Pietsch, J. Xu, O. Voznyy, A. H. Ip, R. Comin and E. H. Sargent, *Nano Lett.*, 2014, **14**, 6281–6286.

50. P. Docampo, F. Hanusch, S. D. Stranks, M. Döblinger, J. M. Feckl, M. Ehrensperger, N. K. Minar, M. B. Johnston, H. J. Snaith and T. Bein, *Adv. Energy Mater.*, 2014, **4**, 1400355.
51. K. Liang, D. B. Mitzi and M. T. Prikas, *Chem. Mater.*, 1998, **10**, 403–411.
52. P. Docampo, F. C. Hanusch, N. Giesbrecht, P. Angloher, A. Ivanova and T. Bein, *APL Mater.*, 2014, **2**, 081508.
53. J. Schlipf, P. Docampo, C. J. Schaffer, V. Körstgens, L. Bießmann, F. C. Hanusch, N. Giesbrecht, S. Bernstorff, T. Bein and P. Muller-Buschbaum, *J. Phys. Chem. Lett.*, 2015, **6**, 1265–1269.
54. L. Zheng, Y. Ma, S. Chu, S. Wang, B. Qu, L. Xiao, Z. Chen, Q. Gong, Z. Wu and X. Hou, *Nanoscale*, 2014, **6**, 8171–8176.
55. S. Ryu, J. H. Noh, N. J. Jeon, Y. Chan Kim, W. S. Yang, J. Seo and S. I. Seok, *Energy Environ. Sci.*, 2014, **7**, 2614–2618.
56. J. H. Heo, S. H. Im, J. H. Noh, T. N. Mandal, C.-S. Lim, J. A. Chang, Y. H. Lee, H.-j. Kim, A. Sarkar, M. K. Nazeeruddin, M. Grätzel and S. I. Seok, *Nat. Photonics*, 2013, **7**, 486–491.
57. P. Docampo, S. Guldin, T. Leijtens, N. K. Noel, U. Steiner and H. J. Snaith, *Adv. Mater.*, 2014, **26**, 4013–4030.
58. A. K. Jena, H.-W. Chen, A. Kogo, Y. Sanehira, M. Ikegami and T. Miyasaka, *ACS Appl. Mater. Interfaces*, 2015, **7**, 9817–9823.
59. C. J. Brabec, S. Gowrisanker, J. J. Halls, D. Laird, S. Jia and S. P. Williams, *Adv. Mater.*, 2010, **22**, 3839–3856.
60. J. Y. Jeng, K. C. Chen, T. Y. Chiang, P. Y. Lin, T. D. Tsai, Y. C. Chang, T. F. Guo, P. Chen, T. C. Wen and Y. J. Hsu, *Adv. Mater.*, 2014, **26**, 4107–4113.
61. J. H. Heo, H. J. Han, D. Kim, T. K. Ahn and S. H. Im, *Energy Environ. Sci.*, 2015, **8**, 1602–1608.
62. J. H. Heo, D. H. Song, H. J. Han, S. Y. Kim, J. H. Kim, D. Kim, H. W. Shin, T. K. Ahn, C. Wolf, T. W. Lee and S. H. Im, *Adv. Mater.*, 2015, **27**, 3424–3430.
63. J. Seo, S. Park, Y. Chan Kim, N. J. Jeon, J. H. Noh, S. C. Yoon and S. I. Seok, *Energy Environ. Sci.*, 2014, **7**, 2642.
64. Q. Xue, Z. Hu, J. Liu, J. Lin, C. Sun, Z. Chen, C. Duan, J. Wang, C. Liao, W. M. Lau, F. Huang, H.-L. Yip and Y. Cao, *J. Mater. Chem. A*, 2014, **2**, 19598–19603.
65. N. J. Jeon, H. G. Lee, Y. C. Kim, J. Seo, J. H. Noh, J. Lee and S. I. Seok, *J. Am. Chem. Soc.*, 2014, **136**, 7837–7840.
66. A. Abrusci, S. D. Stranks, P. Docampo, H.-L. Yip, A. K. Y. Jen and H. J. Snaith, *Nano Lett.*, 2013, **13**, 3124–3128.
67. B. Conings, L. Baeten, C. De Dobbelaere, J. D'Haen, J. Manca and H.-G. Boyen, *Adv. Mater.*, 2014, **26**, 2041–2046.
68. J. Liu, S. Pathak, T. Stergiopoulos, T. Leijtens, K. Wojciechowski, S. Schumann, N. Kausch-Busies and H. J. Snaith, *J. Phys. Chem. Lett.*, 2015, **6**, 1666–1673.

69. G.-W. Kim, J. Kim, G.-Y. Lee, G. Kang, J. Lee and T. Park, *Adv. Energy Mater.*, 2015, **5**, 1500471.
70. H. Li, K. Fu, A. Hagfeldt, M. Grätzel, S. G. Mhaisalkar and A. C. Grimsdale, *Angew. Chem., Int. Ed.*, 2014, **53**, 4085–4088.
71. P. Ganesan, K. Fu, P. Gao, I. Raabe, K. Schenk, R. Scopelliti, J. Luo, L. H. Wong, M. Gratzel and M. K. Nazeeruddin, *Energy Environ. Sci.*, 2015, **8**, 1986–1991.
72. J. Liu, Y. Wu, C. Qin, X. Yang, T. Yasuda, A. Islam, K. Zhang, W. Peng, W. Chen and L. Han, *Energy Environ. Sci.*, 2014, **7**, 2963–2967.
73. M. L. Petrus, T. Bein, T. J. Dingemans and P. Docampo, *J. Mater. Chem. A*, 2015, **3**, 12159–12162.
74. Y. Song, S. Lv, X. Liu, X. Li, S. Wang, H. Wei, D. Li, Y. Xiao and Q. Meng, *Chem. Commun.*, 2014, **50**, 15239–15242.
75. J. A. Christians, R. C. Fung and P. V. Kamat, *J. Am. Chem. Soc.*, 2014, **136**, 758–764.
76. S. K. M. Jönsson, J. Birgerson, X. Crispin, G. Greczynski, W. Osikowicz, A. W. Denier van der Gon, W. R. Salaneck and M. Fahlman, *Synth. Met.*, 2003, **139**, 1–10.
77. M. P. de Jong, L. J. van Ijzendoorn and M. J. A. de Voigt, *Appl. Phys. Lett.*, 2000, **77**, 2255.
78. J. H. Kim, P. W. Liang, S. T. Williams, N. Cho, C. C. Chueh, M. S. Glaz, D. S. Ginger and A. K. Jen, *Adv. Mater.*, 2015, **27**, 695–701.
79. M. C. Beard, J. M. Luther and A. J. Nozik, *Nat. Nanotechnol.*, 2014, **9**, 951–954.
80. M. L. Petrus, R. K. M. Bouwer, U. Lafont, S. Athanasopoulos, N. C. Greenham and T. J. Dingemans, *J. Mater. Chem. A*, 2014, **2**, 9474.
81. M. L. Petrus, F. S. F. Morgenstern, A. Sadhanala, R. H. Friend, N. C. Greenham and T. J. Dingemans, *Chem. Mater.*, 2015, **27**, 2990–2997.
82. M. L. Petrus, R. K. M. Bouwer, U. Lafont, D. H. K. Murthy, R. J. P. Kist, M. L. Böhm, Y. Olivier, T. J. Savenije, L. D. A. Siebbeles, N. C. Greenham and T. J. Dingemans, *Polym. Chem.*, 2013, **4**, 4182.
83. http://abcr.de.
84. http://www.heraeus-clevios.com.
85. http://www.lumtec.com.tw.
86. http://www.sigmaaldrich.com.
87. http://www.trc-canada.com.
88. Z. Xiao, Q. Dong, C. Bi, Y. Shao, Y. Yuan and J. Huang, *Adv. Mater.*, 2014, **26**, 6503–6509.
89. W. Nie, H. Tsai, R. Asadpour, J.-C. Blancon, A. J. Neukirch, G. Gupta, J. J. Crochet, M. Chhowalla, S. Tretiak and M. A. Alam, *Science*, 2015, **347**, 522–525.
90. Z. Xiao, C. Bi, Y. Shao, Q. Dong, Q. Wang, Y. Yuan, C. Wang, Y. Gao and J. Huang, *Energy Environ. Sci.*, 2014, **7**, 2619.

91. J. Yin, D. Cortecchia, A. Krishna, S. Chen, N. Mathews, A. C. Grimsdale and C. Soci, *J. Phys. Chem. Lett.*, 2015, **6**, 1396–1402.
92. S. Ryu, J. Seo, S. S. Shin, Y. C. Kim, N. J. Jeon, J. H. Noh and S. I. Seok, *J. Mater. Chem. A*, 2015, **3**, 3271–3275.
93. S. Kazim, F. J. Ramos, P. Gao, M. K. Nazeeruddin, M. Gratzel and S. Ahmad, *Energy Environ. Sci.*, 2015, **8**, 1816–1823.
94. C. V. Kumar, G. Sfyri, D. Raptis, E. Stathatos and P. Lianos, *RSC Adv.*, 2015, **5**, 3786–3791.
95. Z. Zhu, Y. Bai, T. Zhang, Z. Liu, X. Long, Z. Wei, Z. Wang, L. Zhang, J. Wang, F. Yan and S. Yang, *Angew. Chem., Int. Ed.*, 2014, **53**, 12571–12575.

CHAPTER 3

Characterization of Capacitance, Transport and Recombination Parameters in Hybrid Perovskite and Organic Solar Cells

JUAN BISQUERT,* GERMÀ GARCIA-BELMONTE AND IVAN MORA-SERO

Institute of Advanced Materials (INAM), Universitat Jaume I, 12006 Castelló, Spain
*Email: bisquert@uji.es

3.1 Introduction

The improvement and optimization of solar cells, as well as guidance for the discovery of new photovoltaic materials and their combinations, require an understanding of the physical processes in materials, starting from light absorption and charge generation, leading to the production of electricity in the outer circuit. The traditional methods used to analyze inorganic crystalline solar cells have been mainly based on techniques that determine a specific feature of the semiconductor layer, such as carrier lifetime. With the advent of nanostructured hybrid materials and photoelectrochemical systems for solar cell applications and the resulting widespread research interest, general-purpose frequency techniques have become popular and

RSC Energy and Environment Series No. 16
Unconventional Thin Film Photovoltaics
Edited by Enrico Da Como, Filippo De Angelis, Henry Snaith and Alison Walker

Published by the Royal Society of Chemistry, www.rsc.org

very useful methods. Impedance spectroscopy (IS) has the special advantages of very well established interpretation methodologies and the fact that it probes directly the current voltage characteristics of the device, the key feature of a solar cell. IS has been applied intensively to study dye-sensitized solar cells, and this research has allowed us to develop a more general understanding of the interpretation of IS applied to solar cell measurements. These approaches have been reviewed systematically in papers and book chapters,[1–4] and they will be taken as the starting point for the new developments described in this chapter. The same methods have later been extended to organic bulk heterojunction solar cells[5] and, more recently, to hybrid organic–inorganic halide perovskites. These two last topics will be discussed here.

Organic solar cells constitute an academically interesting and technologically viable alternative to the currently-commercialized energy harvesting systems. Of special interest are organic solar cells made up of an interpenetrating network of donor and acceptor molecules known as bulk-heterojunction solar cells (BHJ). In this technology, selective contacts collect photogenerated carriers transported through bi-continuous organic n- and p-compound networks. Recombination, either before charge separation (geminate) or involving photogenerated charges (non-geminate), and contact losses are the main limiting mechanisms.

The discovery and subsequent rapid development of organic–inorganic lead halide perovskite solar cells has opened up a new dimension in the field of photovoltaic materials. Hybrid perovskites have an ABX_3 structure, where A is commonly an organic cation, B a divalent metallic cation (for photovoltaic applications usually Sn or Pb, especially the latter) and X a halide anion (generally I, Br or Cl or combinations of them). They display a range of extraordinary properties that are both interesting and very challenging to understand. In particular, the methylammonium (MA) lead iodide perovskite has been studied extensively in the last three years and has become the prime example of these materials, although other significant variants have also been analyzed. One of the challenges is to characterize the properties on the basis of IS measurements; a number of impedance studies have been reported,[6–10] but a consolidated interpretation has not been established so far.

This chapter will focus on the basic physical properties and device elements that are related to the interpretation of impedance and capacitance spectroscopy, and what can be learned from such interpretation in terms of materials operation and device mechanisms. Here, we adopt a general stance that gives insight into characteristic physical elements of the solar cells. The main topics of interest are capacitance, transport, and recombination.

Capacitance is a key parameter that provides information on specific operating modes of any kind of solar cell and electronic device. Capacitive effects in organic solar cells may originate from a variety of mechanisms, ranging from those related to purely electrostatic phenomena to others related to the occupancy of the electronic density of states (DOS) and the device energetics. The latter effects are referred to as *chemical capacitance*. In addition, the energetics of the contact between the organic absorbing layer and the transporting layers give rise to carrier depletion zones that introduce additional contact capacitances in bulk-heterojunction organic solar cells. In perovskite solar cells, we encounter the additional possibility of variation of the dielectric properties, since the crystal structure can be modified by different phase transitions that change the local polarity. Therefore, we describe dielectric relaxation in insulators, which is part of the wider class of *relaxation phenomena* that refer generally to the recovery towards an equilibrium state, and include the structural relaxation of glasses and the decay of photoconductivity and photoluminescence. The mechanism of transport of carriers is a necessary consideration for important photovoltaic properties such as the diffusion length. Recombination of photogenerated carriers is also a central feature of solar cells. We aim to connect different ways to measure recombination: in IS, the recombination is characterized by a recombination resistance and, in many time domain methods, the recombination is characterized by a carrier lifetime. We will describe recombination features from very basic radiative phenomena to device physics including the device contacts.

3.2 Dielectric Relaxation, Impedance and Capacitance Spectroscopy

3.2.1 General Definitions and Concepts

Impedance and capacitance spectroscopies provide a great deal of information on carrier accumulation and transport in solar cell materials. The impedance is the measurement of small perturbation voltage ($\hat{V}$) to current ($\hat{I}$):

$$Z(\omega) = \frac{\hat{V}(\omega)}{\hat{I}(\omega)} \tag{3.1}$$

Here, ω is the angular frequency, $\omega = 2\pi f$. For an ohmic process, $Z = R$, the resistance, which is a positive and real quantity. Any kind of relaxation introduces a time delay in the response that brings in an imaginary component in the impedance, associated with capacitive behavior. It is useful to display the measured impedance data in different representations that are equivalent but provide insight into their meaning in terms of physical processes. For a sample of electrode area A and thickness d, the complex impedance $Z(\omega)$ allows to define the complex conductivity as

$$\sigma(\omega) = \frac{d}{AZ(\omega)} \tag{3.2}$$

The complex capacitance $C(\omega)$ is defined from the impedance as

$$C(\omega) = \frac{1}{i\omega Z(\omega)} \tag{3.3}$$

The complex dielectric constant is obtained from the complex capacitance

$$\varepsilon(\omega) = \frac{dC(\omega)}{A\varepsilon_0} \tag{3.4}$$

According to eqn (3.2)–(3.4), there is a direct connection between the complex conductivity and the complex dielectric constant as follows:

$$\sigma(\omega) = i\omega\varepsilon_0\varepsilon(\omega) \tag{3.5}$$

3.2.2 Dielectric Relaxation and Frequency Dispersion

A dielectric material (insulator) has dynamic features that cause time or frequency dependent polarization phenomena generally termed *dielectric relaxation*. These dynamic phenomena produce specific features in the complex permittivity $\varepsilon(\omega)$. These features are due to the local response of different physical elements: the oscillation of intrinsic dipoles in the material, the dipole moments of complex defects (which are formed by an interstitial atom plus a vacancy), and the short range hopping of electronic carriers. In solar cell semiconductor materials, in addition to dielectric relaxation there are also free carriers that cause dc conduction. Conduction and polarization are mutually interconnected and often consist on two facets of a single mechanism.

The dipolar units in the bulk of a dielectric layer produce a polarization vector $P(\omega)$ that influences the dielectric displacement field in response to the applied field $\mathbf{E}(\omega)$ as follows:

$$D(\omega) = P(\omega) + \varepsilon_0\mathbf{E}(\omega) \tag{3.6}$$

For a linear dielectric, we obtain the relation

$$D(\omega) = \varepsilon(\omega)\mathbf{E}(\omega) \tag{3.7}$$

The complex dielectric constant can be separated into real and imaginary parts as

$$\varepsilon(\omega) = \varepsilon'(\omega) - i\varepsilon''(\omega) \tag{3.8}$$

The real part of the dielectric constant indicates the increase of charge accumulated by polarization, while ε'' is denoted "dielectric loss" as it determines the dissipation of energy in excess of the dc dissipation. The frequency dependence of the real part of the permittivity and the dielectric loss are closely related. In general, as dictated by the Kramers–Kronig relationship, a peak in the dielectric loss that corresponds to a specific dielectric relaxation mechanism produces an increase of the real part of the permittivity.[11,12] For one given mechanism, the increment of the complex permittivity from its high frequency value can be expressed as

$$\Delta\varepsilon = \varepsilon_s - \varepsilon_\infty \tag{3.9}$$

Here, ε_s is the "static" value of the real part of the dielectric constant, while ε_∞ is the high frequency part. The latter can be associated to an "instantaneous" relaxation due to the displacement of the electrons with respect to the nuclei, and of ions with respect to their normal equilibrium positions, and it gives the *geometrical capacitance* per unit area:

$$C_g = \frac{\varepsilon_0 \varepsilon_\infty}{d} \tag{3.10}$$

However, we have not yet defined in absolute terms which components of the permittivity should be included in ε_∞. This depends on the frequency range of interest in the measurement.

The simplest relaxation type is described by the Debye expression,

$$\varepsilon(\omega) = \varepsilon_\infty + \frac{\varepsilon_s - \varepsilon_\infty}{1 + i\omega\tau_d} \tag{3.11}$$

The behavior according to eqn (3.11) is illustrated in Figure 3.1a. The imaginary part of the permittivity exhibits a symmetric peak centered at the frequency that corresponds to the angular frequency $\omega_d = \tau_d^{-1}$, where τ_d is the relaxation time. At frequencies higher than $\omega_d/2\pi$, the dipole units cannot follow the ac perturbation and the polarization mechanism is not effective. Thus, the real part of the permittivity, ε', changes in a stepwise fashion from the static value ε_s to the high frequency value (*cf.* eqn (3.9)). The dielectric relaxation of a material over a very broad range of frequencies usually shows a variety of steps in ε' that are coincident with dielectric loss peaks that represent different types of polarization and relaxation mechanisms, as indicated in Figure 3.2a.

The change of the dielectric constant with the frequency is physically very significant for systems with photogenerated carriers. The extent of the electrical screening affecting phenomena such as recombination or exciton dissociation will vary markedly depending on the timescale of the dielectric screening mechanism. Increasing the extent of dielectric shielding due to effective relaxation produces a larger dielectric constant at longer times, thereby progressively decreasing the Coulomb interactions.

A number of phenomenological relaxation functions are used in practice to describe the dielectric relaxation in disordered or entangled systems.[11] These models usually contain one or two added parameters that recover the equation for a single relaxation time for certain parameter values. Among the most widely used models is the Cole–Cole relaxation function,

$$\varepsilon(\omega) = \varepsilon_\infty + \frac{\varepsilon_s - \varepsilon_\infty}{1 + (i\omega\tau_{d0})^{1-\alpha}} \tag{3.12}$$

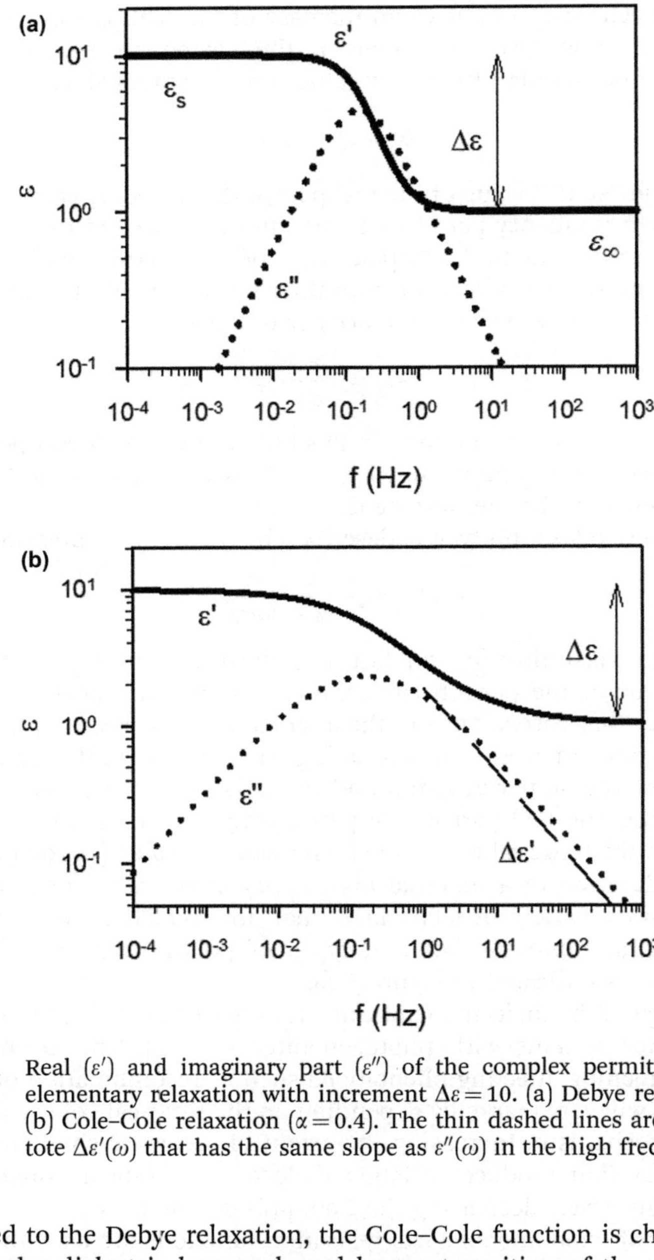

Figure 3.1 Real (ε') and imaginary part (ε'') of the complex permittivity for an elementary relaxation with increment $\Delta\varepsilon = 10$. (a) Debye relaxation and (b) Cole–Cole relaxation ($\alpha = 0.4$). The thin dashed lines are the asymptote $\Delta\varepsilon'(\omega)$ that has the same slope as $\varepsilon''(\omega)$ in the high frequency wing.

Compared to the Debye relaxation, the Cole–Cole function is characterized by a broader dielectric loss peak and longer transition of the capacitance step, as shown in Figure 3.1b. The Cole–Cole model and similar phenomenological functions can be considered a combination of elementary Debye relaxations with a broad distribution of relaxation times. τ_{d0} represents the central characteristic time in the corresponding time distribution. The parameter $\alpha (0 < \alpha < 1)$ describes the broadening of the distribution. In the

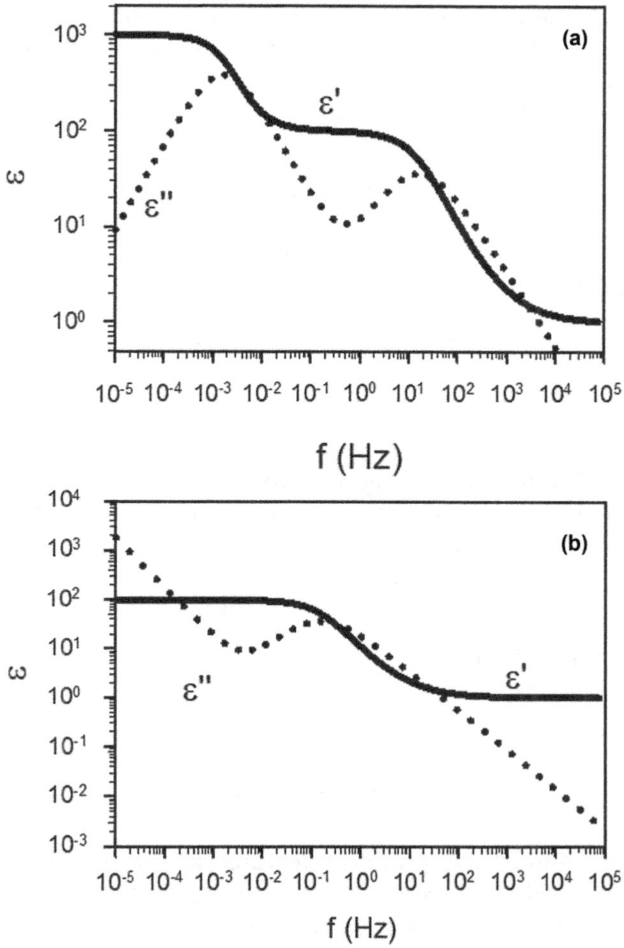

Figure 3.2 Real (ε') and imaginary (ε'') parts of the complex permittivity for (a) two consecutive relaxations with values $\varepsilon_\infty = 1$, $\varepsilon_1 = 100$, $\varepsilon_s = 1000$; and (b) a single dielectric relaxation peak combined with a small dc conduction component.

limiting case of $\alpha = 0$, a peaked delta function is obtained for the time distribution so that eqn (3.12) is reduced to a single relaxation time process described by the Debye expression in eqn (3.11).

However, the occurrence of broad relaxation and dispersive exponents in the relaxation laws is not generally due to a simple combination of independent processes, but rather to cooperative phenomena and entangled relaxation.[13] Indeed, in organic and inorganic solids, and in soft matter formulations of conductors such as ionic liquids, one encounters the general phenomenon of *frequency dispersion*, in which the limiting tails of the relaxation function depart from the idealized case of a Debye relaxation.

For eqn (3.12), the asymptotes of the real and imaginary parts of the permittivity for $\omega \to 0$ are related to the dispersive exponent as follows,

$$\varepsilon'(\omega) \sim \varepsilon_s \tag{3.13}$$

$$\varepsilon''(\omega) \sim \omega^{1-\alpha} \tag{3.14}$$

In the high frequency part (far from the peak region), both real and imaginary parts show the same slope:

$$\Delta\varepsilon'(\omega) = \varepsilon'(\omega) - \varepsilon_\infty - \omega^{\alpha-1} \tag{3.15}$$

$$\varepsilon''(\omega) \sim \omega^{\alpha-1} \tag{3.16}$$

This behavior is illustrated in Figure 3.1b.

The conduction properties are generally described by the ac conductivity, which can be separated into real and imaginary parts as

$$\sigma(\omega) = \sigma'(\omega) + i\sigma''(\omega) \tag{3.17}$$

We have the relationships:

$$\sigma'(\omega) = \omega\varepsilon_0\varepsilon''(\omega) \tag{3.18}$$

$$\sigma''(\omega) = \omega\varepsilon_0\varepsilon'(\omega) \tag{3.19}$$

Obviously, ac conductivity produces features in the permittivity, and *vice versa*. Let us consider this connection in more detail.

In many solids, the real part of the conductivity, taken over a frequency range from mHz to GHz, shows a constant domain, σ_0, at low frequencies and increases rapidly at high frequencies, as described by the expression:[13,14]

$$\sigma'(\omega) = \sigma_0[1 + (\omega/\omega_m)^n] \tag{3.20}$$

where the frequency ω_m marks the onset of the dispersion, which takes the form of a power law $\sigma' \propto \omega^n$, with the exponent $0 < n \leq 1$ lying usually in the range 0.6–0.8. This dependence is shown in Figure 3.3a, and it appears in the experimental data of Figure 3.4,[15] in the high frequency part. The origin of the universal behaviour of the conductivity has been discussed extensively.[16,17] In general, eqn (3.20) indicates that short range motion at high frequency allows displacement pathways of the carrier that are not allowed in the long range excursions that give rise to dc conduction.

In addition to eqn (3.20), Jonscher[12,13] pointed out the existence of two types of "universal relaxation" behavior observed in the complex permittivity for a wide variety of materials that show a large degree of frequency

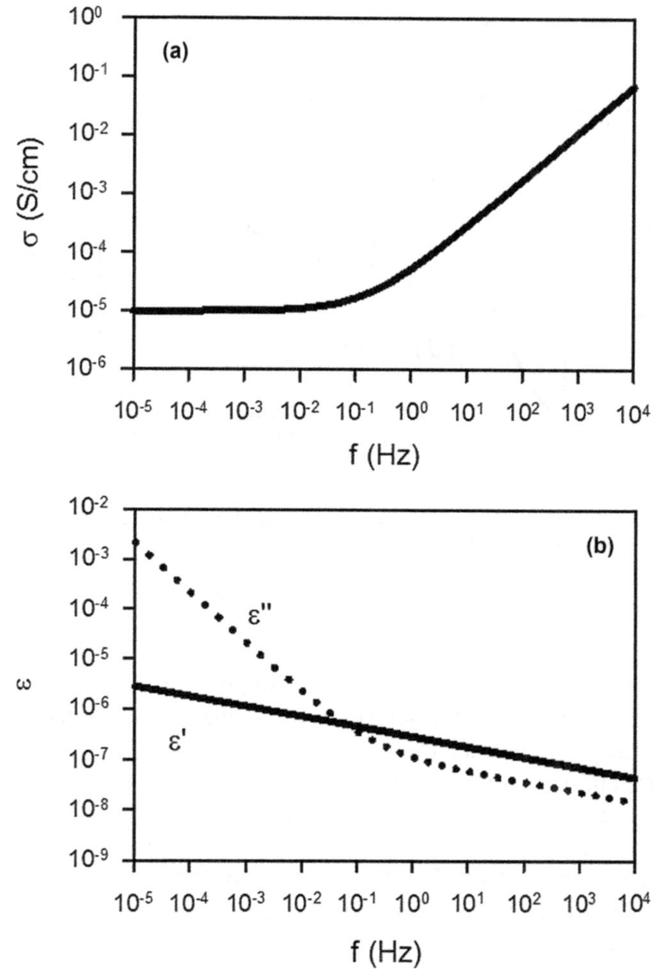

Figure 3.3 (a) Real part of the universal form of dispersive conductivity with dc conduction and (b) real (ε') and imaginary (ε'') parts of the complex permittivity.

dispersion, as shown in Figure 3.5. The first type is the classical dipolar relaxation but with a broad loss peak as in Figure 3.1b. The second type is dominated by the conduction of carriers and includes two different domains of frequency dispersion. The common feature to both types is that the high frequency part of the real and imaginary components of the permittivity show the same dispersive exponent, as remarked in eqn (3.15) and (3.16).

In eqn (3.20), $\sigma' \propto \omega^n$ at high frequency, which implies that $\sigma'' \propto \omega^n$, hence eqn (3.15) and (3.16) are obtained with $\alpha = n$. Both real and imaginary parts of the conductivity rise with the same exponent in a way that is consistent with Figure 3.5a. The rise of the conductivity in the high frequency domain

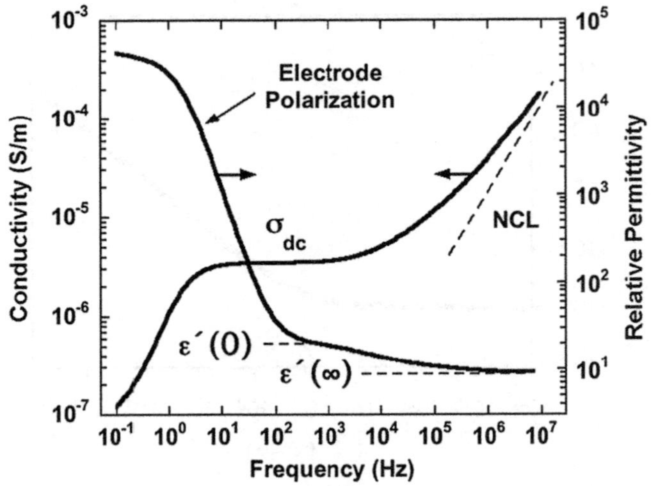

Figure 3.4 Representation of ac conductivity and the real part of the permittivity for a lithium-phosphate glass, indicating the step $\Delta\varepsilon$ and the electrode polarization. © IOP Publishing.
Reproduced with permission. All rights reserved.[15]

in Figure 3.4 is accompanied by a step in the permittivity from $\varepsilon'(\infty)$ to $\varepsilon'(0)$ to. The dc conduction is characterized by a rise of the dielectric loss that increases with f^{-1} towards low frequencies. Figure 3.2b shows a single relaxation mechanism in which the dielectric loss at low frequency exhibits the additional feature due to dc conduction. If the conductivity is constant, as in dc conduction, it does not affect the permittivity, as shown in Figure 3.2b. However, if the low frequency conduction has some frequency dispersion, the same dispersion appears in the permittivity $\varepsilon'(\omega)$ as observed in the dispersive regime of conduction in Figure 3.3b.

So far, the discussion has focused on the conduction and polarization mechanisms for the dielectric capacitance and carrier conduction that are distributed homogeneously in the bulk of the sample. In this case, the permittivity and conductivity are intensive bulk properties of the material and they are independent of the sample thickness. However, there are different types of interfacial effects in electrical measurements, especially those related to the presence of the outer contacts. Considering that different bulk and contact contributions is the domain of the *impedance spectroscopy* technique, $Z(\omega)$ and $C(\omega)$ are used to display and analyze the data. Normally, these features are analyzed using the valuable tool of equivalent circuits. However, dielectric relaxation phenomena are the central topic of interest in some fields of study, and so $\varepsilon(\omega)$ and $\sigma(\omega)$ are the preferred entities to present the data. In this case, the capacitance of the interfaces will also introduce features in $\varepsilon(\omega)$, as defined by eqn (3.4), and the conductivity will be affected as well by features that are no longer related to intensive bulk quantities. Some examples are given below.

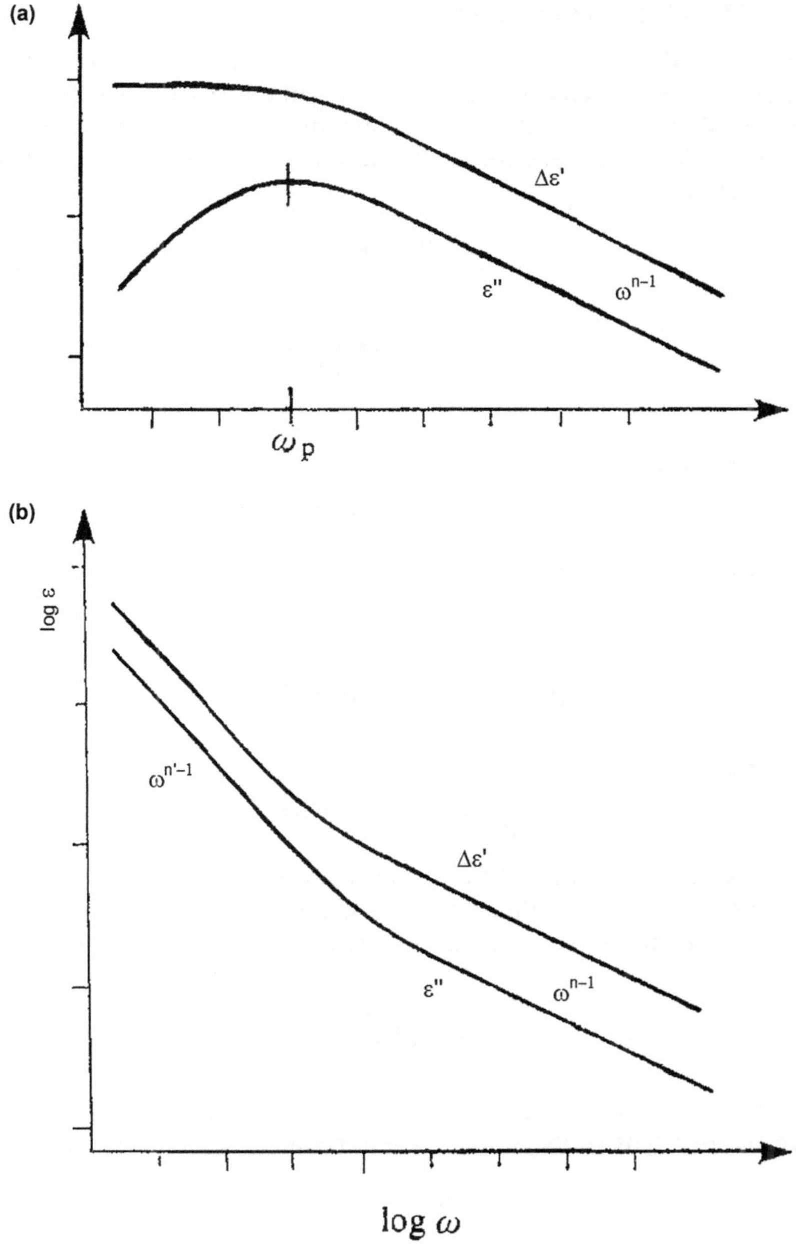

Figure 3.5 Increment of the real part of the complex permittivity (ε') and the imaginary part (ε'') for two types of dielectric response, showing the universal behavior with ω^{n-1} at high frequencies. (a) Broad dipolar relaxation and (b) conductivity dispersion types.
Reproduced from ref. 13 with permission from Nature Publishing Group.

3.2.3 The Chemical Capacitance

We will now discuss the *bulk* capacitance, which is different from the dielectric capacitance considered above. We will follow the notation used in ref. 18. To determine the density of states in the valence and conduction band, we utilize the standard formulae for a parabolic band as a first approximation,

$$N_\mathrm{c} = 2\left[\frac{2\pi m_e{}^{*} k_\mathrm{B} T}{h^2}\right]^{3/2} \tag{3.21}$$

$$N_\mathrm{v} = 2\left[\frac{2\pi m_h^{*} k_\mathrm{B} T}{h^2}\right]^{3/2} \tag{3.22}$$

The intrinsic carrier density is

$$n_i = \sqrt{N_\mathrm{c} N_\mathrm{v}}\, \mathrm{e}^{-E_g/2k_\mathrm{B}T} \tag{3.23}$$

We can express the electron and hole densities as follows:

$$n = n_0 \mathrm{e}^{(E_\mathrm{Fn} - E_\mathrm{F0})/k_\mathrm{B}T} = N_\mathrm{c} \mathrm{e}^{(E_\mathrm{c} - E_\mathrm{Fn})/k_\mathrm{B}T} \tag{3.24}$$

$$p = p_0 \mathrm{e}^{-(E_\mathrm{Fp} - E_\mathrm{F0})/k_\mathrm{B}T} = N_\mathrm{v} \mathrm{e}^{(E_\mathrm{v} - E_\mathrm{Fp})/k_\mathrm{B}T} \tag{3.25}$$

Note the product

$$np = N_\mathrm{c} N_\mathrm{v} \mathrm{e}^{-(E_\mathrm{g} - qV)/k_\mathrm{B}T} \tag{3.26}$$

In the semiconducting light absorber materials used in solar cells, the change of electron carrier density, n, associated to the displacement of E_Fn, the electron Fermi level, gives rise to a chemical capacitance that has the value,[19]

$$c_\mu = q^2 \frac{\partial n}{\partial E_\mathrm{Fn}} \tag{3.27}$$

where q is the elementary charge and $k_\mathrm{B}T$ is the thermal energy.

If the carrier density obeys the Boltzmann statistics (non-degenerate case)

$$n = n_0 \mathrm{e}^{(E_\mathrm{Fn} - E_\mathrm{F0})/k_\mathrm{B}T} \tag{3.28}$$

then the chemical capacitance per unit volume is

$$c_\mu = \frac{q^2 n}{k_\mathrm{B} T} \tag{3.29}$$

In disordered semiconductors, the generation–recombination of carriers determines the population of disorder-induced DOS, g_n, which in fact determines the E_Fn position and consequently the output voltage. In this case, the chemical capacitance mirrors the DOS to a good approximation in such a way that[18]

$$c_\mu = q^2 g_n(E_{Fn}) \tag{3.30}$$

This identity is strictly valid at zero-temperature, when the Fermi–Dirac function converges to the step function.

The total chemical capacitance is directly proportional to the film thickness[18]

$$C_\mu = c_\mu d \tag{3.31}$$

This is in direct contrast to the dielectric capacitance that depends reciprocally on d, see eqn (3.4).

3.2.4 Contact Capacitances

In general, different types of polarization occur at the electrodes, and these are related, for example, to the familiar Helmholtz (or more general double-layer) capacitance in electrochemical systems and to the Schottky barrier leading to the Mott–Schottky capacitance plot used to characterize semiconductor barriers.

When an n-type semiconductor is contacted with an electron-extracting layer with a low work function, Φ_c, a Schottky barrier is formed at the contact. The work function offset between the semiconductor layer and the contacting material provides the origin of the contact barrier height, which determines the flat-band voltage as $qV_{fb} = E_{F0} - \Phi_c$.[20] Under these assumptions, bending of the transporting bands of the bulk absorbing layer appears as a consequence of the presence of immobilized charged defects (space charge). At the semiconductor–metal interface, these depletion zones produce capacitive responses. The depletion zone width, w, controls the capacitance value as

$$C_{sc} = \frac{\varepsilon\varepsilon_0}{w} \tag{3.32}$$

For a uniform distribution of dopants with density N in the space charge region (corresponding to a carrier density in the neutral region $n = N$), $w \propto \sqrt{V_{fb} - V}$, and hence

$$C_{sc}^{-2} = \frac{2}{q\varepsilon\varepsilon_0 N}(V_{fb} - V) \tag{3.33}$$

The representation of C_{sc}^{-2} *vs.* V is commonly known as the Mott–Schottky plot and, in general, the linear behaviour on these plots indicates the presence of a depletion zone. Eqn (3.33) shows that the Mott–Schottky plot allows the determination of the flat-band voltage from the intercept on the voltage axis, and the doping density from the slope of the linear portion of the response.

In ionic conducting systems, one encounters the phenomenon of electrode polarization, which is related to the fact that ions cannot penetrate the metal collector contact, so that conductivity is blocked at low frequencies. Therefore, the general trend of eqn (3.20) will be supplemented by a decrease of the conductivity as shown in Figure 3.4. In a pure ionic conductor, the conductivity must actually decrease to zero as the frequency tends to zero, see Figure 3.4. In the case of a mixed ionic electronic conductor (as expected for any light absorber semiconductor that is an ionic conductor as well), the conductivity will show another plateau at very low frequencies, associated with the electronic conductivity or the charge transfer resistance at the electronic contact.

The accumulation of ions at the contact interface produces a capacitance due to surface space charge.[21,22] This capacitance is located within a short distance from the contact, and it is independent of the film thickness d, if the film is thick enough. However, electrode polarization gives rise to very large values of the apparent permittivity in the $\varepsilon(\omega)$ plot as shown in Figure 3.4. This occurs in addition to the conductivity polarization indicated as $\varepsilon'(0)$. The large increase in ε' is a typical feature of many types of ionic systems.[23] Electrode polarization gives rise to capacitances of the order of 10 μF cm^{-2}, which can be modeled following the classical Gouy–Chapman diffuse double-layer model. At equilibrium (zero bias) and room temperature, excess ion carriers accumulate within a zone (space charge region) equivalent to the ion Debye length[24]

$$\lambda_D = \sqrt{\frac{\varepsilon\varepsilon_0 k_B T}{q^2 N}} \tag{3.34}$$

Here, N accounts for the density of ionic charges. The resulting electrode capacitance is

$$C_{sc} = \frac{\varepsilon\varepsilon_0}{\lambda_D} \tag{3.35}$$

3.2.5 Recombination Parameters

We consider the bimolecular recombination rate with coefficient B:

$$U_n = Bnp \tag{3.36}$$

The voltage is given by the difference of Fermi levels at the selective contacts,

$$qV = E_{Fn} - E_{Fp} \tag{3.37}$$

Therefore, from eqn (3.23) and (3.26),

$$U_n = Bn_i^2 e^{qV/k_B T} \tag{3.38}$$

In general, the current voltage curve is described by the following expression

$$j = j_0 e^{qV/mk_BT} - j_0 \tag{3.39}$$

Here, we have neglected shunt and series resistances. j_0 is the dark current density of the diode and m is the diode quality factor. Under illumination, the photocurrent is added to eqn (3.39). The current density in the solar cell diode can be also written in the form:

$$j = qd(U_n - U_0) \tag{3.40}$$

Here, U_0 is the equilibrium recombination rate that corresponds to generation by thermal radiation in the dark, see Section 3.7.1.

$$j_0 = qd\, Bn_i^2 \tag{3.41}$$

In a high quality solar cell with perfect selective contacts, the current in eqn (3.39) is entirely due to the recombination in the diode.[4] Let us write the recombination model of eqn (3.36) as

$$U_n = k_{rec}\, n^{\beta} \tag{3.42}$$

It is interesting to consider the dependence of the carrier density on the voltage. There are two main types of situation in a solar cell under illumination:

(a) The generated carrier density does not exceed the majority carrier density. Hence $p = p_0$ (for a p-type semiconductor), and the minority carrier density n is variable. Thus we have

$$n = n_0 e^{qV/k_BT} \tag{3.43}$$

Hence, in eqn (3.42), $k_{rec} = Bp_0$ and $\beta = 1$. By eqn (3.39), $m = 1$.

(b) The generated carrier exceeds the majority carrier density. Electroneutrality requires that

$$n = p \tag{3.44}$$

Hence, $k_{rec} = B$ and $\beta = 2$. By eqn (3.38) and (3.40), $m = 1$. The carrier density depends on the voltage as

$$n = n_i e^{qV/2k_BT} \tag{3.45}$$

The recombination resistance can be defined as

$$r_{rec} = \left(\frac{1}{d}\frac{\partial j}{\partial V}\right)^{-1} \tag{3.46}$$

and the total resistance is given by

$$R_{rec} = \frac{r_{rec}}{d} \tag{3.47}$$

Clearly, the recombination resistance is connected to the slope of the current voltage curve:

$$R_{\mathrm{rec}} = \frac{mk_{\mathrm{B}}T}{qj_0}\mathrm{e}^{-qV/mk_{\mathrm{B}}T} \tag{3.48}$$

It can be shown that, by taking a small perturbation of the continuity eqn (3.40), a linear equation[1] is obtained that relates the small perturbation of the voltage ($\hat{\varphi}$, the Fermi level) and the current density at the outer contacts, $\hat{j}$, as follows:

$$c_{\mu}\frac{\partial \hat{\varphi}_n}{\partial t} + \frac{\hat{\varphi}_n}{r_{\mathrm{rec}}} + \frac{\hat{j}}{d} = 0 \tag{3.49}$$

Eqn (3.49) indicates that the equivalent circuit in this case is a parallel connection of the chemical capacitance and recombination resistance. In ref. 25, electrons and hole recombination are considered separately. However, the circuit is the same as indicated previously, with the chemical capacitance being given by a series combination of the chemical capacitances of electrons and holes. The separation of the chemical capacitance electrons and holes is established when Shockley–Read–Hall (SRH) recombination occurs *via* defects in the bandgap.[26]

The recombination resistance can be expressed as

$$r_{\mathrm{rec}} = \frac{1}{c_{\mu}}\left(\frac{\partial \bar{U}_n}{\partial n}\right)^{-1} \tag{3.50}$$

The recombination lifetime is given by[3]

$$\tau_n = r_{\mathrm{rec}}\, c_{\mu} \tag{3.51}$$

Therefore, the connection between the lifetime and recombination resistance is the chemical capacitance. Otherwise, we have

$$\tau_n = \left(\frac{\partial \bar{U}_n}{\partial n}\right)^{-1} \tag{3.52}$$

In the case where $n = p$,

$$\tau_n = \frac{1}{2Bn} \tag{3.53}$$

3.3 Capacitance in Organic Solar Cells

3.3.1 Chemical Capacitance

As mentioned in Section 3.2.3, one of the mechanisms able to produce a capacitive effect, the chemical capacitance, relates to the variation of the electron and hole Fermi levels brought about by illumination or bias voltage-driven carrier injection. The chemical capacitance is a differential monitoring of the charge accumulation in the organic blend. Moreover, chemical capacitance is a very useful parameter for comparing different cells or processing conditions using the same set of materials, because it addresses energetic changes in the operation-determining electronic states of complete devices.

In order to undertake a meaningful interpretation of the observed chemical capacitance, it is important to clarify the change of Fermi levels with respect to the increased carrier density. For solar cells that are not severely limited by carrier transport or contact effects, the separation between the out-of-equilibrium electron and hole Fermi levels is directly related to the available free energy qV, as indicated in eqn (3.37). Previously, we have distinguished two principal modes of variation of the Fermi levels under increasing illumination. For eqn (3.43), only one Fermi level is modified, but in eqn (3.44) both Fermi levels are displaced symmetrically in opposite directions. Organic blends are normally doped compounds,[27] hence, the shift of the separate Fermi levels upon voltage or light application is not symmetric with respect to the equilibrium position E_{F0}. Most commonly, polymer donors undergo some kind of unintentional p-doping during the solar cell processing as a consequence of the appearance of structural or chemical defects brought about by exposure to oxygen or moisture.[28] It has been observed that the amount of photogenerated carriers under 1 sun irradiation intensity roughly equals the preexisting background hole density.[29] This implies that the hole Fermi level constitutes a sort of energy reference because $E_{Fp} \approx E_{F0}$. On the contrary, the electron Fermi energy moves across a large portion of the bandgap upon illumination or application of an external voltage.

A wide variety of compounds making up the absorbing blend of bulk-heterojunction solar cells have been explored using IS.[5] The general trends of the solar cell capacitance as a function of the Fermi voltage can be inferred by examining Figure 3.6. Most of the polymer–fullerene combinations exhibit an increased capacitance that depends approximately exponentially on the voltage of the form $C_\mu \propto \exp(\alpha qV/k_B T)$ with $\alpha \approx 0.3 - 0.4$. This is indeed the case for blends containing [6,6]-phenyl-C61-butyric acid methyl ester (PCBM) and similar fullerene acceptor molecules.[30] Other acceptors, such as 4,40-dihexyloxydiphenylmethano[60]fullerene (DPM6), give much lower capacitance values with a Gaussian-like voltage-dependence.[5] Since materials comprising BHJ solar cells are usually processed by solution methods, a great deal of energetic and structural disorder is expected to occur in complete devices. As indicated in eqn (3.30), the chemical capacitance measurement corresponds to a determination of the relevant DOS actually influencing the final V_{oc} value *via* Fermi level modulation. According to general observations of polymer-fullerene blends, disorder induces electronic gap states (density of states DOS) in addition to LUMO or HOMO levels. Capacitance measurements allow one to distinguish between electronic bands and site tails and to determine their relative position within the effective bandgap.

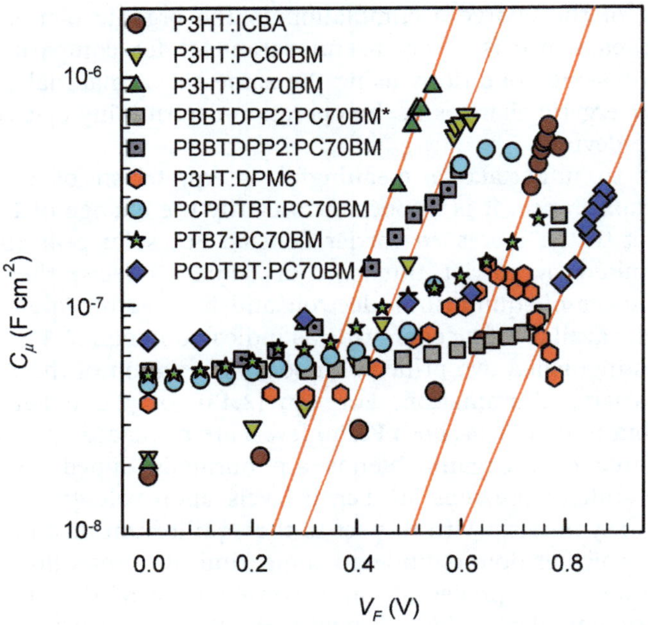

Figure 3.6 Capacitance–voltage extracted from impedance analysis for a variety of different materials forming the blend of BHJ solar cells. The screened fullerenes comprise molecules with different electron affinity. Polymers of wide and low bandgap, and different ionization energy are analyzed. As a general trend, a rise in the capacitance is observed in excess of geometrical values. Straight lines correspond to the exponential behavior when P3HT and PCDTBT are used as donor polymers.
Reproduced with permission from G. Garcia-Belmonte, A. Guerrero and J. Bisquert, *J. Phys. Chem. Lett.*, 2013, **4**, 877–886. Copyright (2013) American Chemical Society.[29]

3.3.2 Mott–Schottky Analysis

By examining Figure 3.6, one can observe that, in addition to the increment originating from the chemical capacitance, there is a much lower capacitance plateau at low voltages (usually <0.4 V). This portion of the dependence of the capacitance on the voltage cannot be understood in terms of the chemical capacitance. At low-voltages and under illumination, the excess charge is so small that other capacitive mechanisms mask the chemical capacitance contribution. These additional mechanisms have a purely dielectric origin. Under reverse bias voltage, the capacitance is dominated by bulk polarization (dielectric) processes that account for the geometrical capacitance of eqn (3.10): see zone I in Figure 3.7a.[2] This observation points to the occurrence of full-depletion conditions that produce a constant capacitor value.[31] This means that the band bending associated to the Schottky barrier extends all the way to the opposite contact and $w=d$. Upon increasing the forward bias, the capacitance rises slightly because of the voltage-driven reduction in the depletion zone that eventually collapses

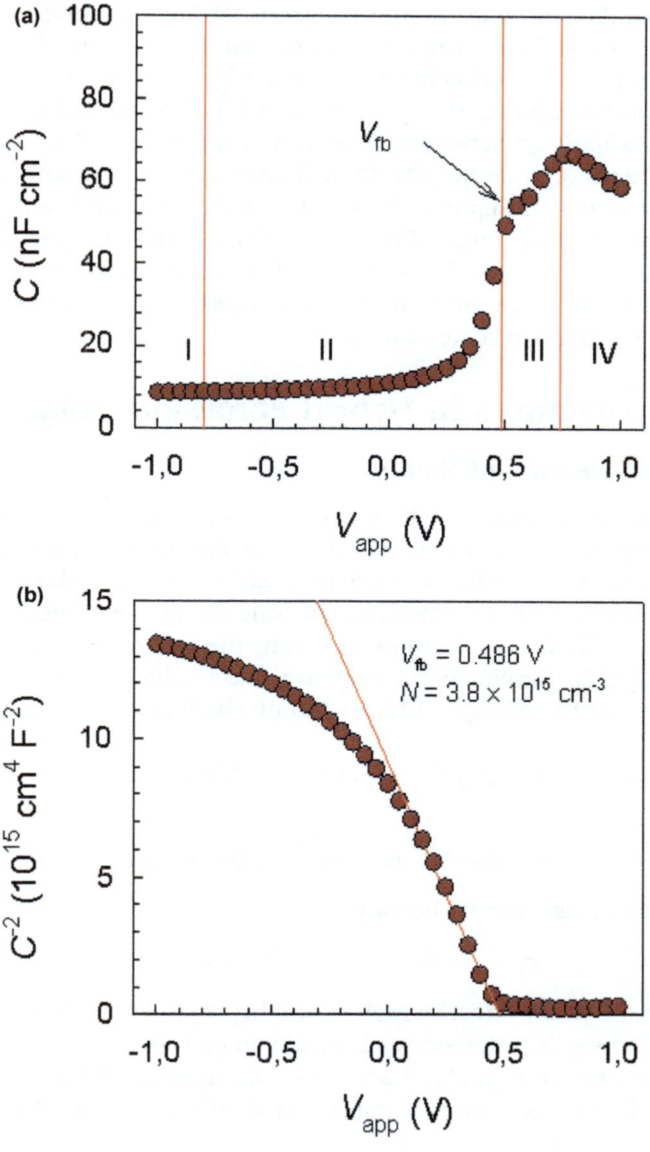

Figure 3.7 (a) Characteristic capacitance response (100 Hz) of a BHJ device with structure ITO/PEDOT:PSS/P3HT:PCBM/Al, as a function of the bias voltage. Vertical lines separate the voltage regions for which the capacitance is determined by the geometrical value C_g (zone I: $V_{app} \ll V_{fb}$), the depletion layer modulation C_{sc} (zone II: $V_{app} < V_{fb}$), the storage of excess carriers C_μ (zone III: $V_{app} > V_{fb}$), and limitation mechanisms (zone IV: $V_{app} \gg V_{fb}$). (b) Mott-Schottky plot (100 Hz) that exhibits a straight line from which the values $V_{fb} = 0.49$ V and $N_A = 3.8 \times 10^{15}$ cm^{-3} are determined, assuming $\varepsilon = 3$ for P3HT:PCBM.
Reproduced from ref. 32 with permission from the PCCP Owner Societies.

when flat-band conditions are achieved (zone II). Zone III corresponds to the previously discussed chemical capacitance that in some cases is observed to decrease by effect of limiting mechanisms at high forward voltages (zone IV).

As mentioned above, the processing of polymers induces p-doping because of oxidation reactions with oxygen or moisture and other structural changes. Figure 3.7b shows the Mott–Schottky plot corresponding to the capacitance values of Figure 3.7a. We note that the typical p-doping levels encountered are of the order of $N \approx 10^{15}$–10^{17} cm^{-3}. Low-voltage capacitance measurements are then an interesting analytical tool to explore doping levels and band bending in complete BHJ solar cells, as well as their variation with processing methods or material changes.

3.4 Capacitances in Hybrid Perovskite Solar Cells

3.4.1 The Density of States

As discussed in previous sections, the chemical capacitance is intimately related to the DOS of the carriers in the semiconductor. Let us estimate the chemical capacitance of free carriers in a lead iodide perovskite. In order to determine the density of states in the valence and conduction band of $MAPbI_3$, we take, in a first approximation, the standard formulae in eqn (3.21) and (3.22). Based on the effective mass values given in the literature,[33,34] $m_e = 0.18$ and $m_p = 0.22$, we obtain the following values:

$$N_c = 2 \times 10^{19}\, m_e^{3/2}\, \text{cm}^{-3} = 1.5 \times 10^{18}\, \text{cm}^{-3} \tag{3.54}$$

$$\text{N}_v = 2 \times 10^{19}\, m_h^{3/2}\, \text{cm}^{-3} = 2.0 \times 10^{18}\, \text{cm}^{-3} \tag{3.55}$$

and for the intrinsic carrier density,

$$n_i = \sqrt{N_c N_v}\, e^{-E_g/2k_B T} = 1.3 \times 10^5\, \text{cm}^{-3} \tag{3.56}$$

It is observed that the effective DOS of $MAPbI_3$ is rather small in comparison with standard semiconductors such as silicon or TiO_2 (10^{19} cm^{-3}). However, the effective DOS in eqn (3.54) and (3.55) are comparable to the value for GaAs, which also possesses a quite small effective mass for electronic carriers.

The effective density of states of the conduction band of the lead halide perovskite can also be obtained from the Burstein Moss (BM) shift.[35] Manser and Kamat have shown that the BM shift occurs at very low carrier densities, about 10^{18} cm^{-3}, that are consistent with the above calculations. For this value of carrier density, the Fermi level would already enter the conduction band. On the other hand, Yan and coworkers[36,37] have reported that the calculations of the transition matrix elements for GaAs and $MAPbI_3$ give similar results, so the larger optical absorption coefficient of $MAPbI_3$ should be due to a larger joint DOS. One study[38] has suggested that the presence of a direct gap between highly dispersed Pb(6s)–I(5p) valence bands and Pb(6p) conduction bands in $MAPbI_3$ is the key ingredient that explains their high absorption coefficient. The work by Amat *et al.*[39] have also analyzed the joint density of states of $MAPbI_3$. Their conclusion is that, although the perovskite CB has a rather low DOS, the VB has a very high DOS due to the I_3-contribution, which – in combination with a good matrix element due to good Pb/I overlap – explains the high absorption coefficient. We note these detailed descriptions are not taken into account in the simple model of eqn (3.54) and (3.55).

One consequence of the low DOS in the CB indicated by these results is that the chemical capacitance of the CB will be one order of magnitude smaller than in many other semiconductors with the same energy separation between the Fermi level and the conduction band edge. Assuming Boltzmann statistics and a perovskite bandgap energy $E_g = 1.5$ eV, one can estimate the charge density at 1 sun illumination giving typically open-circuit voltages around 1 V. For an intrinsic perovskite material, the excess carrier density obtained from eqn (3.45) is $n \approx 10^{14}\ \mathrm{cm^{-3}}$. From the experimental point of view, the chemical capacitance given by eqn (3.29) will be observable only if it becomes larger than the geometrical capacitance. For thin $MAPbI_3$ absorber layers (those that present larger geometrical capacitance values), we find the geometric capacitance $C_g = 0.3\ \mu\mathrm{F\,cm^{-2}}$, as discussed below. The excess carrier limit needed to fulfil the condition $C_\mu \geq C_g$ is $n \approx 5 \times 10^{15}\ \mathrm{cm^{-3}}$. In practice, actual experimental conditions do not allow the chemical capacitance features to be resolved from the geometrical capacitance in the case of thin absorber layers. In addition, C_μ may be masked by additional capacitance effects at low frequencies as discussed below. It is interesting to note that the calculated carrier density value in lead iodide perovskite ($10^{14}\ \mathrm{cm^{-3}}$) is lower than the typical photogenerated carrier densities of around $10^{16}\ \mathrm{cm^{-3}}$ reported for bulk-heterojunction solar cells operating at 1 sun irradiation intensity with a recombination coefficient of the order of $B \approx 10^{-12}\ \mathrm{cm^3\ s^{-1}}$.

3.4.2 Bulk Dielectric Constant

Several authors have measured the dielectric properties of lead halide perovskites experimentally. Poglitsch and Weber[40] in 1987 measured the complex dielectric response of methyl-ammonium lead halides (iodide, chloride, bromide) as a function of the temperature between 100 and 300 K. The effective dielectric constant at 300 K was measured to be 33 for $MAPbI_3$ at a frequency of 90 GHz. Other studies[41] have reported a value of *ca.* 58 at a frequency of 1 kHz. In contrast, the dielectric constant, in the absence of molecular reorientation, is predicted to be 24.1 from electronic structure calculations (PBEsol + QS*GW*),[42] in good agreement with the value of 23.3 determined from a fit of permittivity measurements over 100–300 K to the Kirkwood–Fröhlich equation.

Permittivity measurements on single $MAPbX_3$ (X = Cl or Br) crystals have detected a sharp increase in ε as the material undergoes a phase transformation at $\sim$160 K between orthorhombic (γ-phase with $\varepsilon \approx 24$) and tetragonal (β-phase with $\varepsilon \approx 55$ at 300 K) structures.[43] The phase transformation is also observed when the capacitance spectra of complete perovskite-based solar cells are explored.[44] As observed in Figure 3.8 for measurements in the dark under zero bias, the capacitance spectra of $MAPbI_{3-x}Cl_x$-based planar devices show two steps appearing at low temperatures (120–180 K), and high temperatures (180–320 K), respectively. The low-temperature increment yields a permittivity value for the β-phase of $\varepsilon \approx 32.5$ at 300 K, in good agreement with other reports.[45,46] It should be noted that considerable disparity is found among reported permittivity values ($\varepsilon \approx 24$–55). Considering that perovskite films are polycrystalline and that the TiO_2 layer is mesoporous in some cases, one can expect roughness factors as high as 3–5, as recently pointed out.[6] The high-frequency plateau ($C = 10$ nF cm^{-2}) can be related to the perovskite permittivity at low temperature in addition to the dielectric contribution of contacts layers such as spiro-OMeTAD and TiO_2.

The dielectric properties of $MAPbI_3$ films have been related to the rotational displacement of $CH_3NH_3^+$ cations, which do not take a fixed orientation in the structure (dynamic disorder) and contribute to an increase in the polarizability.[40] The rotation around the *c* axis of the PbI_6 octahedra forming the external perovskite structure also contributes to the polarizability.[47,48] In addition, cooperative ionic off-centering might influence the local polarization, potentially giving rise to long range ion displacement and macroscopic polarization.[49]

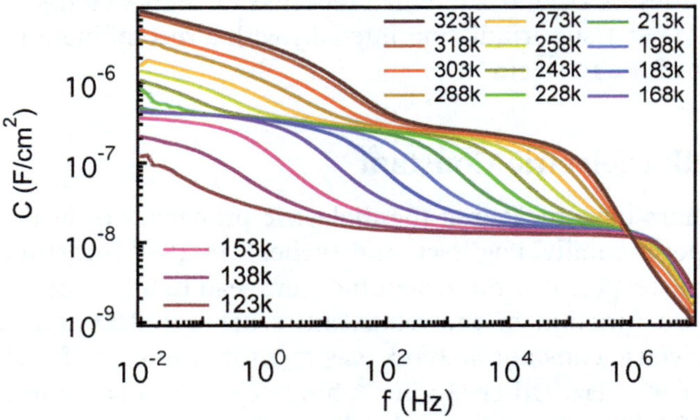

Figure 3.8 Capacitance spectra of planar $FTO/TiO_2/CH_3NH_3PbI_{3-x}Cl_x$/spiro-OMeTAD/Au structures, measured at different temperatures in the dark and at zero bias, showing two capacitance steps as discussed in the text. Reproduced with permission from O. Almora, I. Zarazua, E. Mas-Marza, I. Mora-Sero, J. Bisquert and G. Garcia-Belmonte, *J. Phys. Chem. Lett.*, 2015, **6**, 1645–1652. Copyright (2015) American Chemical Society.[44]

3.4.3 Electronic Contact Capacitance: A Mott–Schottky Analysis

As previously mentioned in the case of bulk-heterojunction solar cells (Section 3.3.2), energy equilibration at the outer interfaces between the absorbing and contacting layers depends on the relative doping character of the materials. In the most studied perovskite-based solar cell structure (FTO/TiO_2/$MAPbI_3$/spiro-OMeTAD/Au), the TiO_2 and spiro-OMeTAD layers are highly doped n- and p-compounds, respectively. If the perovskite layer is intrinsic, it is expected that the device operates under full-depletion conditions as a p-i-n structure.[50] However, $MAPbI_3$ can undergo either n- or p-doping as a consequence of complex defect chemistry. For p-doped compounds, a space charge region is formed in the vicinity of the cathode contact which accommodates the work function offset. *C–V* signaling indicates that the built-in voltage V_{bi} is always quite large ($\sim$1 V) indicating that, after equilibration, the absorbing layer accommodates a large potential drop,[51] as observed in the energy band diagram of Figure 3.9. The depletion layer plays an important role in charge collection if the diffusion transport is insufficient.[52] The capacitance has been measured in the dark varying the applied voltage and using intermediate measuring frequencies ($\sim$1 kHz) that avoid the influence of low-frequency capacitive effects (see Figure 3.8).

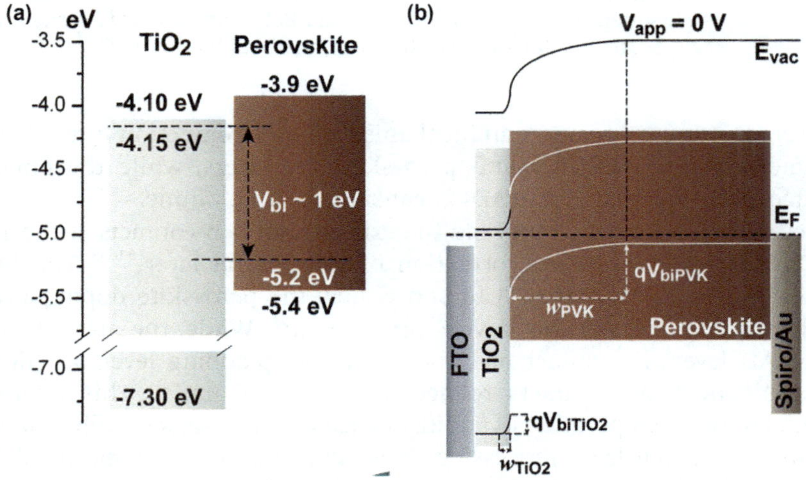

Figure 3.9 (a) Illustrative energy diagram of the separate materials, TiO_2 and $MAPbI_3$ (PVK). The position of the respective work function is based on the doping level encountered. A built-in voltage approximately equal to 1 eV is expected. (b) Band diagram in equilibrium showing the formation of a p–n (perovskite–TiO_2) type heterojunction. The extension of the respective depletion zones and built-in voltages are indicated at each side.
Reproduced with permission from A. Guerrero, E. J. Juarez-Perez, J. Bisquert, I. Mora-Sero and G. Garcia-Belmonte, *Applied Physics Letters*, 2014, **105**, 133902. Copyright (2014). AIP Publishing LLC.[51]

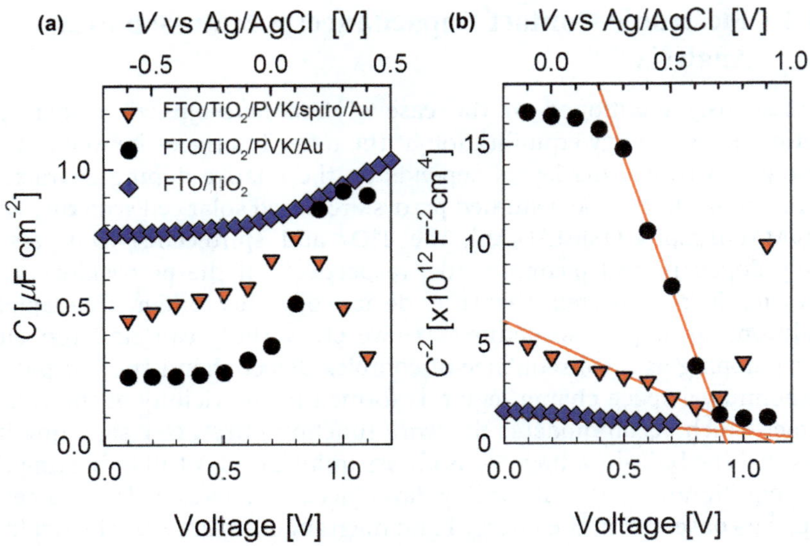

Figure 3.10 (a) Capacitance *vs.* voltage response of $MAPbI_3$ (PVK) films contacted by different anode structures, and (b) Mott–Schottky analysis of the devices in (a). For FTO/TiO_2, an electrochemical three electrode setup was used for a separate determination of the TiO_2 permittivity, with Ag/AgCl as the reference electrode.
Reprinted with permission from A. Guerrero, E. J. Juarez-Perez, J. Bisquert, I. Mora-Sero and G. Garcia-Belmonte, *Applied Physics Letters*, 2014, **105**, 133902. Copyright (2014). AIP Publishing LLC.[51]

The contact model assumes an ideal anisotype heterojunction formed at the contact between the TiO_2 and perovskite flat layers, while the opposite contact ($MAPbI_3$/spiro-OMeTAD) is considered to be ohmic.

Several analyses have proposed the existence of p–n contacts at the TiO_2/$MAPbI_3$ interface with the formation of a depletion layer.[51,53] The Mott–Schottky analysis of Figure 3.10 shows how the perovskite doping-level is highly dependent on the device processing.[51] While the use of spiro-OMeTAD layers as contact materials produces p-doping levels as high as $1.8\times10^{17}\,cm^{-3}$, Au contacts reduce it to $3.2\times10^{16}\,cm^{-3}$. This might be related to the incorporation of additives such as the Li salts usually employed during contacting layer processing. It is concluded that the electrical field profile throughout the $MAPbI_3$ bulk is controlled by the quantity of defects responsible for the perovskite doping.

3.4.4 Electrode Capacitance

As mentioned in Section 3.2.4, ionic conductors exhibit a significant increment in the capacitance as a consequence of the ion accumulation in the vicinity of the outer contacts.[21] Since perovskite materials are known to possess non-negligible ionic conductivity,[54] one can expect similar electrode polarization phenomena as those observed for purely ion conductors. Recently, the low-frequency capacitance increment exhibited by both planar and mesoporous $CH_3NH_3PbI_{3-x}Cl_x$-based solar cells, shown in Figure 3.8 (high temperatures curves), has been related to an electrode polarization effect.[44] It was shown that the low-frequency capacitance is independent of the thickness of the $MAPbI_3$ layer. This fact underlines the need to interpret the capacitance in terms of outer interface mechanisms as in eqn (3.35). Other explanations of the capacitance, either dielectric ($C \propto 1/d$) or chemical ($C \propto d$), can be safely discarded because they are inconsistent with the test of the absorbing-layer thickness. Ref. 9 also attributes the low frequency arc observed in IS to an ionic blocking effect.

Using the simple model of eqn (3.35), Figure 3.8 at room-temperature yields $\lambda_D \approx 6$–13 nm and $N \sim 10^{17}$ cm^{-3}. It is important to note that eqn (3.35) assumes only electrostatic interactions and Boltzmann statistics. The actual electrode capacitances may exhibit more complex features when the electrode reacts with the contacting perovskite film in such a way that both ionic and electronic charges are involved. In the latter case, the electrode reactivity can determine the low-frequency capacitance trends.

3.4.5 Dependence of Capacitance on Illumination

Summarizing the previous discussions, the capacitance of a $MAPbI_3$ solar cell in the dark under zero bias voltage contains at least three major features, *cf.* Figure 3.8:

(1) A low temperature, high frequency plateau ($C = 10$ nF cm^{-2}).
(2) A high temperature, high frequency plateau with permittivity value of $\varepsilon = 32.5$.
(3) A high temperature, low frequency apparent giant dielectric constant.

The capacitance in (1) contains contributions from contacting films in addition to the perovskite layer, because the reduction in the conductivity of TiO_2 and spiro-OMeTAD at low temperatures moves their respective responses into the measuring frequency window. It has been observed[29] that capacitance component (2) is inversely proportional to the film thickness as $C \propto 1/d$ and, consequently, it is dominated by the permittivity of the perovskite film in the β-phase. On the other hand, (3) is independent of the film thickness and it is therefore attributed to electrode polarization.

Under illumination, or for a forward voltage bias that injects carriers in the solar cell, a strong modification of the dielectric constant is observed, as shown in Figure 3.11.[55] There is a large increase on both the real and imaginary parts of the permittivity, and these changes are clearly connected to the presence of excess carriers, as shown by the large increase of the conductivity in Figure 3.12. The evolution of the real part of the permittivity with increasing illumination is shown in Figure 3.13, and the inset shows that the pseudo-static dielectric constant depends linearly on the

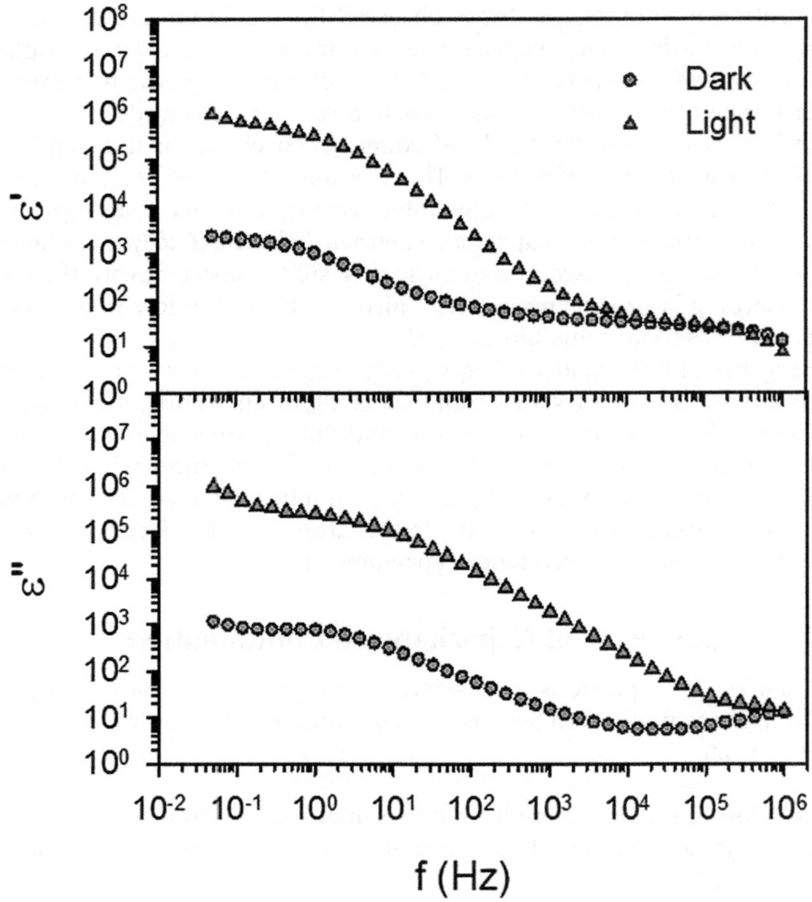

Figure 3.11 Bode plot of the real and imaginary part of the dielectric permittivity under dark and 1 sun illumination conditions for TiO_2/*meso*-Al_2O_3/$CH_3NH_3PbI_{3-x}Cl_x$/OMeTAD/Au perovskite solar cell. Bode plot of the real part of the conductivity under dark and 1 sun illumination conditions for measurements that were carried out at room temperature and 0 V applied bias.

illumination intensity. Figure 3.14 shows that a very large increase of the permittivity can also be obtained by voltage-injected carriers in the dark. In addition, the effect becomes even bigger if both illumination and forward bias are applied.

There are different interpretations of the physical origin of the increase of permittivity with illumination, and they can be summarized as follows, depending on the central property that is changed by the illumination:

(a) A change in the electrode polarization capacitance. This will be associated with an increased concentration of mobile ions, caused by an excess of electronic carriers.

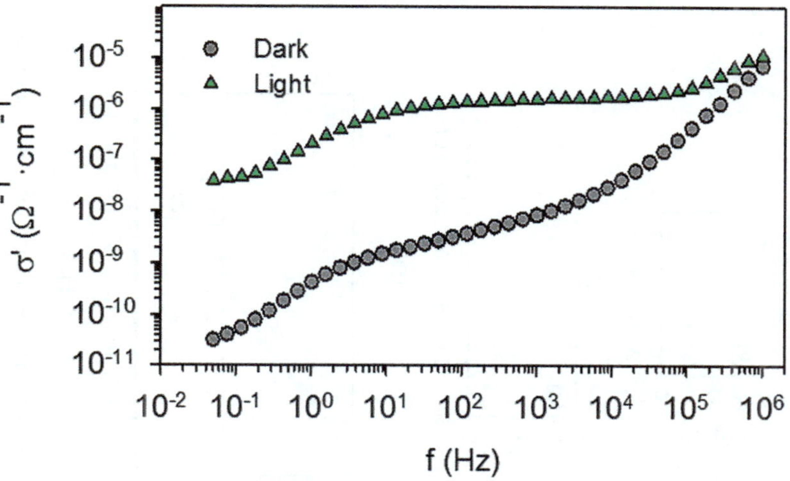

Figure 3.12 Bode plot of the real part of conductivity under dark and 1 sun illumination conditions for TiO_2/*meso*-Al_2O_3/$CH_3NH_3PbI_{3-x}Cl_x$/OMeTAD/Au perovskite solar cell. Measurements were carried out at room temperature and 0 V applied bias.

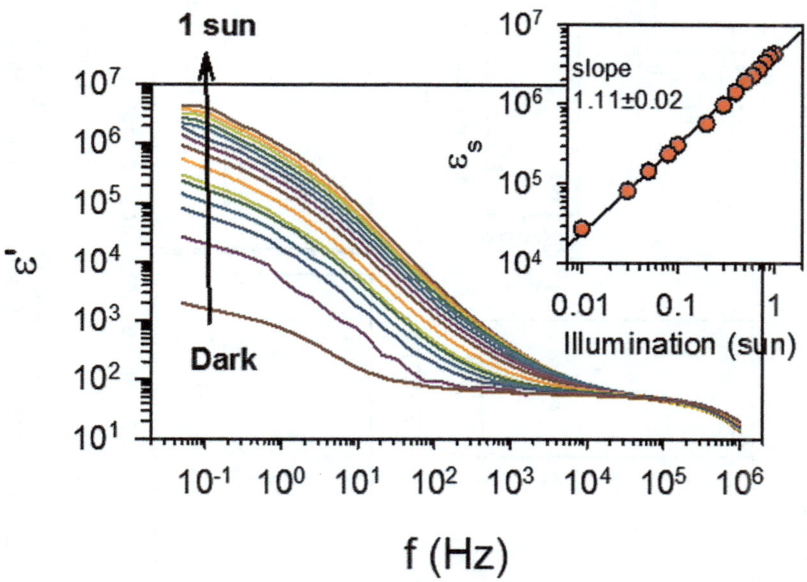

Figure 3.13 Plot of the real permittivity as a function of the frequency for different incident light intensities (Φ_0) from dark to 1 sun, for a TiO_2/*meso*-Al_2O_3/$CH_3NH_3PbI_{3-x}Cl_x$/OMeTAD/Au perovskite solar cell. Measurements were carried out at room temperature and 0 V applied bias. Inset, linear regression of the dielectric constant at $f = 50$ mHz *vs.* illumination intensity, observing a close to linear dependence between ε_s and the intensity of illumination.

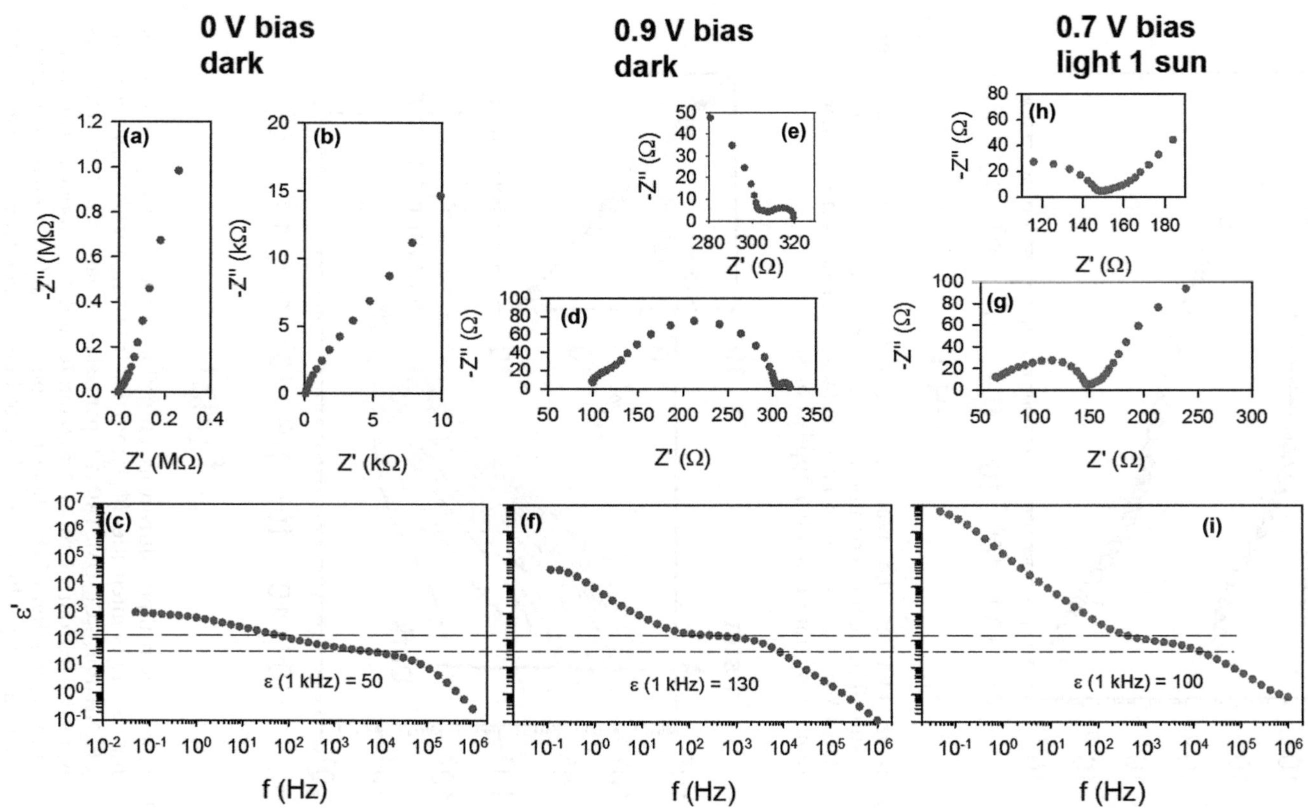

Figure 3.14 Complex plane impedance plot and plot of the real permittivity as a function of the frequency for TiO_2/*meso*-TiO_2/$CH_3NH_3PbI_{3-x}Cl_x$/OMeTAD/Au perovskite solar cell, in different conditions.
Reprinted with permission from E. J. Juarez-Perez *et al.*, *J. Phys. Chem. Lett.*, 2014, **5**, 2390–2394. Copyright (2014) American Chemical Society.[55]

(b) A change in the conductivity.[56] This explanation attributes the increase of ε' to the conduction polarization.
(c) A change in the bulk polarizability. In this model, an enhancement of the elementary dipole in the perovskite octahedral cage is caused by light-generated carriers.[29,57]
(d) A change in the chemical capacitance.[58]

The measurements shown in Figure 3.11 can be understood better by looking at the temperature variations in Figure 3.8. The latter indicate that the observed response in room temperature measurements (at present, data are not available for measurements at low temperature under illumination) is composed of a middle range frequency plateau and a low frequency giant dielectric constant, labeled (2) and (3) above. The mechanism (a) is for the low frequency feature (3), which seems a consistent explanation in view of the ionic conductivity that has been highlighted in ref. 58. It appears that the middle frequency plateau of ε' is not affected by the illumination, but the data available at present are not conclusive. It has been previously shown that such a plateau is a dielectric capacitance, in the dark at 0 bias; hence, mechanism (c) would require an increase of capacitance under illumination. Explanation (d) assumes that a chemical capacitance would take control of the middle frequency range, but so far there is no evidence for this. In ref. 58, it is suggested that the electronic chemical capacitance induces the feature (3), but this seems unlikely in view of the fact that feature (3) corresponds to a contact capacitance independent of the film thickness, whereas the chemical capacitance is proportional to the film thickness.

In regard to explanation (b), it is based on the general remark that if σ' has some frequency dispersion, this dispersion is translated to ε'' and, consequently, to ε', as explained in Section 3.2.2. This observation is interesting as, in fact, a small degree of frequency dispersion is observed in the middle frequency range of Figure 3.12, hence the increase of ε' is correlated with an increase of conductivity. One problem is that the morphological model (based on a porous structure) suggested in ref. 56 is rather implausible, as some perovskite solar cells showing these dispersion features are formed by large crystal grains. In addition, there is at present considerable uncertainty about the interpretation of the different parts of the impedance spectra,[6–8] and hence the conductivity pattern is more complex than it looks at first sight. The conductivity should contain both ionic and electronic conductivity components, but we need to remember that Figure 3.12 is for asymmetric solar cell structures, in which the low frequency resistance is controlled by a recombination resistance. Therefore, at the present time, it seems rather difficult to explain the increase of ε' in terms of a specific dispersive conduction mechanism. In general, it is evident that the identification of the main experimental trends and the origin of the dependence of conduction and polarizability with light, voltage and temperature in lead halide perovskite solar cells require further investigation.

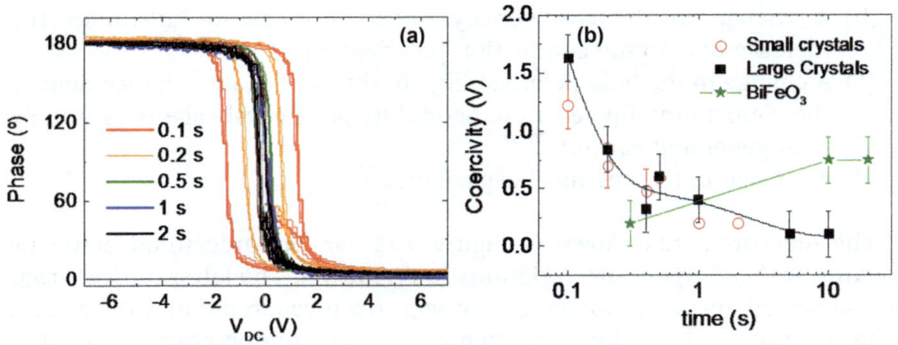

Figure 3.15 (a) Piezo-phase hysteresis loops of $MAPbI_3$ thin films with small crystals performed at different acquisition times. (b) Coercivity dependence with time of $MAPbI_3$ films with small (red open circles) and large crystals (black squares) and 25 nm $BiFeO_3$ thin films (green stars). Reproduced with permission from ref. 59. http://pubs.acs.org/doi/pdf/10.1021/acs.jpclett.5b00502.

An alternative approach to the study of the polarization is *via* piezoelectric force microscopy (PFM), which is a variant of AFM that is widely used to image polarization structures and local switching in ferroelectric materials. In fact, a study of $MAPbI_3$ perovskite films[59] has shown piezo-phase hysteresis loops as seen in Figure 3.15a. In principle, this observation provides evidence for ferroelectric polarization but, in contrast to a standard material such as $BiFeO_3$ (Figure 3.15b), the polarization in the lead halide perovskite thin film vanishes in a matter of seconds. It is also observed that the coercivity is greatly enhanced under illumination, which would agree with the light-enhanced dipole hypothesis (c). However, it seems unclear at present whether the ionic polarization effects described above may interfere with the observed polarization. Furthermore, the results of several reports on the ferroelectric properties of lead halide perovskites are contradictory,[9,60,61] and the matter needs more careful investigation.

3.5 Carrier Transport

Transport properties play a secondary role in the case of organic solar cells, at least for thin absorber layers of thickness around 100 nm. Therefore, this section will be directed mainly to transport in perovskite solar cells. In some cases, transport properties, such as the diffusion coefficient or diffusion length, can be accessed through measurements of the impedance, provided that the characteristic features of transport processes are not camouflaged by other phenomena. However, there are several works that point to an important role of ionic transport in the case of perovskite solar cells. The current situation of perovskite solar cells, in which the different processes occurring under working conditions are not completely understood, makes it difficult to provide a complete description of the analysis of transport processes by impedance measurements. However, significant advances in this area can be expected in the near future. In this section, we provide some insight into studies of electron and ionic transport properties measured on perovskite solar cells. Electronic transport properties have been measured following classical characterization techniques as time-of-flight (TOF) or the Hall effect.

Several studies on resistivity measurements on hybrid halide perovskites as a function of temperature were carried out during the 1990s. Knop *et al.* measured the resistivity, ρ, of $CH_3NH_3PbI_3$, obtaining values of 1.3×10^7 Ω cm at 0 °C and 2.6×10^7 $\Omega\cdot$cm at 94 °C, and they did not observe any effect of the presence or absence of ambient light.[62] The wide-reaching study of 2D and 3D hybrid perovskites carried out by Mitzi and coworkers is especially interesting.[63–65] The resistivity of hybrid perovskite based materials can be controlled by a layered structure, see Figure 3.16a. In a 2D perovskite layered structure, one or more perovskite layers are separated by organic layers as shown in Figure 3.16a for the $[NH_2C(I){=}NH_2]_2(CH_3NH_3)_mSn_mI_{3m+2}$ family. The resistivity for single crystals is highly dependent on the number of perovskite layers, see Figure 3.16b. The lowest resistivity is obtained for the 3D structure ($m=\infty$). This fact explains partially the superior performance of photovoltaic devices with 3D structures, such as $CH_3NH_3PbI_3$,[66] in comparison with their 2D $CH_3CH_2NH_3PbI_3$ counterparts.[67]

For 3D hybrid perovskites, the resistivity is also sensitive to the different elements composing the ABX_3 perovskite structure. Kanatzidis and coworkers carried out an extensive study of the structural and electrical properties of different hybrid halide perovskites, see Figure 3.17.[68] The electrical properties are affected by the type of the organic cation, methylammonium or formamidinium, used in the A site, as can be observed for $CH_3NH_3SnI_3$ in Figure 3.17a, and $HC(NH_2)_2SnI_3$ in Figure 3.17b. In addition, the use of different metallic cations also affects the final performance, as can be verified by comparing $CH_3NH_3SnI_3$ (Figure 3.17a) to $CH_3NH_3PbI_3$ (Figure 3.17c). Finally, it is also important to highlight that the resistivity is also affected by the crystallinity of the material. $CH_3NH_3PbI_3$ single crystals, Figure 3.17c, exhibit a resistivity more than four orders of magnitude lower than that observed for pellets prepared from $CH_3NH_3PbI_3$, Figure 3.17d.

Good conductivity is beneficial for long diffusion lengths, L_d, a key parameter for outstanding photovoltaic performance. Recent results on methylammonium lead perovskite single crystals have reported L_d values exceeding 10 μm,[69] or even surpassing 175 μm, although in this case only a single carrier was present in the layer.[70] Moreover, the exceptionally high extinction coefficient of the perovskite[71] means that only sub-micrometer perovskite layers are needed for complete light absorption of the solar spectrum above the band edge. The combination of low thickness and large L_d highlights the enormous potential of these materials for photovoltaic applications. Nevertheless, the absorber material employed in perovskite solar cells is polycrystalline due to the low cost deposition methods commonly used. In this context, the study of carrier diffusion lengths in a polycrystalline layer gives direct information on the parameters of the photovoltaic device. L_d has been measured on thin perovskite layers by

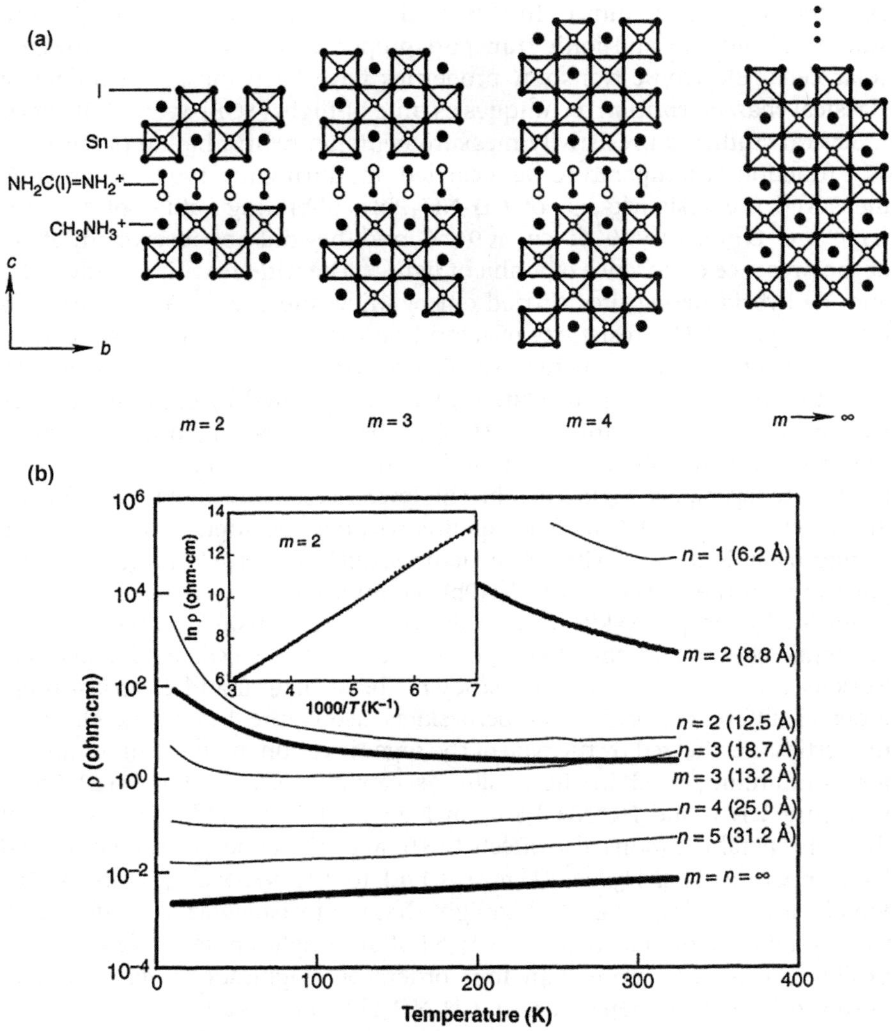

Figure 3.16 (a) Schematic presentation of the $NH_2C(I){=}NH_2]_2(CH_3NH_3)_mSn_mI_{3m+2}$ layered perovskite as a function of m; for $m = \infty$, a 3D perovskite is obtained. (b) Resistivity of different layered perovskites of the families $NH_2C(I){=}NH_2]_2(CH_3NH_3)_mSn_mI_{3m+2}$ and $(C_4H_9NH_3)_2(CH_3NH_3)_{n-1}Sn_nI_{3n+1}$, including the resistivity for the corresponding 3D perovskites $m = n = \infty$.[63] Reproduced with permission from D. B. Mitzi, S. Wang, C. A. Feild, C. A. Chess, and A. M. Guloy, *Science* 1995, **267**(5203), 1473.

analysis of photoluminescence transients.[72,73] For $MAPbI_3$ prepared from the precursor PbI_2, diffusion lengths of ~100 nm for both electrons and holes have been measured independently by different groups.[72,73] Even larger diffusion lengths of ~1 μm have been observed for $MAPbI_3$ prepared with the precursor $PbCl_2$.[73] These exceptionally high and balanced L_d for electrons and holes in polycrystalline materials have their origin in

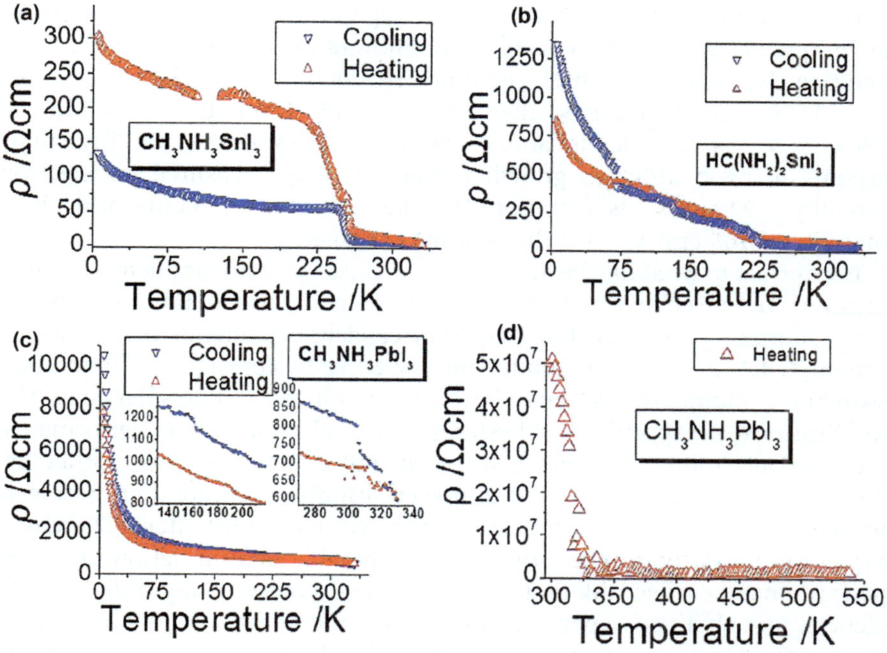

Figure 3.17 Resistivity of single crystal perovskites measured by the four-probe method. (a) $CH_3NH_3SnI_3$, (b) $HC(NH_2)_2SnI_3$, (c) $CH_3NH_3PbI_3$ in the 5–330 K temperature range and (d) CH3NH3PbI3 pellets in the 300–540 K temperature range.
Reproduced with permission from C. C. Stoumpos, C. D. Malliakas and M. G. Kanatzidis, *Inorg. Chem.*, 2013, **52**, 9019–9038. Copyright (2013) American Chemical Society.[68]

ambipolar diffusion with high electron and hole diffusion coefficients of 0.017 and 0.011 $cm^2\ s^{-1}$, respectively, when the iodine precursor is used, and 0.042 and 0.054 $cm^2\ s^{-1}$, respectively, when the chlorine precursor is used. It has also been observed that transport properties, such as diffusion length, depend strongly on the perovskite growth conditions and on the cell structure (planar device or nanostructured scaffold, and also on the type of scaffold).[74]

Several studies using THz spectroscopy have also provided quantitative results on carrier mobilities. Hertz and coworkers have highlighted the effect of different preparation methods on the final electric parameters of the perovskite layers.[75,76] High-frequency charge mobilities of 11.6 $cm^2\ V^{-1}\ s^{-1}$ and 8 $cm^2\ V^{-1}\ s^{-1}$ have been measured for solution-deposited perovskite layers prepared from chlorine and iodine lead precursors, respectively. In the case of the perovskite grown from the chorine precursor, even higher mobilities (>33 $cm^2\ V^{-1}\ s^{-1}$) are observed when the layer is prepared by evaporation. It is very significant that these values obtained on polycrystalline perovskite layers are almost the same as those reported for perovskite single crystals. For a $MAPbBr_3$ single crystal, Bakr and coworkers measured very similar mobilities for electron and holes of 115 $cm^2\ V^{-1}\ s^{-1}$

and 20–60 $cm^2 V^{-1} s^{-1}$, respectively, depending on the measurement technique: TOF and Hall effect.[69] The authors attributed the discrepancy between the measured values to surface effects that are negligible for TOF. For $MAPbI_3$ single crystals, they conservatively estimated a mobility of 2.5 $cm^2 V^{-1} s^{-1}$.[69] Huang and coworkers also produced $MAPbI_3$ single crystals using a different growth method.[70] They measured an electron mobility of 24 $cm^2 V^{-1} s^{-1}$ (by both Hall and TOF measurements) and a hole mobility of 105 $cm^2 V^{-1} s^{-1}$ (by Hall measurement).

In contrast to what has been observed in classical semiconductors, several unconventional effects have been observed for hybrid halide perovskites. One example is the giant low frequency capacitance already mentioned in Section 3.4.5. In addition, anomalous *J–V* curve hysteresis[49,77–80] and giant switchable photocurrents[81] have been reported in perovskite solar cells. The implications of capacitive *J–V* hysteresis in final photovoltaic performance are relatively minor, as solar cells work under DC conditions. Other behaviours of *J–V* curve associated to preconditioning indicate that hysteresis in perovskite solar cells is not only a transient capacitive effect. Therefore, hysteresis has strong implications for the reliable determination of device performance because different *J–V* curves, and consequently different efficiencies, are obtained depending on the direction of the voltage sweep (from short circuit to open circuit or *vice versa*). This hysteresis phenomenon has introduced some controversy in the field, since reported efficiencies depend strongly not only on the direction and voltage scan used for *J–V* measurements, but also on preconditioning illumination or poling.[82]

Many metal halide perovskites show ionic conductivity.[54,83–85] The internal migration of ions may have important consequences for the electrical characteristics of the solar cell and for the modification of interfacial properties under bias or illumination like the giant switchable photocurrent.[86,87] The ionic conductivity has been highlighted in different types of experiments.[49,80,81,88] It has also been suggested that ionic mobility plays an important role in electrode polarization (as discussed in Section 3.4.4) and in hysteresis phenomena.[49,80]

3.6 Recombination in Organic Solar Cells

3.6.1 Recombination Mechanisms

The rate of recombination of photogenerated electrons and holes is an essential factor determining the power conversion efficiency. In an organic cell, after charge separation at the polymer–fullerene interface, carriers may either be collected at the outer contacts or recombine *via* electron–hole annihilation.[89] The rate of carrier loss is widely believed to obey a bimolecular law of the form of eqn (3.36). In these systems, the bimolecular recombination coefficient is defined for recombination involving a reduced fullerene molecule and an oxidized polymer unit. The physics underlying the recombination process determines the coefficient B. The most widely accepted picture considers recombination as a Langevin-type process in which each individual recombination event is limited by the velocity by which carriers reach the recombination location.[90] Thus, the rate of carrier loss is governed by the carrier mobility. This model may reasonably be applied in the case of single phase materials but, in the case of heterogeneous blends, the Langevin theory imposes a drastic correlation between transport properties and recombination rates. Such a strong correlation between the carrier mobility and the recombination coefficient has rarely been observed in detailed experiments. Alternatively, recombination may be limited by the rate constant for charge transfer reactions at the internal Donor/Acceptor (D:A) interface of the blend, while the supply of electrons and holes is not thought to be a limiting factor.[91] According to this view, the fullerene:polymer reaction controls the recombination process, which depends consequently on the chemistry of the D:A molecular environment.[92] Recent papers have stressed that charge recombination properties are closely related to molecular interfacial properties.[93,94] Theoretical studies predict that relative molecular orientations and intermolecular distances play an important role in the charge transfer event kinetics.[95,96]

An issue that is related to the previous discussion on recombination mechanisms is the role of the distribution of electronic states (DOS) on the overall recombination rate. A useful way to explore a recombination process is by measuring the open-circuit voltage as a function of light intensity. Since the photogeneration rate G is proportional to the light intensity, it is expected that[97]

$$V_{oc} \propto \frac{k_B T}{q} \ln G \tag{3.57}$$

In some cases, a slope equal to k_BT/q has been reported for plots of $V_{oc} \propto \ln G$, signaling a bimolecular-like recombination process with a charge density-independent B value.[98] In other cases, a change in the recombination mechanisms has been proposed, since larger slopes $k_BT/\gamma q$ with $\gamma < 1$ have been observed. Instead of purely bimolecular recombination, a trap-assisted recombination process was suggested in terms of the SRH mechanism[99,100] or a density-dependent recombination coefficient.[101] Based on these approaches, it has been claimed that the slope of $V_{oc} \propto \ln G$ plots can be considered as an indication of the specific recombination process ($\gamma = 1$ for band-to-band bimolecular, or $\gamma < 1$ for trap-assisted) governing the solar cell performance. An alternative explanation relates the slope of the $V_{oc} \propto \ln G$ plot to the disorder in the electronic DOS, keeping the individual microscopic recombination events between electron and hole DOS of bimolecular type.[102] A more general model has been able to establish the crossover between the two regimes, those of density- and velocity-dependent recombination, according to the fundamental hopping and charge transfer rates.[91]

3.6.2 Determination of Recombination Kinetics

Recombination kinetics in bulk-heterojunction solar cells have been studied by several experimental techniques, including time-resolved charge extraction (TRCE), transient absorption spectroscopy (TAS), photoinduced charge extraction by linearly increasing voltage (photo-CELIV), transient photovoltage, and IS;[103] but a general picture and physical interpretation has not yet been established. The use of IS is particularly interesting because it allows the simultaneous measurement of charge density and electronic DOS (from the capacitance) and the recombination kinetics (from the recombination resistance). By integrating the chemical capacitance in eqn (3.29), one can obtain the electron carrier density involved in the photovoltaic process at a given voltage V as

$$n = \frac{1}{qd}\int_0^V C_\mu dV \tag{3.58}$$

The effective carrier lifetime can be calculated easily from the product r_{rec} C_μ, as indicated in eqn (3.51). These two parameters can be used to explore the validity of the bimolecular rate law eqn (3.36) assuming electroneutrality, eqn (3.44), which leads to $U = Bn^2$ ($\gamma = 2$). Under small perturbation conditions (IS), one can readily derive eqn (3.53). Bimolecular recombination then entails a reciprocal relationship between the excess carrier density and the effective lifetime.

An example of the relationship given by eqn (3.53) can be viewed in Figure 3.18 for the widely-studied combination between the donor P3HT and the acceptor PCBM.[25] For carrier densities within the interval 10^{16}–10^{17} cm^{-3}, it is observed that the effective lifetime is reduced from 100 μs to 10 μs, following eqn (3.53). This observation indicates that recombination is dominated by processes within the blend bulk, with outer contact recombination pathways being only a secondary influence. Deviations from the linear relationship at low carrier densities are explained by the dominance of contact over chemical capacitance, leading to time constants that are not related to true lifetime values. Recombination coefficients resulting from different D:A blends have been compared following similar procedures.[92] Figure 3.19 shows that poly(thiophenes) (P3HT) exhibit less recombination compared to (poly)carbazoles (PCDTBT) or diketopyrrolopyrroles (PDPP5T). The recombination coefficients encountered are always larger than $B > 10^{-12}$ cm^3 s^{-1}, with lifetimes of the order of μs at 1 sun irradiation intensity.

The analysis of the recombination kinetics in P3HT–PCBM organic solar cells has been used to provide input information for the reconstruction of J–V curves under varying illumination.[104] Here, the recombination resistance of eqn (3.46) is observed to be independent of the irradiation intensity after correction of the series resistance potential drop. This allows for full reconstruction of the photovoltaic response by integration of the recombination resistance to recover eqn (3.39), with the current entirely dominated by bimolecular recombination mechanisms. Related approaches based on the analysis of transient responses have been suggested.[105]

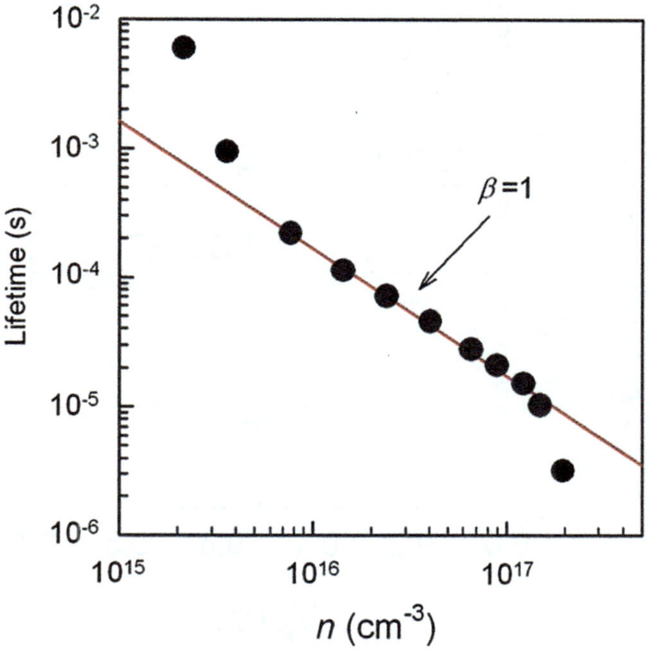

Figure 3.18 Recombination time (effective lifetime) as a function of the charge-carrier density, which is obtained by DOS integration, of bulk-heterojunction solar cells comprising P3HT:PCBM. A linear relationship is obtained $\gamma = 2$.
Reprinted from G. Garcia-Belmonte *et al.*, Simultaneous determination of carrier lifetime and electron density-of-states in P3HT:PCBM organic solar cells under illumination by impedance spectroscopy, *Sol. Energy Mater. Sol. Cells*, **94**, 366–375, Copyright (2010), with permission from Elsevier.[25]

3.7 Recombination in Perovskite Solar Cells

3.7.1 Radiative Recombination Coefficient of $MAPbI_3$

The study of radiative recombination can give important clues for the optimization of solar cells. In fact, efficient external luminescence is a necessity in order to approach the Shockley–Queisser efficiency limit for maximum photoconversion efficiency.[106] In this sense, the excellent emissive behavior of hybrid halide perovskites, with photoluminescence quantum efficiencies as high as 70% at high excitation densities,[107] is another indication of the enormous potential of these materials in photovoltaic applications.

As it is possible to quantify the radiation absorbed by a semiconductor in dark equilibrium, one can determine the fundamental radiative recombination rate of a semiconductor.[108,109] The emission of photons of energy E by radiative recombination occurs at a rate

$$U_{\mathrm{rad}}\,(E) = B_{\mathrm{rad}}\,(E) n_i^2 \tag{3.59}$$

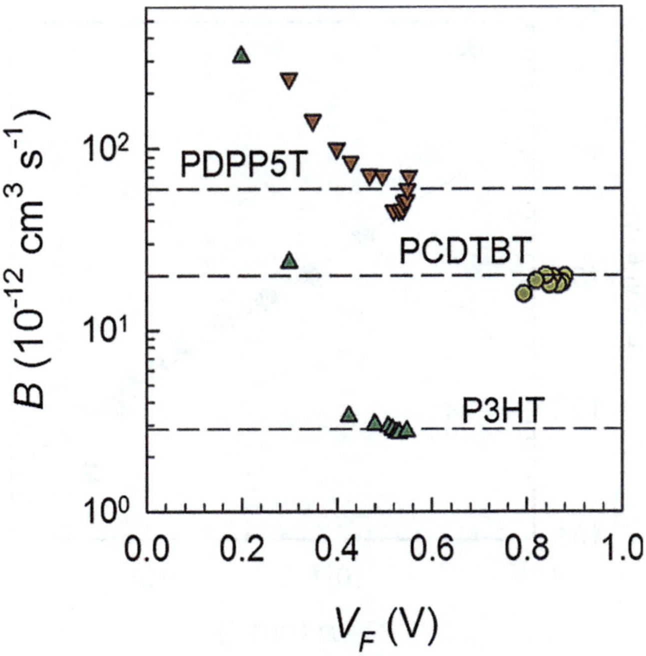

Figure 3.19 Recombination coefficient B calculated using the parameters extracted from the impedance analysis for diketopyrrolopyrrole (PDPP5T), carbazole (PCDTBT), and thiophene (P3HT) polymers. Horizontal dashed lines mark the average values of data points.
Reproduced with permission from ref. 92 http://pubs.acs.org/doi/pdf/10.1021/jp402751v.

where we have used the relationship $n_0 p_0 = n_\mathrm{i}^2$. Taking into account the blackbody radiation absorbed per volume element and emitted at a solid angle Ω, we obtain

$$B_\mathrm{rad}(E) = \frac{2\Omega n_r^2 \alpha(E)}{n_i^2 h^3 c^2} \frac{E^2}{\mathrm{e}^{E/k_\mathrm{B}T} - 1} \tag{3.60}$$

Here, n_r is the index of refraction, $\alpha(E)$ is the absorption coefficient, and the rest are standard physical constants. Integrating over the entire energy spectrum and total solid angle, we arrive at the radiative recombination constant

$$B_\mathrm{rad} = \frac{8\pi n_r^2}{n_i^2 h^3 c^2} \int_0^\infty \alpha(E) E^2 \mathrm{e}^{-E/k_\mathrm{B}T} \mathrm{d}E \tag{3.61}$$

In order to establish an analytical expression for the absorption coefficient of $MAPbI_3$, we have adopted the following expressions from ref. 106. The absorption coefficient is nearly constant at $E > E_g$ and it is well described by the form:

$$\alpha = \alpha_0 \left(1 + \frac{E - E_g}{E_1}\right) E > E_g \tag{3.62}$$

where E_1 is a constant. The optical absorption coefficient displays an Urbach tail due to perturbation of the bands and shows the exponential spectral dependence

$$\alpha(E) = \alpha_0 \exp\left(\frac{E - E_g}{E_0}\right) E \leq E_g \tag{3.63}$$

Here, E_0 is a tailing parameter. The following values have been obtained from the experimental data in ref. 71: $\alpha_0 = 2\times10^4\,\text{cm}^{-1}$, $E_0 = 0.015$ eV, $E_1 = 0.356$ eV.

Taking into account these values, $n_i = 1.3\times10^5\,\text{cm}^{-3}$ from eqn (3.56) and $n_r = 2.3$, it is now possible to use eqn (3.61) to calculate the recombination coefficient for the $MAPbI_3$ perovskite, which results in $B_{rad} = 2.3\times10^{-9}\,\text{cm}^3\,\text{s}^{-1}$ at 300 K. For comparison, the recombination coefficient for GaAs at 300 K is $B_{rad} = 7.2\times10^{-10}\,\text{cm}^3\,\text{s}^{-1}$.[110]

3.7.2 Radiative and Non-radiative Recombination in Perovskite Solar Cells

The recombination of photogenerated electron–hole pairs limits the efficiency of photovoltaic devices. It is important to distinguish between the different kinds of recombination processes, especially between radiative and non-radiative recombination. If it is physically allowed to absorb one photon to promote an electron from the valence band to the conduction band, the reverse process, the recombination of photoexcited electrons with holes emitting photons, is also allowed. By this reciprocity principle, radiative recombination is unavoidable. This is not the case for the non-radiative recombination processes that, in an ideal situation, can be completely removed. In fact, the Shockley–Queisser efficiency limit, the maximum theoretical efficiency for any photovoltaic device, is calculated assuming that the only recombination present is the unavoidable radiative recombination.[111] It follows that the presence of good emissivity of a photovoltaic device is a clear indicator of low losses by non-radiative recombination pathways and, consequently, of high photovoltaic performances.[112]

The relative weights of radiative and non-radiative recombination processes can be quantified by photoluminescence analysis.[113] Yamada *et al.* have studied the photoluminescence decay to analyze the different recombination pathways in perovskite $MAPbI_3$ for solar cell applications.[114] The rate of decay of photoluminescence is proportional to the density of photoexcited carriers. Considering an intrinsic material, the electron and hole densities, n and p respectively, are equal, and so the decay in carrier density can be described by the equation:

$$\frac{dn}{dt} = -A_{nr}n - B_{rad}n^2 \tag{3.64}$$

The first term in the right hand side of the equation describes single carrier non-radiative recombination *via* trap states, where A_{nr} is the average of electron and hole trapping rates. The second term describes the radiative recombination of electron–hole pairs. Note that, for the sake of simplicity, eqn (3.64) does not take into account other recombination pathways as SRH recombination[115] or third order Auger recombination, since they have a small weight in the total recombination process.[35] Herz *et al.* observed $B_{rad} = 9 \times 10^{-10}$ cm^{-3} s^{-1}.[76] The value of B_{rad} is of similar order than the calculated one in the previous section, and also of the same order than the radiative recombination coefficient of GaAs, as it has been previously highlighted. This fact, together with the high oscillator strength, are clear indicators of the potential of hybrid halide perovskites for the development of LEDs[116,117] and lasers.[107,118] Moreover, the low measured A_{nr} value indicates that recombination occurs mainly *via* radiative processes. This observation contrasts with the behavior of organic solar cells and dye-sensitized solar cells, where recombination is usually governed by non-radiative processes.

Recombination has an especially significant effect on the open circuit potential, V_{oc}. Tress, Grätzel and coworkers have studied the role of radiative and non-radiative recombination in the final open circuit potential by photoluminescence analysis.[113] They have analyzed three types of perovskite devices: with a TiO_2 scaffold, with an Al_2O_3 scaffold and spiro-OMeTAD in both cases as the hole transporting layer (HTL), and a third device without the HTL. The predicted V_{oc}, considering just the recombination resistance, was in all the cases ~1.3 V, while the measured values were ~1 V for the devices with scaffold and 0.8 V for the device without HTL. The difference between the theoretical and obtained values is due to non-radiative recombination, indicating that there is room for further optimization in perovskite solar cells. In addition, these results point to the presence of interfacial recombination that reduces the efficiency of samples without a carrier-selective contact.

Impedance spectroscopy is an effective characterization technique to evaluate recombination processes through the recombination resistance, R_{rec}. Recombination is a process always present in photovoltaic devices and the coupling with a capacitive process produces a semi-arc feature in the complex plane plot of the impedance spectra.[120] The diameter of the low frequency semicircle (highest value of Z' in the complex plane impedance plot, see Figure 3.20a), is associated with R_{rec}, which is inversely proportional to the recombination rate. Figure 3.20 shows an example in which a complete perovskite solar cell using a TiO_2 scaffold and compact TiO_2 and spiro-OMeTAD as the electron and hole selective contacts (EPH sample) is compared to samples without HTL (EP sample), without electron transporting layer (PH sample), and without both (P sample).[119] At a high applied bias, close to V_{oc}, the EPH and PH samples have similar R_{rec} values, see Figure 3.20b, and similar V_{oc} values are obtained for both samples, see Figure 3.20c. By contrast, devices without an electron selective contact layer show a significantly lower R_{rec} value, *i.e.* a higher recombination rate, pointing to an important recombination pathway at the transparent conductive oxide/perovskite interface that is blocked effectively by the presence of the TiO_2 electron selecting contact. The effect of the contacts on the

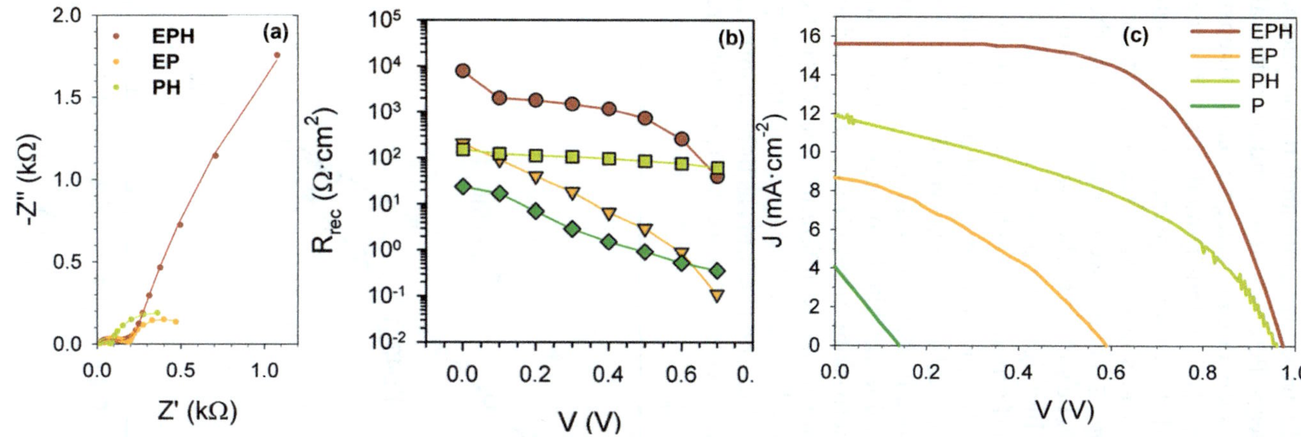

Figure 3.20 Perovskite solar cell using both selective contacts, EPH sample, with TiO_2 as selective electron contact and spiro-OMeTAD as selective hole contact, compared to samples without hole selective contact (EP), without electron selective contact (PH), or without both selective contacts (P). (a) Nyquist plot at DC bias of $V = 0.1$ V under 1 sun illumination; (b) R_{rec} and (c) J–V curves. Reproduced with permission from E. J. Juarez-Perez *et al.*, *J. Phys. Chem. Lett.*, 2014, **5**, 680–685. Copyright (2014) American Chemical Society.[119]

impedance response and the role of the recombination impedance have been studied systematically by Cheng and coworkers,[121] who concluded that the main recombination pathway is interfacial recombination at the perovskite/contact interfaces. IS results were also compared to the photoluminescence and open circuit voltage decays. Peter, Cameron and coworkers also compared IS results to those from open circuit voltage decays, intensity-modulated photovoltage spectroscopy and intensity-modulated photocurrent spectroscopy.[122] They analyzed 12 samples prepared in the same way, and observed two different behaviors in the R_{rec} measurements, see Figure 3.21. Different ideality factors, $m = 2.5$ and 5, were extracted from the slopes of the R_{rec} plots and were found to be similar to the ones obtained from open circuit voltage decays. This result highlights the power of IS for the characterization of perovskite solar cell performance as well as the difficulty in achieving reproducibility, even when the preparation conditions were nominally identical for all the samples. The origin of the observed difference in the R_{rec} slope is not clear, but has a significant effect on the sample behavior since the group of cells with the lowest m value exhibited a persistent photovoltage effect.

The effect of different perovskite materials as light absorbers, the use of different lead precursors or the use of different solar cell configurations can be analyzed in the same way.[123] Figure 3.22a presents the J–V plots obtained for different cells using $MAPbI_3$ and $MAPbBr_3$ as light-absorbing materials in two different configurations, with a TiO_2 or Al_2O_3 scaffold, respectively. It

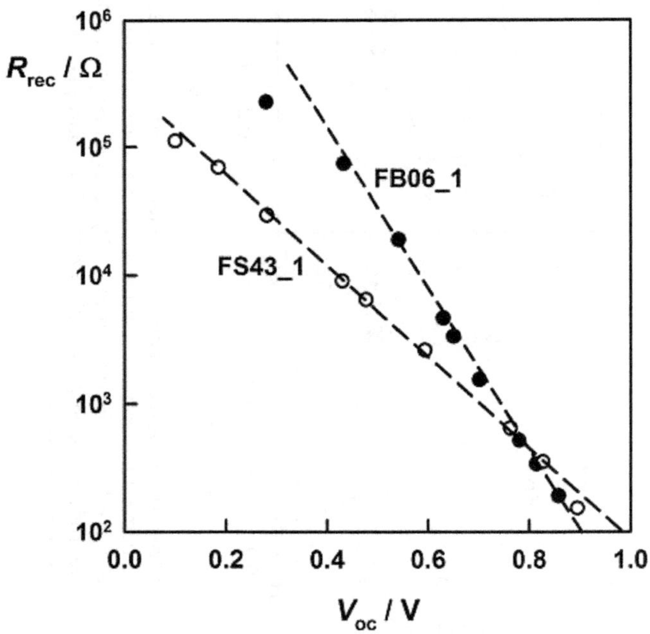

Figure 3.21 Recombination resistance measured for the two types of samples reported, where different ideality factors, $m = 2.5$ and 5 are observed. Reproduced from ref. 122 (http://pubs.acs.org/doi/pdf/10.1021/jp510837q).

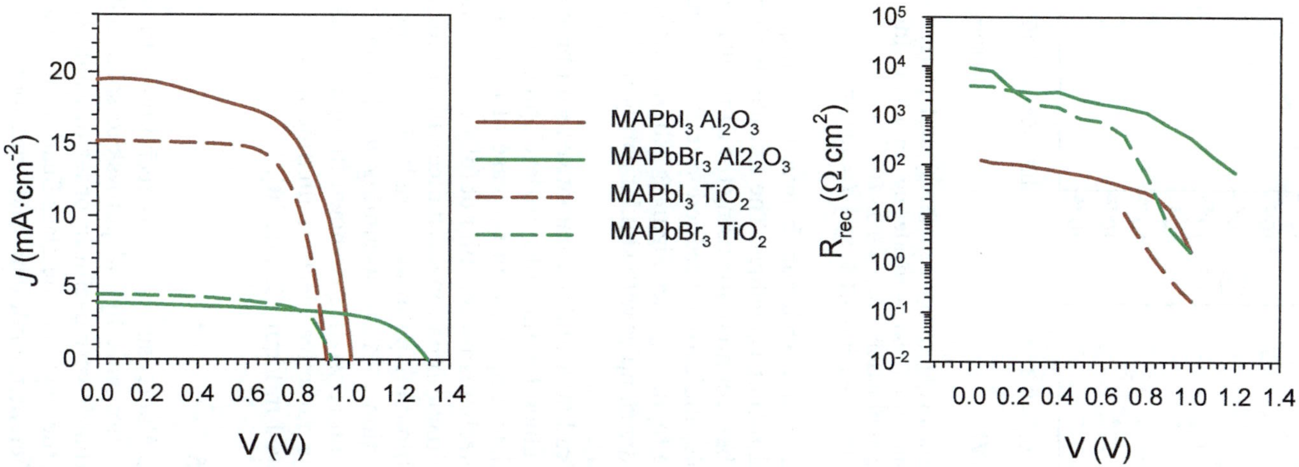

Figure 3.22 (a) *J–V* curves and (b) R_{rec} of solar cells prepared with $MAPbI_3$ and $MAPbBr_3$ as light absorbing materials, using two different configurations with titania or alumina mesoporous scaffolds.[123]
Reproduced with permission from B. Suarez *et al.*, *J. Phys. Chem. Lett.*, 2014, **5**, 1628–1635. Copyright (2014) American Chemical Society.

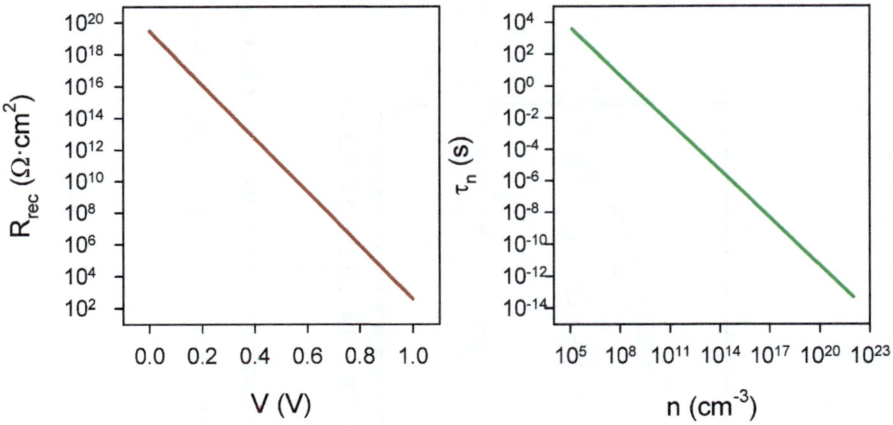

Figure 3.23 (Left) R_{rec}, calculated using eqn (3.48) and (3.41), and (right) τ_n, obtained from eqn (3.53) considering an ideal situation just with radiative recombination. $B_{rad} = 9.9 \times 10^{-10}\ cm^3\ s^{-1}$ (Section 3.7.1), $n_i = 1.3 \times 10^5\ cm^{-3}$ (Section 3.4.1) and $d = 300$ nm.

can be seen that there is a correlation between the higher R_{rec} (lower recombination rate) and higher V_{oc} values. The samples with alumina have higher R_{rec} values than their counterparts using the same absorbing material but with a titania scaffold, indicating a reduction in the recombination rate and, as a consequence, a higher V_{oc}, see Figure 3.22b. An analogous analysis can be made by comparing samples with similar configuration and different light absorbing materials.

The determination of radiative and non-radiative recombination pathways can provide important clues for the further development of photovoltaic devices, and it is especially important in the case of perovskite solar cells because their working mechanisms are still not fully understood. It is useful to compare the values of recombination resistance with the ideal situation in which the only recombination pathway is radiative, using the radiative coefficient derived in Section 3.7.1 and assuming $m = 1$. The resulting upper limit of the recombination resistance is shown in Figure 3.23(left). In addition, the corresponding radiative electron lifetime as a function of the carrier density is displayed in Figure 3.23(right).

3.8 Conclusions

The application of impedance spectroscopy techniques to study organic solar cells and organic–inorganic hybrid perovskite solar cells provides a variety of information that is very useful in order to establish the physical models of operation of the devices. The classification of capacitances indicates that there are different phenomena that contribute to the capacitive response: bulk and interfacial, dielectric and chemical capacitance, ionic and electronic. By using suitable variations such as thickness, type of contact and temperature, one can distinguish the physical nature of the measured capacitances and thus obtain a picture of the distribution of the bulk energy levels and energy barriers at the contacts. Great control of the kinetics and energetics of recombination has been achieved by IS in recent studies on organic solar cells. The understanding of transport and recombination with respect to the specific electronic and ionic features in perovskite solar cells is so far incomplete, but an initial understanding of their critical aspects has been achieved. Impedance spectroscopy, in combination with time transient and other experimental methods, will be key elements in the evolution towards the quantitative control of the photophysical behavior of hybrid organic–inorganic perovskite solar cells.

Acknowledgements

We acknowledge financial support by MINECO of Spain (project MAT2013-47192-C3-1-R) and by Generalitat Valenciana (project PROMETEO/2014/020).

References

1. J. Bisquert and F. Fabregat-Santiago, in *Dye-sensitized Solar Cells*, ed. K. Kalyanasundaram, CRC Press, Boca Raton, 2010.
2. F. Fabregat-Santiago, G. Garcia-Belmonte, I. Mora-Seró and J. Bisquert, *Phys. Chem. Chem. Phys.*, 2011, **13**, 9083–9118.
3. J. Bisquert, F. Fabregat-Santiago, I. Mora-Seró, G. Garcia-Belmonte and S. Giménez, *J. Phys. Chem. C*, 2009, **113**, 17278–17290.
4. J. Bisquert and R. A. Marcus, *Top. Curr. Chem.*, 2014, **352**, 325–396.
5. G. Garcia-Belmonte, A. Guerrero and J. Bisquert, *J. Phys. Chem. Lett.*, 2013, **4**, 877–886.
6. A. Pockett, G. E. Eperon, T. Peltola, H. J. Snaith, A. B. Walker, L. M. Peter and P. J. Cameron, *J. Phys. Chem. C*, 2015, **119**, 3456–3465.
7. A. R. Pascoe, N. W. Duffy, A. D. Scully, F. Huang and Y.-B. Cheng, *J. Phys. Chem. C*, 2015, **119**, 4444–4453.
8. V. Gonzalez-Pedro, E. J. Juarez-Perez, W.-S. Arsyad, E. M. Barea, F. Fabregat-Santiago, I. Mora-Sero and J. Bisquert, *Nano Lett.*, 2014, **14**, 888–893.
9. J. Beilsten-Edmands, G. E. Eperon, R. D. Johnson, H. J. Snaith and P. G. Radaelli, *Appl. Phys. Lett.*, 2015, **106**, 173502.
10. E. Guillén, F. J. Ramos, J. A. Anta and S. Ahmad, *J. Phys. Chem. C*, 2014, **118**, 22913–22922.
11. C. J. F. Böttcher and P. Bordewijk, *Theory of Electric Polarization*, Elsevier, Amsterdam, 1978.
12. A. K. Jonscher, *Dielectric Relaxation in Solids*, Chelsea Dielectrics Press, London, 1983.
13. A. K. Jonscher, *Nature*, 1977, **267**, 673.
14. D. P. Almond, G. K. Duncan and A. R. West, *J. Non-Cryst. Solids*, 1985, **74**, 285–301.

15. J. C. Dyre, P. Maass, B. Roling and D. L. Sidebottom, *Rep. Prog. Phys.*, 2009, **72**, 046501.
16. J. C. Dyre and T. B. Schroder, *Rev. Mod. Phys.*, 2000, **72**, 873.
17. D. L. Sidebottom, *Rev. Mod. Phys.*, 2009, **81**, 999–1014.
18. J. Bisquert, *Nanostructured Energy Devices: Equilibrium Concepts and Kinetics*, CRC Press, Boca Raton, 2014.
19. J. Bisquert, *Phys. Chem. Chem. Phys.*, 2003, **5**, 5360–5364.
20. A. Guerrero, L. F. Marchesi, P. P. Boix, S. Ruiz-Raga, T. Ripolles-Sanchis, G. Garcia-Belmonte and J. Bisquert, *ACS Nano*, 2012, **6**, 3453–3460.
21. J. H. Beaumont and P. W. M. Jacobs, *J. Phys. Chem. Solids*, 1967, **28**, 657.
22. M. Tomozawa and D.-W. Shin, *J. Non-Cryst. Solids*, 1998, **241**, 140–148.
23. P. Lunkenheimer, V. Bobnar, A. V. Pronin, A. I. Ritus, A. A. Volkov and A. Loidl, *Phys. Rev. B: Condens. Matter Mater. Phys.*, 2002, **66**, 052105.
24. C. Kim and M. Tomozawa, *J. Am. Chem. Soc.*, 1976, **59**, 127–130.
25. G. Garcia-Belmonte, P. P. Boix, J. Bisquert, M. Sessolo and H. J. Bolink, *Sol. Energy Mater. Sol. Cells*, 2010, **94**, 366–375.
26. L. Bertoluzzi, P. P. Boix, I. Mora-Sero and J. Bisquert, *J. Phys. Chem. C*, 2014, **118**, 16574–16580.
27. B. A. Gregg and Z. Liang, *Adv. Mater.*, 2012, **24**, 3258–3262.
28. L. G. Kaake, P. F. Barbara and X.-Y. Zhu, *J. Phys. Chem. Lett.*, 2010, **1**, 628–635.
29. G. Garcia-Belmonte, A. Guerrero and J. Bisquert, *J. Phys. Chem. Lett.*, 2013, **4**, 877–886.
30. A. Guerrero, L. F. Marchesi, P. P. Boix, J. Bisquert and G. Garcia-Belmonte, *J. Phys. Chem. Lett.*, 2012, **3**, 1386–1392.
31. J. Bisquert and G. Garcia-Belmonte, *J. Phys. Chem. Lett.*, 2011, **2**, 1950–1964.
32. F. Fabregat-Santiago, G. Garcia-Belmonte, I. Mora-Seró and J. Bisquert, *Phys. Chem. Chem. Phys.*, 2011, **13**, 9083–9118.
33. E. Menendez-Proupin, P. Palacios, P. Wahnon and J. C. Conesa, *Phys. Rev. B: Condens. Matter Mater. Phys.*, 2014, **90**, 045207.
34. G. Giorgi, J.-I. Fujisawa, H. Segawa and K. Yamashita, *J. Phys. Chem. Lett.*, 2013, **4**, 4213–4216.
35. J. S. Manser and P. V. Kamat, *Nat. Photonics*, 2014, **8**, 737–743.
36. W.-J. Yin, T. Shi and Y. Yan, *Adv. Mater.*, 2014, **26**, 4653–4658.
37. W.-J. Yin, T. Shi and Y. Yan, *J. Phys. Chem. C*, 2015, **119**, 5253–5264.
38. A. Filippetti and A. Mattoni, *Phys. Rev. B: Condens. Matter Mater. Phys.*, 2014, **89**, 125203.
39. A. Amat, E. Mosconi, E. Ronca, C. Quarti, P. Umari, M. K. Nazeeruddin, M. Grätzel and F. De Angelis, *Nano Lett.*, 2014, **14**, 3608–3616.
40. A. Poglitsch and D. Weber, *J. Chem. Phys.*, 1987, **87**, 6373–6378.
41. N. Onoda-Yamamuro, T. Matsuo and H. Suga, *J. Phys. Chem. Solids*, 1992, **53**, 935–939.
42. F. Brivio, K. T. Butler, A. Walsh and M. van Schilfgaarde, *Phys. Rev. B: Condens. Matter Mater. Phys.*, 2014, **89**, 155204.
43. M. Masaki, M. Hattori, A. Hotta and I. Suzuki, *J. Phys. Soc. Jpn.*, 1997, **66**, 1508–1511.

44. O. Almora, I. Zarazua, E. Mas-Marza, I. Mora-Sero, J. Bisquert and G. Garcia-Belmonte, *J. Phys. Chem. Lett.*, 2015, **6**, 1645–1652.
45. J. M. Frost, K. T. Butler and A. Walsh, *APL Mater.*, 2014, **2**, 081506.
46. F. Brivio, A. B. Walker and A. Walsh, *APL Mater.*, 2013, **1**, 042113.
47. Y. Kawamura, H. Mashiyama and K. Hasebe, *J. Phys. Soc. Jpn.*, 2002, **71**, 1694–1697.
48. H. Mashiyama, E. Magome, Y. Kawamura and Y. Kubota, *J. Korean Phys. Soc.*, 2003, **42**, 1026–1029.
49. E. L. Unger, E. T. Hoke, C. D. Bailie, W. H. Nguyen, A. R. Bowring, T. Heumuller, M. G. Christoforo and M. D. McGehee, *Energy Environ. Sci.*, 2014, **7**, 3690–3698.
50. J. M. Ball, M. M. Lee, A. Hey and H. J. Snaith, *Energy Environ. Sci.*, 2013, **6**, 1739–1743.
51. A. Guerrero, E. J. Juarez-Perez, J. Bisquert, I. Mora-Sero and G. Garcia-Belmonte, *Appl. Phys. Lett.*, 2014, **105**, 133902.
52. T. Kirchartz, J. Bisquert, I. Mora-Sero and G. Garcia-Belmonte, *Phys. Chem. Chem. Phys.*, 2015, **17**, 4007–4014.
53. W. A. Laban and L. Etgar, *Energy Environ. Sci.*, 2013, **6**, 3249–3253.
54. J. Mizusaki, K. Arai and K. Fueki, *Solid State Ionics*, 1983, **11**, 203–211.
55. E. J. Juarez-Perez, R. S. Sanchez, L. Badia, G. Garcia-Belmonte, V. Gonzalez-Pedro, Y. S. Kang, I. Mora-Sero and J. Bisquert, *J. Phys. Chem. Lett.*, 2014, **5**, 2390–2394.
56. D. P. Almond and C. R. Bowen, *J. Phys. Chem. Lett.*, 2015, **6**, 1736–1740.
57. X. Wu, H. Yu, L. Li, F. Wang, H. Xu and N. Zhao, *J. Phys. Chem. C*, 2014, **119**, 1253–1259.
58. T.-Y. Yang, G. Gregori, N. Pellet, M. Grätzel and J. Maier, *Angew. Chem., Int. Ed.*, 2015, **54**, 7905–7910.
59. M. Coll, A. Gomez, E. Mas-Marza, O. Almora, G. Garcia-Belmonte, M. Campoy-Quiles and J. Bisquert, *J. Phys. Chem. Lett.*, 2015, 1408–1413.
60. H.-S. Kim, S. K. Kim, B. J. Kim, K.-S. Shin, M. K. Gupta, H. S. Jung, S.-W. Kim and N.-G. Park, *J. Phys. Chem. Lett.*, 2015, **6**, 1729–1735.
61. B. Chen, J. Shi, X. Zheng, Y. Zhou, K. Zhu and S. Priya, *J. Mater. Chem. A*, 2015, **3**, 7699–7705.
62. O. Knop, R. E. Wasylishen, M. A. White, T. S. Cameron and M. J. M. V. Oort, *Can. J. Chem.*, 1990, **68**, 412–422.
63. D. B. Mitzi, S. Wang, C. A. Feild, C. A. Chess and A. M. Guloy, *Science*, 1995, **267**, 1473.
64. D. B. Mitzi, C. A. Feild, W. T. A. Harrison and A. M. Guloy, *Nature*, 1994, **369**, 467–469.
65. D. B. Mitzi, *Synthesis, Structure, and Properties of Organic-Inorganic Perovskites and Related Materials*, 1999.
66. J.-H. Im, C.-R. Lee, J.-W. Lee, S.-W. Park and N.-G. Park, *Nanoscale*, 2011, **3**, 4088–4093.
67. J.-H. Im, J. Chung, S.-J. Kim and N.-G. Park, *Nanoscale Res. Lett.*, 2012, **353**, 1–7.

68. C. C. Stoumpos, C. D. Malliakas and M. G. Kanatzidis, *Inorg. Chem.*, 2013, **52**, 9019–9038.
69. D. Shi, V. Adinolfi, R. Comin, M. Yuan, E. Alarousu, A. Buin, Y. Chen, S. Hoogland, A. Rothenberger, K. Katsiev, Y. Losovyj, X. Zhang, P. A. Dowben, O. F. Mohammed, E. H. Sargent and O. M. Bakr, *Science*, 2015, **347**, 519–522.
70. Q. Dong, Y. Fang, Y. Shao, P. Mulligan, J. Qiu, L. Cao and J. Huang, *Science*, 2015, **347**, 967–970.
71. S. De Wolf, J. Holovsky, S.-J. Moon, P. Löper, B. Niesen, M. Ledinsky, F.-J. Haug, J.-H. Yum and C. Ballif, *J. Phys. Chem. Lett.*, 2014, **5**, 1035–1039.
72. G. Xing, N. Mathews, S. Sun, S. S. Lim, Y. M. Lam, M. Gratzel, S. Mhaisalkar and T. C. Sum, *Science*, 2013, **342**, 344–347.
73. S. D. Stranks, G. E. Eperon, G. Grancini, C. Menelaou, M. J. P. Alcocer, T. Leijtens, L. M. Herz, A. Petrozza and H. J. Snaith, *Science*, 2013, **342**, 341–344.
74. A. Listorti, E. J. Juarez-Perez, C. Frontera, V. Roiati, L. Garcia-Andrade, S. Colella, A. Rizzo, P. Ortiz and I. Mora-Sero, *J. Phys. Chem. Lett.*, 2015, **6**, 1628–1637.
75. C. Wehrenfennig, M. Liu, H. J. Snaith, M. B. Johnston and L. M. Herz, *Energy Environ. Sci.*, 2014, **7**, 2269–2275.
76. C. Wehrenfennig, G. E. Eperon, M. B. Johnston, H. J. Snaith and L. M. Herz, *Adv. Mater.*, 2014, **26**, 1584–1589.
77. H.-S. Kim and N.-G. Park, *J. Chem. Phys. Lett.*, 2014, **5**, 2927–2934.
78. R. S. Sanchez, V. Gonzalez-Pedro, J.-W. Lee, N.-G. Park, Y. S. Kang, I. Mora-Sero and J. Bisquert, *J. Chem. Phys. Lett.*, 2014, **5**, 2357–2363.
79. H. J. Snaith, A. Abate, J. M. Ball, G. E. Eperon, T. Leijtens, N. K. Noel, S. D. Stranks, J. T.-W. Wang, K. Wojciechowski and W. Zhang, *J. Phys. Chem. Lett.*, 2014, **5**, 1511–1515.
80. W. Tress, N. Marinova, T. Moehl, S. M. Zakeeruddin, M. K. Nazeeruddin and M. Gratzel, *Energy Environ. Sci.*, 2015, **8**, 995–1004.
81. Z. Xiao, Y. Yuan, Y. Shao, Q. Wang, Q. Dong, C. Bi, P. Sharma, A. Gruverman and J. Huang, *Nat. Mater.*, 2015, **14**, 193–198.
82. B. C. O'Regan, P. R. F. Barnes, X. Li, C. Law, E. Palomares and J. M. Marin-Beloqui, *J. Am. Chem. Soc.*, 2015, **137**, 5087–5099.
83. K. Yamada, K. Isobe, E. Tsuyama, T. Okuda and Y. Furukawa, *Solid State Ionics*, 1995, **79**, 152–157.
84. T. A. Kuku, *Solid State Ionics*, 1987, **25**, 105–108.
85. T. A. Kuku and A. M. Salau, *Solid State Ionics*, 1987, **25**, 1–7.
86. Z. Xiao, Y. Yuan, Y. Shao, Q. Wang, Q. Dong, C. Bi, P. Sharma, A. Gruverman and J. Huang, *Nat. Mater.*, 2015, **14**, 193–198.
87. J. M. Azpiroz, E. Mosconi, J. Bisquert and F. De Angelis, *Energy Environ. Sci.*, 2015, **8**, 2118–2127.
88. E. T. Hoke, D. J. Slotcavage, E. R. Dohner, A. R. Bowring, H. I. Karunadasa and M. D. McGehee, *Chem. Sci.*, 2015, **6**, 613–617.

89. A. Maurano, R. Hamilton, C. G. Shuttle, A. M. Ballantyne, J. Nelson, B. O'Regan, W. Zhang, I. McCulloch, H. Azimi, M. Morana, C. J. Brabec and J. R. Durrant, *Adv. Mater.*, 2010, **22**, 4987–4992.
90. C. Deibel, A. Wagenpfahl and V. Dyakonov, *Phys. Rev. B: Condens. Matter Mater. Phys.*, 2009, **80**, 075203.
91. M. Ansari-Rad, G. Garcia-Belmonte and J. Bisquert, *Appl. Phys. Lett.*, 2015, **107**, 073301.
92. T. Ripolles-Sanchis, S. R. Raga, A. Guerrero, M. Welker, M. Turbiez, J. Bisquert and G. Garcia-Belmonte, *J. Phys. Chem. C*, 2013, **117**, 8719–8726.
93. C. W. Schlenker and M. E. Thompson, *Chem. Commun.*, 2010, **47**, 3702–3716.
94. S. Yamamoto, A. Orimo, H. Ohkita, H. Benten and S. Ito, *Adv. Energy Mater.*, 2012, **2**, 229–237.
95. Y. Yi, V. Coropceanu and J.-L. Bredas, *J. Am. Chem. Soc.*, 2009, **131**, 15777–15783.
96. S. Ko, E. T. Hoke, L. Pandey, S. Hong, R. Mondal, C. Risko, Y. Yi, R. Noriega, M. D. McGehee, J.-L. Brédas, A Salleo and Z Bao, *J. Am. Chem. Soc.*, 2012, **134**, 5222–5232.
97. L. J. A. Koster, V. D. Mihailetchi, R. Ramaker and P. W. M. Blom, *Appl. Phys. Lett.*, 2005, **86**, 123509.
98. S. R. Cowan, A. Roy and A. J. Heeger, *Phys. Rev. B: Condens. Matter Mater. Phys.*, 2010, **82**, 245207.
99. M. M. Mandoc, F. B. Kooisrta, J. C. Hummelen, B. de Boer and P. W. M. Blom, *Appl. Phys. Lett.*, 2007, **95**, 263505.
100. M. Kuik, H. T. Nicolai, M. Lenes, G.-J. A. H. Wetzelaer, M. Lu and P. W. M. Blom, *Appl. Phys. Lett.*, 2011, **98**, 093301.
101. T. Kirchartz, B. E. Pieters, J. Kirkpatrick, U. Rau and J. Nelson, *Phys. Rev. B: Condens. Matter Mater.*, 2011, **83**, 115209.
102. G. Garcia-Belmonte, *Solid-State Electron.*, 2013, **79**, 201–205.
103. T. M. Clarke, C. Lungenschmied, J. Peet, N. Drolet and A. J. Mozer, *Adv. Energy Mater.*, 2015, **5**, 01345.
104. P. P. Boix, A. Guerrero, L. F. Marchesi, G. Garcia-Belmonte and J. Bisquert, *Adv. Energy Mater.*, 2011, **1**, 1073–1078.
105. C. G. Shuttle, R. Hamilton, B. C. O'Regan, J. Nelson and J. R. Durrant, *Proc. Nat. Acad. Sci. U. S. A.*, 2010, **107**, 16448–16452.
106. O. D. Miller, E. Yablonovitch and S. R. Kurtz, *IEEE J. Photovoltaics*, 2012, **2**, 303–311.
107. F. Deschler, M. Price, S. Pathak, L. E. Klintberg, D.-D. Jarausch, R. Higler, S. Hüttner, T. Leijtens, S. D. Stranks, H. J. Snaith, M. Atatüre, R. T. Phillips and R. H. Friend, *J. Phys. Chem. Lett.*, 2014, **5**, 1421–1426.
108. W. van Roosbroeck and W. Shockley, *Phys. Rev.*, 1954, **94**, 1558–1560.
109. T. Trupke, M. A. Green, P. Wurfel, P. P. Altermatt, A. Wang, J. Zhao and R. Corkish, *J. Appl. Phys.*, 2003, **94**, 4930–4937.
110. Y. P. Varshni, *Phys. Status Solidi*, 1967, **19**, 459–514.
111. W. Shockley and H. J. Queisser, *J. Appl. Phys.*, 1961, **32**, 510–519.

112. O. D. Miller, E. Yablonovitch and S. R. Kurtz, *IEEE J. Photovoltaics*, 2012, **2**, 303–311.
113. W. Tress, N. Marinova, O. Inganäs, M. K. Nazeeruddin, S. M. Zakeeruddin and M. Graetzel, *Adv. Energy Mater.*, 2015, **5**, 1400812.
114. Y. Yamada, T. Nakamura, M. Endo, A. Wakamiya and Y. Kanemitsu, *J. Am. Chem. Soc.*, 2014, **136**, 11610–11613.
115. L. Bertoluzzi, P. P. Boix, I. Mora-Sero and J. Bisquert, *J. Phy. Chem. C*, 2014, **118**, 16574–16580.
116. O. A. Jaramillo-Quintero, R. S. Sanchez, M. Rincon and I. Mora-Sero, *J. Phys. Chem. Lett.*, 2015, **6**, 1883–1890.
117. Z.-K. Tan, R. S. Moghaddam, M. L. Lai, P. Docampo, R. Higler, F. Deschler, M. Price, A. Sadhanala, L. M. Pazos, D. Credgington, F. Hanusch, T. Bein, H. J. Snaith and R. H. Friend, *Nat. Nano*, 2014, **9**, 687–692.
118. G. Xing, N. Mathews, S. S. Lim, N. Yantara, X. Liu, D. Sabba, M. Grätzel, S. Mhaisalkar and T. C. Sum, *Nat. Mater.*, 2014, **13**, 476–480.
119. E. J. Juarez-Perez, M. Wuβler, F. Fabregat-Santiago, K. Lakus-Wollny, E. Mankel, T. Mayer, W. Jaegermann and I. Mora-Sero, *J. Phys. Chem. Lett.*, 2014, **5**, 680–685.
120. I. Mora-Seró, G. Garcia-Belmonte, P. P. Boix, M. A. Vázquez and J. Bisquert, *Energy Environ. Sci.*, 2009, **2**, 678–686.
121. A. R. Pascoe, N. W. Duffy, A. D. Scully, F. Huang and Y.-B. Cheng, *J. Phys. Chem. C*, 2015, **119**, 4444–4453.
122. A. Pockett, G. E. Eperon, T. Peltola, H. J. Snaith, A. Walker, L. M. Peter and P. J. Cameron, *J. Phys. Chem. C*, 2015, **119**, 3456–3465.
123. B. Suarez, V. Gonzalez-Pedro, T. S. Ripolles, R. S. Sanchez, L. Otero and I. Mora-Sero, *J. Phys. Chem. Lett.*, 2014, **5**, 1628–1635.

CHAPTER 4

Photophysics of Hybrid Perovskites

AJAY RAM SRIMATH KANDADA,[a] VALERIO D'INNOCENZO,[a,b] GUGLIELMO LANZANI[a,b] AND ANNAMARIA PETROZZA*[a]

[a] Center for Nano Science and Technology @Polimi, Istituto Italiano di Tecnologia, via Giovanni Pascoli 70/3, 20133, Milano, Italy;
[b] Dipartimento di Fisica, Politecnico di Milano, Piazza Leonardo da Vinci, 32, 20133, Milano, Italy
*Email: annamaria.petrozza@iit.it

4.1 Introduction

Hybrid perovskites are an emergent class of materials for optoelectronic applications.[1] However, understanding of their photophysics is lagging behind, hampering the systematic engineering of the best performing devices. There are open questions regarding the nature of elementary excitations:[2–5] the complex role of lattice dynamics in dielectric screening[6] and the energetics and dynamics at the interfaces. In order to answer these questions, it is necessary to deconvolute many interfering phenomena such as ion migration, carrier trapping and crystal phase transitions, to mention a few. Indeed, all these phenomena contribute to the ultimate behavior of the interfaces that lie at the heart of the devices and thus affect the final performance. Disentangling the different contributions and identifying the intrinsic properties are pre-requisites for the development of this technology on a market-scale.

In this chapter, we consider the physical basis for the optical properties of these semiconductors. We focus in particular on the absorption and emission properties of methyl-ammonium lead iodide $CH_3NH_3PbI_3$,

RSC Energy and Environment Series No. 16
Unconventional Thin Film Photovoltaics
Edited by Enrico Da Como, Filippo De Angelis, Henry Snaith and Alison Walker

Published by the Royal Society of Chemistry, www.rsc.org

which is the active semiconductor in champion solar cells, as well as on $CH_3NH_3PbBr_3$. We discuss specific photophysical observables and phenomena, such as the bandgap, carrier lifetime, trapping dynamics, carrier diffusion/mobility and Coulomb correlation effects. Attention is given to the role of the microstructure in the photophysics of the material. We also review various experimental methods used to perform photophysical investigations on perovskites, and highlight the precautions to be kept in mind when employing such techniques.

4.2 Linear Absorption

One of the primary reasons for the outstanding performance of solar cells based on hybrid perovskites is their strong optical absorption over a broad spectral range. Excellent light-harvesting properties with mere 500 nm thick films enables a good charge collection efficiency by reducing the solar cell's series resistance.[7] The optical behavior exhibited by hybrid perovskites is very similar to that of conventional inorganic direct-bandgap semiconductors.[8,9] Hence, it is appropriate to begin by reviewing the theoretical framework of the semi-classical theory of light–matter interaction in semiconductors in order to understand the photophysical properties of these materials.

4.2.1 3D Semiconductor Absorption

Hybrid perovskites, especially those employed in photovoltaics, are three-dimensional inorganic semiconductors. Within the *independent electron approximation*, the optical transition at the absorption edge for such systems is described by a band to band electronic transition, from the filled *valence* (E_v) band to the empty *conduction* (E_c) band (see Figure 4.1a). An absorbed photon gives rise to a pair of free particles, an electron in the conduction band and a hole in the valence band. Due to the negligibly small photon wave vector compared to the reciprocal space unit of the crystal lattice, only vertical transitions are optically allowed, and the nascent particles have opposite momentum.

The breakdown of the kinetic energy between electron and hole depends on the band curvature or, in other terms, on the effective mass of the two particles, according to $\varepsilon = h\nu - E_{gap} \approx \frac{\hbar k^2}{2m_s{}^*} = E_e(k_e)$ where "s" is either "electron" or "hole". In most semiconductors, $m_h{}^* \gg m_e{}^*$ and the photon energy exceeding the gap goes mainly into the kinetic energy of the electrons. In hybrid perovskites however, since $m_h{}^* \approx m_e{}^*$, the photon energy is equally shared by the two particles.

Since electrons are fermions, they follow Fermi–Dirac statistics and therefore, if we consider the semiconductor in thermal equilibrium, the occupation probability of a given level of energy E will be:

$$f(E) = \frac{1}{1 + e^{(E - E_F)/k_b T}} \tag{4.1}$$

where k_b is the Boltzman constant and E_F is the Fermi level. In the case of an intrinsic semiconductor, E_F will lie in the middle of the bandgap giving rise to a fully occupied valence band and an empty conduction band (Figure 4.1a). The initially photo-excited carriers are however out of thermal equilibrium, and in some situations may also have a specific polarization in the k space due to the light polarization. Electron–electron collisions re-distribute the momentum in few hundred femtoseconds by elastic scattering, while inelastic scattering establishes on a similar time scale a thermal distribution of the carrier population in energy, corresponding to a carrier temperature that is higher than that of the lattice. Such a hot carrier distribution relaxes further by electron–phonon interactions, cooling down the carrier temperature in a few ps. In perovskites, the intraband relaxation time[6,10] is $\approx$1 ps, much faster than electron–hole recombination, which occurs in the 1–100 ns time range.[1,11,12] As a consequence, we can safely assume that thermal equilibrium at the band edges is established after absorption, as represented in Figure 4.1b.

The strength of absorption, or *absorption coefficient*, at a photon frequency ν is given by the product: $d\alpha = \sigma(\nu - \nu_0)dN(\nu_0)$, where $\sigma(\nu - \nu_0)$ corresponds to the transition probability within that frequency range as given by Fermi's golden rule, and $dN(\nu_0)$ is the elemental number of transitions whose energies lie between $h\nu_0$ and $h(\nu_0 + d\nu_0)$. This number can be evaluated as $dN = \rho_j(E_0)dE_0$ where ρ_j is the density of possible transitions satisfying the energy and momentum conservation, referred as the *joint density of states*. For a 3D semiconductor, the frequency dependent absorption coefficient can be evaluated as

$$\alpha(\nu) \simeq \alpha_0 = \frac{\pi^3 \nu}{n\varepsilon_0 ch^3} \frac{\mu^2}{3} (2m_r)^{3/2} \sqrt{h\nu - E_G} \tag{4.2}$$

The absorption line shape is predominantly dictated by the joint density of states of the band to band transition, which monotonically increases

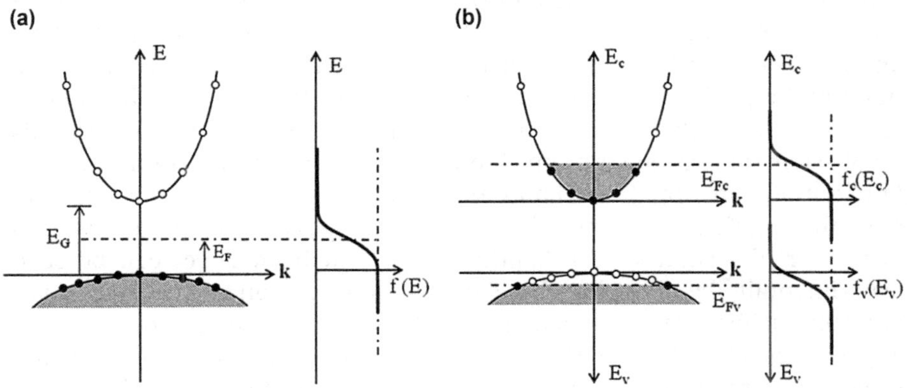

Figure 4.1 Dispersion relations and occupation probability function in (a) thermal equilibrium and (b) quasi-thermal equilibrium condition after photo-excitation.

above the band edge. In three dimensional semiconductors it follows the form: $\alpha \propto (\hbar\omega - E_{gap})^{1/2}$, while in 2D it is a step function.

The theory predicts that the absorption of a 3D direct bandgap semiconductor below the bandgap is null due to the lack of absorbing states. However, energetic disorder within the lattice due to various defects results in sub-bandgap states that contribute to the absorption. These are usually modeled as an exponential tail of the conduction band density of states that is described by a simple expression: $\exp\left(\frac{h\nu - E_g}{E_u}\right)$, where E_u is the Urbach energy. This empirical rule was first proposed by Urbach[13] in 1953, who employed it to describe alkali halide absorption. Later works[14,15] demonstrated that a random distribution of impurities in a semiconductor can introduce fluctuations in the lattice potential, thereby allowing the presence of some energy levels within the forbidden region of the energy gap. The Urbach energy, and thus the steepness of absorption close to the bandgap, is a quantitative measure of the crystallinity of a material and, in the case of perovskites, E_u has been found to be around 15 meV, which is comparable to the values for CIGS and GaAs.[1] It is remarkable that hybrid perovskites, unlike the other two, are solution processed and yet exhibit such high crystallinity. We return in Section 4.3 to the discussion of the defect states that can induce carrier trapping and strongly influence the recombination dynamics.

4.2.2 Bandgap

In the previous section we provided the formalism that defines the shape and extent of the free carrier absorption spectrum that also characterizes the hybrid perovskites. The success of $CH_3NH_3PbI_3$ as an efficient PV material arises from its appropriate bandgap value, a crucial parameter that dictates the open circuit voltage of a single-junction solar cell.[7] In the case of hybrid perovskites, theoretical studies have shown that the bandgap value is predominantly dictated by the Pb–X lattice, where the valence band maximum is composed of antibonding hybrid orbitals between the Pb-s and X-p orbitals, while the conduction band minimum is determined by a non-bonding hybrid state between Pb-p and X-p orbitals.[16–20] The organic cation, on the other hand, does not contribute to the frontier orbitals involved in optical absorption. However, its steric interaction with the lattice modulates the Pb–X bond angle, affecting the electronic structure and thus influencing the electronic and optical properties and, in particular, the value of the bandgap.[16,21,22] Such an effect can be seen experimentally by changing the chemical composition of the cation. For example, using a relatively smaller cation such as Cs increases the bond angle, thus blue-shifting the band edge. Such a strong structural and electronic correlation also manifests itself in the modulation of the optical properties of films with the same chemical composition, but different microstructures.

The value of the bandgap of any semiconductor is a temperature-dependent quantity.[23] There are two main reason for this dependence: (a) the thermal expansion of the lattice, which alters the periodic potential experienced by the electrons and thus the band structure and E_G, and (b) the effects of lattice vibrations (phonons), which are considered in more detail in Section 2.3.1. The combination of these effects leads the energy gap (E_g) dependence on the temperature to follow the empirical Varshni law:[24]

$$E_G(T) = E_0 - \alpha \frac{T^2}{T + \beta} \tag{4.3}$$

where E_0 represents the material energy gap at $T = 0$, while α and β are constant parameters characteristic of a given material. For most semiconductors (*e.g.* GaAs, Si), the gap shifts to lower energies with the increasing temperature ($dE_G/dT < 0$). However, Pb-based bulk semiconductors have an exceptional behavior, since the lattice dilation contributes to an increase of the bandgap energy with the temperature ($dE_G/dT > 0$).[25,26]

Hybrid perovskites are usually characterized by an excitonic transition in the absorption (which will be discussed in the next section), making it difficult to properly evaluate the E_G value from a simple UV-Vis absorption measurement. For this reason, in order to evaluate the temperature dependence of the energy gap for these materials, an effective approach is to monitor the edge of the absorption in a high temperature regime where the excitonic transition should not play a role (see Section 4.2.3). In this region, eqn (4.3) can be approximated by a linear function of $T\left(E_G(T)_0 + \frac{dE_G}{dT}T\right)$. Figure 4.2a shows the model case of $CH_3NH_3PbI_3$ prepared from PbCl and CH_3NH_3I precursors.[2] Figure 4.2b is a plot of the band edge (taken as the energy at which the first derivative of the absorption spectra is maximum) *versus* T, from which a linear fit gives the temperature dependence $\frac{dE_G}{dT} \simeq 262\ K^{-1}$. The temperature at which a PV cell operates can vary depending on the time of the day or the location. In a single-junction cell, temperature-induced variations may not be relevant. However, in the case of tandem cells, the use of two active materials with two different Varshini coefficients can strongly affect the overall efficiency.

4.2.3 Excitonic Absorption

The preceding discussion does not take into account any kind of interaction between the photo-generated carriers and assumes that the crystal states are totally unaffected by the transition of one electron from the valence band to the conduction band. The lowest order correction to this approximation is the inclusion of the Coulomb interaction between electrons and holes. The two particle problem can be reduced to that of the hydrogen atom, considering the effective masses of the two particles and the dielectric

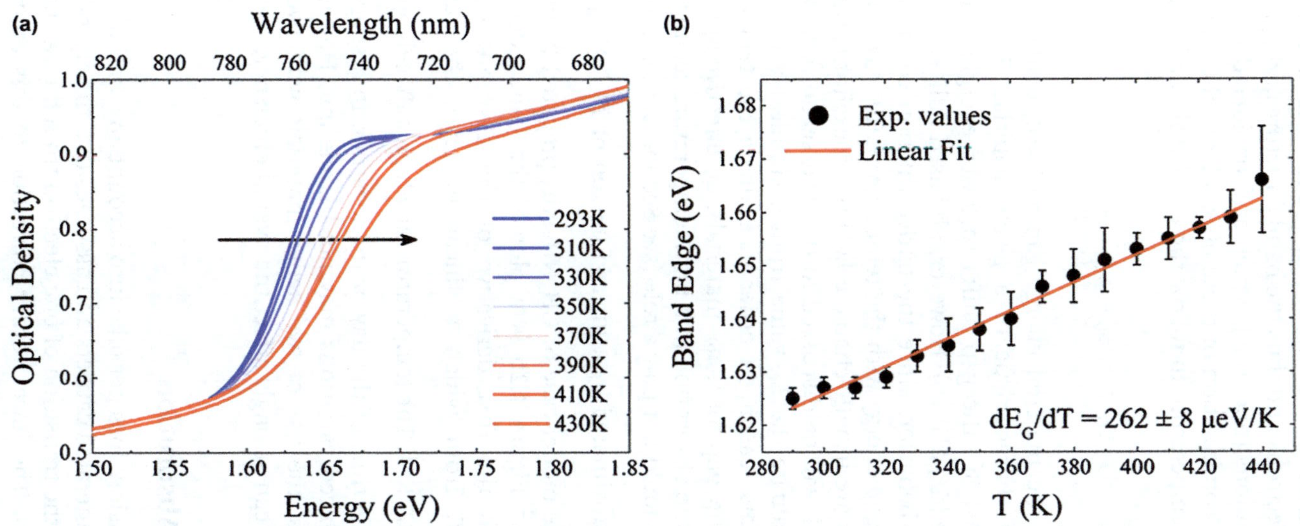

Figure 4.2 (a) Absorption spectra of Cl doped $CH_3NH_3PbI_3$ deposited on a glass substrate at different temperatures (290–440 K). (b) Plot of the band edge *vs.* temperature, T. The experimental data are fitted according to the linear approximation of eqn (4.3), yielding $\frac{dE_G}{dT} = 262 \mp 8\ \mu eV\,K^{-1}$.
Data reproduced from ref. 26.

function of the lattice,[27] which leads to the well known Wannier–Mott (W–M) exciton model.[28,29] The exciton is a wavelike excitation in the periodic crystal, whose real states are wavepackets characterized by their position and momentum. Such quasi particles propagate freely in the crystal, with effective mass $\mu=\left(\frac{1}{m_e}+\frac{1}{m_h}\right)^{-1}$ and parabolic dispersion. However, the exciton has another degree of freedom, associated with the relative motion of the electron and hole. This leads to a discrete spectrum of levels, representing the bound states of the hydrogen-like system. Accordingly, the total energy of the W–M exciton is

$$E_{K_X,n}=E_g-\frac{\mathrm{R}}{n^2}+\frac{\hbar^2 K_X^2}{(m_e{}^*+m_h{}^*)} \tag{4.4}$$

where: E_g is the minimum energy required to generate a free electron and hole (*i.e.* the semiconductor bandgap); and $-\frac{\mathrm{R}}{n^2}$ (with $n=1, 2, \ldots$) is the n-eigenvalue energy of the bound e–h pair, the solution of the hydrogen-model Schrödinger equation with potential energy $U(\bar{r})=\frac{1}{4\pi\varepsilon}\frac{e^2}{r}$. The bound states are located below the conduction band minimum, essentially lowering the optical gap of the material. R is the modified Rydberg constant $\mathrm{R}=\mathrm{R}_0\frac{\mu}{m_0}\frac{1}{\varepsilon^2}$ with $\mathrm{R}_0=13.6$ eV, ε is the dielectric constant of the semiconductor and $\mu^{-1}=\frac{1}{m_e{}^*}+\frac{1}{m_h{}^*}$ is the reduced mass of the system. The energy spectrum has a series of discrete states for increasing n until it merges into the conduction band continuum with a rapidly decreasing oscillator strength ($\propto n^{-3}$).

The (Bohr) radius of the hydrogenoid, electron–positron atom in the semiconductor medium is $a_B=a_B^0\varepsilon\left(\frac{m_0}{\mu}\right)$. Using $a_B^0=0.05$ nm, $\varepsilon=20$, $\mu=0.1$ m_0 $a_B\approx 10$ nm $\gg a$, where a is the lattice constant. The binding energy of the exciton is defined by $E_b=E_1-E_g$.

The presence of Coulomb correlations between electrons and holes strongly modifies the absorption spectrum of the semiconductor.[30] A series of hydrogen-like lines appear below the bandgap, and the line shape above the gap is renormalized. The resultant absorption coefficient, as formulated by Elliott, is described by the equation:[28,30]

$$\alpha(\nu)=\alpha_0^{3D}\left[\sum_{n=1}^{\infty}\frac{4\pi E_b^{3/2}}{n^3}\delta\left(h\nu-E_G-\frac{E_b}{n^2}\right)+\frac{2\pi\sqrt{E_b}}{1-\mathrm{e}^{-2\pi\sqrt{\frac{E_b}{h\nu-E_G}}}}\theta(h\nu-E_G)\right] \tag{4.5}$$

where $\alpha_0^{3D} = \frac{\pi^3 \nu}{n\varepsilon_0 ch^3}\frac{\mu^2}{3}(2m_r)^{3/2}$ is the pre-factor related to the 3D joint density of states and n is a positive integer.

By comparing this continuum contribution with eqn (4.2), we can write a relation for the free-carrier absorption α_{free} with excitonic effects as

$$\alpha_{cont} = \alpha_{free} C(\nu) \tag{4.6}$$

where

$$C(\nu) = \frac{2\pi}{1 - e^{-2\pi\sqrt{\frac{E_b}{h\nu - E_G}}}}\sqrt{\frac{h\nu - E_G}{E_b}} \tag{4.7}$$

is the *Sommerfeld correction* factor and α_{free} is the free carrier absorption without any excitonic effects. Notice that, for $h\nu \to E_G$, $C(\nu) \to 2\pi\sqrt{\frac{E_b}{(h\nu - E_G)}}$, showing a striking difference compared to the square root behavior displayed by eqn (4.2).

Figure 4.3 shows the simulated absorption coefficient for a 3D semiconductor according to eqn (4.5) (solid line) and eqn (4.2) (dashed line). In both cases, the thermal broadening has been accounted for by the convolution of $\alpha(h\nu)$ with a hyperbolic secant of width Γ. For the sake of discussion, we show a system with a very high binding energy ($E_b = 400$ meV) and low thermal broadening ($\Gamma = 10$ meV). As it is evident from the simulation, the presence of Coulomb interactions in a 3D semiconductor

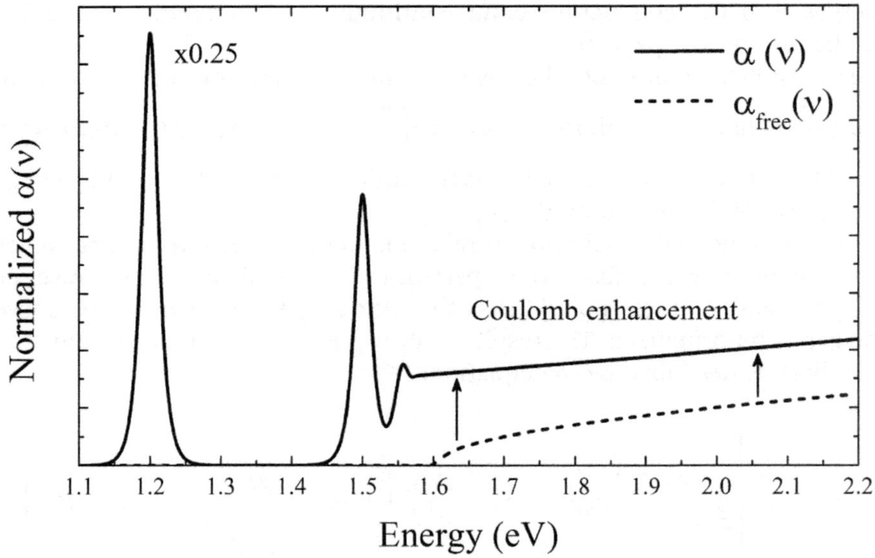

Figure 4.3 Representative band edge absorption for a 3D semiconductor with (solid line) and without (dashed line) excitonic effects.

increases the absorption cross section, not only close to the band edge, where there are the resonance lines of the excitonic states, but also over the entire free carrier absorption spectrum *via* Sommerfeld correction.

The presence of excitonic effects in a PV material is very beneficial for its light harvesting capability due to the general increase of the absorption cross section across the whole spectrum. Nevertheless, creation of bound electron–hole pairs implies that an energy equivalent to E_b must be spent in order to generate the free carriers within the conduction band that give rise to the current. In order to properly assess the role of excitons in hybrid perovskite solar cells, it is extremely important to estimate the value of E_b and comprehend the dissociation mechanism of the bound states at solar excitation densities.

4.2.3.1 Exciton Line-broadening

The primary mechanism of exciton dissociation is exciton–phonon scattering. The dissociation rate is determined by the available phonon density and it contributes to the finite line-width of the excitonic resonance in the absorption spectrum Γ in eqn (4.5). Note, however, that the major cause of line broadening is optical dephasing (*t2*). This is still due to phonon scattering and, consequently, it depends on the phonon density. Among the different mechanisms governing the exciton–phonon interaction, the Fröhlich mechanism is the most relevant here.[31] The physical principle underlying this process comprises the Coulomb interaction between the e–h pairs and the local electric field generated by the lattice vibration. In fact, the metal cation (B^{2+}) and the halide anion (X^{-}) within the perovskite crystal lattice lie on alternate planes, acting as capacitor plates. As a consequence, LO phonons (with k vector orthogonal to such planes) will give rise to a vibrating capacitor with constant charge density but an oscillating plate separation, inducing a local oscillating electric field whose amplitude will increase with T (Figure 4.4). Note that TO modes do not contribute due to their relative orientation.

According to the above discussion, the degree of exciton linewidth broadening can be approximated by considering only the interaction with the LO phonons as:[32]

$$\Gamma = \Gamma_0 + \frac{\Gamma_B}{e^{\hbar\omega_{LO}/kT} - 1} \tag{4.8}$$

where Γ_0 is an inhomogeneous broadening parameter, which is T independent, and Γ_B is the exciton–phonon coupling parameter.

4.2.3.2 Estimation of the Binding Energy of Excitons

The simplest approach to estimate the binding energy E_b of a Wannier–Mott exciton is by considering the equation[33] $E_b = \dfrac{m^* e^4}{2\hbar^2 \epsilon^2}$, where m^* is the exciton effective mass, e is the unit charge, and ϵ is the effective dielectric constant. The effective mass has a relatively small impact on the estimate, because of

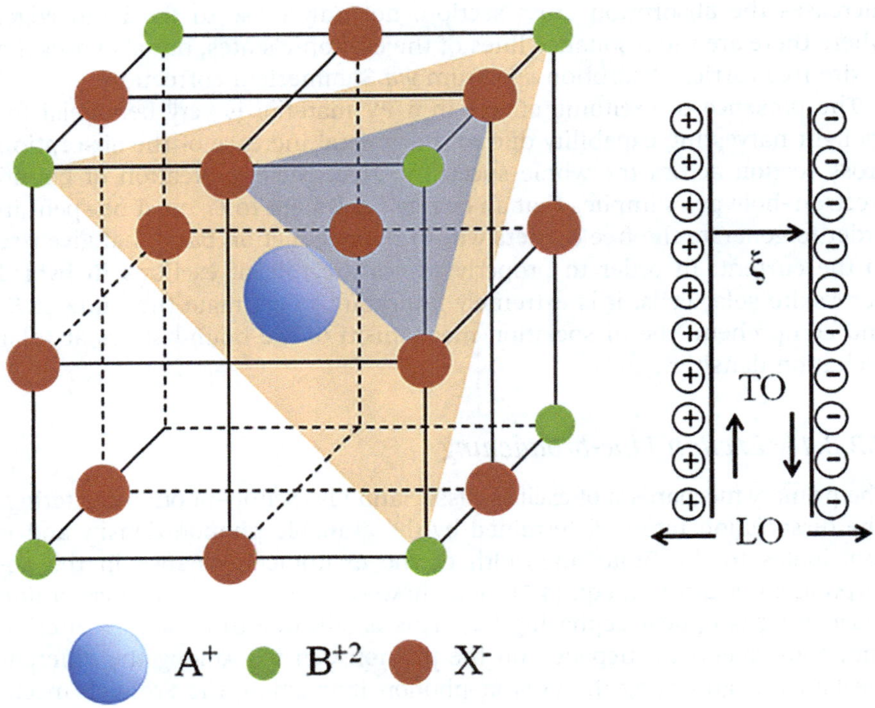

Figure 4.4 Perovskite cubic structure. The blue plane is defined by negative X^- ions while the orange one is defined by positive B^{2+} ions, generating a local capacitor.

the relatively small range of variation and the reliable theoretical prediction. The dielectric function, on the contrary, can be crucial. In general, the dielectric constant, which describes the ability of the material to screen an electric field, is frequency-dependent due to the different time scales of the relaxation phenomena involved. In simple terms, ϵ is built up by adding the contributions from lattice displacements (10^0–10^{10} s^{-1}), molecular reorientations (10^{10}–10^{12} s^{-1}) and electronic polarizability (10^{15} s^{-1}).

The relevant frequency of the electron motion, E_b/h, can be adopted as a figure of merit for the time scale of the screening mechanism. Alternatively, the Bohr radius can be adopted. For instance, a very large radius, which is generally associated to a small binding energy, would justify the use of the static dielectric constant. In this case, the ionic and molecular motions happen on the time scale of the relative motion of electrons and holes and, hence, contribute to the screening. On the other hand, if the binding energy is large, the frequency of the exciton is so high that the electron–hole interactions are not affected by the lattice vibrations, and the relevant dielectric constant should be the value at optical frequencies (ϵ_∞). In the case of most 3D semiconductors, especially covalent solids, the static dielectric constant which includes all the phonon contributions ($\epsilon_0 = \epsilon_\infty + \epsilon_{\text{lattice}} + \epsilon_{\text{rotations}}$) is commonly used. Indeed for such materials, the dielectric permittivity changes little from static to optical frequencies (*e.g.* GaAs $\epsilon_\infty = 10.86$, $\epsilon_0 = 12.90$ at 300 K[34]). On the contrary, hybrid perovskites exhibit a strong frequency dependence of the dielectric constant, which varies from approximately 10 at optical frequencies to 100 under static conditions.[33] The choice of the exact value of the dielectric constant will strongly affect the estimation of E_b. Clearly, a direct experimental measurement of the relevant exciton characteristics should allow us to resolve the ambiguities between cause and effect. Several attempts at exciton characterization have been reported in the literature, following different methods. We review them briefly in the following section.

4.2.3.2.1 Exciton Binding Energy from Elliott Fitting. The Elliott formula stated in eqn (4.5) relates the absorption cross section to specific semiconductor parameters that include the exciton binding energy. Hence, by a simple numerical fitting of the absorption line shape with the formula allows one to extract the value of the binding energy as one of the fitting parameters.[4] Figure 4.5 shows, as a real example, a comparison of the experimental UV-Vis absorption spectrum of a $CH_3NH_3PbBr_3$ perovskite thin film (open circles) with the spectrum simulated using the Elliott formula (red line: free parameters reported in the box). It follows that one can deduce the exciton binding energy value by simply inspecting the UV-Vis absorption spectrum. The model deviates from the experimental data below the bandgap due to a non-negligible absorption arising from the Urbach tails discussed in the previous section. The divergence between the experiment and the theoretical model can be corrected by adding an exponential sub-bandgap absorption term $\exp\left(\frac{h\nu - E_g}{E_u}\right)$, where σ and E_u are fitting parameters.

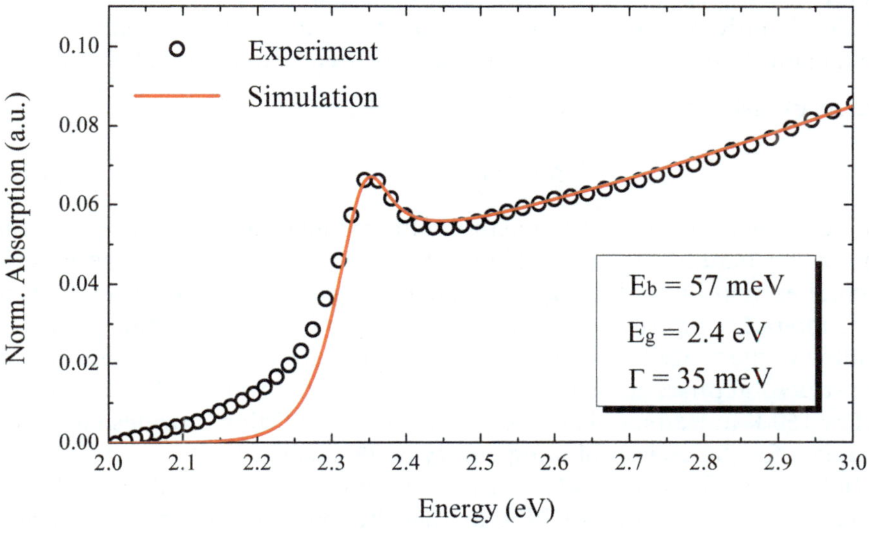

Figure 4.5 Experimental UV-Vis absorption spectrum of a $CH_3NH_3PbBr_3$ perovskite thin film together with the simulated spectrum using eqn (4.5).

Of course this fitting method requires the experimental determination of an accurate absorption spectral shape, which is non-trivial in highly-absorbing samples. For the application of Elliott's theory, it is also imperative that the binding energy is independent of the temperature,[28,30] which is not necessarily true for hybrid perovskites (see later sections). Nevertheless, this method does enable the prediction of the approximate range of binding energy.

4.2.3.2.2 Exciton Binding Energy from T Dependence. Assuming that disorder-induced broadening and pure dephasing (γ) are independent of the temperature, one can account for the temperature dependence of the exciton line-width just by taking into account thermal dissociation.[29] This is particularly suitable for ionic semiconductors at high temperatures (*e.g.* $T > 100$ K). In the frequency domain, the resonance linewidth of a transition is given by $\Delta\nu = 1/\pi T_2$, where T_2 is the excited state coherence lifetime, or dephasing. This value can be affected both by the natural population decay of the excited state population (with time T_1) and by elastic scattering events (with rate γ), according to the following equation:

$$\frac{1}{T_2} = \frac{1}{2T_1} + \gamma \tag{4.9}$$

According to the model, T_1 is the sum of the *intrinsic* decay rate (k_0) and the thermal dissociation rate (k_T), $\frac{1}{T_1} = k_0 + k_T$. k_T is modeled with an Arrhenius-like law:

$$k_T = \nu_T e^{-\varepsilon_B/kT} \tag{4.10}$$

where ε_B is the upper estimate for the exciton binding energy assuming that predominantly phonon interactions lead to exciton dissociation, k is the Boltzmann constant and ν_T is the ionization attempt frequency. Substituting the relationship into the linewidth, $\Delta\nu = \frac{1}{\pi T_2}$, we obtain:

$$\Delta\nu = \Delta\nu_0 + \nu_T e^{-\varepsilon_B / kT} \tag{4.11}$$

where $\Delta\nu_0$ accounts for the temperature-independent broadening. Figure 4.6 shows the natural logarithm of only the T-dependent part of the experimental exciton linewidth ($\ln(h\nu - h\nu_0)$), collected as reported in ref. 2, as a function of $1/T$ together with the linear fitting $\ln(h\nu_T) - \varepsilon_B/kT$. Note that T is limited to the range between 160 K and 290 K to avoid complications due to the structural phase transition (from tetragonal to orthorhombic) occurring below 160 K in lead-iodide based perovskites. The fitting provides an upper estimate for the exciton binding energy of 55 ± 20 meV.

Notice that for $\hbar\omega_{LO} \gg kT$, eqn (4.8) resembles the formulation of eqn (4.11). This observation is telling us that the ionization phenomenon, mediated by LO phonons, is involving phonons of energy ε_B. Thus, we can safely say that the real binding energy of our system will be $E_B < \varepsilon_B$.

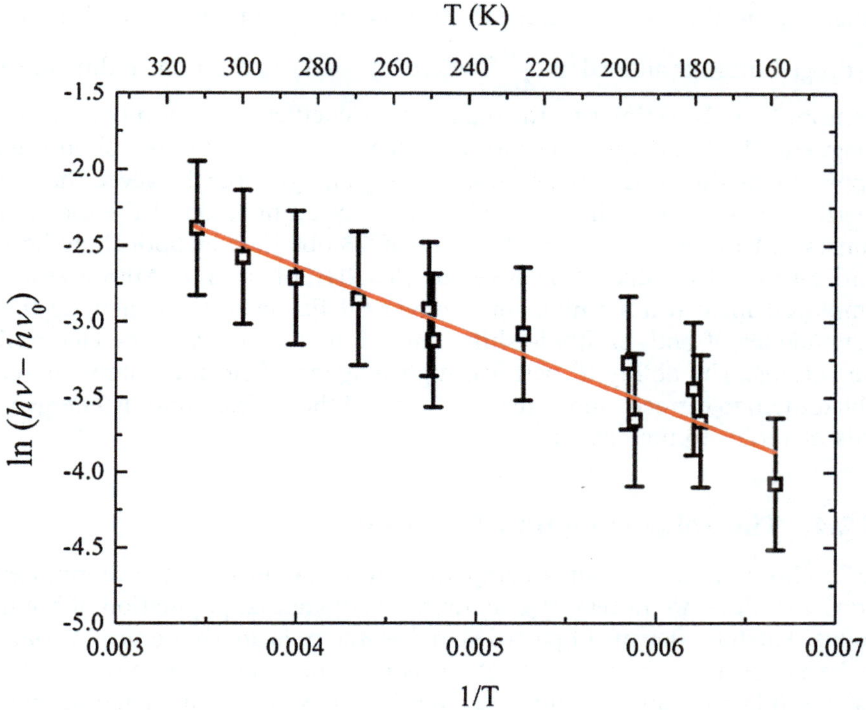

Figure 4.6 Plot of the exciton absorption band broadening $\Delta\nu_0$ in a $CH_3NH_3PbI_3$ thin film *vs.* inverse temperature (150 K$<T<$290 K). The error bars indicate the maximum variation of the experimental value due to a $\pm$10 K uncertainty in the determination of the sample temperature. The solid line is the fit to $\ln(h\nu_T) - \varepsilon_B/kT$. Parameters: $\varepsilon_B = (55 \pm 20)$ meV, $\nu_T = (7\ fs)^{-1} \cdot \chi^2 = 0.92$.
Figure reproduced from ref. 2.

4.2.3.2.3 Magneto-absorption. External perturbation of the exciton either by electric or magnetic fields can provide deep insight into its characteristics. In fact, the effect of the magnetic field on the excitonic absorption is used as a standard means to estimate the binding energy. The magnetic field has two predominant effects on spectroscopically observable energies: (a) Zeeman splitting, which varies linearly with the field, and (b) the diamagnetic shift, a blue-shift of the exciton energy. For certain approximations, the field-dependent exciton energy can be expressed as:

$$E = E_0 \mp \frac{1}{2} g\mu_B B + c_0 B^2 \tag{4.12}$$

where E_0 is the energy at zero field, μ_B is the Bohr mangeton and c_0 is the diamagnetic coefficient. The diamagnetic coefficient is related to the exciton binding energy by the relation:

$$\frac{E_b}{\mathrm{Ry}} = \left[\varepsilon \frac{c_0}{c_H}\right]^{\frac{1}{3}} \tag{4.13}$$

where Ry is the Rydberg constant and c_H is the diamagnetic shift of a hydrogen atom evaluated as $\frac{e^2 a_H^2}{4m_0}$ where $a_H = 0.53$ Å, the Bohr radius of the exciton. Thus, by estimating the diamagnetic coefficient experimentally from magnetic field dependent absorption measurements, one can obtain and approximate the value of the exciton binding energy. Note, however, that this method also requires the knowledge of the relevant value of the dielectric constant. Hirasawa *et al.*[35] used a value of 6.5 obtained by optical methods, and estimated a value of 37 meV for $CH_3NH_3PbI_3$ at 4 K. Miyata *et al.*[36] employed instead a simultaneous numerical fitting of the magnetic field dependence of both excitonic shifts and Landau energies of the electronic transitions. The ability to perform high magnetic field measurements enabled them to extract a more accurate value of the exciton binding energy for this material: 16 meV at 4 K.

4.2.4 The Saha–Langmuir Equation

An excitonic binding energy comparable to or larger than the thermal energy, $k_B T$, does not necessarily guarantee a substantial population of bound states. The total density of particles in the system is another crucial variable to be taken into account. The reason is that the system composed by free carries (plasma) and bound pairs (excitons) is in thermodynamic equilibrium. This implies a balance between dissociation and recombination events. In simple terms, if the free carrier density is very low, the probability of regenerating an exciton from the dissociated state is very small (assuming non-zero T), and most of the population will be in the plasma state in spite of a binding energy exceeding $k_B T$. However, if the binding energy is much larger than $k_B T$, there will be a stable exciton population even at a low free carrier density while, conversely, a negligibly small binding energy will see a prevailing plasma state even at a high density or low T. In between these limits, there is a full phase diagram that has been described by Saha:

$$\frac{x^2}{1-x} = \frac{1}{n}\left(\frac{2\pi\mu kT}{h^2}\right)^{3/2} e^{-E_B/kT} \tag{4.14}$$

where $x = n_{eh}/n$ is the fraction of free carriers, and n is the total number of particles. μ is the exciton reduced mass $\left(\mu = \frac{m_e^* + m_h^*}{m_e^* m_h^*}\right)$ and k is the Boltzmann constant. The form of this equation is quite intuitive:[37,38] the higher the temperature, the greater e–h plasma ratio due to improved ionization processes but, as the excitation density increases, the possibility of e–h interactions with subsequent exciton formation prevails and the ratio is reduced.

Figure 4.7 illustrates the dependence of the fraction x of free charges with respect to the excitation density n, varying either the temperature of the overall system at fixed E_B or its binding energy ($E_B = 20$ meV to 200 meV) at

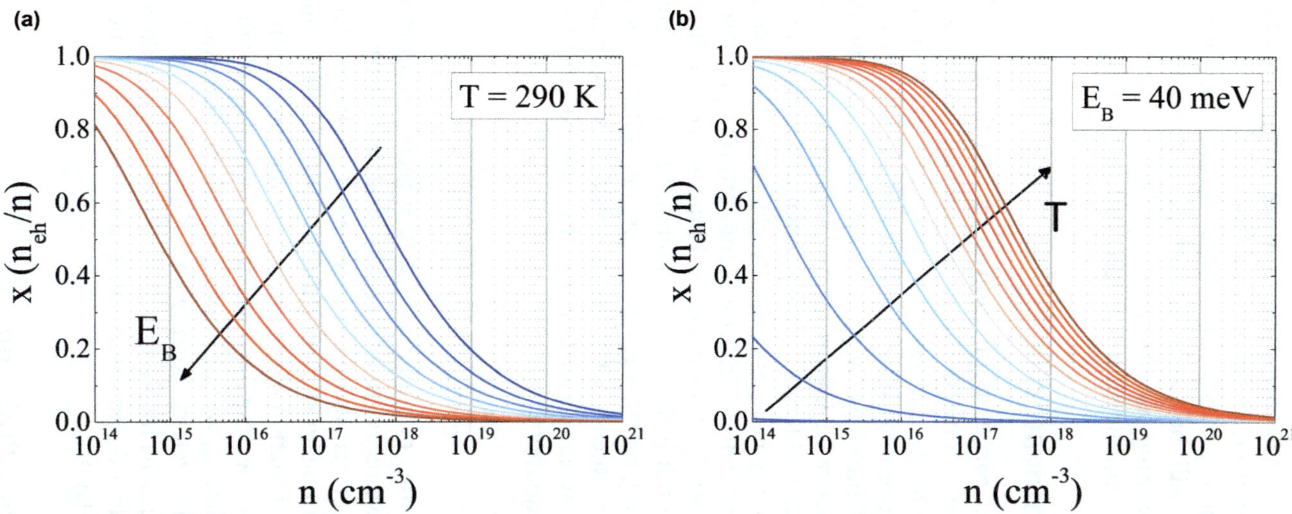

Figure 4.7 Saha simulations of the fraction of free charge carriers as a function of the total photo-excitation density: (a) with various binding energies ranging from 20 meV to 200 meV (20 meV steps), at 290 K; (b) over the temperature range 0 K to 290 K with a constant binding energy of 40 meV. The effective mass for the above simulations was set to 0.2 m_e.

room temperature. It is interesting to notice that, in order to have a consistent fraction of excitons in conventional experimental conditions such as in transient absorption and photoluminescence measurements ($n = 10^{17}$ to $10^{18}\,cm^{-3}$), relatively high binding energies are needed (<50% for $E_B > 50$ meV). Moreover, we can predict, from this simple analysis, that in the case of hybrid perovskites meant for photovoltaic applications such as $CH_3NH_3PbI_3$, there will be no dominant population of excitons under photovoltaic cell operating conditions ($n < 10^{16}\,cm^{-3}$ at room temperature), since the established equilibrium is largely in favor of free carriers.[2]

4.3 Photoluminescence

The photo-generated carrier population decays to the ground state *via* a number of relaxation channels. The decay channels can be separated into *radiative* and *non-radiative* processes.[31] In the former, at least part of the excess energy is released by light emission; in the later, excess energy is eventually converted into heat by phonon emission. Both processes can be affected by the presence of impurities and defects behaving as recombination centers. Overall, the lifetime of the population is determined by the sum of all the decay paths. For an efficient PV cell, it is desirable to have long carrier lifetimes, small radiative rates and minimum trapping so that carriers can be collected efficiently at the contacts. Photo-generated carriers in hybrid perovskites evidently fulfill these requirements, having long lifetime and an apparently low trapping rate that results in very long diffusion lengths. This is another reason for the success of hybrid perovskites as PV materials.[1]

The excited state dynamics in $CH_3NH_3PbI_3$ can be effectively described with a simple rate equation:[4,5]

$$\frac{dn}{dt} = -R(n) - k_t n - \gamma n^3 \tag{4.15}$$

which accounts for the spontaneous radiative recombination rate $R(n)$, an extrinsic non-radiative recombination rate k_t describing carrier trapping and a non-radiative Auger recombination process (γ). Notice that the photoluminescence (PL) intensity is proportional to the radiative rate ($I_{PL} \propto R(n)$), which in general is a function of the excited population density. Thus, all the different decay paths affecting the population, regardless of whether they are radiative or non-radiative, will also affect the photo-emission properties of the material. In this section, we present a review of the three main recombination mechanisms involved in the excited state population decay in a direct-bandgap semiconductor material, and in relation of the specific case of perovskite $CH_3NH_3PbI_3$.

4.3.1 Spontaneous Emission

In the simple case where radiative recombination of electrons and holes is the only recombination mechanism after photo-excitation, one will have

$n(t) = p(t)$. Thus, the general rate equation will be $\frac{dn}{dt} = R_G - \beta_{rad}n^2$, where R_G will be the charge photo-generation rate. Solving this equation with $n(0) = 0$, one obtains the time evolution of the excited state population density $n(t)$. At early times ($t \approx 0$), the result is that $n(t) \propto R_G t$. Since $R_G \propto I_{ex}$, the PL intensity at early times is a quadratic function of the light excitation intensity:

$$I_{PL}(0) \propto \beta_{rad}n^2(0) \approx I_{ex}^2 \tag{4.16}$$

Once the excited state population has been built up, we can apply the initial condition $n(t=0) = n(0)$, and consider the simplified rate equation $dn/dt = -\beta_{rad}n^2$ in order to obtain the PL decay behavior. This is the so called *bimolecular* recombination limit, which can be readily integrated giving

$$n(t) = \frac{n(0)}{(\beta_{rad}n(0)t + 1)} \tag{4.17}$$

and

$$I_{PL}(t) \propto \beta_{rad}n^2(t) = \frac{\beta_{rad}n^2(0)}{(\beta_{rad}n(0)t + 1)^2} \tag{4.18}$$

The picture above changes in the presence of excitons. Excitons may decay radiatively according to a monomolecular rate law. However, this does not necessary imply a linear dependence of PL on the excitation intensity, I_{ex}, because the formation of excitons is a bimolecular reaction following the above gap photo-excitation. The role of excitons in the emission in perovskites is not entirely clear at present, especially because one needs to consider carefully, one by one, hybrid perovskites with different chemical composition. We show below that the exciton-less rate equations reported above can account fairly well for the observed intensity dependence of the PL.

4.3.2 Shockley–Read–Hall Recombination

When carrier recombination occurs through localized states, such as intra-gap traps, we can replace the valence (conduction) band DOS with that of the localized states in the above treatment. The term $p(n)$ is replaced by the density of trapped electrons $N_t(1 - f_t)$ (holes) according to:

$$R_{nt} = \beta_n n N_t(1 - f_t) \tag{4.19}$$

where N_t is the density of trap states and f_t is the probability that a trap state is occupied.

If we now consider a semiconductor in which both electrons and holes can be trapped, we can define two distinct coefficients $\beta_{n,p}$ and two de-trapping rates $G_{n,p}$. Considering that the trapping and de-trapping rates should be balanced at equilibrium ($R_{nt,pt} = G_{n,p}$), the expressions for the four above mentioned rates as a function of the trap occupation probability f_t are

$$\begin{aligned} R_{nt} &= \beta_n n N_t (1 - f_t) G_n = \beta_n n_t N_t f_t \\ R_{pt} &= \beta_p p N_t f_t G_p = \beta_p p_t N_t (1 - f_t) \end{aligned} \tag{4.20}$$

where n_t and p_t are the electron and hole densities when the quasi-Fermi level matches the trap energy E_t. One can eliminate the trap states occupation probability f_t by imposing the steady-state condition that the net trapping rate of electrons must match the net recombination rate for the holes, *i.e.* $R_{nt} - G_n = R_{pt} - G_p$. This leads eventually to the Shockley–Read–Hall expression for recombination through a trap state:[7]

$$R_{SRH} = \frac{np}{\tau_n (p + p_t) + \tau_p (n + n_t)} \tag{4.21}$$

where we have defined the average lifetime for electron/hole capture as $\tau_{n,p} = \dfrac{1}{\beta_{n,p} N_t}$.

If we consider the simple case in which the density of available trap states is always greater than the free e–h pair, so that $n(t) = p(t) \gg n_t$ and $\tau_n = \tau_p$, we obtain

$$R_{SRH} \approx \frac{n^2}{2\tau_n n} = \beta_n N_t n \tag{4.22}$$

To sum up, we see that, in contrast to the previous case, here the excited state de-population is described by a *monomolecular* recombination mechanism ($dn/dt = R_G - \beta_n N_t n$). In this case $R_G \propto n(0)$ and thus, remembering that $R_G \propto I_{ex}$, one obtains a linear dependence of the PL emission with respect to the excitation density ($I_{PL}(0) \propto I_{ex}$) and *mono-exponential* kinetics:

$$I_{PL}(t) \propto \beta_n N_t n(t) \approx n(0) e^{-t/\tau_n} \tag{4.23}$$

4.3.3 Auger Recombination

Auger recombination is a three-particle process in which two carriers of the same sign undergo an elastic scattering event resulting in the excitation of one carrier to a higher kinetic energy, and the recombination of the second with an available carrier of the opposite sign. In the case of a band-like semiconductor, the Auger recombination will involve either two electrons and one hole (R_{Aug}^{eeh}) or two holes and one electron (R_{Aug}^{hhe}). Thus, following an argument similar to the one presented above, one retrieves a recombination rate proportional to the electron and hole densities:

$$R_{Aug}^{eeh} = A_p n^2 p R_{Aug}^{hhe} = A_n n p^2 \tag{4.24}$$

Taking, for simplicity, $n(t) = p(t)$ yields the well-known cubic behavior of the Auger rate according to:

$$R_{\mathrm{Aug}} = \gamma n^3 \tag{4.25}$$

This recombination mechanism is usually more relevant in indirect bandgap semiconductors since the radiative recombination is hindered, but it can also be observed in direct bandgap semiconductors in high carrier density regimes.

4.3.4 Model for PL Dynamics in $CH_3NH_3PbI_3$

Within the plethora of available hybrid perovskites, $CH_3NH_3PbI_3$ is the most widely studied and the most commonly employed material for efficient device fabrication. It therefore represents a model case for the description of fundamental PL properties.[3–5,39] Here, we discuss the photo-luminescence properties for the particular case of thin films of perovskite $CH_3NH_3PbI_3$ deposited in a two-step process.[40] This material is a direct bandgap semiconductor characterized by the presence of a weakly bound exciton. Due to the low binding energy of the exciton, in a first approximation, we can neglect the excitonic contribution to the PL. It is worth noting that this approximation no longer holds for cooled samples since the exciton population becomes significant, as predicted by the Saha equation.

As reported by Saba and coworkers,[4] the integrated PL emission intensity as a function of the pump laser fluence follows a 3/2 power law over more than three orders of magnitude in excitation intensity. This power law indicates that the electron–hole population recombination is governed by the existence of competition between a radiative recombination channel and at least one non-radiative path, possibly controlled by mid-gap traps of electrons or holes. Referring back to eqn (4.26), we consider an intragap trap state accounting for a SRH-like recombination path, an Auger non-radiative term and a spontaneous emission which, for simplicity, we take as bimolecular band-to-band recombination (*cf.* eqn (4.16)). For the purposes of illustration, the sketch in Figure 4.8a shows an electron trap.

The resulting rate equations for the electron n_e, hole n_h and filled trap n_t population densities are:

$$\begin{aligned}
\frac{dn_e}{dt} &= -\beta_{\mathrm{rad}} n_e n_h - k_t n_e (N_t - n_t) - \gamma n_e n_h^2 + R_G \\
\frac{dn_h}{dt} &= -\beta_{\mathrm{rad}} n_e n_h - G_n n_h n_t - \gamma n_e n_h^2 + R_G \\
\frac{dn_t}{dt} &= k_t n_e (N_t - n_t) - G_n n_h n_t
\end{aligned} \tag{4.26}$$

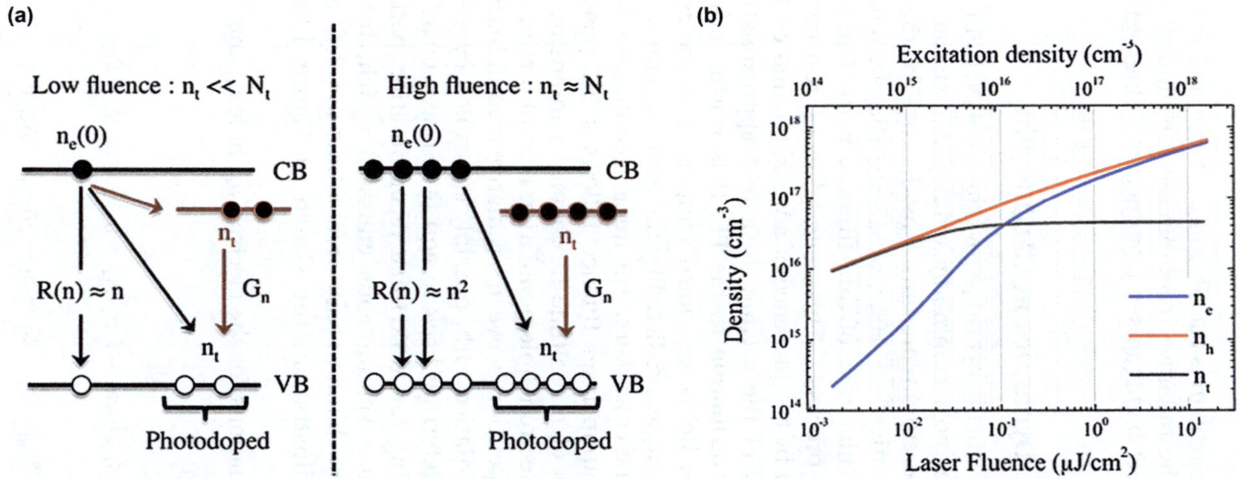

Figure 4.8 (a) Sketch illustrating the recombination mechanisms for the low- and high-fluence regimes. (b) Population density of electrons (blue), holes (red) and filled traps (grey) obtained as the steady-state solution of eqn (4.26). As the excitation intensity increases, the traps are filled and the bimolecular recombination becomes faster than the trapping rate, preventing the accumulation of trapped electrons.

where N_t is the density of available trapping sites, G_n is the rate of recombination of a trapped electron with a hole in the valence band, and R_G is the photo-generation rate.

4.3.4.1 Steady-state Solution

Figure 4.8b shows the steady-state solutions of the three rate equations ($dn_i/dt = 0$) as a function of the impinging laser fluence $F(\mu J\,cm^{-2})$. As a consequence of the two competing recombination pathways, the solution is characterized by two different regimes.

- At low-excitation fluencies, trapping is the most efficient recombination path for electrons. So, for $n_t \ll N_t$, electron trapping is a monomolecular process and n_e will scale linearly with the laser fluence, $n_e \propto F$. However, the dependence of the valence band hole density with the fluence is limited by the rate of non-radiative recombination with trapped electrons, which is a bimolecular process ($\approx n_h n_t$), inducing the hole population to scale as the square root of the laser fluence, $n_h \propto \sqrt{F}$. The resulting steady-state PL intensity then reads $I_{PL} \propto \beta_{rad} n_e n_i \propto \beta_{rad} F^{3/2}$.
- At high fluences, the trapping rate slows down since $n_t \approx N_t$ and the bimolecular radiative recombination of e–h pairs becomes dominant. The intensity dependence of the steady-state PL therefore changes from superlinear to linear, since $n_h n_e \propto F$, as shown in Figure 4.9a.

At even higher laser intensities, the Auger processes are no longer negligible, resulting in a sublinear intensity dependence of the integrated PL and a reduction in the PL quantum yield (red dots in Figure 4.9a).

4.3.4.2 Dynamic Solution

The same rate equation can be solved in order to give the dynamic solution and thus fit the PL decays, as shown in Figure 4.9b. For the sake of simplicity, we have removed the generation rate term and simply used the number of absorbed photons as the initial state for n_e and n_h, evaluated as:

$$n_{phot} = \frac{P_{CW}\lambda}{RRhcA_p\alpha(\lambda)^{-1}} \tag{4.27}$$

where P_{CW} is the average impinging laser power, RR and A_p are the repetition rate of the pulsed laser and the focal spot on the sample, respectively, and λ and $\alpha(\lambda)^{-1}$ are the excitation wavelength and the light penetration depth at that particular wavelength. In this case, the two different regimes are also clearly observable. The PL decay is nearly mono-exponential when the sample is excited at low fluences, while it deviates more and more from an exponential decay as the pump fluence increases. Notice that when one electron is trapped, a hole will be left in the valence band and, in a

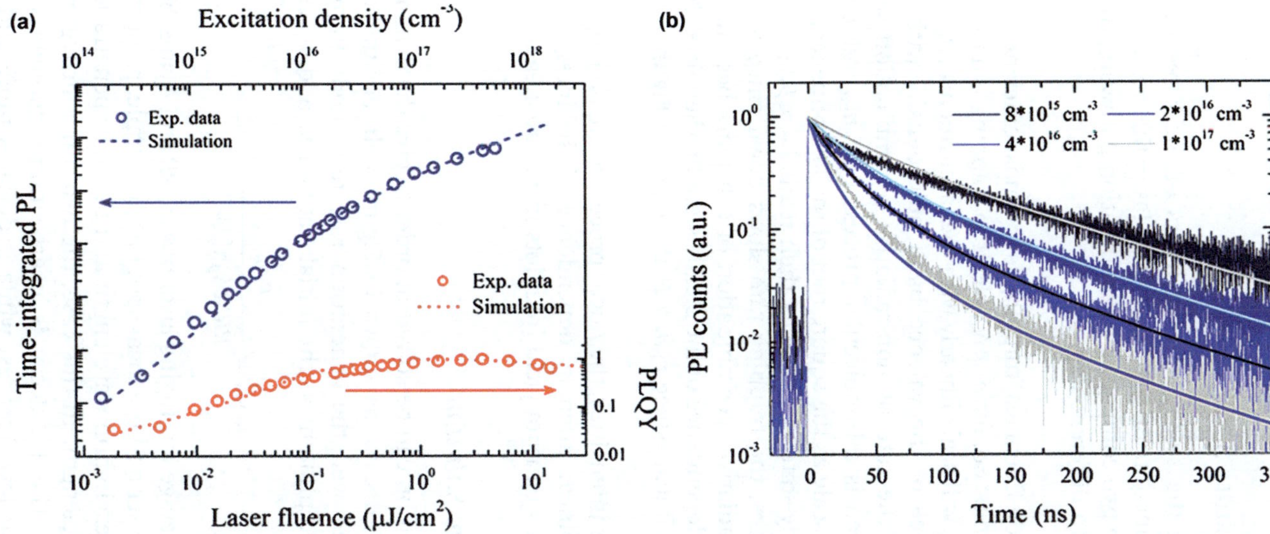

Figure 4.9 (a) Experimental and simulated time integrated PL and PL quantum yield, PLQY, of a representative $CH_3NH_3PbI_3$ film plotted as a function of the excitation density. (b) Experimental and fitted PL decays of the same sample at different excitation densities.

steady-state, a certain population density of *photodoped* holes n_t builds up as the electron trap states are filled. This photodoped hole density can be very high, even at low fluences, due to a very slow de-trapping rate G_n (left of Figure 4.9a).

In the dynamic solution, the two regimes can be physically explained as follows:

- Under low fluence excitation, the photo-excited electron density n_e is much lower than the total concentration of free holes in the valence band, since $n_h(t) = n_e(t) + n_t(t)$ where $n_e(0) \ll n_t$. In this case, since the additional photo-excited charge does not noticeably alter the hole population density, the system will be well described by the monomolecular recombination case considered in the previous section, and the PL decay will be monoexponetnial (*cf.* eqn (4.23)).
- When the excitation fluence is high enough, such that $n_e(0) > n_t$, electron–hole bimolecular recombination is achieved (as in eqn (4.18)), resulting in a power-law decay. When the free-electron concentration drops below the photo-doped hole density, the decay again becomes mono-exponential.

4.4 Micro-structure and Optical Properties

One of the major differences between hybrid perovskites and other conventional semiconductors is the presence of organic–inorganic interactions and the resultant structural disorder which may strongly affect the material's properties.[18] Even though the organic cation is not involved in the electronic structure at low energies, it is far from being negligible in photo-excitation dynamics. Its steric hindrance strongly affects the inorganic cage geometry and hence the electronic structure of the material. Hydrogen bonding between the amine group and the halide network and the strong orientational disorder of the organic cation within the inorganic cage, given by its relatively high rotational mobility, induce strong lattice deformations that perturb the electronic landscape within the Pb–X cage.[16,20,21,41] This phenomenon can be studied experimentally by changing the chemical nature of the cation.[20] The role of the cation also seems to be correlated to the type of structural and morphological order. We discovered that, depending on the processing conditions, lead-iodide perovskite thin films are formed with variable mesoscale morphologies, with grain size ranging from tens to thousands of nanometers and with varying degrees of structural disorder, which strongly influence the optical properties even though the chemical composition is unchanged.[12] Using Raman spectroscopy, we have established the effect of increasing the polycrystallinity of $CH_3NH_3PbI_3$ thin films on the semiconductor structural properties at the molecular level.[26,42] The primary effects are the modulation of the lattice strain and the orientational order of the organic cation within the lattice, which can be ascribed to the quality of crystallization and/or the presence of grain boundaries and/or dangling bonds. In addition to such intrinsic material properties, the interaction of the crystalline grain surface with external agents such as water or oxygen molecules should be considered as a source of structural disorder, as we have recently demonstrated.[42] In summary, $CH_3NH_3PbX_3$ is a fluctuating structure where tilting and distortion of the octahedral networks and the rotation and polarizability of the molecular dipole strongly affect the optoelectronic properties of the semiconductor.

4.4.1 Bandgap and Photoluminescence

Figure 4.10a–d shows top-view SEM images of four representative $CH_3NH_3PbI_3$ samples. These films were fabricated using a two-step sequential deposition technique involving dipping a spin-coated film of PbI_2 into a CH_3NH_3I (MAI) solution with variable temperature and concentration, resulting in an observed average crystallite size ranging from 250 nm to 2 μm. The top panel of Figure 4.10e shows the positions of the optical UV-Vis absorption spectra band edges as well as the positions of the PL peaks, while the bottom panel shows the average PL lifetime as a function of the average crystallite size. Notably, the optical absorption edge, along with the PL peak position, shifts to longer wavelengths (lower energies) moving from 1.63 eV to 1.61 eV as the average crystallite size increases. At the same time, the PL dynamics get slower with an average lifetime of about 2 ns for the smallest crystallites (<250 nm) and of more than 100 ns for the largest crystallites (>1 μm).

Figure 4.10f shows the PL lifetime, τ_{PL}, as a function of the initial excitation density n_0 for two representative samples (corresponding to Figure 4.10a and d, respectively). A simple theoretical relation between these quantities can be retrieved from the rate equation considered above. Working in a low excitation regime, we can discard the Auger contribution. Thus, simplification of the rate equation to $dn = -\beta_{rad}n^2 - An$ gives:

$$\tau_{PL} = \frac{1}{A + \beta_{rad}n_0} \tag{4.28}$$

By fitting eqn (4.28) to the experimental data, one obtains the intrinsic bimolecular radiative recombination coefficient β_{rad}. Notably, while the non-radiative coefficient A is not affected much (1.3×10^7 s^{-1} to 0.72×10^7 s^{-1}), a reduction of β_{rad} is observed going from small to big crystallites ($\beta_{rad}\approx0.62\times10^{-9}$ cm^3 s^{-1} and 3.7×10^{-9} cm^3 s^{-1}, respectively), demonstrating a correlation between the optical bandgap shift and the radiative recombination rate.

In Section 4.2.2, we already commented on the red-shift of the optical bandgap in hybrid lead-halide perovskites arising from lattice distortion (the *Varshni shift*). According to our work,[2] as the temperature decreases, the Varshni shift follows the reduction of the bimolecular radiative coefficient β_{rad}. Hence, the correlation between E_G and β_{rad} observed by increasing the

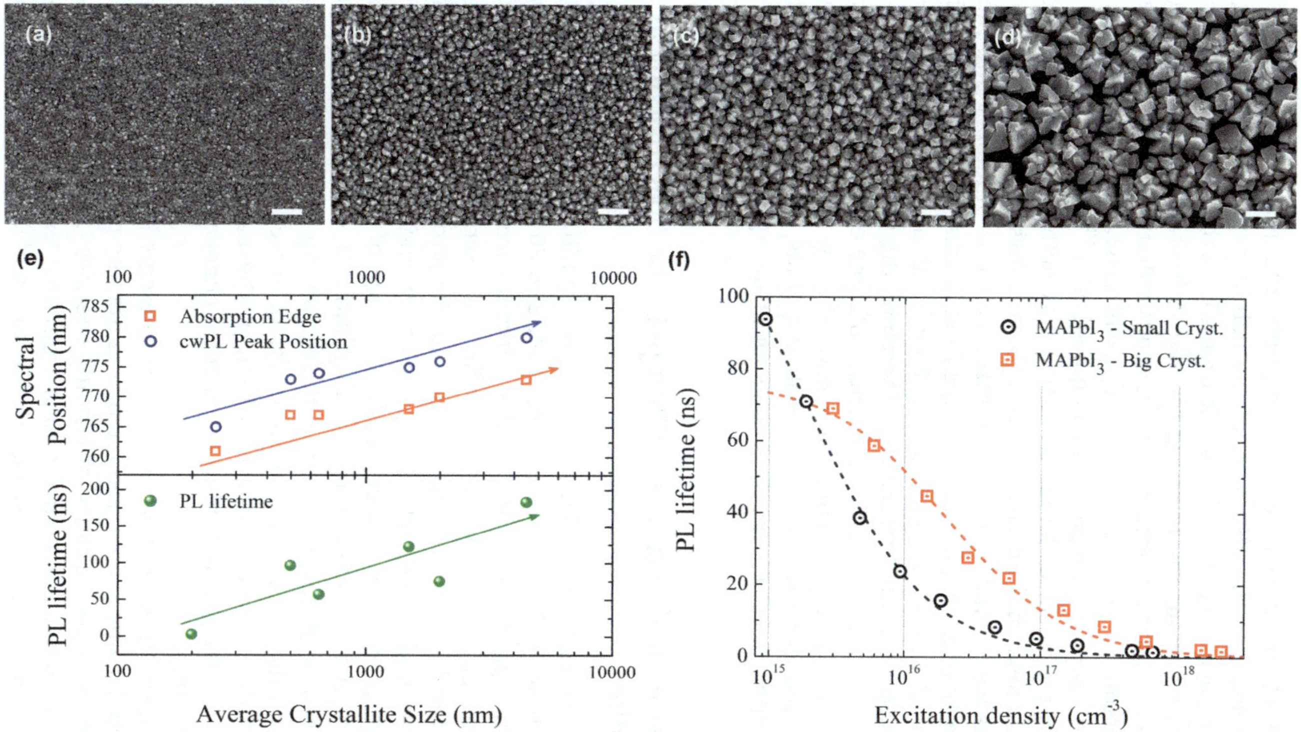

Figure 4.10 (a–d) Top-view SEM images of $CH_3NH_3PbI_3$ prepared using (a) [MAI] = 0.063 M, T = 25 °C; (b) [MAI] = 0.045 M, T = 25 °C; (c) [MAI] = 0.031 M, T = 25 °C; (d) [MAI] = 0.045 M, T = 70 °C. Scale bars are 2 μm. (e) Spectral position of the UV-Vis absorption band edges and cw-PL peaks (top) and PL average lifetime as a function of the crystallite size (bottom). (f) PL lifetime as a function of the excitation density for samples with small (black circles) and large (red squares) crystallites. Dashed lines represent the curves obtained after fitting the data to eqn (4.26).
Figure adapted from ref. 12. Copyright (2014) American Chemical Society.

crystal dimension can also be seen by lowering the temperature, which notably affects the lattice strain, suggesting a direct correlation between the two.

In fact, the observed change in absorption and photo-luminescence properties is related to the distortion of the halide–metal–halide bonding angle. The rearrangement of the inorganic cage affects the overall electronic band-structure, consequently inducing the red-shift of the energy gap. In turn, due to the relation between the spontaneous radiative rate R_{rad} and the electronic structure of the semiconductor, β_{rad} will also be affected indirectly by the lattice distortion, as reported by Filippetti *et al.*[43]

Interestingly, the effect of lattice distortion was both predicted through *ab initio* calculations, and observed when hybrid perovskites with different organic cations were considered. In that case, the steric hindrance and the local arrangement of the organic cation play a fundamental role in the ground state crystalline structure by tuning the Pb–I–Pb bond angle, thus indirectly affecting the electronic band structure.[20] Therefore, the possibility of engineering the bandgap and the intrinsic radiative lifetime of the compound by simply tuning the crystallization procedure paves the way for the optimization of the emissive properties of hybrid perovskite thin films, especially in view of the potential employment of these materials for lasing applications.

4.4.2 Electron–Hole Screening: Pump–Probe Spectroscopy as a Probe

There is a wide spread in the values of the exciton binding energies reported in the literature for $CH_3NH_3PbI_3$, ranging from 2 to 50 meV.[2,9,33,35,36] This variability comes partly from the fact that different techniques have been used to estimate E_b. Another possible cause for this discrepancy could be the variability of the local microstructure. Due to the strong contribution of various phonon modes to the effective permittivity, and hence to electron–hole screening within these materials, the variation of the degree of local order/disorder (and therefore phonon distributions) over different film architectures can result in the tuning of the exciton binding energies, even for samples with the same chemical composition. In fact, we have observed that the strength of the excitonic transition at the band edge, which is representative of the exciton binding energy, changes with the microstructure of the film.[2,6]

As it is evident from the Saha analysis presented in Figure 4.7, at solar excitation densities and at room temperature, the predominant photoexcitations for the reported range of binding energies (2–50 meV) are free-carriers, with insignificant contributions from real exciton populations. However, at reasonably higher excitation densities, which can be reached in other applications such as lasers, contributions from excitonic dynamics have to be appropriately assessed. For this, we need optical probes to observe exciton dynamics, and pump–probe spectroscopy, in particular transient absorption, offers the essential tools.

The transient absorption spectrum represents the change in the absorbance induced by a photo-exciting pump pulse, typically measured as a transmittance change measured using a delayed white-light probe pulse through the sample. In the case where primary photo-excitations are free carriers, due to the state-filling close to the band edges by the photo-excited species and the resultant Pauli blocking, the absorption probability is reduced at the associated photon-energies. This induces transparency at the energies corresponding to the band edge of the excited sample, measured as a positive $\Delta T/T$ (normalized transmittance change) signal, commonly indicated as photo-bleach (PB), see Figure 4.11a. If there is a considerable fraction of excitons, a new derivative-like feature (giving a sharp S-like shape, see illustration in Figure 4.11b) appears. This is due to the blue-shift of the exciton energy induced by the stable exciton population and as perceived by the probe pulse.

The physical origin of such a blue shift can be understood by considering the two kinds of many-body interactions that arise within the excitonic population: a van der Waals-like attractive interaction between excitons and a repulsive force due to the Pauli exclusion principle acting on the charge carriers that make up the exciton.[44,45] The attractive interaction results in a red-shift of the excitonic energy (*i.e.* a more stable exciton), while the latter effect creates a blue-shift. In 3D semiconductors, these effects usually balance each other out, leaving the exciton energy constant after the photo-excitation. On the other hand, in 2D semiconductors, a clear blue-shift can be observed due to the reduction of the attractive component.[46] The overall effect is the appearance of a negative $\Delta T/T$ signal on the blue side of the PB signal, which we indicate as the MA* feature. Interestingly, as we have demonstrated,[6] hybrid perovskites with a stable exciton population show this MA* feature.

The MA* feature can also arise due to the presence of carrier screening effects. However, in such a scenario, we would not expect the formation time for the MA* feature that we observe ($\approx$1 ps time window[6]). In spite of the ambiguities surrounding whether this feature is an excitonic population signature, it remains a sensitive probe for the strength of excitonic transitions and thus the electron–hole correlation.

As an example, Figure 4.11c shows the linear and transient absorption spectra at an excitation density $>10^{17}$ cm^3, of two samples of $CH_3NH_3PbBr_3$, one composed of small crystallites grown within a mesoporous scaffold and the other made of micron size crystallites grown on a flat substrate. In the case of small crystallites, there is no excitonic transition at the band edge, and the TA spectrum shows only the PB feature. The sample with large crystals shows an obvious excitonic peak, with an excitonic binding energy around 60 meV (see the Elliott fitting in Figure 4.5), and the TA spectrum exhibits the MA* feature, characteristic of excitonic population.

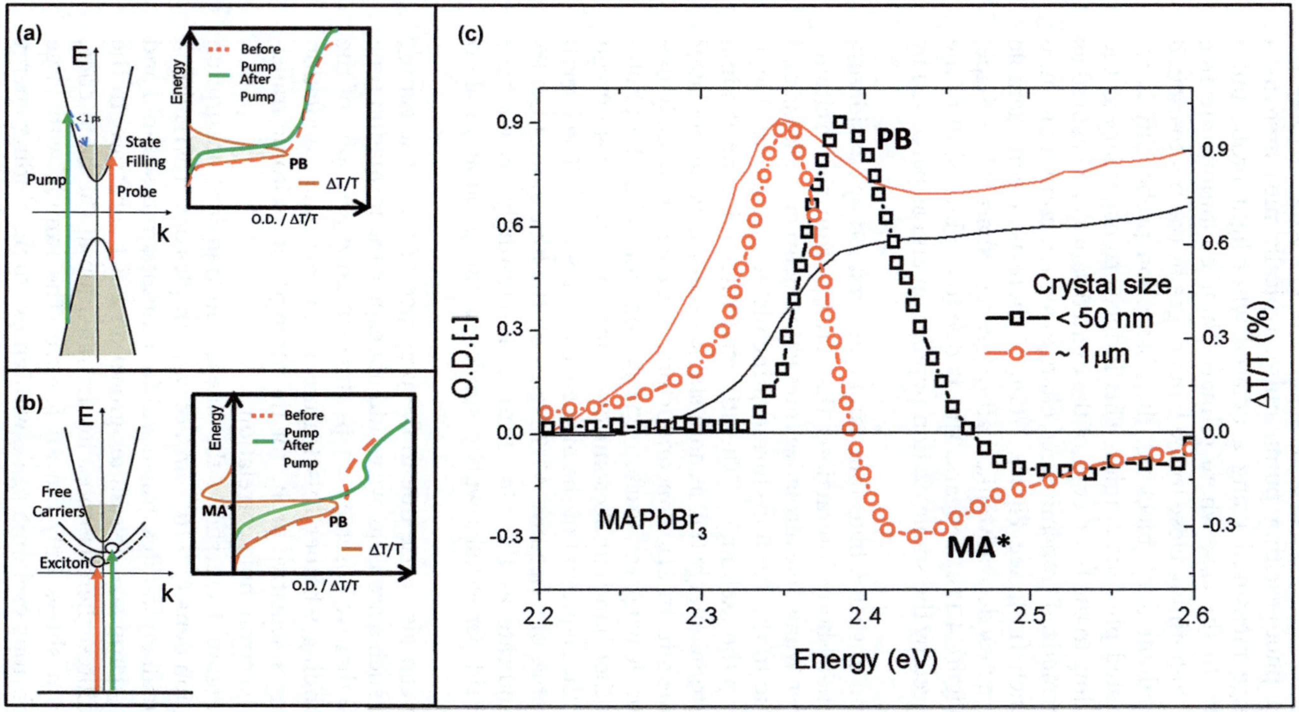

Figure 4.11 Schematic representation of TA spectra: (a) free carrier picture: after the pump excitation, hot carriers are generated which thermalize to the bottom of the conduction band and, due to the state filling effect, those states are no longer available for absorption. The result is a blue shift of the absorption edge, experimentally seen as a positive $\Delta T/T$ signal at the band edge (photo-bleach (PB)). (b) Excitonic picture: in the presence of an excitonic transition and its subsequent population after photo-excitation, the excitonic energy blue shifts due to exciton–exciton interactions, which result in the appearance of the MA* feature along with PB. (c) UV-Vis absorption spectra and TA spectra (pump–probe delay: 1 ps) of $CH_3NH_3PbBr_3$ films composed of small crystals (grown within an Al_2O_3 scaffold) and large crystals (grown on bare glass). Data reproduced from ref. 6.

We can correlate the variability of the exciton binding energy with the structural disorder and the resultant modulation of the effective dielectric permittivity. As already pointed out, the total ε is a cumulative sum that includes dielectric contributions from all the phonon modes. By tuning the polycrystallinity of the film, the values of these respective contributions change due to the modulation of the phonon energies and lattice interactions, thus giving rise to the dependence of excitonic stability on polycrystallinity.[42]

Alternatively, the role of the organic cation, which has a permanent dipole moment, can also be considered. The rotational degree of freedom of the cations can generate a gradient of electrostatic potential that results in local electric fields that dissociate the exciton. Walsh and coworkers[6] calculated such a variance in the potential and found that it decreases when the temperature is lowered, because the rotational motion of the organic dipole is reduced. This is in full agreement with the enhancement of the exciton binding energy observed experimentally by Nicholas *et al.*[36] and Herz *et al.*[47] All rotational disorder is quenched, leading to the formation of fully twinned dipole domains. The twinned dipoles within neighboring unit cells interact *via* dipole–dipole coupling, and this will increase the cumulative energy needed for their individual rotation, and lead to the "locking" of the cations in an orderly fashion. Apart from thermal fluctuations, when moving from a perfect single crystal to a polycrystalline sample, the dipole twinning can also be disrupted by any type of defect, which can be a vacancy, a dangling bond at a grain boundary or the intercalation of water or oxygen molecules at the surface. The density of such defects can become increasingly important as the average crystallite size is reduced. Therefore, the variance of the electrostatic potential, and thus the local screening, can be controlled by the local order within the crystal. While larger and less defective crystals provide lower screening, the electron–hole separation due to electrostatic disorder can be significant in smaller crystals. The effect on the excitonic stability of disorder induced by lattice strain has been investigated in other semiconductors such as GaAs, where the localized electric fields created by microscopic internal strains can dissociate the exciton.

4.5 Transport Properties

So far, we have dealt with absorption and emission dynamics in hybrid perovskites. Another crucial topic for any opto-electronic device optimization is carrier transport. It is important to understand the transport mechanism in hybrid perovskites and to properly quantify carrier diffusion lengths and mobilities. Optical spectroscopy offers easy tools for such purposes, which have been applied extensively to the study of $CH_3NH_3PbI_3$ during the last couple of years. In this section, we discuss two of these techniques, one uses PL lifetimes to extract carrier diffusion lengths, while the other employs optical pump–terahertz probe spectroscopy (OPTP) to extract photo-generated carrier mobilities.

4.5.1 Carrier Diffusion from Photoluminescence Decays

There is a variety of experimental methods available to extract carrier diffusion lengths. Some of them are based on device measurements that include time-of-flight techniques with photo-current as the probe. Alternatively, pure optical techniques also exist, among which the most popular method is PL quenching, which involves comparing the photoluminescence in the presence and absence of quenching layers.

$CH_3NH_3PbI_3$ films exhibit PL lifetimes of hundreds of nanoseconds as discussed in Section 4.3. In the presence of electron or hole accepting layers, the PL decay is quenched to a few nanoseconds owing to efficient carrier transfer. In this case, the PL decay behavior can be modelled by calculating the excited carrier density according to a one dimensional diffusion equation:[11]

$$\frac{\partial n(x,t)}{\partial t} = D\frac{\partial^2 n(x,t)}{\partial x^2} - k(t)n(x,t) \tag{4.29}$$

where D is the diffusion coefficient and $k(t)$ is the PL decay rate in the absence of a quencher, which can be obtained by fitting the PL decay of the pristine film with a stretched exponential. The effect of the quenching layer is considered by employing the boundary condition: $n(L,t) = 0$, where L is the perovskite film thickness and $x = 0$ at the glass/perovskite. This amounts to the assumption that the quenching efficiency, in other words, the charge transfer efficiency, is 100%. The equation can be solved either analytically or numerically to obtain the value of the diffusion coefficient, D. The average diffusion length can then be extracted from the simple relationship: $L_D = \sqrt{D\tau}$, where τ is the carrier recombination lifetime. Following such a methodology, a diffusion length of around a micrometer has been obtained for Cl-doped $CH_3NH_3PbI_3$ films.[11]

Based on the discussion on PL dynamics presented earlier in this chapter, a few cautionary points should be kept in mind while employing this method. We have shown that the PL lifetime of the pristine material is dependent on two parameters: the intrinsic bimolecular recombination coefficient and the trap density. We have observed that both of these parameters, and in particular the trapping rate, change from sample to sample. Hence, care must be taken to evaluate the role of these parameters in the PL dynamics at the excitation densities used in the experiment. Moreover, when fitting the experimental data to the diffusion equation, the effect of changes in the sample morphology/trap density on the PL dynamics must be differentiated from the quenching effect across the heterogeneous interface. The assumption of 100% quenching at the interface implicit with the boundary condition is not necessarily true for the various quenching layers. A more detailed method that includes the quenching rate has been developed recently by Lee and Tisdale[48] for another heterogeneous interface, and it can also be extended to the case of perovskites.

4.5.2 Optical Pump–Terahertz Probe (OPTP) Experiments

Among the various techniques to estimate the charge carrier mobility, the most widely used, especially in the context of optical spectroscopy, is OPTP. This technique is based on the fact that the transmission of a terahertz pulse in a semiconductor is modulated in the presence of carriers. The relative change in the transmission is directly proportional to the carrier conductivity, and hence one can obtain the charge carrier mobility directly from such an experiment.[5,47] In addition, the spectral shape of the photo-induced transmission change of the terahertz pulse is representative of the nature of the photo-excitation, and hence can act as an unambiguous probe of any excitonic state. For $CH_3NH_3PbI_3$ at room temperature, THz measurements suggest that the predominant population is that of free charge carriers,[5,49] while excitonic signatures are observed at lower temperatures, where the organic cation motion is frozen.[47] This fits well with the dielectric argument presented in Section 4.3.2 of this chapter. It is intriguing that, in a recent paper, Cooke and coworkers[49] have reported a fraction of stable exciton population in single crystals and at high excitation densities.

In the case where only free carriers are present, the relative decay of the THz transients enables one to observe the free carrier population directly, and hence the method can act as a complementary technique to monitor the photo-excitation dynamics. Herz and co-workers used this fact to model the THz transients using eqn (4.15) and extract the relevant trapping rates and bimolecular coefficients.[5] They also obtained a room temperature charge carrier mobility of about 15 $cm^2\ V^{-1}\ s^{-1}$.[5] As the temperature was lowered, they observed an increase of the mobility, reaching about 50 $cm^2\ V^{-1}\ s^{-1}$ at the phase-transition temperature, below which there was a further rapid increase to 150 $cm^2\ V^{-1}\ s^{-1}$.[47] It is also intriguing to note that the authors observed a substantial deviation from Langevin-like recombination, and this highlights the need for further efforts aimed towards understanding transport mechanisms in these materials.

4.6 Summary and Outlook

In this chapter, we have outlined the physical basis for understanding the optical properties of hybrid perovskites. While, at a first approximation, this class of semiconductors can be pictured within the established inorganic crystalline semiconductor framework, their "soft" character strongly influences their electronic functionality and leads to many remarkable properties. At room temperature, in particular, the fluctuating nature of the crystalline unit makes the phyotophysics of hybrid perovskites extremely susceptible to minor changes in the molecular geometry. This calls for further efforts to achieve awareness and control of electron–vibronic coupling phenomena, and to fully elucidate the role of chemical and structural modifications.

Acknowledgements

The authors acknowledge the financial support from the EU 7th Framework Program under the grant agreements No. 604032 (MESO), No. 316494 (DESTINY), EU Horizon 2020 Research and Innovation program under the grant agreement No. 643238 (SYNCHRONICS), Fondazione Cariplo Project GREENS (No. 2013-0656). The authors also thank all their collaborators.

References

1. S. D. Stranks and H. J. Snaith, *Nat. Nanotechnol.*, 2015, **10**, 391–402.
2. V. D'Innocenzo, G. Grancini, M. J. P. Alcocer, A. R. S. Kandada, S. D. Stranks, M. M. Lee, G. Lanzani, H. J. Snaith and A. Petrozza, *Nat. Commun.*, 2014, **5**, 3486.
3. S. D. Stranks, V. M. Burlakov, T. Leijtens, J. M. Ball, A. Goriely and H. J. Snaith, *Phys. Rev. Appl.*, 2014, **2**, 034007.
4. M. Saba, M. Cadelano, D. Marongiu, F. Chen, V. Sarritzu, N. Sestu, C. Figus, M. Aresti, R. Piras, A. Geddo Lehmann, C. Cannas, A. Musinu, F. Quochi, A. Mura and G. Bongiovanni, *Nat. Commun.*, 2014, **5**, 5049.
5. C. Wehrenfennig, G. E. Eperon, M. B. Johnston, H. J. Snaith and L. M. Herz, *Adv. Mater.*, 2014, **26**, 1584–1589.
6. G. Grancini, A. R. Srimath Kandada, J. M. Frost, A. J. Barker, M. De Bastiani, M. Gandini, S. Marras, G. Lanzani, A. Walsh and A. Petrozza, *Nat. Photonics*, 2015, **9**, 695–701.
7. J. Nelson, *The Physics of Solar Cells*, Imperial College Press, 2003.
8. P. Umari, E. Mosconi and F. De Angelis, *Sci. Rep.*, 2014, **4**, 4467.
9. J. Even, L. Pedesseau and C. Katan, *J. Phys. Chem. C*, 2014, **118**, 11566–11572.
10. J. S. Manser and P. V. Kamat, *Nat. Photonics*, 2014, **8**, 737–743.
11. S. D. Stranks, G. E. Eperon, G. Grancini, C. Menelaou, M. J. P. Alcocer, T. Leijtens, L. M. Herz, A. Petrozza and H. J. Snaith, *Science*, 2013, **342**, 341–344.
12. V. D'Innocenzo, A. R. Srimath Kandada, M. De Bastiani, M. Gandini and A. Petrozza, *J. Am. Chem. Soc.*, 2014, **136**, 17730–17733.
13. F. Urbach, *Phys. Rev.*, 1953, **92**, 1324.
14. P. Van Mieghem, *Rev. Mod. Phys.*, 1992, **64**, 755–793.
15. L. Bai, C. Xu, P. G. Schunemann, K. Nagashio, R. S. Feigelson and N. C. Giles, *J. Phys.: Condens. Matter*, 2005, **17**, 549–558.
16. A. Amat, E. Mosconi, E. Ronca and C. Quarti, *Nano Lett.*, 2014, **14**, 3608–3616.
17. F. Brivio, K. T. Butler, A. Walsh and M. van Schilfgaarde, *Phys. Rev. B: Condens. Matter Mater. Phys.*, 2014, **89**, 155204.
18. J. Even, L. Pedesseau, C. Katan, M. Kepenekian, J.-S. Lauret, D. Sapori and E. Deleporte, *J. Phys. Chem. C*, 2015, **119**, 10161–10177.
19. A. Filippetti and A. Mattoni, *Phys. Rev. B: Condens. Matter Mater. Phys.*, 2014, **89**, 125203.

20. M. R. Filip, G. E. Eperon, H. J. Snaith and F. Giustino, *Nat. Commun.*, 2014, **5**, 5757.
21. C. Quarti, G. Grancini and E. Mosconi, *J. Phys. Chem. Lett.*, 2014, **5**, 279–284.
22. C. C. Stoumpos and M. G. Kanatzidis, *Acc. Chem. Res.*, 2015, **48**, 2791–2802.
23. N. W. Ashcroft and N. D. Mermin, *Solid State Physics*, Holt, Rinehart and Winston, 1976.
24. Y. P. Varshni, *Physica*, 1967, **34**, 149–154.
25. A. Olkhovets, R.-C. Hsu, A. Lipovskii and F. Wise, *Phys. Rev. Lett.*, 1998, **81**, 3539–3542.
26. C. Quarti, E. Mosconi, J. M. Ball, V. D'Innocenzo, C. Tao, S. Pathak, H. J. Snaith, A. Petrozza and F. De Angelis, *Energy Environ. Sci.*, 2015, **9**, 155–163.
27. H. Kalt and M. Hetterich, *Optics of Semiconductors and Their Nanostructures*, Springer Science & Business Media, 2004, vol. 5.
28. H. Haug and S. W. Koch, *Quantum Theory of the Optical and Electronic Properties of Semiconductors*, World Scientific, 2004.
29. G. Lanzani, *The Photophysics Behind Photovoltaics and Photonics*, Wiley, 2012.
30. R. J. Elliott, *Phys. Rev.*, 1957, **108**, 1384–1389.
31. I. Pelant and J. Valenta, *Luminescence Spectroscopy of Semiconductors*, OUP, Oxford, 2012.
32. D. A. B. Miller, D. S. Chemla, D. J. Eilenberger, P. W. Smith, a. C. Gossard and W. T. Tsang, *Appl. Phys. Lett.*, 1982, **41**, 679–681.
33. Q. Lin, A. Armin, R. C. R. Nagiri, P. L. Burn and P. Meredith, *Nat. Photonics*, 2014, **9**, 106–112.
34. W. J. Moore and R. T. Holm, *J. Appl. Phys.*, 1996, **80**, 6939.
35. M. Hirasawa, T. Ishihara, T. Goto, K. Uchida and N. Miura, *Phys. Rev. B: Condens. Matter Mater. Phys.*, 1994, **201**, 427–430.
36. A. Miyata, A. Mitioglu, P. Plochocka, O. Portugall, J. T.-W. Wang, S. D. Stranks, H. J. Snaith and R. J. Nicholas, *Nat. Phys.*, 2015, **11**, 582–587.
37. R. Cingolani, L. Calcagnile, G. Col'i, R. Rinaldi, M. Lomoscolo, M. DiDio, A. Franciosi, L. Vanzetti, G. C. LaRocca and D. Campi, *J. Opt. Soc. Am. B*, 1996, **13**, 1268.
38. M. N. Saha, *Proc. R. Soc. A*, 1921, **99**, 135–153.
39. Y. Yamada, T. Nakamura, M. Endo, A. Wakamiya and Y. Kanemitsu, *J. Am. Chem. Soc.*, 2014, **136**, 11610–11613.
40. J. Burschka, N. Pellet, S.-J. Moon, R. Humphry-Baker, P. Gao, M. K. Nazeeruddin and M. Grätzel, *Nature*, 2013, **499**, 316–319.
41. G. Grancini, S. Marras, M. Prato, C. Giannini, C. Quarti, F. De Angelis, M. De Bastiani, G. E. Eperon, H. J. Snaith, L. Manna and A. Petrozza, *J. Phys. Chem. Lett.*, 2014, **5**, 3836–3842.
42. G. Grancini, V. D'Innocenzo, E. R. Dohner, N. Martino, A. R. Srimath Kandada, E. Mosconi, F. De Angelis, H. I. Karunadasa, E. T. Hoke and A. Petrozza, *Chem. Sci.*, 2015, **6**, 7305–7310.

43. A. Filippetti, P. Delugas and A. Mattoni, *J. Phys. Chem. C*, 2014, **118**, 24843–24853.
44. D. Hulin, A. Mysyrowicz, A. Antonetti, A. Migus, W. T. Masselink, H. Morkoc, H. M. Gibbs and N. Peyghambarian, *Phys. Rev. B: Condens. Matter Mater. Phys.*, 1986, **33**, 4389–4391.
45. S. Schmitt-Rink, D. Chemla and D. Miller, *Phys. Rev. B: Condens. Matter Mater. Phys.*, 1985, **32**, 6601–6609.
46. N. Peyghambarian, H. Gibbs, J. Jewell, A. Antonetti, A. Migus, D. Hulin and Mysyrowicz, *Phys. Rev. Lett.*, 1984, **53**, 2433–2436.
47. R. L. Milot, G. E. Eperon, H. J. Snaith, M. B. Johnston and L. M. Herz, *Adv. Funct. Mater.*, 2015, DOI: 10.1002/adfm.201502340.
48. E. M. Y. Lee and W. A. Tisdale, *J. Phys. Chem. C*, 2015, **119**, 9005–9015.
49. D. A. Valverde-Chávez, C. Ponseca, C. Stoumpos, A. Yartsev, M. G. Kanatzidis, V. Sundström and D. G. Cooke, *Energy Environ. Sci.*, 2015, **8**, 3700–3707.

CHAPTER 5

The Role of Nanostructured Metal Oxides in Hybrid Solar Cells

JAMES A. DORMAN† AND LUKAS SCHMIDT-MENDE*

Department of Physics, University of Konstanz, P.O. Box M680, 78457 Konstanz, Germany
*Email: Lukas.Schmidt-Mende@uni-konstanz.de

5.1 Introduction

As an alternative to the standard p–n junction photovoltaic cell, many researchers have investigated the combination of organic and inorganic materials to harness photoelectrochemical conversion for energy generation. The first attempt to combine an organic dye with a metal oxide occurred in the late 1970s,[1] which resulted in relatively low conversion efficiencies due to the limited total light absorption by the monolayer of dye adsorbed on the surface (roughly 1% of the total light). The overall device performance was found to increase proportionally with the surface area of the metal oxide layer, with a first reported conversion efficiency of 1.5% under monochromatic light for sintered ZnO particles and rose Bengal dye.[2] This is the first reported implementation of a dye-decorated sintered metal oxide structure for solar energy conversion. Of course, the field of structured electrodes for dye sensitized solar cells (DSSC) emerged with the use of a

†Present address: Department of Chemical Engineering, Louisiana State University, Baton Rouge, LA 70802, USA.

RSC Energy and Environment Series No. 16
Unconventional Thin Film Photovoltaics
Edited by Enrico Da Como, Filippo De Angelis, Henry Snaith and Alison Walker

Published by the Royal Society of Chemistry, www.rsc.org

mesoporous TiO_2 electrode as a high surface area light harvesting material by O'Regan and Grätzel[3] in 1991, with an impressive conversion efficiency of ~7.9% under solar light.

This mesoporous TiO_2 thin film is still the benchmark material for second generation solar cells due to the high performance, chemical stability, and commercial availability of the metal oxide.[4] Particularly interesting is the range of materials that have been used in conjunction with the mesoporous film, including dyes,[5] quantum dots,[6–8] solid absorbers,[9] and more recently the metal-organic material methylammonium lead iodide perovskite.[10–12] What makes mesoporous TiO_2 unique is the combination of many physical properties that are inherent to the metal oxide and its particular structuring. In addition to the chemical stability mentioned previously, TiO_2 has a relatively high electron mobility (10 $V cm^{-2} s^{-1}$) and a wide bandgap (~3.2 eV), allowing for the separation of the optical absorption and electrical conductivity.[13] More important for the functionality of the DSSCs is the extremely high surface area of the film, resulting in the almost complete absorption of light by a monolayer or less of dye molecules.[3] However, it has been proposed that the performance of mesoporous films is limited for several reasons, *e.g.* the sintering of nanoparticles introduces numerous grain boundaries that limit the flow of electrons, and it is not necessarily ideal to completely decouple the optical and electrical properties.[14,15]

More recently, these second generation solar cells have reach impressive efficiencies and appear as a competitive alternative to the standard silicon solar cell. The most impressive of these devices is a perovskite solar cell with a reported conversion efficiency of over 20%, as certified by the National Renewable Energy Laboratory (NREL).[16] While the exact structure of this record cell is unknown, the most recent report by the same group (the Seok group at the Korean Research Institute of Chemical Technology (KRICT)), describes the use of a mesoporous TiO_2–perovskite layer to achieve an average of 16.46% conversion without significant hysteresis.[17] As a comparison, the highest confirmed AM 1.5 efficiency of a DSSC is 11.9% for a cell using the YD2-o-C8 dye developed at the École Polytechnique Fédérale de Lausanne (EPFL).[18] The Sargent group at the University of Toronto has reported air-stable PbS quantum dot solar cells with efficiencies of 8% on a mesoporous TiO_2 layer with certified NREL efficiencies of 9.9%.[6,16] Finally, hybrid solar cells of nanostructured TiO_2–dye–polymer layers have been reported to have efficiencies between 2.7–3.1% depending on the thickness of the nanostructured layers.[19,20] These materials will be discussed in further detail later in the chapter.

After a general introduction to hybrid solar cells with emphasis on the role of the metal oxide layer, we focus on the use of metal oxides as electrode materials in hybrid solar cells. This aspect of nanostructuring of the metal oxide is discussed with the goal of highlighting the relationship between electron mobility, light management, and high surface area structures. The following sections are divided based on the role of the metal oxide, from transparent conducting oxides to blocking layers (Section 5.3), and as active layers (Section 5.4). The chapter concludes with an outlook on future nanostructuring for next generation solar cells.

5.2 Fundamentals of Hybrid Solar Cells

This section presents the working mechanisms of hybrid solar cells in order to explain the motivation for the incorporation and nanostructuring of metal oxides. First, a brief introduction into the working mechanism is presented to highlight the role of n- and p-type metal oxides as membranes for charge carriers. Once this has been established, the loss mechanisms and methods of mitigation are discussed. Finally, the use of nanostructured electrode materials is presented along with a discussion of the underlying physical mechanisms that occur with their incorporation into solar cells.

Hybrid solar cells can consist of a variety of organic–inorganic material combinations and therefore must be clearly defined. A hybrid solar cell as defined here is a solar cell in which generated excitons are separated into free charges across an inorganic/organic interface, as opposed to organic devices where an organic/organic interface is used. Hybrid solar cells can be further separated into subgroups based on where charges are being generated, as illustrated in Figure 5.1. This figure also shows the evolution of the devices with respect to each other. Hybrid solar cells were developed from organic bulk heterojunction (BHJ) and liquid electrolyte dye-sensitized solar cells (DSSCs). The hybrid solar cell architecture is based on the BHJ structure that is employed in organic devices, *i.e.*, n-type and p-type polymers or small molecules are thoroughly mixed, and domains are formed to transport the electrons and holes, respectively. In the hybrid devices, the n-type material is traditionally replaced with a metal oxide with the aim of producing a high surface area structure similar to that of the BHJ for efficient charge separation. In a DSSC, for instance, the n-type polymer is replaced with TiO_2 nanoparticles that are sintered together to form a mesoporous film. In these devices, the excitons are generated and separated at the dye/TiO_2 interface and holes are transported *via* a redox couple shuttle, such as the liquid I_3^-/I^- electrolyte. Replacing the liquid electrolyte with a solid organic hole-conductor has led to solid-state DSSCs (ssDSSCs), which are one class of hybrid devices, where the dye is solely responsible for charge generation, and the commonly transparent hole transporter material does not contribute to charge generation. The limited light absorption in a monolayer of dye is one of the major drawbacks of such structures and the reason why the high surface area supplied by the mesostructured TiO_2 is necessary. In ssDSSCs, the optimal film thickness of this active layer is in the range of one or two micrometers. In contrast with traditional hybrid solar cells, the generation of excitons in hybrid polymer cells takes place in an absorbing polymer, and excitons are subsequently separated at the polymer/metal oxide interface. High interfacial area, as supplied by metal oxide nanostructures, is necessary for efficient exciton separation. As in BHJs, the strong absorption of the polymer allows efficient light harvesting in films only 100 to a few hundreds of

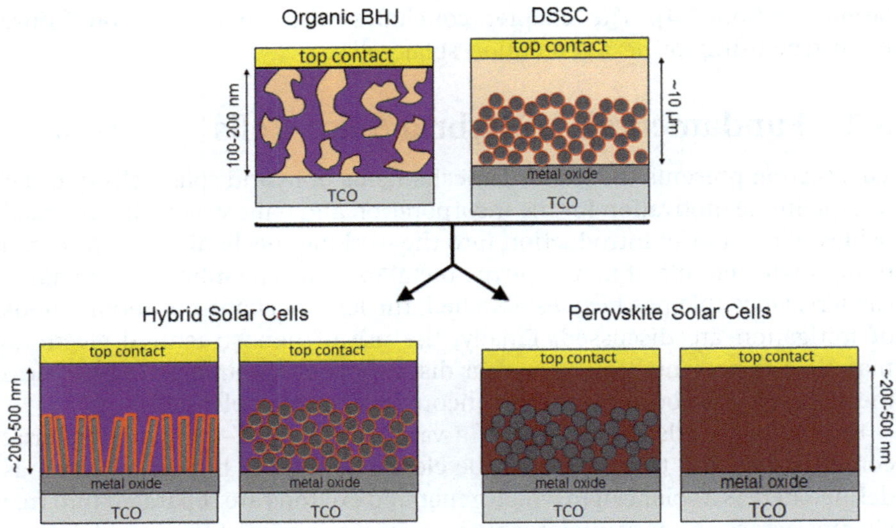

Figure 5.1 The different classes of organic and hybrid solar cell structures are shown along with their respective architectures. The organic bulk heterojunction (BHJ) consists of intermixed n-type and p-type polymers that separate into various domains to transport electrons and holes, respectively. In the DSSC, the n-type polymer is replaced with a high surface area TiO_2 film that is decorated with a dye and the p-type material is replaced with a hole shuttle. The hybrid structure incorporates the p-type polymer from the BHJ with the high surface area TiO_2 film for direct injection from the polymer to the metal oxide. Finally, the perovskite replaces the dye absorber with a high absorption organometallic structure and a hole shuttle. The active material is sandwiched between electrodes and additional layers that aid in the selective transport of the charges.

nanometers in thickness. In many cases, a dye is still used to increase the charge injection from the polymer to the metal oxide *via* chemical bonding with the inorganic structure and orbital overlap with the organic material. Perovskite solar cells, as covered in the other chapters of this book, originally replaced the dye molecule in the DSSCs with an organometallic $CH_3NH_3PbI_3$ perovskite. Light absorption in perovskite solar cells is believed to directly form free charges that are transported to the respective electrodes.[12] A nanostructured metal oxide is not obligatory in perovskite solar cells, and planar device architectures have also shown impressive device performances.

The standard architecture for hybrid solar cells (shown in Figure 5.2(a)) consists of a transparent conducting oxide (TCO), a metal oxide blocking layer, metal oxide scaffolding, an absorbing layer/hole transport layer, and a reflective metal top contact. In a traditional DSSC, the absorbing layer is a dye molecule, however, the hole transport layer can simultaneously act as the absorbing material, as is the case for poly(3-hexythiophene) (P3HT). As opposed to fully organic solar cells, the n-type metal oxide acts as the electron transport material due to the position of the conduction band (CB) with

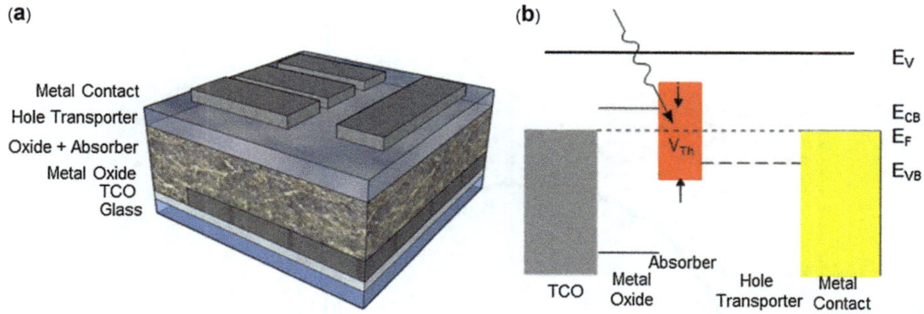

Figure 5.2 (a) The standard solar cell design is shown. The device consists of a transparent conducting layer on a glass substrate, a hole blocking layer, a high surface area metal oxide scaffold that is decorated with an absorber material, a hole transport layer and a metal top contact. (b) The energy landscape for this architecture is shown. The absorber material converts absorbed photons to excitons that are separated at the interface. The photogenerated electrons travel through the metal oxide and the holes through the respective transport material. The theoretical V_{OC} is highlighted and is determined from the HOMO of the hole transport layer and the CB of the metal oxide.

respect to the lowest unoccupied molecular orbital (LUMO) of the polymer, transporting charges to the TCO cathode, as seen in the energy level diagram (Figure 5.2(b)). Additionally, as the absorber medium is typically an organic molecule, excitons are generated instead of free charges, since the exciton binding energy is larger than the thermal energy:[21]

$$\varepsilon_{exc} = \frac{m_{red}e^4}{2(4\pi\epsilon\epsilon_0)^2\hbar^2} \tag{5.1}$$

where m_{red} is the reduced exciton mass, ϵ is the dielectric permittivity, ϵ_0 is the dielectric permittivity of free space, and $\hbar$ is the reduced Planck's constant. For inorganic materials where ϵ is greater than 10, the absorbed light is directly converted to free charges at room temperature. In order to separate and harness these light generated charges, the exciton must diffuse to an interface before recombination (typical exciton diffusion length $l = 10$ nm).[21,22] The interface facilitates the separation of the exciton into free charges due to the offset of the LUMO/CB energies, forming either free charges or charge transfer states.[21,23] The delocalization of the charge transfer states varies for each material, as seen in Figure 5.3, with the electron distributed over many lattice planes and the hole confined to within ~1 nm of the interface.[23] Finally, the separated charges move toward the respective contacts and are harnessed as power.

In order to achieve the best possible performance, an in-depth understanding of the metal oxide–polymer interface is required. To facilitate charge separation, a type-II staggered junction is formed allowing for the selective charge transport in each material.[22] In this structure (Figure 5.2(b)), the electrons travel in the direction of the lowest energy CB/LUMO and the

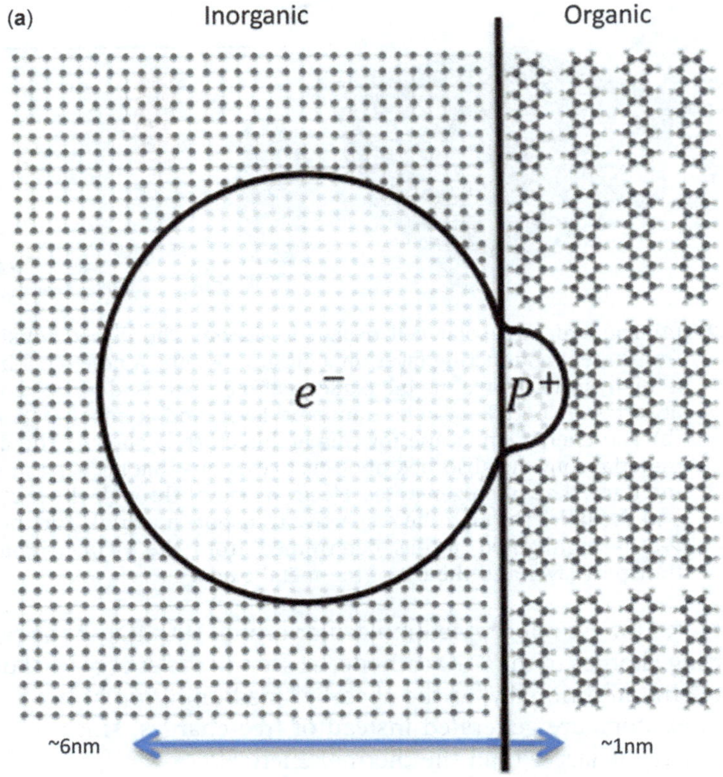

Figure 5.3 A conceptual illustration of a charge transfer state in a hybrid solar cell. The electron is delocalized over many crystal lattice sites due to high dielectric screening. This electron is Coulombically bound to the hole which is located within a nm of the interface. The delocalization in the metal oxide can be controlled by using materials with high dielectric constants, while the hole is also dependent on the crystallinity of the polymer.
Reprinted with permission from Renshaw and Forrest, *Phys. Rev. B: Condens. Matter Mater. Phys.*, 2014, **90**, 045302. Copyright 2014 American Physics Society.[23]

holes toward the highest energy valence band/highest occupied molecular orbital (HOMO). However, as discussed by many groups, the actual energy landscape at the interface is non-trivial.[23–26] An exact schematic of the alignment is difficult to produce as the type of alignment changes with the material, *i.e.*, the Fermi level pinning at the inorganic interfaces and the vacuum level alignment within the organics.[27,28] Additionally, the presence of dipole moments and major offsets between Fermi levels at the inorganic/organic interfaces results in a mixture of both Fermi pinning and vacuum alignment, adding extra parameters to the discussion. When the materials are contacted, a quasi-Fermi level is formed near the respective energy levels. The alignment of the energy levels is used to predict the maximum open circuit voltage (V_{OC}) possible with two materials based on the difference between the metal oxide CB and the HOMO level of the organic under ambient conditions. Under working conditions, the measured V_{OC} is determined instead by the quasi-Fermi levels, and quantification of these values is not trivial. Care must be taken when designing the system to maximize both the V_{OC} and the offset between LUMO and CB levels, which is considered to create the driving force of charge injection. When materials are chosen solely to maximize V_{OC}, charge injection may be inhibited, thereby preventing charge separation and allowing the excitons to recombine. The resulting discrepancy between theoretical and measured V_{OC} values can be attributed to the recombination of excitons and free charges.

The three main recombination pathways in hybrid solar cells are: exciton recombination, trap state-mediated recombination (monomolecular), and recombination of free charge carriers (bimolecular).[29] In order to limit exciton recombination, the solar cell must be designed so that any photon absorbed is within the exciton diffusion length of an interface. A prime example of this structuring is observed in organic bulk heterojunction blends, where the polymer and C_{60} molecules separate into intermixed but mostly continuous domains.[30,31] Photoluminescence quenching studies offer a powerful alternative to the common electronic measurements for characterization of exciton recombination.[32,33] The other two recombination mechanisms are probed during the measurement of the hybrid solar cell and can be deconvoluted by investigating three different operating regimes, *i.e.*, the short circuit current density (J_{SC}), V_{OC} and the maximum power point (MPP). The most widely used method to elucidate the mechanisms under short circuit (J_{SC}) and open circuit (V_{OC}) conditions is through light intensity-dependent current density–voltage (J–V) measurements.[34] An ideality factor is extracted from these measurements at V_{OC} using the ideal diode equation:[21]

$$V'_{OC} = \frac{nkT}{q} \ln\left(\frac{XJ_{SC}}{J_0}\right) \tag{5.2}$$

where V'_{OC} is the intensity dependent V_{OC}, n is the ideality factor, k is Boltzmann's constant, T is the temperature, q is the electron charge, X is the light intensity, and J_0 is the dark saturation current density. In high quality devices, the ideality factor is between 1–2 and indicates the type of recombination.[35] The more ideal the device, the closer n is to 1; the more bimolecular recombination, the better the device functions. Alternatively, as n approaches 2, trap state recombination increases, indicating the presence of defects near the inorganic–organic interfaces. Additional probing of these trap state defects is possible by plotting J_{SC} as a function of light intensity based on the following relationship:[36]

$$J_{SC} \propto X^{\alpha} \tag{5.3}$$

where α is ≤ 1, and the deviation is strictly due to trap states near the donor–acceptor interface, as all the charges that are generated are either extracted or trapped in positions where recombination can occur.[37] To prevent the occurrence of these trap states, high quality metal oxide particles are needed, including single crystal structures to prevent the formation of grain boundaries which limit electron transport. However, as with other aspects of the solar cell discussed below, it is important to consider both sides of the interface. Recently, Burke and McGehee[38] discussed the role of disorder on the energy landscape of bulk heterojunction solar cells and concluded that polymer disorder increases at the interface, resulting in a slight increase of the bandgap and conduction band levels. A by-product of this shift in the conduction band energy is that a barrier forms at the interface between the metal oxide and the polymer, hindering direct electron injection across this interface. A similar mechanism is believed to occur in hybrid systems, with many groups focusing on the crystallinity of P3HT within porous matrices.[26,39] Without a uniform crystal structure, the probability of excitons diffusing to the interfaces decreases, resulting in increased recombination and lower device performance.

Furthermore, the mobility of charges in the various hole and electron conductors is important for solar cell operation. It is commonly stated that the materials must have similar mobilities in order to prevent recombination of highly mobile species with their slowly moving counterparts.[40] However, this does not mean that the inorganic and organic materials are optimized simultaneously in a single cell, as most groups are experts in either organic or inorganic materials. For example, there are many naturally occurring n-type inorganic materials that can be used as electron transport materials, as highlighted in Table 5.1. Commercially available anatase mesoporous TiO_2 has an extremely low electron mobility (10^{-4}–10^{-2} cm^2 V^{-1} s^{-1})[41] compared to the value in the table due to the numerous grain boundaries present. Fortunately, it is possible to augment the physical properties of metal oxides through doping without degrading their natural properties. This approach is used in the formation of commercially available transparent conducting oxides such as tin doped indium oxide (ITO), fluorine doped tin oxide (FTO), and aluminium doped zinc oxide (AZO). There has been relatively little work done on the doping of the electrode material in hybrid solar cells; most work has focussed on the engineering of the absorption of dyes and polymer materials. When the electrode is the focus of the research, groups tend to focus on nanostructuring to maximize the interfacial area between polymers and metal oxides.

Table 5.1 Crystal structure, charge mobilities, and lattice parameters for p- and n-type metal oxides commonly used as scaffolding materials in hybrid-type solar cells.[76–82]

Material	Crystal structure	Mobilities (cm^2 V^{-1} s^{-1})	Lattice parameters (Å)
TiO_2	Tetragonal (A)	10	3.8, 3.8, 9.5
TiO_2	Tetragonal (R)	1	4.6, 4.6, 3.0
SnO_2	Tetragonal (R)	250	4.7, 4.7, 3.2
ZnO	Hexagonal	200	3.3, 3.3, 5.2
NiO	Cubic	0.02–0.06	4.2, 4.2, 4.2

The functionality of the electrodes is not solely limited to the interfacial contact area with the polymer donor. While complete light absorption is important for device performance, increasing the thickness of the donor and acceptor layers can result in increased recombination and difficulties in device fabrication such as pore filling and electrode deposition.[42] It is possible to engineer light coupling and interaction in the devices by carefully designing the dimensionality and orientation of the nanostructures.[43–48] A common approach to manipulate the role of light within solar cells is to selectively pattern the metal contact, allowing for amplified electric fields due to plasmonic interactions.[49,50] However, the location of the metal particles must be considered carefully to avoid creating recombination centres within the active material, while simultaneously adjusting the field.[47] Alternative approaches, which are discussed more thoroughly below, consist on the creation of optical cavities (TCOs), light in-coupling (blocking layers), and light scattering/trapping (metal oxide scaffolding). While the formation of an optical cavity and light in-coupling are highly dependent on the materials used and must be systematically modelled, the scattering and trapping of light can be easily achieved by controlling the dimensions and structuring of the metal oxide electrode.

5.3 Transparent Electrodes and Blocking Layers

As discussed previously, the ability to selectively transport electrons and holes determines the quality of a hybrid solar cell device. This anisotropic nature of the device begins at the TCO/blocking layer interface in the inverted solar cell structure employed for hybrid devices. Common TCOs contain either expensive components (ITO), have low transparencies or mobilities (FTO), or are chemically and temperature unstable (AZO, and ITO). Therefore, a large subset of work has been placed on finding suitable alternatives to these materials, which also maintain a high TCO figure of merit (Φ_{TE}) as, *e.g.*, defined by Haacke:[51]

$$\Phi_{TE} = \frac{T_{tot}^{10}}{R_S} \tag{5.4}$$

where T_{tot} is the transmission at 550 nm, normally between 400–900 nm, and R_S is the measured sheet resistance in $\Omega\,cm^{-1}$. These corresponding conductance values range between 10^{-2}–10^{-3} Ω^{-1} for the standard TCOs mentioned previously.[52] Unfortunately, the materials have ambipolar charge mobility requiring the addition of a blocking layer. Suitable replacements are required in order to achieve the goal of low cost and chemically stable materials.

More recently, there has been a bi-directional push for commercialization of TCO replacements, one in the direction of networks of 1D metallic nanostructures[53,54] and the other with 2D films and metallic sandwiches.[53,55,56] Depending on the architecture of the device, these films can be free-floating, *i.e.*, on top of the organic polymer, or must be encased in a metal oxide.

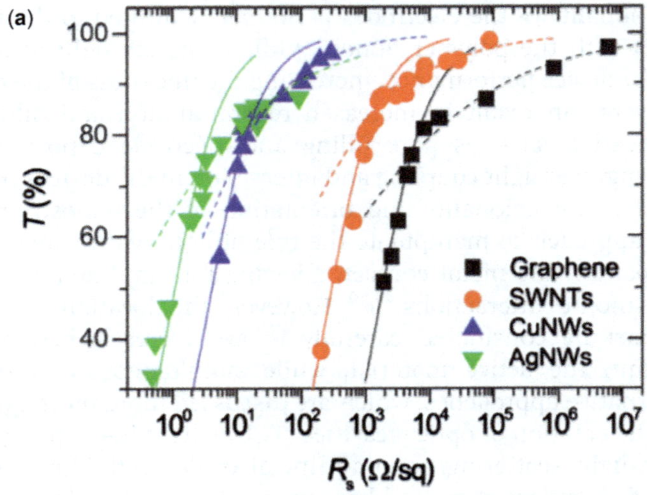

Figure 5.4 The optical transparency *vs.* sheet resistance is shown for a range of metallic 1D and 2D nanostructures. Each material is divided into two regions: the bulk regime, shown as a solid curve, and the percolation regime, plotted as a dashed line.
Reprinted with permission from De and Coleman, The effects of percolation in nanostructured transparent conductors, **36**(10), 774–781.[66] Copyright 2011 Cambridge University Press.

In addition to the price and abundance of the materials, these thin metallic films are suitable for the fabrication of flexible electrodes. The nanowire arrays, fabricated from Ag,[57] Au,[58] Cu,[59,60] Cu/Ni composites,[61,62] carbon nanotubes,[63] graphene[64] *etc.*, are deposited in both ordered and random networks. Generally, Ag is the material of choice for nanowire arrays due to its high bulk electrical conductivity compared to the other materials.[57] Figure 5.4 directly compares the electrical conductivity with the transmittance of nanowire films, exemplifying the properties of Ag as the ideal thin film material.[57] A range of transparent electrode materials are tabulated below for a direct comparison between the different structures (Table 5.2). Ordered nanowire arrays are used to produce direct percolation pathways across the surface of the substrate.[57,65] This requires less overall metal to be deposited across the surface of the substrate for metal-like properties, allowing for a high transmission across the visible range due to voids in the mesh. However, the mesh results in voids in transport regions where generated charges are not within the diffusion range of the electrode and recombine before collection can occur. The effect of these regions can be limited through the use of a highly conductive layer such as poly(3,4-ethylenedioxythiophene)-poly(styrenesulfonate) (PEDOT:PSS). Most research has tended to focus on random nanowire networks to prevent these "dead spaces", as the disorder produces much smaller voids on the surface, depending on the surface coverage and desired transmission.[57] As with ordered arrays, there are certain drawbacks with these random films. Since the random networks are deposited *via* solution

Table 5.2 Thickness (nm), transparency (%), sheet resistance ($\Omega\,cm^{-1}$), and figure of merit (Φ_{TE}) for assorted types of transparent electrodes implemented in current organic and hybrid electronics.[53,54,67–72]

Transparent electrode	Thickness (nm)	Transparency at 550 nm (%)	Sheet resistance ($\Omega\,cm^{-1}$)	Φ_{TE} ($10^{-3}\ \Omega^{-1}$)
ITO	200	86	11	20
AZO	—	86	10	22
FTO	1100	85	2	98
ITO/Ag/ITO	35/8/35	78	10	8
ITO/Au/ITO	50/10/40	72	5.6	7
ZnO/Ag/ZnO	35/12/35	75	10	6
ZnO/Cu/ZnO	30/6/30	88	100	3
MoO_3/Ag/MoO_3	2/6/16	75	6.2	9
TiO_2/Ag/TiO_2	5/12/20	78	6.7	12
TiO_2/Ag/TiO_2	30/9.5/30	86	5.7	39

processing, the films can suffer from poor junction conductivity and high surface roughness. These issues can be addressed through certain post processing techniques, such as thermal/electrical annealing, high pressures, and optical sintering, but care must be taken due to the sensitivity of the nanostructures.[57] Additionally, the wires can be encapsulated in metal oxide hosts, allowing for annealing treatments up to 750 °C without destruction of the wires.[57]

The alternative to nanowire networks is to use a 2D film encapsulated directly between two metal oxides. It is possible to fabricate thin metal films (<15 nm) with high transmittance. This can be enhanced further by carefully selecting materials with suitable refractive indices (n). One of the first reported fabrications of a metal oxide–Ag–metal oxide sandwich structure was by Lewis *et al.* in 2004, who used an ITO–Ag–ITO composite on polyethylene terephthalate (PET) as a flexible electrode.[54] Addition of the metal layer between the 35 nm thick ITO layers resulted in ductile films that were much thinner than commercially available ITO (roughly 200 nm). The results showed that a 4 nm thick Ag layer is required before a conductivity increase can be observed; below this point the metal forms individual islands that have no effect on charge transport. An optimal layer thickness of around 10 nm resulted in a sheet resistance of 10 $\Omega\,cm^{-1}$ and transmittance of 80% at 550 nm, compared to 20 $\Omega\,cm^{-1}$ and 82% for the commercial substrate. Comparison of the two substrates for OLED applications resulted in a higher current density for the sandwich layer above 5 V and an overall increase of 22% in lum W^{-1} efficiency.

The next step in this sandwich structure was to replace ITO with a benign metal oxide, such as ZnO, as first reported by Sahu *et al.*[73] and Sahu and Huang[74] in 2006. This replacement allows for the selective transport of electrons in the transparent electrode, eliminating the hole blocking layer that is required on traditional TCOs. Their studies showed a peak in transmission around 600 nm that increased proportionally with the Ag layer thickness between 8–12 nm. These results were confirmed by Han *et al.*[72] Additionally, the paper went on to explain the mechanism behind the formation of ZnO as a transparent conductor, as seen in Figure 5.5. Since the

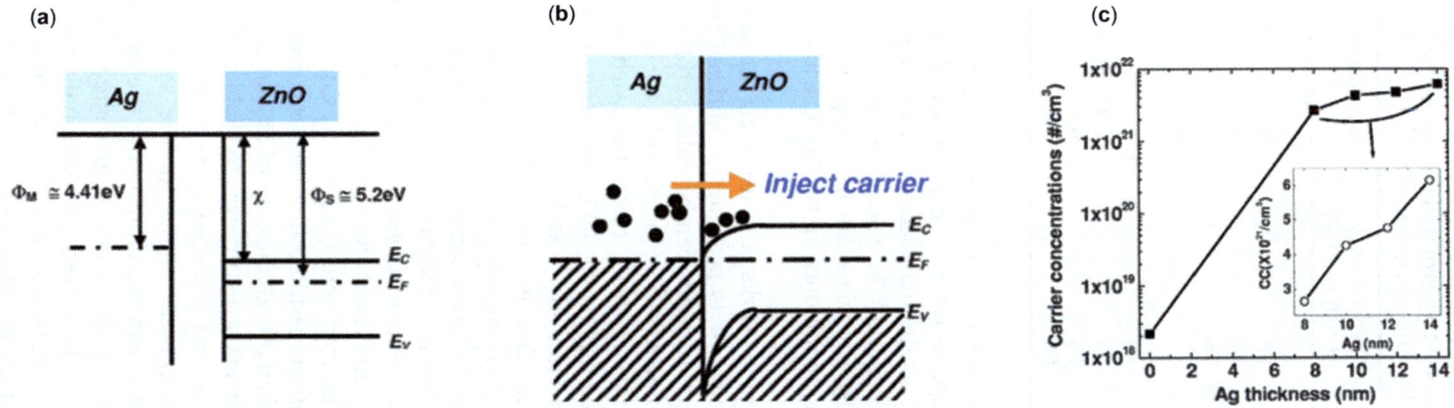

Figure 5.5 (a) The energy levels are shown for Ag and ZnO with respect to the vacuum level. (b) The Fermi level pins after the two films are contacted causing the ZnO CB and VB levels to move toward the vacuum level. The resulting bending allows electrons to be injected into the ZnO film resulting in the increased conductivity. (c) A critical thickness of 8 nm significantly increases the carrier concentration. This is due to the formation of an Ag film instead of the original island structure.
Reprinted with permission from H. Han, N. D. Theodore and T. L. Alford, Improved conductivity and mechanism of carrier transport in zinc oxide with embedded silver layer. *J. Appl. Phys.*, 2008. **103**(1), 013708.[72] Copyright 2008 AIP Publishing LLC.

work function of silver (4.4 eV) is lower than that of ZnO (5.2 eV), the Fermi level is pinned at the metal, forming a continuous band structure which allows for direct electron injection between the two materials. As such, the measured charge carrier density is shown to increase significantly with the Ag interlayer, with an overall increase of three orders of magnitude for a 14 nm thick layer. However, an opposite trend was observed in Hall mobility measurements, indicating a decrease in mobility with Ag layers below 10 nm. After ruling out other mechanisms for this phenomenon, such as grain boundary scattering, surface scattering, and phonon scattering, the authors suggested that Ag forms islands below 10 nm. As the film thickness increases, the islands coalesce into a dense film which allows the electrons to flow freely across the surface. Additionally, it is believed that coalescence is also responsible for the increase in transmission around this critical thickness. Unfortunately, no figure of merit was published for this study, although it can be estimated to be around 0.004 Ω^{-1}.

In an alternative approach, Sergeant *et al.*[71] designed a p-type transparent electrode material using molybdenum oxide (MoO_3) as the encapsulation material in order to produce a non-inverted type organic solar cell. A goal in this work was to control the deposition parameters to reduce the island threshold of the Ag film using thermal evaporation, as opposed to physical sputtering. It proved possible to limit the island growth, and a near-homogenous 6 nm thin film was achieved by cooling the substrate to -5 °C with a deposition rate of 6 Å s^{-1} on 2 nm of MoO_3. While the film was not completely conformal, the voids in the film were no larger than 20 nm with a total surface area of approximately 5.6%. Despite the small voids, the MoO_3–Ag film had a direct percolation pathway and an overall sheet resistance of 6.2 $\Omega\ cm^{-1}$, compared to 18 $\Omega\ cm^{-1}$ for the ITO reference. A top layer was not included during these measurements in order to allow the study of the relationship between thickness and the optical cavity-induced electric field. Variation of the top layer thickness from 5 to 20 nm resulted in a red shift in the transmission maximum ($\sim$80%) from roughly 400 to 475 nm, respectively. The group simulated the electric field for a bulk heterojunction device of poly[*N*-9″-hepta-decanyl-2,7-carbazole-*alt*-5,5-(4′,7′-di-2-thienyl-2′,1′,3′-benzothiadiazole) (PCDTBT) and [6,6]-phenyl-C_{71}-butyric acid methyl ester ($PC_{71}BM$). An optimum thickness of 10 nm was found to maximize the overall absorption between 500–700 nm, where the PCDTBT polymer displays the highest absorption. The effect of the engineered electric field is directly observable in the relative external quantum efficiency (EQE) value for both the simulated and experimental data, with an overall improvement of 25% centered around 650 nm, and a slightly higher total conversion efficiency than the commercial ITO substrates. While this substrate is a suitable alternative to ITO for organic devices, there are very few p-type metal oxides that are utilized for hybrid solar cells.

The electric field engineering concept was extended by members of our own group to a TiO_2–Ag–TiO_2 (TAT) transparent electrode for inverted hybrid type solar cell devices, where n-type metal oxides were employed as the scaffolding material. To test the TAT electrode, a high efficiency poly[[4,8-bis[(2-ethylhexyl)oxy]benzo[1,2-*b*:4,5-*b*′]dithiophene-2,6-diyl][3-fluoro-2-[(2-ethylhexyl)-carbonyl] thieno[3,4-*b*]thiophenediyl]] (PTB7):[6,6]-phenyl-C_{71}-butyric acid methyl ester ($PC_{71}BM$) combination was used as the initial reference bulk heterojunction and compared to standard ITO films. The fabricated TAT electrode showed an inverse relationship between the top film thickness and Φ_{TE}, with the latter varying between 0.01–0.04 Ω^{-1} (ITO $\Phi_{TE} = 0.01\ \Omega^{-1}$) with TiO_2 layers of 30 nm. While solar cells based on this substrate showed an impressive performance for the PTB7:$PC_{71}BM$ combination of 8.73% (an improvement of 16%), it also displays the capability of being used as a hybrid substrate due to its high temperature stability. In order to process hybrid solar cells, high temperature annealing is required, specifically for the sintering of mesoporous TiO_2 at 450 °C. To quantify the temperature stability, the sheet resistance was measured for the substrates and compared to that of ITO references for temperatures up to 550 °C. The ITO sheet resistance increased two-fold from 11 $\Omega\ cm^{-1}$ during this annealing process as compared to the TAT substrate, which had a maximum value of 8.4 $\Omega\ cm^{-1}$. Maintaining a low sheet resistance over this temperature range opens the way to a new class of transparent electrodes, possible for hybrid solar cells with reduced loss mechanisms.

In addition to the layered architecture for transparent electrodes, Pfadler *et al.* have recently reported the structuring of the TiO_2 blocking layer for light coupling within the solar cell.[75] Using direct laser interference patterning (DLIP), an asymmetric TiO_2 blocking layer was fabricated with controllable periods based on the laser angle of incidence, as seen in Figure 5.6. It is possible to increase the total absorption of light within an

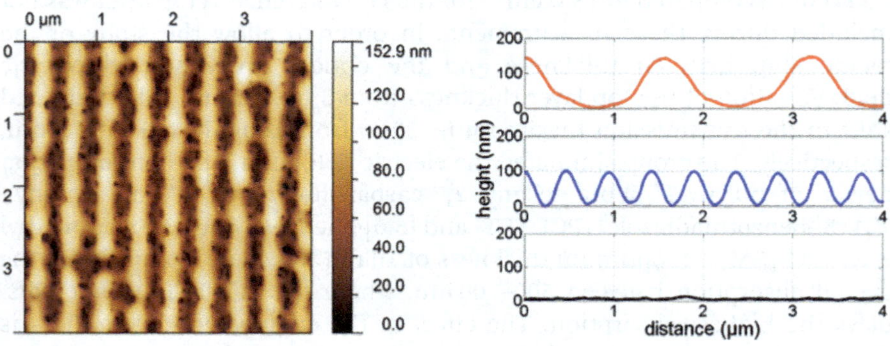

Figure 5.6 (Left) The atomic force microscopy (AFM) image of the surface structure of a TiO_2 film after a Direct Laser Interference Patterning (DLIP) treatment with a 266 nm laser is shown. (Right) The periodicity of the structuring is directly dependent on the angle of incidence, which in this case varied between no structuring (bottom), 470 nm (middle), and 1600 nm (top).
Reprinted with permission from T. Pfadler *et al.*, A comparison of light-coupling into high and low index nanostructured photovoltaic thin films. *APL Mater.*, 2015, **3**(6), 066101. Copyright 2015 AIP Publishing LLC.[75]

absorption layer through differences in refractive indices *via* selective structuring of the blocking layer. This is particularly beneficial for solid absorbers such as antimony sulfide (Sb_2S_3) and methylammonium lead halide ($CH_3NH_3PbI_3$) perovskites, where the refractive index is higher than that of TiO_2. However, for standard hybrid solar cells, where the scaffolding layer is a high surface area metal oxide film, the light in-coupling will be lost due to the small difference in refractive indices. In the case of hybrid and dye-sensitized solar cells, the metal oxide scaffolding should be nanostructured to increase the interfacial area and charge separation, as discussed in the next section.

5.4 Nanostructured Active Layers

As discussed in Section 5.3, the metal oxide plays a vital role in the operation of hybrid solar cells by selectively transporting charges to their respective electrodes. Table 5.1 lists the charge mobilities for ZnO, TiO_2, SnO_2, and NiO, highlighting the difference in physical performance of each class of material. This table also includes the different crystal structures and lattice parameters. It can be seen that many materials listed in the table, such as TiO_2 and SnO_2, have similar crystal structures and lattice parameters that are not necessarily important when used as a single component, but must be considered when creating multilayer structures where stress and strain can play a role in the conformity of the different layers. It is important to consider these issues when designing new nanostructures to take advantage of certain electronic and physical properties that may be lacking in commercially available materials. The focus of this section will be on TiO_2 nanostructures, but other metal oxides will be discussed where appropriate.

5.4.1 Physical Structuring

The system often referred to for reference materials is the high surface area mesoporous TiO_2 film that has changed little from the original Grätzel paper when introduced for DSSCs.[3] The main driving force behind the continued use of the mesoporous films is the simplicity of the fabrication process, which involves dissolving the commercial paste in ethanol and spin-coating or doctor blading the solution for layers up to 20 μm thick. There have been plenty of reports on various crystal structures and synthetic methods for the mesoporous layer, including self-assembly of the film from a diblock copolymer, hydrothermal growth of core–shell structures, chemical vapour deposition, *etc.*[83–85] While these methods have resulted in interesting film morphologies and extremely high surface areas, >400 $m^2\ g^{-1}$ compared to 45 $m^2\ g^{-1}$, the pore sizes between 5–10 nm make polymer infiltration difficult.[86] An alternative approach that has been implemented by Alvaro *et al.*[86] and Zukalová *et al.*[87] is the use of the Pluronic P123 triblock copolymer to synthesize ordered mesoporous films with periodic pores. The thickness of these layers can be controlled *via* subsequent layering and annealing, reaching thicknesses up to 1 μm. As the layer thickness is increased, the pore size distribution is reduced from around 8 nm to 6 nm. While it might be expected that the decreasing pore size would adversely affect the performance of the device, a near 2-fold increase in efficiency was observed compared to the traditional mesoporous film (2.21% to 4.04%). These ordered mesoporous films also resulted in an increase in the photocatalysis turn over frequency (TOF) when the periodic pore size was controlled. Finally, Crossland *et al.*[88] used the diblock copolymer poly(4-fluorostyrene)-*b*-poly(D,L-lactide) (PFS-*b*-PLA) to create a gyroid structure that was treated with a mild base to selectively etch the PLA block (Figure 5.7(a)). Once the PLA was etched, TiO_2 was electrochemically deposited in the void, and the remaining polymer was removed, resulting in the gyroid structure shown in Figure 5.7(b). The unit cells were simulated to be 47 nm and were measured to be 50.5 nm using grazing incidence small angle X-ray scattering (GISAXS) spectroscopy. The GISAXS patterns were consistent with nanostructures with an average crystallite diameter of 9 nm. The 4 μm structures were used to

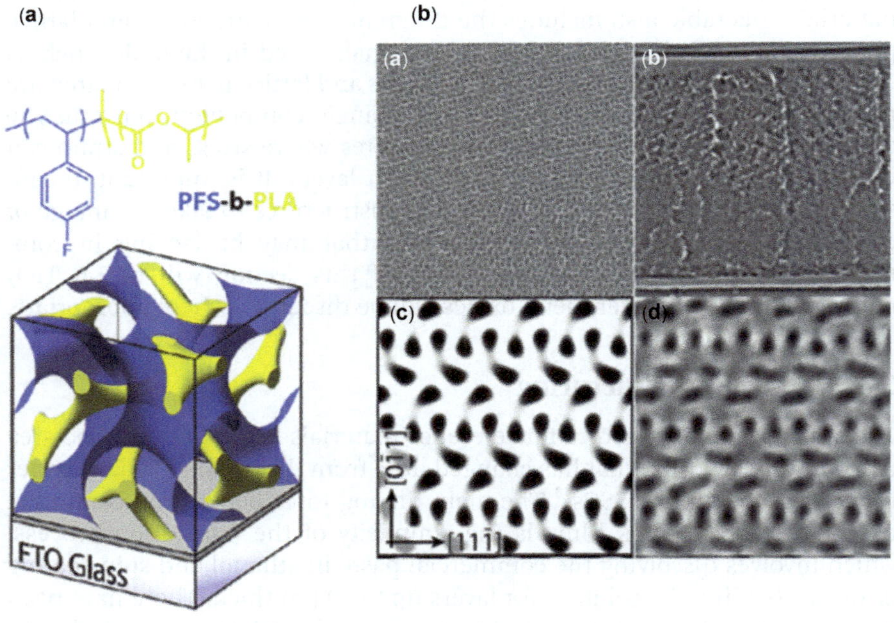

Figure 5.7 (a) A poly(4-fluorostyrene)-*b*-poly(D,L-lactide) (PFS-*b*-PLA) diblock copolymer was used to form a double gyroid blend. (b) The PLA can be selectively removed and the template can be used for the electrodeposition of TiO_2. The resulting TiO_2 structure displays impressive ordering with direct transport pathways and high surface areas that are thought to be ideal for hybrid solar cells. The resulting structure was modelled (left portion of (b)) and directly compared to the experimental results (right portion of (b)).
Reprinted with permission from E. J. W. Crossland *et al.*, A Bicontinuous Double Gyroid Hybrid Solar Cell. *Nano Lett.*, 2008. **9**(8), 2807–2812. Copyright (2008) American Chemical Society.[88]

fabricate solid state DSSCs using the indoline dye D149, resulting in conversion efficiencies of 1.7%. The device performance was attributed to the direct pathways for both the electrons and holes which are not available in the traditional mesoporous structure, but clearly more work is required to match the state-of-the-art devices.

Several drawbacks occur with mesoporous films. Specifically, grain boundaries form when the particles are sintered together, reducing the electron mobility, and the charge collection pathways consist of randomly assembled particles relying on diffusion pathways for extraction. One approach that has been suggested to avoid this is the fabrication of ordered arrays of nanotubes, thereby maintaining a large surface area while producing a direct conduction pathway for electrons and holes. A commonly used top-down approach to produce ordered arrays is the electrochemical anodization of valve metals (Figure 5.8(a)).[89] By controlling the anodization process, it is possible to produce ordered and nanostructured metal oxides (Figure 5.8(b)). Anodization is frequently used in commercial applications for the passivation of surfaces with a thick oxide layer. In commercial applications, the structuring is less of a concern than the thickness of the oxide layer and anodization can be conducted at high voltages. Anodization can be

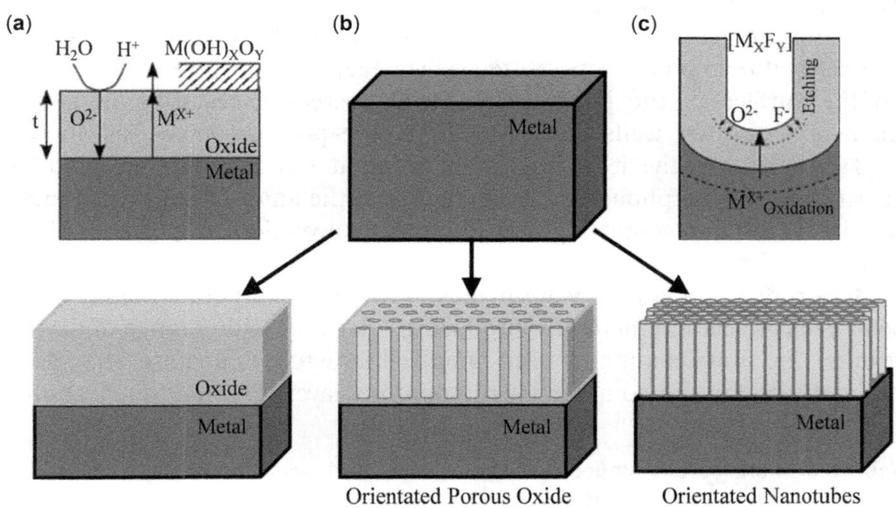

Figure 5.8 (a) The oxide formation of a valve metal film is shown. Water reacts on the surface and an oxygen ion diffuses through the oxide layer until meeting a metal ion. (b) The anodization of the valve metal can follow one of several pathways resulting in an oxide film (left), an ordered porous structure (middle), or a high surface area nanotube structure (right). (c) The nanoporous/tube anodization mechanism is shown. In this process oxygen diffuses through the surface and reacts with the metal ions while fluorine ions react with the film producing a soluble M_XF_Y ion that is shuttled away from the film.
Adapted from ref. 89 with permission from the Royal Society of Chemistry.

performed on most metals but requires optimization of both the chemistry and processing parameters. Many groups have studied the anodization of Ti to TiO_2 due to its practicality.[89–92] In order to anodize Ti, a fluoride electrolyte is required to solvate Ti^{4+} in the form of $TiF_6{}^{2-}$ (Figure 5.8(c)). During the anodization process, Ti^{4+} formed by oxidation of Ti at the metal oxide interface is transported to the surface due to an applied electric field.[89] Under the same applied field, O^{2-} ions migrate through the electrolyte toward the metal oxide interface. The tubes are formed as defects within the film that allow the build-up of charge, creating a high density of Ti^{4+} which is dissolved by reaction with the fluoride ions, producing a void at the surface where the charges can continue to build up and begin the formation of a tube structure. A randomly orientated metal oxide capping layer is formed before an ordered tubular structure begins to grow, which is visible in the time-dependent current response, as discussed thoroughly by Weickert *et al.*,[90] based on the observed variations in the current. This paper explains the steps involved in the anodization process as follows: (1) growth of an oxide barrier layer, (2) formation of a tubular structure through the barrier layer, (3) elongation of tubes within the valve metal, and (4) a final drop in current due to the complete consumption of the valve Ti. The quality of the nanotube is controlled by the processing conditions such as the electrolyte pH, temperature, anodization voltage, and quality of the valve metal (*i.e.* impurities, defects, and surface roughness). As the anodization speed increases, due to both temperature and voltage, the distribution of charges on the surface of the initial valve metal decreases, resulting in larger diameters, thinner walls, and increased pore–pore distances, as shown in Table 5.3. Finally, it is important to point out that the initial tube structures are amorphous due to the nature of the anodization process and require additional annealing steps for the conversion to polycrystalline anatase and rutile.

Many different anodized valve metals have been incorporated into hybrid solar cells as scaffold materials, producing solar cells with a range of device results. The most commonly used anodized structure is anatase TiO_2, as a replacement for the commercial mesoporous layer, as mentioned above.

Table 5.3 Pore–pore distance, pore diameter, and wall thickness for various anodization voltages and temperatures.[78]

Voltage (V)	Temperature (K)	Pore–pore distance (nm)	Pore diameter (nm)	Wall thickness (nm)
10	298	29.1 ± 4.0	17.1 ± 2.9	6.8 ± 0.9
15	298	38.3 ± 4.0	21.2 ± 2.9	10.5 ± 2.4
20	298	51.6 ± 5.5	28.3 ± 4.8	13.8 ± 1.4
25	298	63.0 ± 9.8	34.7 ± 3.4	14.7 ± 2.4
20	273	34.6 ± 8.1	20 ± 3.2	N/A
20	298	37.1 ± 5.1	23.9 ± 1.9	13.8 ± 1.4
20	308	42.8 ± 5.3	23.6 ± 2.5	11.2 ± 1.2
20	318	44.7 ± 5.5	26 ± 3.5	9.4 ± 1.2

This technique was utilized extensively in the mid-2000s by the Grimes group,[93–95] resulting in numerous papers discussing the performance in hybrid and dye sensitized solar cells. Most notable is their work with an unsymmetrical squaraine dye (SQ1) and P3HT on TiO_2 for a hybrid solar cell. The combination is noteworthy as it promotes both energy transfer and charge transfer, resulting in a truly hybrid device in which both layers generate charge as opposed to other common dyes, where the energy alignment prevents both processes from occurring simultaneously. Their structures generally consisted of nanotube diameters between 20–60 nm and lengths of 500 nm–1.2 μm. At the same time, the Schmuki group[96–98] investigated anodized TiO_2 nanotube arrays for dye sensitized solar cells with lengths up to 20 μm. The devices with 20 μm length TiO_2 nanotubes resulted in a conversion efficiency of 2.61%, with a wavelength dependent conversion (EQE) of up to 90% at the peak dye absorption of 550 nm. The study investigated the charge transport and trapping within these extremely long nanotubes in order to elucidate the origin of this extremely high EQE value, and concluded that an electron diffusion length of 100 μm was responsible for the obtained results.

However, while the nanotubes exhibit a direct charge pathway, they are limited by the dye adsorption due to a limited surface area. While most anodization is carried out at constant voltage, it is possible to vary this parameter during structure formation. Albu *et al.*[99] demonstrated the ability to control the structuring and surface area of the anodized metal oxide by selectively alternating the voltage resulting in a "bamboo-like" nanotube. By alternating between 120 and 40 V, the etch rate rapidly increases while the electrolyte diffusion is limited, requiring a sufficient incubation time in order to maintain the growth profile. An illustration of the voltage profile and subsequent anodized structure can be seen in Figure 5.9. If the cycling process is continued for too long, the chemistry changes to $TiO_2 + 6HF \rightarrow TiF_6^{2-} + 2H_2O + 2H^+$ and results in the dissolution of the metal oxide. When this process is allowed to continue for an extended period, the narrow portions of the bamboo structure are etched away and a nanolace is formed. As opposed to the nanolace structures, the "bamboo-like" films have been implemented in DSSCs by Kim *et al.*[98] By controlling the location of the tube interconnections, an increase in dye loading of 50% was observed over the smooth nanotubes (52 mM to 78 mM for Ru based N719 dye). This increase resulted in an improvement in the conversion efficiency of over 1% absolute (1.9% to 2.96%). The improvement is attributed to the high V_{OC}, J_{SC}, and FF, with an EQE maximum of 89% for the "bamboo-like" structure compared to that of 50% for the reference.

Similar problems are observed with anodized samples as with mesoporous structures, *i.e.*, because of the post-annealing process, the anatase TiO_2 is a composite of nanocrystals and contains numerous grain boundaries that inhibit electrons transport. In order to allow for direct electron transport, a single crystal 1D structure is required. A simple and elegant method to achieve these structures is the direct growth of TiO_2 on FTO substrates *via* a

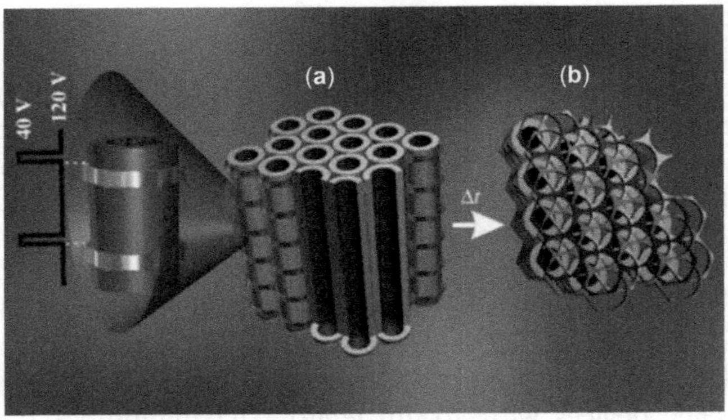

Figure 5.9 An illustration of the formation of the TiO_2 bamboo structures is shown. (a) Alternation between voltages changes the anodization rate and results in areas of increased etching. (b) These regions can be completely etched by allowing the substrate to rest in the electrolyte for the production of nanolaces.
Reprinted from S. P. Albu, D. Kim and P. Schmuki, Growth of Aligned TiO_2 Bamboo-Type Nanotubes and Highly Ordered Nanolace. *Angew. Chem., Int. Ed.*, 2008, **47**(10), 1916–1919 with permission from John Wiley and Sons. Copyright © 2008 WILEY-VCH Verlag GmbH & Co. KGaA, Weinheim.[99]

hydrothermal method.[100] The hydrothermal process is a semi-versatile method that can be used to produce nanowires of varying dimensions and crystal structures, with the typical process producing single crystal rutile nanowires.[100,101] Liu has given a comprehensive report of the growth of orientated rutile TiO_2 nanowires on FTO, focusing on the role of acidity, temperature, Ti precursor, and buffer salts on the nanowire density and dimensionality for applications in DSSCs.[100] Normal growth conditions for high density, sub-100 nm diameter wires consist of a H_2O–HCl 50 : 50 mixture with Ti *tert*-butoxide as the precursor in the presence of pure FTO substrates. These wires have since been extensively used for all types of hybrid solar cells due to their ease of fabrication and possible nanowire length up to 20 μm. Unfortunately, the performances of the resulting devices have typically trailed behind those of devices fabricated from the standard mesoporous paste, with many reports pointing to reduced dye loading as the main problem.

Recently, Wisnet *et al.*[19] have reported the formation of a "finger" structure that propagates throughout the length of the nanowire due to a high concentration of hydroxyl groups at the (001) plane that prevents the diffusion and reaction of $TiCl_4$ molecules at this interface *via* electrostatic repulsion (Figure 5.10). As the growth continues on the (111) surface, slip dislocations form to compensate for the different growth rates. It is believed that these fingers limit the radial diffusion of electrons, confining the separated charges toward the interface where recombination occurs. To counter this, annealing at 600 °C was proposed as a way to allow the crystal to

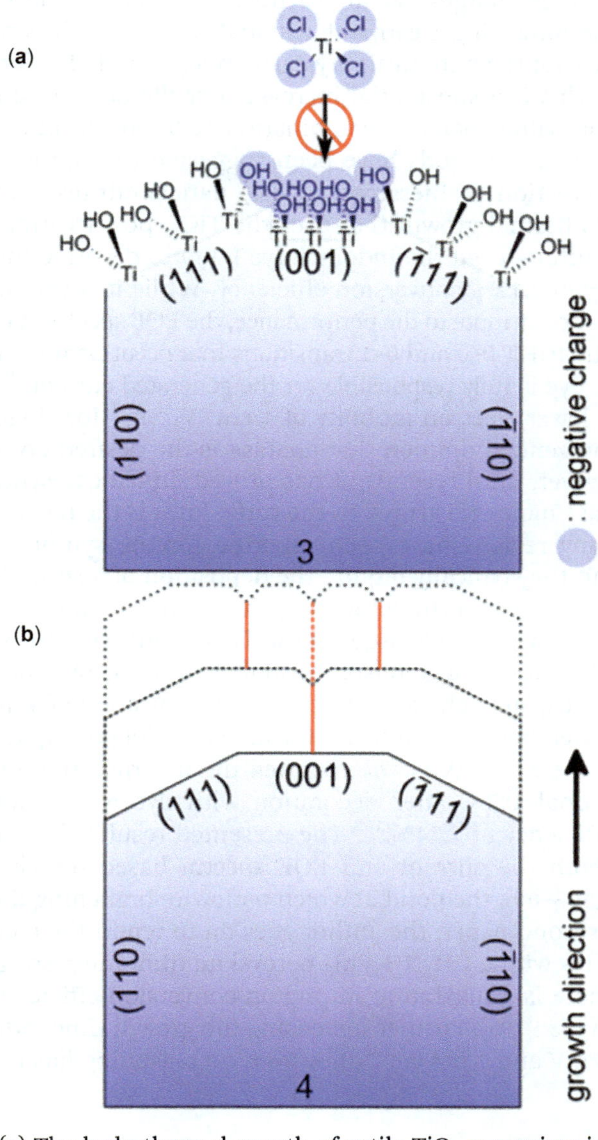

Figure 5.10 (a) The hydrothermal growth of rutile TiO_2 nanowires is facilitated by the diffusion/reaction of $TiCl_4$ molecules with surface hydroxide groups. The available surface facets change the density of the hydroxide groups, which electrostatically limit the diffusion toward the (001) plane. (b) The electrostatic repulsion will eventually be overcome by the surface energy *via* defect planes that propagate throughout the length of the nanowire resulting in "finger" structures.
Reprinted with permission from A. Wisnet *et al.*, Model for Hydrothermal Growth of Rutile Wires and the Associated Development of Defect Structures. *Cryst. Growth Des.*, 2014, Copyright (2014) American Chemical Society.[101]

reorganize so that charges can move freely throughout the whole of the nanowire. The annealing treatment resulted in an overall conversion efficiency of 2.58% for the pure rutile crystal with SQ2 and P3HT for sub-micron wire lengths. This is a substantial increase over the optimized mesoporous films with the same material combination (2.05%), despite a lower dye loading. Additional methods have been suggested to improve dye loading through the creation of hierarchical TiO_2 nanodendrites through a subsequent hydrothermal growth that uses the $TiCl_4$ post treatment as a seed layer.[20] The authors used the indoline dye D149 to decorate the nanostructure, resulting in 3.12% conversion efficiency. While it is claimed that both D149 and P3HT contribute to the performance, the EQE spectrum is missing the signature of the P3HT 0–0 and 0–1 transitions that occur between 600–700 nm, indicating the dye is fully responsible for the generated current.[39]

Due to the lower electron mobility of 1 $cm^2\ V^{-1}\ s^{-1}$ for single crystalline rutile, it is the general opinion that anatase is the desired crystal structure for TiO_2. However, until recently, it has proved difficult to synthesise semi-ordered anatase nanowire arrays, as the rutile form is the most stable crystal structure. Many recipes have been reported for the synthesis of anatase nanowires, but they typically require the deposition of a seed/blocking layer before the wires are grown, something that is not required for the rutile form.[102–104] The most widely-used synthesis generally consists of a solution of water, diethylene glycol, and isopropanol at a volume ratio of 2 : 3 : 3 and a titanium oxide oxalate dehydrate precursor concentration of 3 mM. Reaction times were varied between 1.5 to 9 h for increased length (up to 20 μm) and branching of the wires. Wu *et al.* reported the incorporation of these wires into a traditional DSSC after decoration with dye N719 with an overall conversion efficiency of 7.34%.[105] The presented results show a substantial increase in both the current and EQE spectra based on TiO_2 electrodes grown between 3–6 h, the point at which nanowire branching dominates the growth. In a second paper, the author goes on to report the combination of these nanowires with a $CH_3NH_3PbI_3$ perovskite film. Interestingly, the work shows that there is a maximum in photon conversion efficiency with wires grown for 6 h (13.97%). Further increasing the growth time reduces the V_{OC} and J_{SC} by 70 mV and 1 mA cm^{-2}, however, no reasoning has been given for this effect.

5.4.2 Nanostructures for Light Manipulation

As discussed with the metal oxide physical structuring, the amount of light that is absorbed has a direct effect on the extracted J_{SC}. The traditional approach, as discussed above, is to increase the absorbed light *via* higher dye adsorption. However, an alternative method to improve light absorption is the manipulation of the photon pathways within the active material, either due to light scattering or light trapping. Scattering of light is rather straightforward and is based on the interaction of the photon with structures with dimensions (r) approximately: $r \leq \lambda$.[106] Two regimes are defined for light scattering within the active area, either Rayleigh scattering ($r \ll \lambda$) or Mie scattering ($r \approx \lambda$).[106] For mesoporous hybrid and solid state DSSCs, the size of the nanoparticles ($d \sim 25$ nm) falls between the two regimes with scattering below 300 nm. Ferber and Luther simulated the interaction between spherical particles with 500 nm light in a liquid electrolyte to quantify the maximum light scattering efficiency, as shown in Figure 5.11(a).[107] The simulation shows that scattering is negligible for the 25 nm TiO_2 particles, as expected, with a maximum in light scattering for the 100 nm particles in the forward direction. Binary particle blends were also simulated as a function of the larger particle diameter, with an increasing volume fraction of the larger particles (Figure 5.11(b)). The binary mixtures exhibited a maximum at 5 vol% 250–300 nm particles. As the volume fraction increases, the peak loses its Gaussian shape and splits into a bimodal distribution at 20 vol%. Solutions of commercially available particles blends are typically a mixture of 20 and 450 nm and are commonly assumed to 5 vol% of the latter, as simulated in the paper.

An alternative approach to the binary particle blend suggested by Chou *et al.*[108] was to create mesoporous particles from 20 nm ZnO spheres. The nanoparticles were synthesized from a diethylene glycol solution with 0.01 mol of dissolved zinc acetate, which was refluxed at 160 °C for 24 h. The resulting precipitate was observed to consist of ~300 nm monodispersed colloids of 20 nm particles. The colloids were compared to a commercial ZnO paste of 25 nm particles, both of which were annealed at 450 °C. The amount of light scattered can be quantified using the haze factor, which is measured using a UV-Vis system with an integrating sphere to calculate ratio of photons scattered to transmitted. Unfortunately, no haze data was reported for these samples. Instead, the UV-Vis data showed an increase in light absorption between 400–500 nm, and nearly one order of magnitude increase in the generated current over the commercial ZnO particles. Similar results have been reported by numerous other groups for TiO_2 mesostructures, with the light scattering quantified by diffuse reflection. A particularly interesting study was performed by Hwang *et al.*[109] in which the mesoporous TiO_2 structures were fabricated by electrostatically spraying commercial P25 particles dispersed in ethanol directly onto FTO with a 15 V potential applied across the solution and substrate. The solution was passed through a nozzle at 30 μl min^{-1}. Again, the magnitude of light scattering arising from the 260 nm diameter particles was not quantified, but the size is close to that simulated by Ferber and Luther. It was shown that the dye loading does not change with the formation of the mesostructures, and the increase in device performance (from 6.49% to 10.57%) is attributed to two main factors: light scattering and a more direct electron collection pathway.

Incorporation of large light scattering particles has been taken one step further with the deposition of a scattering overlayer on top of transparent

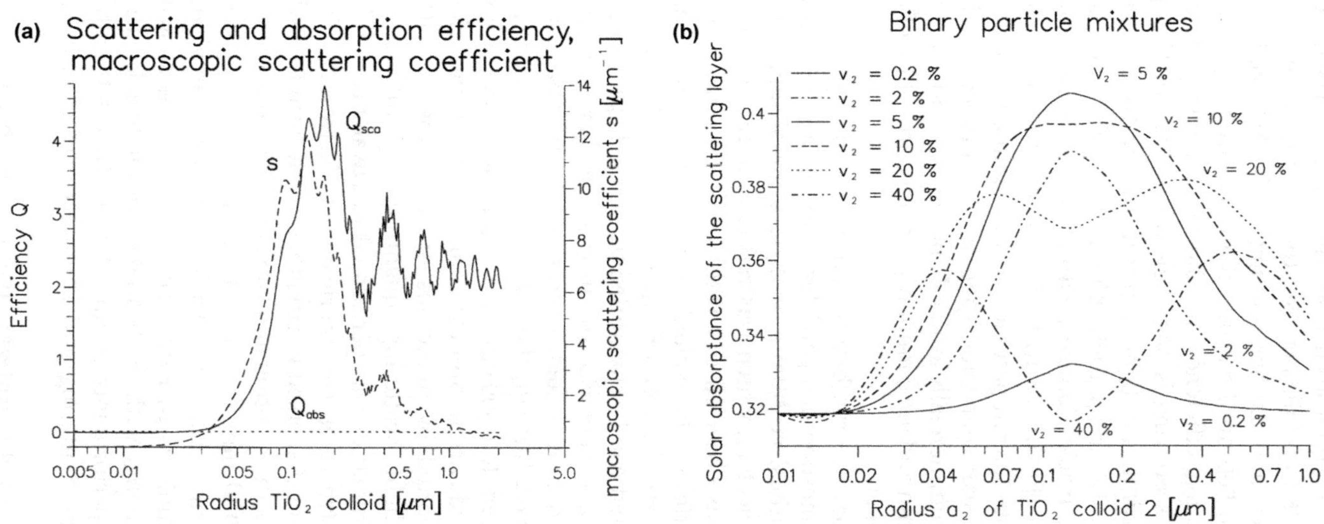

Figure 5.11 (a) The scattering efficiency (Q_{sca}), absorption efficiency (Q_{abs}), and scattering coefficient (s) are simulated for TiO_2 spheres with 500 nm light illumination. (b) Solar absorption of binary TiO_2 colloid mixtures with increasing volume fractions. The solar absorption is dependent on the size of a_2 with the radius of a_1 fixed at 10 nm.
Reprinted with permission from Ferber and Luther, Computer simulations of light scattering and absorption in dye-sensitized solar cells, *Sol. Energy Mater. Sol. Cells*, **54**, 265–275. Copyright (1998) Elsevier Limited.[107]

mesoporous films for application in DSSCs. Huang *et al.*[110] did exactly this with the combination of the previously studied 400 nm TiO_2 mesostructures mixed with 20 wt% 20 nm particles on a 6.3 μm transparent mesoporous film. The film thicknesses of the scattering layers were carefully controlled to maintain the dye absorption at approximately 1.2 $(\pm 0.2)\times 10^{-7}$ mol cm^{-2}. Additionally, diffuse reflectance and transmission data indicate that the beads act as part of the photoanode, scattering light back into the transparent layer while simultaneously absorbing light. While dye loading on the scattering overlayer is on the low side due to the large voids, larger currents were obtained due to the increased light pathway and absorption probability. The photon conversion efficiency was increased to 8.35% (13% improvement) due to the improved absorption above 600 nm caused by light scattering. These results are consistent with similar studies conducted with ZnO particles, anatase TiO_2 nanowires, rutile TiO_2 nanowires, and ZnO nanowires. In order to further increase the applicability of this approach, it is important to optimize the solar cell with a quantitative focus on light scattering in conjunction with the optimization of the direct pathway for electron transport, as discussed above for nanowire arrays.

Light-trapping nanostructures, *i.e.* photonic crystals, are commonly used due to their ability to selectively tune the wavelength trapped to the absorber material. There is a plethora of available literature on this topic, which will not be covered here. A few interesting reviews on the application of inverse opal structures and the role of these photonic crystals in solar cells can be found in ref. 111 and 112. The wavelength of light that is confined within the electrodes is dependent on the inter-particle/void separation as defined by:

$$\lambda_{\text{max}} = 1.633D(\varphi n_{\text{wall}} + (1-\varphi)n_{\text{voids}}) \tag{5.5}$$

where 1.633 is a geometrical factor, D is the average pore size, φ is the solid fraction, and n_x are the refractive indices of the wall and the voids, respectively. It is important to fully consider all aspects of the solar cell when designing the photonic crystal electrode material due to the limited surface area available and the change in the confined wavelength brought about by infiltration with the hole transport/donor material.

5.4.3 Nanostructures for Energy Landscape Engineering

In the previous two subsections, the idea of controlling the morphology for increased light absorption has been discussed. As the light absorption increases, the number of generated charges subsequently increases as well. However, when the motion of these charges is not controlled, an increase in the extracted current is not guaranteed due to recombination taking place at the interface. To reduce this recombination, one must control the position of the electrons with respect to the holes. Separation of the charges must be larger than the electron tunneling distance in order to reduce the electron emission, as in the case of Fowler–Nordheim emission:[113]

$$\frac{J}{F^2} = Ce^{-\beta/F} \tag{5.6}$$

where J is the current density, F is the field across the oxide, and C and β are defined by:

$$C = \frac{q^3 m}{16\pi^2 \hbar m_{ox} \phi_0} \tag{5.7}$$

and

$$\beta = \frac{4}{3} \frac{(2m_{ox})^{1/2}}{q\hbar} \phi_0^{3/2} \tag{5.8}$$

In these two equations, q is the elementary charge, $\hbar$ is the reduced Planck's constant, m and m_{ox} are the electron mass in free space and the metal oxide, and ϕ_0 is the barrier height.[113] From eqn (5.6)–(5.8), it is seen that the tunnelling distance is dependent on the metal oxide used at the interface, but typically it is a few nanometers, with the probability decreasing as the separation increases. While the theory for hybrid solar cell interfaces will differ from that of the field emission devices considered above, the general principle of increasing the separation distance between charges is germane. Additionally, this physical phenomenon can be achieved easily with correctly designed nanoarchitectures, in which a cascade of energy levels directs the electrons away from the interface.

The idea of using a core–shell structure to form a cascading conduction band was first proposed by Law *et al.* to compare ZnO–Al_2O_3 and ZnO–TiO_2 nanoarchitectures.[114] The paper discusses the thickness of both the Al_2O_3 and TiO_2 shell layers on the device performance using dye N719 and the I_3/I^- electrolyte. It was found that alumina layers up to 2.2 nm slightly increased the V_{OC} of the cell but resulted in a drastic decrease in J_{SC} as a consequence of poor charge injection. These results are consistent with the Al_2O_3 layer acting as a tunnelling barrier inhibiting recombination. It was noted that an increase in V_{OC} is only obtained when the photoinjected charge density is greater than the recombination, after which point it decreases. The TiO_2 shell layer resulted in a lower fill factor and a roughly 2.5 mA cm^{-2} decrease in current density with shell thicknesses below 5 nm, but an increased V_{OC} by more than 100 mV. Further deposition of TiO_2 between 10–35 nm recovered the J_{SC} and FF and further increased the V_{OC} by up to 300 mV. An optimum device performance of 2.1% conversion efficiency was achieved after deposition of a 21 nm TiO_2 shell with a V_{OC} of 0.77 V, a J_{SC} of 4.5 mA cm^{-2}, and a FF around 60%. The decrease in current density for thinner TiO_2 shells was attributed to dye loading since the N719 dye is slightly acidic (etching ZnO) and binds poorly with amorphous TiO_2 (below 4 nm). Once the TiO_2 is crystallized, the dye can readily attach itself, which

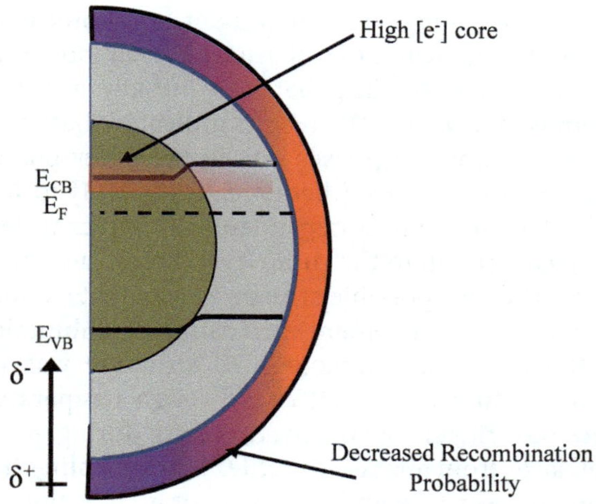

Figure 5.12 The general energy level alignment is shown for a cascading conduction band system. The core material has a lower CB energy than the shell layer. When the two are in contact, the energy levels pin to the Fermi level which produces a natural dipole across the interface which forces the electrons toward the core. The low concentration of electrons at the oxide-organic interface decreases the overall recombination probability.

can be seen *via* UV-Vis, with an ultimate increase in J_{SC} over the ZnO reference cell. The core–shell structure produces a cascading energy landscape (Figure 5.12) with the higher TiO_2 conduction band producing an internal dipole moment that forces the electrons away from the interface and reduces recombination. This hypothesis was initially tested by depositing a thin TiO_2 blocking layer below the ZnO nanowires, resulting in a decrease in device performance which was linked to the higher conduction band of TiO_2 preventing injection into the TCO layer.

Other groups have investigated the cascading conduction band strategy with metal oxide–TiO_2 core–shell structures with similar improvements in device performances.[115] However, as discussed previously, the combination of several metal oxides begins to become problematic when matching crystal structures are not available and stress builds up in the shell layer. The ideal combination would consist of a single metal oxide material that has multiple conduction band levels. One of the simplest combinations that meets this criterion is that of anatase–rutile TiO_2. Combining these two crystal structures still provides some difficulties due to lattice mismatch (18–21%) and the alignment of the band structures. Pfeifer *et al.* recently discussed the band alignment of both crystal morphologies with respect to RuO_2 and pointed out that the rutile form has been reported to have a similar valance band positions but a 200 meV smaller band gap.[116] In this work, photoemission spectroscopy (XPS) and density functional theory (DFT) showed that the two morphologies are in fact staggered with a 700 meV difference in the valence band. Additionally, Scanlon *et al.* probed the pinning of the two crystals to elucidate the energy level alignment for photocatalysis.[117] Again, XPS and quantum-mechanical/molecular-mechanical modelling was used to characterize a rutile–anatase film that was initially grown on NaCl and subsequently removed in water. The results indicated that the electrons flow from the rutile to the anatase phases due to the latter possessing a lower conduction band of $\Delta E = 0.3$ eV (theoretical, QM/MM) to 0.22 eV (experimental, XPS).[117] However, this is dependent on the Fermi level pinning to the anatase level as opposed to the rutile level. From these two conclusions, it would seem that the best possible structure is an anatase core with a rutile shell, with the opposite arrangement increasing recombination.

In practice, the rutile–anatase core–shell structure was found to be a suitable nanoarchitecture by Wisnet *et al.*[19] High temperature processing not only removes the "finger" structures that form during growth, it can also convert the shell layer from anatase to rutile. Two annealing treatments were investigated with respect to standard processing conditions. Annealing at 600 °C directly after wire growth produced a rutile–anatase inorganic interface, while annealing after $TiCl_4$ treatment resulted in a higher surface area rutile nanowire array. The electronic and physical structuring shown in Figure 5.13 produced higher efficiencies (2.71%) with an anatase shell, in spite of a decrease in both photoluminescence and transient photovoltage decay lifetimes. This phenomenon was explained by an increase in defect states at the inorganic interface which act as recombination centres under open circuit conditions. When the device is operated at short circuit, the charges are rapidly extracted through the high mobility core, and so a decrease in surface trap states does not play a significant role. Finally, the cascading conduction band was shown to assist in the transfer of charges from the P3HT layer to the TiO_2, leading to a further increase in the conversion efficiency.

In some cases, high temperature annealing is not always feasible, *e.g.* when using ITO as a substrate, and lower temperature alternatives must be developed. This was the case in the study performed by Dorman *et al.*, in which Sn^{4+} was doped into the TiO_2 nanowire array.[118] The conduction band level of SnO_2 has been reported to be 500 meV lower in energy than that of TiO_2, and the electron mobility in SnO_2 was 250 times higher.[41] The doping of TiO_2 has shown increased conductivity for applications in photoanodes but had not previously been used in solar cells. The study used the squaraine dye SQ2 in combination with P3HT in order to extract charges from both light absorbing layers. A thick anatase layer was once again deposited using a high concentration $TiCl_4$ treatment, which resulted in a full shell layer with thickness between 1–5 nm and a lower overall dye absorption. While the Sn^{4+} doping was below detectable limits (<0.5 at.%), an increase in carrier density was measured under dark conditions with impedance spectroscopy, confirming its presence.[119] A significant performance increase of 33% was observed in device measurements due to the formation of a cascading conduction band. Interestingly, the increase in J_{SC}, which is responsible for the improved performance was attributed to increases in both dye and polymer charge injection, and the formation of charges in the TiO_2 region

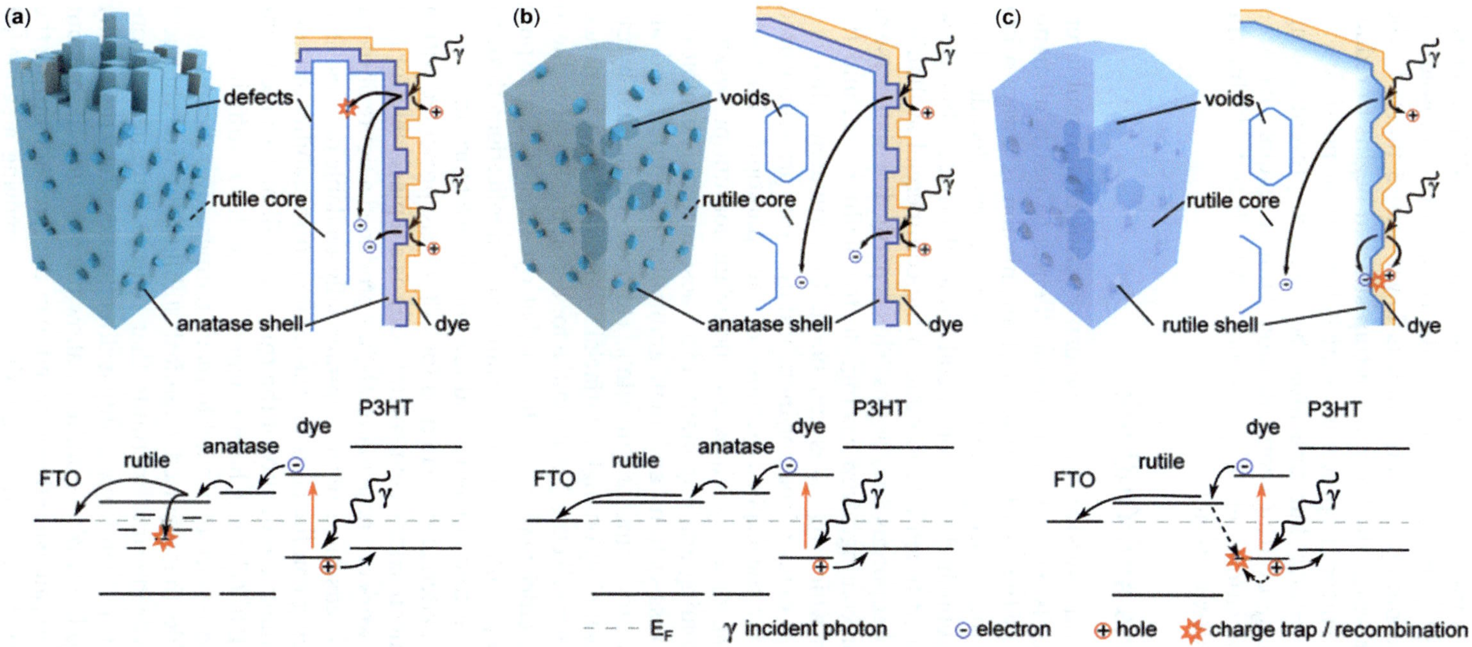

Figure 5.13 The morphology and energy level alignments are shown for hydrothermally grown rutile TiO_2 nanowires with different annealing treatments. (a) No high temperature annealing was performed, maintaining the finger structure that limits the radial electron mobility and increases recombination through defects. (b) A 600 °C anneal is performed before the $TiCl_4$ treatment for the formation of a rutile–anatase core–shell structure. This structure has the beneficial cascading conduction band without the defects produced with the fingers. (c) The 600 °C anneal is performed after fully processing the nanowires eliminating the anatase shell and the finger structures.
Reprinted from A. Wisnet *et al.*, Defeating Loss Mechanisms in 1D TiO-Based Hybrid Solar Cells, *Adv. Funct. Mater.*, 2015, **25**(17), 2601–2608 with permission from John Wiley and Sons. Copyright © 2015 WILEY-VCH Verlag GmbH & Co. KGaA, Weinheim.[19]

that are not present without the dopant (apparent from IPCE spectra). Measured transient photovoltage and impedance lifetimes showed enhancements when the doping concentration is lower than 5 mol% in the growth solution, after which the Sn^{4+} acts a trap site and begins to reduce the charge extraction. Further improvements are believed to be possible with high temperature annealing, but care must be taken to prevent the agglomeration of the dopant during reorganization. Similar results have been shown with various transition metal dopants such as W and Ta,[120,121] but the exact energy level landscape is not yet clear.

5.5 Conclusions and Outlook

This chapter has given an overview on the nanostructuring of different layers of hybrid type solar cells. One of the common themes to nearly all nanostructuring studies was that light absorption/transmission is always a key component affecting the device performance. Over the past decade, there has been plenty of published work on metal oxide electrodes, but these projects have not gained much traction due to the commercial availability and performance of mesoporous TiO_2 pastes. This is partly due to the ease of processability *via* spin-coating, drop casting, doctor blading, *etc.*, and the high surface area on which the dyes can bond. Unfortunately, this is a double-edged sword, as many of the polymers and dyes that are synthesized have been designed for the explicit purpose of incorporation with these materials. With new absorbers, such as the methylammonium perovskites, the structuring is becoming less important since these materials exhibit impressive absorption cross-sections and high ambipolar mobilities, and are amenable to solution processing of thin films. However, traditional hybrid type solar cells still rely mainly on high interfacial areas for their performance. At the moment, this solar cell class has potential for decorative surfaces due to its colour tuning and efficiency. It is believed that dye-sensitized and pure hybrid solar cells have potential for broader commercial applications through correct design of the electrode material and low cost device fabrication.

In the third section, alternative materials for TCOs were presented with a focus on the layered metal oxide–metal–metal oxide as a replacement for ITO and FTO. The current work on these materials has produced a figure or merit (Φ_{TE}) that meets or exceeds those of the commercially available devices. One area where work is being conducted is on the temperature stability of these films in order to maintain the quality of the thin metal film, with a current temperature limit of 450 °C having been reached. While this is sufficient for mesoporous films, more work is needed to reach temperatures greater than 600 °C and to provide the required chemical stability for wet chemical processes such as hydrothermal methods that require extreme pH conditions (pH < 1 or > 12). Chemical stability is directly affected by the quality of the metal oxide top layer, where pinholes allow the diffusion of ions but not long chain polymers, allowing the destruction of the electrode during the growth of the nanostructure. At the same time, by employing a sandwich structure, it is possible to localize the electric field within the absorbing material for maximum current generation, which is not possible with mass-produced ITO. Furthermore, combination of these optical cavities with directly patterned blocking layers offers the potential to produce unique light scattering and trapping within the device, although careful modelling is necessary for successful implementation.

There are many possibilities for improving the scaffolding metal oxide materials, and only a few highlights were mentioned in the fourth section. Many electronic and optical fields are working with similar TiO_2 based nanostructures, which require higher mobilities and wavelength-dependent responses. In the opinion of the authors, the large variation in electron and hole mobilities can be overcome with careful engineering of the structure, *i.e.* a cascading conduction band, which limits the flow of electrons toward the interface. Recently, it was shown that the mobilities of high performance devices do not have to be equal, as seen with perovskite solar cells, in which the lower mobility 2,2′,7,7′-tetrakis-(*N*,*N*-di-4-methoxyphenylamino)-9,9′-spirobifluorene (spiro-OMeTAD) is the hole transport medium of choice (10^{-3} cm^2 V^{-1} s^{-1}). There are many cases of inorganic titanium-based perovskites with mobilities between 20–30 cm^2 V^{-1} s^{-1}, *e.g.* $La_xSr_{1-x}TiO_3$.[122] These electrical improvements are also obvious in TiS_2, which is used frequently as the cathode material in Li^+ batteries.[123] Moreover, nanostructuring of these new materials in order to increase light scattering and optimize the energy landscape will allow better charge transport, increased light absorption, and reduced recombination, all of which are vital for high efficiency devices. Additionally, single crystal structures are ideal as they allow for the direct collection and rapid extraction of charges. Due to the multitude of materials and synthetic approaches, there is no distinct pathway to follow. While some may view this as unfortunate, benefits can also be seen in the fact that it allows a wide range of disciplines to tackle the task of high surface area, single crystal, and engineered conduction band metal oxide electrodes. Finally, while it is outside the scope of this chapter, attention must also be paid to the electronic coupling between the inorganic and organic materials which are involved in charge transport across the interfaces.

Acknowledgements

The authors would like to thank the German Research Foundation and the Alexander von Humboldt Foundation for their generous funding. We acknowledge the fruitful discussions with current and former group member of the "Hybrid Nanostructures" group at the University of Konstanz.

References

1. M. Matsumura, Y. Nomura and H. Tsubomura, *Bull. Chem. Soc. Jpn.*, 1977, **50**, 2533–2537.
2. H. Tsubomura, M. Matsumura, Y. Nomura and T. Amamiya, *Nature*, 1976, **261**, 5559.

3. B. O'Regan and M. Grätzel, *Nature*, 1991, **353**, 737–740.
4. J. Bisquert, D. Cahen, G. Hodes, S. Rühle and A. Zaban, *J. Phys. Chem. B*, 2004, **108**, 8106–8118.
5. Y. Ooyama and Y. Harima, *Eur. J. Org. Chem.*, 2009, **2009**, 2903–2934.
6. Z. Ning, O. Voznyy, J. Pan, S. Hoogland, V. Adinolfi, J. Xu, M. Li, A. R. Kirmani, J.-P. Sun, J. Minor, K. W. Kemp, H. Dong, L. Rollny, A. Labelle, G. Carey, B. Sutherland, I. Hill, A. Amassian, H. Liu, J. Tang, O. M. Bakr and E. H. Sargent, *Nat. Mater.*, 2014, **13**, 822–828.
7. B. Ehrler, K. P. Musselman, M. L. Böhm, F. S. Morgenstern, Y. Vaynzof, B. J. Walker, J. L. MacManus-Driscoll and N. C. Greenham, *ACS Nano*, 2013, **7**, 4210–4220.
8. T. Englman, E. Terkieltaub and L. Etgar, *.J. Phys. Chem. C*, 2015, **119**, 12904–12909.
9. E. Zimmermann, T. Pfadler, J. Kalb, J. A. Dorman, D. Sommer, G. Hahn, J. Weickert and L. Schmidt-Mende, *Adv. Sci.*, 2015, **2**, 1500059.
10. W.-Q. Wu, F. Huang, D. Chen, Y.-B. Cheng and R. A. Caruso, *Adv. Funct. Mater.*, 2015, **25**, 3264–3272.
11. M. He, D. Zheng, M. Wang, C. Lin and Z. Lin, *J. Mater. Chem. A*, 2014, **2**, 5994–6003.
12. S. A. Bretschneider, J. Weickert, J. A. Dorman and L. Schmidt-Mende, *APL Mater.*, 2014, **2**, 040701.
13. L. Forro, O. Chauvet, D. Emin, L. Zuppiroli, H. Berger and F. Lévy, *J. Appl. Phys.*, 1994, **75**, 633–635.
14. F. Sauvage, D. Chen, P. Comte, F. Huang, L.-P. Heiniger, Y.-B. Cheng, R. A. Caruso and M. Graetzel, *ACS Nano*, 2010, **4**, 4420–4425.
15. Y. J. Kim, M. H. Lee, H. J. Kim, G. Lim, Y. S. Choi, N.-G. Park, K. Kim and W. I. Lee, *Adv. Mater.*, 2009, **21**, 3668–3673.
16. N. R. E. Laboratory, *Journal*, 2015.
17. N. J. Jeon, J. H. Noh, Y. C. Kim, W. S. Yang, S. Ryu and S. I. Seok, *Nat. Mater.*, 2014, **13**, 897–903.
18. A. Yella, H. W. Lee, H. N. Tsao, C. Yi, A. K. Chandiran, M. K. Nazeeruddin, E. W. G. Diau, C. Y. Yeh, S. M. Zakeeruddin and M. Grätzel, *Science*, 2011, **334**, 629–634.
19. A. Wisnet, K. Bader, S. B. Betzler, M. Handloser, P. Ehrenreich, T. Pfadler, J. Weickert, A. Hartschuh, L. Schmidt-Mende, C. Scheu and J. A. Dorman, *Adv. Funct. Mater.*, 2015, **25**, 2601–2608.
20. W.-P. Liao, S.-C. Hsu, W.-H. Lin and J.-J. Wu, *J. Phys. Chem. C*, 2012, **116**, 15938–15945.
21. P. Würfel, *Physics of Solar Cells – From Basic Principles to Advanced Concepts*, Wiley-VCH, 2nd edn, 2009.
22. J. Nelson, *The Physics of Solar Cells*, Imperial College Press, 2003.
23. C. K. Renshaw and S. R. Forrest, *Phys. Rev. B: Condens. Matter Mater. Phys.*, 2014, **90**, 045302.
24. A. Panda, C. K. Renshaw, A. Oskooi, K. Lee and S. R. Forrest, *Phys. Rev. B: Condens. Matter Mater. Phys.*, 2014, **90**, 045303.
25. G. Wu, Z. Li, X. Zhang and G. Lu, *J. Phys. Chem. Lett.*, 2014, **5**, 2649–2656.

26. A. R. S. Kandada, S. Guarnera, F. Tassone, G. Lanzani and A. Petrozza, *Adv. Funct. Mater.*, 2014, **24**, 3094–3099.
27. M. T. Greiner, M. G. Helander, W.-M. Tang, Z.-B. Wang, J. Qiu and Z.-H. Lu, *Nat. Mater.*, 2012, **11**, 76–81.
28. R. J. Davis, M. T. Lloyd, S. R. Ferreira, M. J. Bruzek, S. E. Watkins, L. Lindell, P. Sehati, M. Fahlman, J. E. Anthony and J. W. P. Hsu, *J. Mater. Chem.*, 2011, **21**, 1721–1729.
29. J. Weickert, R. B. Dunbar, H. C. Hesse, W. Wiedemann and L. Schmidt-Mende, *Adv. Mater.*, 2011, **23**, 1810.
30. M. Campoy-Quiles, T. Ferenczi, T. Agostinelli, P. G. Etchegoin, Y. Kim, T. D. Anthopoulos, P. N. Stavrinou, D. D. C. Bradley and J. Nelson, *Nat. Mater.*, 2008, **7**, 158–164.
31. F. C. Jamieson, E. B. Domingo, T. McCarthy-Ward, M. Heeney, N. Stingelin and J. R. Durrant, *Chem. Sci.*, 2012, **3**, 485–492.
32. X. Wang, Z. Feng, J. Shi, G. Jia, S. Shen, J. Zhou and C. Li, *Phys. Chem. Chem. Phys.*, 2010, **12**, 7083–7090.
33. M. Hallermann, E. D. Como, J. Feldmann, M. Izquierdo, S. Filippone, N. Martin, S. Juchter and E. v. Hauff, *Appl. Phys. Lett.*, 2010, **97**, 023301.
34. W. J. E. Beek, M. M. Wienk and R. A. J. Janssen, *Adv. Mater.*, 2004, **16**, 1009.
35. R. A. Street, M. Schoendorf, A. Roy and J. H. Lee, *Phys. Rev. B: Condens. Matter Mater. Phys.*, 2010, **81**, 205307.
36. L. Koster, V. Mihailetchi, R. Ramaker and P. Blom, *Appl. Phys. Lett.*, 2005, **86**, 123509.
37. J. van de Lagemaat, N. G. Park and A. J. Frank, *J. Phys. Chem. B*, 2000, **104**, 2044–2052.
38. T. M. Burke and M. D. McGehee, *Adv. Mater.*, 2014, **26**, 1923–1928.
39. P. Ehrenreich, T. Pfadler, F. Paquin, L.-I. Dion-Bertrand, O. Paré-Labrosse, C. Silva, J. Weickert and L. Schmidt-Mende, *Phys. Rev. B: Condens. Matter Mater. Phys.*, 2015, **91**, 035304.
40. L. J. A. Koster, V. D. Mihailetchi and P. W. M. Blom, *Appl. Phys. Lett.*, 2006, **88**, 052104.
41. P. Tiwana, P. Docampo, M. B. Johnston, H. J. Snaith and L. M. Herz, *ACS Nano*, 2011, **5**, 5158–5166.
42. R. Ferragut, S. Aghion, F. Moia, M. Binda, E. V. Canesi, G. Lanzani and A. Petrozza, *J. Phys.: Conf. Ser.*, 2013, **443**, 012051.
43. J. S. King, E. Graugnard and C. J. Summers, *Adv. Mater.*, 2005, **17**, 1010–1013.
44. C.-Y. Cho and J. H. Moon, *Langmuir*, 2012, **28**, 9372–9377.
45. P. Sun, X. Zhang, C. Wang, Y. Wei, L. Wang and Y. Liu, *J. Mater. Chem. A*, 2013, **1**, 3309–3314.
46. K. Zhu, N. R. Neale, A. Miedaner and A. J. Frank, *Nano Lett.*, 2006, 7, 69–74.
47. H. A. Atwater and A. Polman, *Nat. Mater.*, 2010, **9**, 205–213.
48. S. Pillai and M. A. Green, *Sol. Energy Mater. Sol. Cells*, 2010, **94**, 1481–1486.
49. R. B. Dunbar, T. Pfadler, N. N. Lal, J. J. Baumberg and L. Schmidt-Mende, *Nanotechnology*, 2012, **23**, 385202.

50. R. B. Dunbar, H. C. Hesse, D. S. Lembke and L. Schmidt-Mende, *Phys. Rev. B: Condens. Matter Mater. Phys.*, 2012, **85**, 035301.
51. G. Haacke, *J. Appl. Phys.*, 1976, **47**, 4086–4089.
52. C. Guillén and J. Herrero, *Thin Solid Films*, 2011, **520**, 1–17.
53. A. Dhar and T. L. Alford, *APL Mater.*, 2013, **1**, 012102.
54. J. Lewis, S. Grego, B. Chalamala, E. Vick and D. Temple, *Appl. Phys. Lett.*, 2004, **85**, 3450–3452.
55. S. Das, P. Sudhagar, Y. S. Kang and W. Choi, *J. Mater. Res.*, 2014, **29**, 299–319.
56. J. Wu, H. A. Becerril, Z. Bao, Z. Liu, Y. Chen and P. Peumans, *Appl. Phys. Lett.*, 2008, **92**, 263302.
57. L. Daniel, G. Gaël, M. Céline, C. Caroline, B. Daniel and S. Jean-Pierre, *Nanotechnology*, 2013, **24**, 452001.
58. P. E. Lyons, S. De, J. Elias, M. Schamel, L. Philippe, A. T. Bellew, J. J. Boland and J. N. Coleman, *J. Phys. Chem. Lett.*, 2011, 2, 3058–3062.
59. A. R. Rathmell, S. M. Bergin, Y. L. Hua, Z. Y. Li and B. J. Wiley, *Adv. Mater.*, 2010, **22**, 3558–3563.
60. A. R. Rathmell and B. J. Wiley, *Adv. Mater.*, 2011, **23**, 4798–4803.
61. A. R. Rathmell, M. Nguyen, M. Chi and B. J. Wiley, *Nano Lett.*, 2012, **12**, 3193–3199.
62. D. Ghosh, R. Betancur, T. Chen, V. Pruneri and J. Martorell, *Sol. Energy Mater. Sol. Cells*, 2011, **95**, 1228–1231.
63. K. Sierros, D. Hecht, D. Banerjee, N. Morris, L. Hu, G. Irvin, R. Lee and D. Cairns, *Thin Solid Films*, 2010, **518**, 6977–6983.
64. D. S. Hecht, L. Hu and G. Irvin, *Adv. Mater.*, 2011, **23**, 1482–1513.
65. D. S. Ghosh, T. L. Chen and V. Pruneri, *Appl. Phys. Lett.*, 2010, **96**, 041109.
66. S. De and J. N. Coleman, *MRS Bull.*, 2011, **36**, 774–781.
67. A. Rakhshani, Y. Makdisi and H. Ramazaniyan, *J. Appl. Phys.*, 1998, **83**, 1049–1057.
68. S. Flickyngerova, M. Netrvalova, L. Prusakova, I. Novotny, P. Sutta and V. Tvarozek, *Vacuum*, 2009, **84**, 215–217.
69. D. Kim, *Appl. Surf. Sci.*, 2010, **256**, 1774–1777.
70. K. Sivaramakrishnan and T. Alford, *Appl. Phys. Lett.*, 2009, **94**, 052104.
71. N. P. Sergeant, A. Hadipour, B. Niesen, D. Cheyns, P. Heremans, P. Peumans and B. P. Rand, *Adv. Mater.*, 2012, **24**, 728–732.
72. H. Han, N. D. Theodore and T. L. Alford, *J. Appl. Phys.*, 2008, **103**, 013708.
73. D. Sahu, S.-Y. Lin and J.-L. Huang, *Appl. Surf. Sci.*, 2006, **252**, 7509–7514.
74. D. Sahu and J.-L. Huang, *Thin Solid Films*, 2006, **515**, 876–879.
75. T. Pfadler, M. Stärk, E. Zimmermann, M. Putnik, J. Boneberg, J. Weickert and L. Schmidt-Mende, *APL Mater.*, 2015, **3**, 066101.
76. S. Sasaki, K. Fujino and Y. Takéuchi, *Proc. Jpn. Acad., Ser. B*, 1979, **55**, 43–48.
77. H. F. McMurdie, M. C. Morris, E. H. Evans, B. Paretzkin, W. Wong-Ng, L. Ettlinger and C. R. Hubbard, *Powder Diffr.*, 1986, **1**, 64–77.

78. W. H. Baur and A. A. Khan, *Acta Crystallogr., Sect. B: Struct. Crystallogr. Cryst. Chem.*, 1971, **27**, 2133–2139.
79. T. Mitsuhashi and O. Kleppa, *J. Am. Ceram. Soc.*, 1979, **62**, 356–357.
80. D. C. Look, D. C. Reynolds, J. Sizelove, R. Jones, C. W. Litton, G. Cantwell and W. Harsch, *Solid State Commun.*, 1998, **105**, 399–401.
81. Z. Jarzebski and J. Marton, *J. Electrochem. Soc.*, 1976, **123**, 299C–310C.
82. E. Yagi, R. R. Hasiguti and M. Aono, *Phys. Rev. B: Condens. Matter Mater. Phys.*, 1996, **54**, 7945.
83. B. Smarsly, D. Grosso, T. Brezesinski, N. Pinna, C. Boissière, M. Antonietti and C. Sanchez, *Chem. Mater.*, 2004, **16**, 2948–2952.
84. H. Yoshitake, T. Sugihara and T. Tatsumi, *Chem. Mater.*, 2002, **14**, 1023–1029.
85. Y. Cui, L. Liu, B. Li, X. Zhou and N. Xu, *J. Phys. Chem. C*, 2010, **114**, 2434–2439.
86. M. Alvaro, C. Aprile, M. Benitez, E. Carbonell and H. García, *J. Phys. Chem. B*, 2006, **110**, 6661–6665.
87. M. Zukalová, A. Zukal, L. Kavan, M. K. Nazeeruddin, P. Liska and M. Grätzel, *Nano Lett.*, 2005, **5**, 1789–1792.
88. E. J. W. Crossland, M. Kamperman, M. Nedelcu, C. Ducati, U. Wiesner, D. M. Smilgies, G. E. S. Toombes, M. A. Hillmyer, S. Ludwigs, U. Steiner and H. J. Snaith, *Nano Lett.*, 2008, **9**, 2807–2812.
89. A. Ghicov and P. Schmuki, *Chem. Commun.*, 2009, 2791–2808, DOI: 10.1039/b822726h.
90. J. Weickert, C. Palumbiny, M. Nedelcu, T. Bein and L. Schmidt-Mende, *Chem. Mater.*, 2011, **23**, 155–162.
91. R. Salazar, M. Altomare, K. Lee, J. Tripathy, R. Kirchgeorg, N. T. Nguyen, M. Mokhtar, A. Alshehri, S. A. Al-Thabaiti and P. Schmuki, *ChemElectroChem*, 2015, **2**, 824–828.
92. G. K. Mor, S. Kim, M. Paulose, O. K. Varghese, K. Shankar, J. Basham and C. A. Grimes, *Nano Lett.*, 2009, **9**, 4250.
93. K. Shankar, J. Bandara, M. Paulose, H. Wietasch, O. K. Varghese, G. K. Mor, T. J. LaTempa, M. Thelakkat and C. A. Grimes, *Nano Lett.*, 2008, **8**, 1654–1659.
94. K. Shankar, G. K. Mor, H. E. Prakasam, O. K. Varghese and C. A. Grimes, *Langmuir*, 2007, **23**, 12445–12449.
95. G. K. Mor, K. Shankar, M. Paulose, O. K. Varghese and C. A. Grimes, *Nano Lett.*, 2006, **6**, 215–218.
96. J. R. Jennings, A. Ghicov, L. M. Peter, P. Schmuki and A. B. Walker, *J. Am. Chem. Soc.*, 2008, **130**, 13364–13372.
97. R. P. Vitiello, J. M. Macak, A. Ghicov, H. Tsuchiya, L. F. P. Dick and P. Schmuki, *Electrochem. Commun.*, 2006, **8**, 544.
98. D. Kim, A. Ghicov, S. P. Albu and P. Schmuki, *J. Am. Chem. Soc.*, 2008, **130**, 16454–16455.
99. S. P. Albu, D. Kim and P. Schmuki, *Angew. Chem., Int. Ed.*, 2008, **47**, 1916–1919.
100. B. Liu and E. S. Aydil, *J. Am. Chem. Soc.*, 2009, **131**, 3985–3990.

101. A. Wisnet, S. B. Betzler, R. V. Zucker, J. A. Dorman, P. Wagatha, S. Matich, E. Okunishi, L. Schmidt-Mende and C. Scheu, *Cryst. Growth Des.*, 2014, **14**, 4658–4663.
102. W.-Q. Wu, Y.-F. Xu, H.-S. Rao, C.-Y. Su and D.-B. Kuang, *J. Am. Chem. Soc.*, 2014, **136**, 6437–6445.
103. B. Tan and Y. Wu, *J. Phys. Chem. B*, 2006, **110**, 15932–15938.
104. J.-S. Yang, W.-P. Liao and J.-J. Wu, *ACS Appl. Mater. Interfaces*, 2013, **5**, 7425–7431.
105. W.-Q. Wu, B.-X. Lei, H.-S. Rao, Y.-F. Xu, Y.-F. Wang, C.-Y. Su and D.-B. Kuang, *Sci. Rep.*, 2013, **3**, 1352.
106. R. G. Newton, *Scattering Theory of Waves and Particles*, Courier Corporation, 1982.
107. J. Ferber and J. Luther, *Sol. Energy Mater. Sol. Cells*, 1998, **54**, 265–275.
108. T. P. Chou, Q. Zhang, G. E. Fryxell and G. Z. Cao, *Adv. Mater.*, 2007, **19**, 2588–2592.
109. D. Hwang, H. Lee, S.-Y. Jang, S. M. Jo, D. Kim, Y. Seo and D. Y. Kim, *ACS Appl. Mater. Interfaces*, 2011, **3**, 2719–2725.
110. F. Huang, D. Chen, X. L. Zhang, R. A. Caruso and Y.-B. Cheng, *Adv. Funct. Mater.*, 2010, **20**, 1301–1305.
111. A. Stein, B. E. Wilson and S. G. Rudisill, *Chem. Soc. Rev.*, 2013, **42**, 2763–2803.
112. C. Aprile, A. Corma and H. Garcia, *Phys. Chem. Chem. Phys.*, 2008, **10**, 769–783.
113. Z. A. Weinberg, *J. Appl. Phys.*, 1982, **53**, 5052–5056.
114. M. Law, L. E. Greene, A. Radenovic, T. Kuykendall, J. Liphardt and P. Yang, *J. Phys. Chem. B*, 2006, **110**, 22652.
115. S. Gubbala, V. Chakrapani, V. Kumar and M. K. Sunkara, *Adv. Funct. Mater.*, 2008, **18**, 2411–2418.
116. V. Pfeifer, P. Erhart, S. Li, K. Rachut, J. Morasch, J. Brötz, P. Reckers, T. Mayer, S. Rühle, A. Zaban, I. Mora Seró, J. Bisquert, W. Jaegermann and A. Klein, *J. Phys. Chem. Lett.*, 2013, **4**, 4182–4187.
117. D. O. Scanlon, C. W. Dunnill, J. Buckeridge, S. A. Shevlin, A. J. Logsdail, S. M. Woodley, C. R. A. Catlow, M. J. Powell, R. G. Palgrave, I. P. Parkin, G. W. Watson, T. W. Keal, P. Sherwood, A. Walsh and A. A. Sokol, *Nat. Mater.*, 2013, **12**, 798–801.
118. J. A. Dorman, J. Weickert, J. B. Reindl, M. Putnik, A. Wisnet, M. Noebels, C. Scheu and L. Schmidt-Mende, *J. Phys. Chem. C*, 2014, **118**, 16672–16679.
119. M. Xu, P. Da, H. Wu, D. Zhao and G. Zheng, *Nano Lett.*, 2012, **12**, 1503–1508.
120. Y. Wang, Y.-Y. Zhang, J. Tang, H. Wu, M. Xu, Z. Peng, X.-G. Gong and G. Zheng, *ACS Nano*, 2013, **7**, 9375–9383.
121. X. Feng, K. Shankar, M. Paulose and C. A. Grimes, *Angew. Chem., Int. Ed.*, 2009, **48**, 8095.
122. R. Moos and K. H. Härdtl, *J. Appl. Phys.*, 1996, **80**, 393–400.
123. M. S. Whittingham, *Chem. Rev.*, 2004, **104**, 4271–4302.

CHAPTER 6

Structural, Electronic, and Optical Properties of Lead Halide Perovskites

WAN-JIAN YIN, TINGTING SHI AND YANFA YAN*

Department of Physics and Astronomy, and Wright Center for Photovoltaics Innovation and Commercialization, The University of Toledo, Toledo, Ohio 43606, USA
*Email: yanfa.yan@utoledo.edu

6.1 Introduction

Halide perovskites generally refer to a class of materials with chemical formula ABX_3 (A = 1+ ions or molecules, B = Ge, Sn, Pb, and X = I, Br, Cl) and exhibit sought-after physical properties, such as magnetic, ferroelectric, and two-dimensional electronic conductivity.[1,2] Recently, halide perovskites have attracted substantial interest in the fields of materials, chemistry and physics.[3–15] This is due to the rapid increase of the reported conversion efficiency of solar cells based on halide perovskites, which reached 20.1% in 2014,[16] up from 3.8% in 2009.[17] Studies have shown that this rapid improvement of cell efficiency is mainly due to the fact that lead halide perovskite absorbers exhibit superior photovoltaic properties, including an extremely high optical absorption coefficient,[18] super long carrier diffusion lengths,[19,20] extraordinary power conversion efficiencies,[16,21,22] and extremely high open circuit voltages (V_{OC}).[7,21–23] For example, the reported absorption coefficient of a $CH_3NH_3PbI_3$ thin film[24] is around $10^5\,cm^{-1}$, which is nearly one order of magnitude higher than that of GaAs.[25] This high optical

RSC Energy and Environment Series No. 16
Unconventional Thin Film Photovoltaics
Edited by Enrico Da Como, Filippo De Angelis, Henry Snaith and Alison Walker

Published by the Royal Society of Chemistry, www.rsc.org

absorption coefficient enables high efficiency lead halide-based thin film solar cells to be fabricated with rather thin absorbers, typically less than 500 nm. Furthermore, polycrystalline lead halide perovskite thin films have shown exceptionally long carrier diffusion lengths, which in the very best cases can be as high as 1 μm,[19,20] even for films grown at low temperature by wet chemical synthesis. It seems that the grain size and grain boundaries (GBs) have no significantly adverse effects on the carrier diffusion length,[26–28] indicating that polycrystalline lead halide perovskite thin films may achieve performances similar to those of epitaxial single-crystal films. As a result, thin-film solar cells based on $CH_3NH_3PbI_{3-x}Cl_x$ perovskites have achieved V_{OC} values as high as 1.13 V. The V_{OC} deficit, $E_g/q - V_{OC}$, for $CH_3NH_3PbI_{3-x}Cl_x$ based solar cells is smaller than those of the best CIGS cells, and is approaching those of the best c-Si and epitaxial single-crystal GaAs solar cells.[29]

In this chapter, we summarize our recent theoretical investigations on the structural and electronic properties of halide perovskites[30–33] using density functional theory (DFT) calculations. The unique properties of the halide perovskites have been investigated systematically, and they explain the origin of the high performance of perovskite-based solar cells.[32] The dominant defects, including point defects and grain boundaries, in halide perovskites have been found to be benign, since they do not create detrimental deep levels within the bandgap that could act as carrier traps and centers for electron–hole recombination in the solar cells.[33,34] We attribute the superior performance of these perovskite cells to the antibonding coupling between Pb lone-pair s- and I p-orbitals, the ionicity of halide perovskites and the large lattice constant of the halide perovskite. We also discuss the possibility of Cl or O atoms existing at the GBs, and how this may improve the photovoltaic properties of the perovskites.

6.2 Crystal Structure

Perovskites, with formula ABX_3, typically have unit cells composed of five atoms in a cubic structure (α phase), where cation B has six nearest neighbor anions X, and cation A has twelve (Figure 6.1(a)). Under ideal conditions, to maintain a high-symmetry cubic structure, the ion radii of A, B, and X should satisfy the requirement that the tolerance factor t should be close to 1, where $t = (R_A + R_X)/\{\sqrt{2}(R_B + R_X)\}$, and R_A, R_B, and R_X are the ionic radii of the corresponding ions. Otherwise, the cubic structure will be distorted and the crystal symmetry is reduced. It should be noted that the A ion must be much larger than the B ion to satisfy $t \approx 1$. In halide perovskites, the B site is usually occupied by a large Pb or Sn atom, so A must be extremely large. Cs is almost the largest group-I element in the periodic table. However, Cs is still not large enough to hold the stable cubic perovskite structure, so it must be replaced with a larger molecule. This may be the reason why $CH_3NH_3PbI_3$ is more stable and performs better than $CsPbI_3$, and $HC(NH_2)_2PbI_3$ is more stable than $CH_3NH_3PbI_3$.

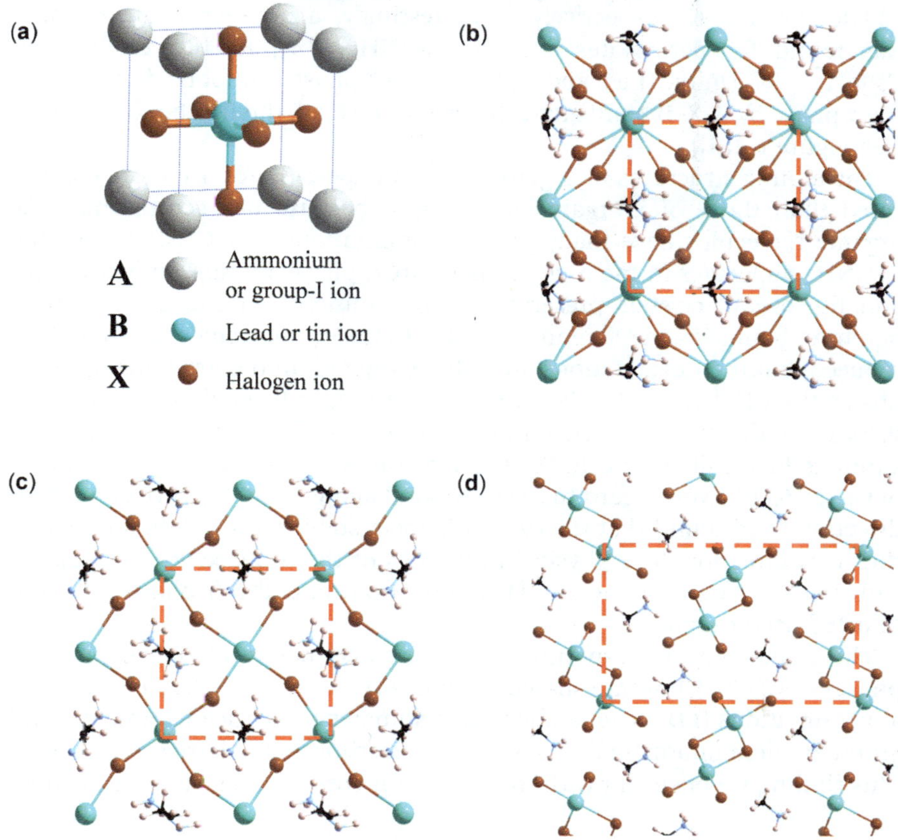

Figure 6.1 The atomic structure of the (a) α phase, (b) β phase, (c) γ phase, and (d) δ phase of $CH_3NH_3PbI_3$. The unit cells of the β and γ phases are $\sqrt{2} \times \sqrt{2} \times \sqrt{2}$, based on the α phase.
Reprinted from ref. 18 with permission from John Wiley and Sons © 2014 WILEY-VCH Verlag GmbH & Co. KGaA, Weinheim.

At finite temperatures, a cubic structure may exist when t lies between 0.89 and 1.[35] Generally, smaller t values could lead to the low-symmetry tetragonal (β phase) or orthorhombic (γ phase) structures, whereas larger t values ($t > 1$) could destabilize the three-dimensional (3D) B–X network, leading to a two-dimensional (2D) layer structure. Notably, in DFT calculations at zero temperature, the γ phase is always the most stable, and the α phase is always the least stable, because it is difficult to satisfy the condition $t = 1$. However, transitions between those structures at finite temperatures often happen in most perovskites.[36,37] In fact, the soft transverse acoustic phonon mode results in easy displacement of the X anion from the B–B midpoint of the ideal cubic phase. At finite temperatures, such displacements are likely to be dynamic in nature.[36] Due to the different ion sizes and structural stabilities, the transition temperatures of different perovskites vary. For the archetypal halide perovskite $CH_3NH_3PbI_3$, the α to β to γ phase transitions happen at

330 K and 160 K, respectively.[38] Interestingly, a non-perovskite δ phase was found for perovskites such as $HC(NH_2)_2PbI_3$, $FAPbI_3$, $CsPbI_3$ and $CsSnI_3$.[36,37,39] Unlike the β and γ phases, the δ phase cannot be derived from the α phase by B–X–B bond angle distortion. Instead, the B–X bond is broken (see Figure 6.1(d)).

The atomic structures of organic–inorganic perovskites are more complicated than those of inorganic perovskites, because non-centrosymmetric organic molecules have many possible orientations. For $CH_3NH_3PbI_3$, the CH_3NH_3 molecules in the high-temperature α phase are randomly oriented, and the overall crystals preserve the O_h symmetry. In the medium temperature β phase, the freedom of the CH_3NH_3 molecules is somewhat reduced, but their orientations are still disordered. In the low-temperature γ phase, the CH_3NH_3 molecules are frozen and have fixed alignments.[36] The disorder of the molecule orientations in the α and β phases is impossible to simulate in a finite unit cell. For the α phase, we compared the total energy of $CH_3NH_3PbI_3$ with regard to the C–N bond along the [001], [110], and [111] directions and found that the system is most stable with a C–N bond along the [111] direction. The physical properties of the α phase were calculated with the C–N bond along the [111] direction, and the β phase was constructed accordingly.

The calculated and experimental lattice parameters of the α, β and γ phases of $CH_3NH_3PbI_3$ are shown in Table 6.1. We can see that local density approximations (LDAs) generally underestimates the lattice parameters and grand gradient approximations (GGAs) overestimates the lattice parameters. This finding is consistent with the results for conventional semiconductors.

Table 6.1 The calculated lattice constants and bandgaps of the three phases of $CH_3NH_3PbI_3$ compared to the experimental values. Because the CH_3NH_3 molecule is non-centrosymmetric, in the α phase, the calculated pseudocubic lattice constant is defined by $a = V^{1/3}$ and, for the β phase, $a = 1/2(a + b)$. The relative total energy (meV/formula), relative to the α phase, is also given.

Phase	Symmetry	Calculated lattice constant (Å)	Experimental lattice constant (Å)	Relative total energy	PBE Calculated bandgap (eV)	Experimental bandgap (eV)
$CH_3NH_3PbI_3$						
α	$Pm\bar{3}m$	$a = 6.39$	$a = 6.31$,[37] 6.28[36]	0	1.53	
β	$I4/mcm$	$a = 8.80$, $c = 12.99$	$a = 8.85$,[36,37] 8.88,[43] $c = 12.64$,[37] 12.67[43]	−85	1.57	1.52,[37] 1.5[17]
γ	$Pbnm$	$a = 8.84$, $b = 8.77$, $c = 12.97$	$a = 8.84$,[36] $b = 8.56$,[36] $c = 12.58$[36]	−119	1.46	

However, for $CH_3NH_3PbI_3$, the overestimation in the GGA(PBE)-calculated lattice parameters has been attributed to the neglect of van der Waals interactions between the CH_3NH_3 and [Pb–I] frameworks.[40] When dispersive interactions with a particular form are considered, the calculated lattice parameters match the experimental results well.[40–42] Based on this, it has been claimed that van der Waals interactions are an important aspect in hybrid halide perovskites.[40] However, we posit that the importance of van der Waals interactions should be revisited carefully for two reasons. One, GGA(PBE) overestimates the lattice parameters not only for $CH_3NH_3PbI_3$ but also for other compounds, as this is an intrinsic problem of exchange-correlation methods. Two, for inorganic $CsPbI_3$, when considering dispersive interactions, the calculated lattice parameters can also be in agreement with the experimental ones.[42] Therefore, it may be that GGA under-binds the compound. Adding dispersive interactions increases the binding of the compound and corrects the GGA errors. This does not mean that dispersive interactions are intrinsic to halide perovskites. Corrections can also be achieved by mixing LDA and GGA, since LDA generally over-binds compounds. Following this reasoning, CH_3NH_3 behaves like an ion, and the interaction between CH_3NH_3 and [Pb–I] is similar to that of Cs and [Pb–I], which exhibits ionic character.

6.3 Electronic Structure

We calculated the band structure, (partial) density of states (DOS), and the partial charge density of the valence band maximum (VBM) and conduction band minimum (CBM) states of the α phase $CH_3NH_3PbI_3$ using DFT-GGA (Figure 6.2). The direct bandgap is at the *R* point. Unlike most cations whose outer s-orbitals are empty, Pb has an occupied 6s orbital, which is below the top of the valence band of perovskites. This so-called lone pair of s-electrons in Pb often gives rise to unusual properties.[44–46] The DOS and partial charge density plots show that the valence band maximum (VBM) has a strong Pb s- and I p-antibonding character; whereas the conduction band minimum (CBM) is almost all contributed from the Pb p-state, which reflects the unique dual nature (ionic and covalent) of halide perovskite electronic structures. In conventional semiconductors such as GaAs and CdTe, the CBM primarily has s-orbital character, whereas the VBM has primarily p-orbital character. In contrast, halide perovskites exhibit an inverted band structure. The state contributed by the organic molecule is far from the band edges, which means that the organic molecules do not play a direct role in determining the basic electronic structure of halide perovskites. As discussed above, an important role of the organic molecules in halide perovskites is stabilizing the perovskite structure and changing the lattice constant.

The band structure of the β, γ, and δ phases of $CH_3NH_3PbI_3$ is shown in Figure 6.3. The band structure of the α phase, with a $\sqrt{2} \times \sqrt{2} \times \sqrt{2}$ supercell, is also shown for comparison. These calculations suggest that β and γ phases have band structures similar to that of the α phase, which indicates

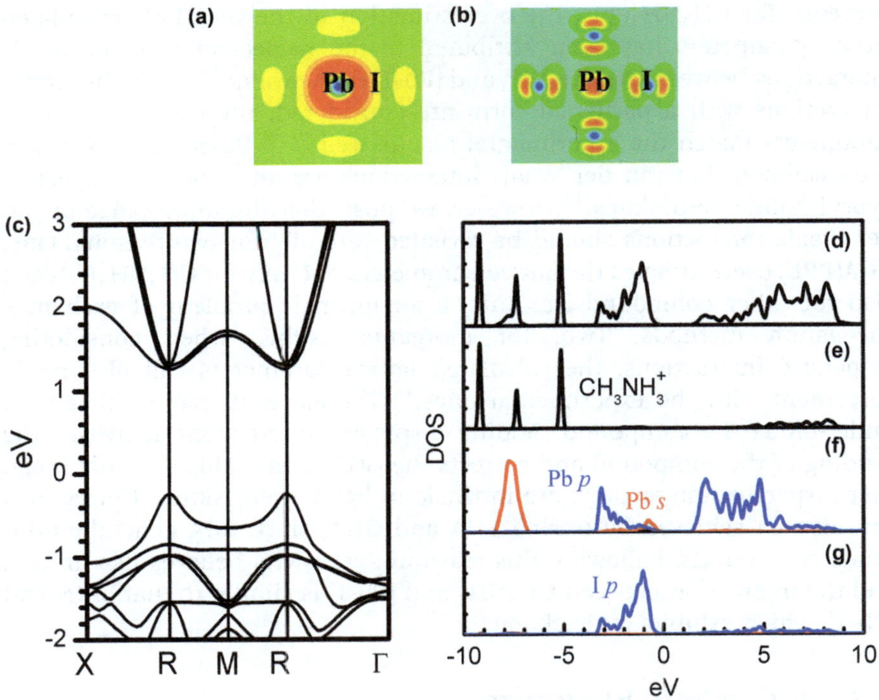

Figure 6.2 The (a) partial charge density of the VBM, (b) partial charge density of the CBM, (c) band structure, and (d–g) DOS of $CH_3NH_3PbI_3$ using DFT-PBE calculations.
Modified by permission from W.-J. Yin, T. Shi and Y. Yan, *Appl. Phys. Lett.*, 2014, **104**, 063903. Copyright (2014), AIP Publishing LLC.[33]

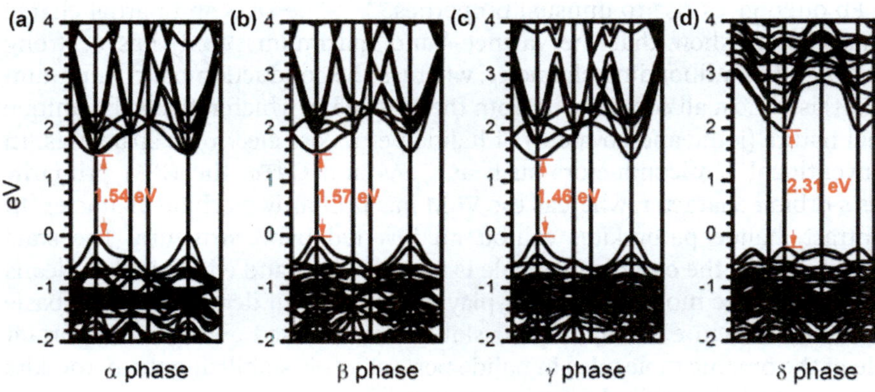

Figure 6.3 Band structure of the (a) α phase, (b) β phase, (c) γ phase, and (d) δ phase of $CH_3NH_3PbI_3$. The α phase was calculated using the same tetragonal supercell as for the β and γ phases.
Reprinted from ref. 18 with permission from John Wiley and Sons © 2014 WILEY-VCH Verlag GmbH & Co. KGaA, Weinheim.

that the Pb–I–Pb bond angle distortions do not significantly change the electronic structure. However, the band structure of the δ phase is different. Partial Pb–I bonds are broken and the 3D [Pb–I] framework is destroyed in the δ phase. In this phase, the coupling between Pb s- and I p-orbitals is weakened. As a result, the VBM, which is in the s-p antibonding state, drops in energy, leading to an enlarged bandgap and a flat upper valence band. In experiments, the δ phase is usually yellow and has a bandgap of over 2 eV,[36,37] *i.e.*, more than 0.5 eV larger than that of the other phases. Its hole effective mass is expected to be much larger than that of the other perovskite phases.

The DFT-PBE calculated bandgap for the β phase of $CH_3NH_3PbI_3$ is 1.57 eV, which is in good agreement with experiments. However, this result is a fortuitous one, due to the accidental cancellation between the underestimation from GGA and the neglect of the spin orbit coupling (SOC).[47] This is not observed in other perovskites such as $CH_3NH_3PbBr_3$, $CH_3NH_3PbCl_3$ and $CsPbI_3$, whose GGA-calculated bandgaps are lower than experimental values. In those cases, advanced computational methods such as PBE + SOC,[47,48] HSE + SOC,[49] and GW + SOC,[50–52] which take SOC into account, have been used to study the electronic structure of $CH_3NH_3PbI_3$ and other perovskites. Notably, the key features of GGA-derived electronic structures, such as the strong s–p coupling and dual nature of the bond character, do not change with the advanced methods. In fact, the SOC effect is not a unique finding for $CH_3NH_3PbI_3$. Heavy atoms such as Pb and Bi have large SOC effects, which also exist in semiconductors like PbTe[46] and Bi_2Se_3.[53] For those compounds, even though the band degeneracy and symmetry will change when SOC effects are included in the calculations, the electronic structure and bond character can be accurately described by non-SOC calculations.

The effective mass (m^*) of electrons (holes) around the bottom of the conduction band (top of the valence band) can be approximately fitted by $m^* = \hbar^2 \left[\frac{\partial^2 \varepsilon(k)}{\partial^2 k}\right]^{-1}$, where $\varepsilon(k)$ are energy dispersion relation functions, which are typically described by the band structures. Intuitively, the more dispersive (flat) the band near the band edges is, the lighter (heavier) the effective mass is. In conventional thin-film solar cell absorbers, such as GaAs and CdTe, the CBM is mostly contributed by the cation s- and the anion s-orbitals, and the VBM is mostly contributed by the anion p-character (p-s semiconductor). High-energy-level s-orbitals are more delocalized than low-energy-level p-orbitals, and thus the lowest conduction band is more dispersive than the highest valence band. Consequently, the electron effective mass is much smaller than the hole effective mass. That is why electrons transport much faster than holes, and why second-generation thin-film solar cells must use p-type absorbers.

The existence of lone-pair Pb s-electrons makes the situation dramatically different in perovskite solar cells. The electronic structure of $CH_3NH_3PbI_3$ is inverted compared to conventional p-s semiconductors. Its CBM is derived from the Pb p-orbitals, and the VBM is a mixture of Pb s- and I p-orbitals (s-p semiconductor). Even though the CBM comes from p-orbitals, a cation Pb p-orbital has a much higher energy level than anion p-orbitals, as in p-s semiconductors. Therefore, the lower conduction band of $CH_3NH_3PbI_3$ is more dispersive than the upper valence band in p-s semiconductors. On the other hand, the upper valence band of $CH_3NH_3PbI_3$ is dispersive due to strong s–p coupling around the VBM. The calculated effective masses of electrons and holes in the α phase $CH_3NH_3PbI_3$ are very similar, $0.35m_e$ and $0.31m_e$ respectively. These balanced effective masses may lead to ambipolar conductivity, which enhances the performance of the p-i-n junction solar cell.

6.4 Optical Absorption

One of the aspects of perovskite solar cells that helps explain their superior performance is their high optical absorption, such that the absorber layer, which is less than 500 nm thick, can absorb enough sunlight to achieve efficiencies above 15%. In fact, the photoactive layer is effectively even thinner since the mesoscopic absorber layer includes additional scaffold oxides (TiO_2, ZrO_2, or Al_2O_3). In comparison, the thickness of the absorber layers in first- and second-generation solar cells are about 300 μm and 2 μm, respectively.

First-principles calculations have been used to obtain the optical absorption of halide perovskites and to unravel the physics underlying this property. The optical absorption of a semiconductor is fundamentally determined by two factors. One is the transition matrix elements between valence band (VB) states and conduction band (CB) states. The other is their joint density of states (JDOS), defined as the product of the DOS between the VB and CB states. The former measure the probability of each photoelectric transition, and the latter measures the total number of possible photoelectric transitions. Thus, the optical absorption coefficient of a material is closely related to its electronic structure.

The mechanisms of optical absorption for first-generation, second-generation, and perovskite-based solar cell absorbers are depicted schematically in Figure 6.4. For the first-generation absorber (silicon), the optical absorption close to the band edge is from the Si p-orbital to the Si p- and s-orbitals. However, silicon is an indirect bandgap semiconductor, and its transition probability between band edges is two orders lower than that of its direct bandgap, making it necessary to employ absorber layers two orders thicker, thereby increasing the cost of materials. The bandgap of second-generation absorbers (GaAs for example) and of halide perovskites ($CH_3NH_3PbI_3$ for example) are direct, and consequently their optical absorption is much stronger than that of silicon. However, the electronic structure of GaAs and $CH_3NH_3PbI_3$ is different. The lower part of the GaAs CB is derived from the dispersive s band (delocalized s-orbitals), whereas the lower part of the $CH_3NH_3PbI_3$ CB is mainly composed of degenerate

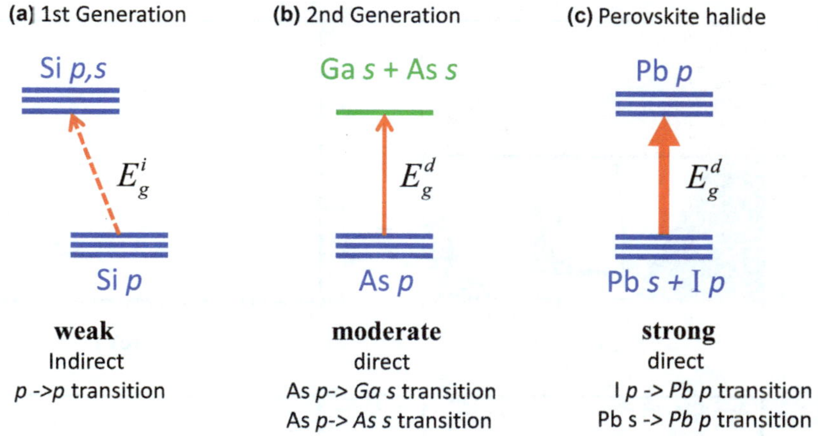

Figure 6.4 The schematic optical absorption of (a) first-generation, (b) second-generation, and (c) halide perovskite solar cell absorbers. GaAs has been chosen as the prototypical second-generation solar cell absorber. Reprinted by permission from ref. 74 with permission from the Royal Society of Chemistry.

Pb p-bands. The atomic p-orbital exhibits less dispersion than s-orbitals. As a result, the DOS in the lower CB of the halide perovskites is significantly higher than that of GaAs, as shown in Figure 6.5(a), leading to a higher JDOS (Figure 6.5(b)). Additionally, the edge transition for $CH_3NH_3PbI_3$ comes from mixed-(Pb s, I p) to Pb p-orbitals. The intra-atomic Pb s to Pb p transition probability is high, which makes the VBM–CBM transition probability of $CH_3NH_3PbI_3$ comparable to that of GaAs. Therefore, halide perovskites show stronger optical absorption than GaAs mainly as a consequence of the higher DOS. The calculated optical absorption coefficients of $CH_3NH_3PbI_3$ and GaAs are shown in Figure 6.5(c). It is seen that the optical absorption coefficient of $CH_3NH_3PbI_3$ is up to one order of magnitude higher than that of GaAs within the visible light range. Because visible light accounts for the major usable portion of the full solar spectrum, high absorption in this range is critical in order to achieve high-efficiency cells.

Achieving high optical absorption is crucial for realizing economic and efficient solar cells with extremely thin absorbers. However, the effect of the optical absorption coefficient is not considered in the calculation of the well-known Shockley–Queisser limit.[54] In a real-world solar cell, the theoretical maximum efficiency depends on the thickness of the absorber layer. Recently, Yu *et al.*[55,56] took the absorption coefficient and absorber layer thickness into consideration and developed a method to calculate the maximum theoretical efficiency based on the absorber thickness. Using this approach, we have calculated the maximum efficiencies of perovskites and other mainstream thin-film absorbers. The results are shown in Figure 6.5(d).[32] The halide perovskites ($CH_3NH_3PbI_3$, $CsSnI_3$, and $CsPbI_3$) exhibit much higher conversion efficiencies than GaAs, CIS, and CZTSe for

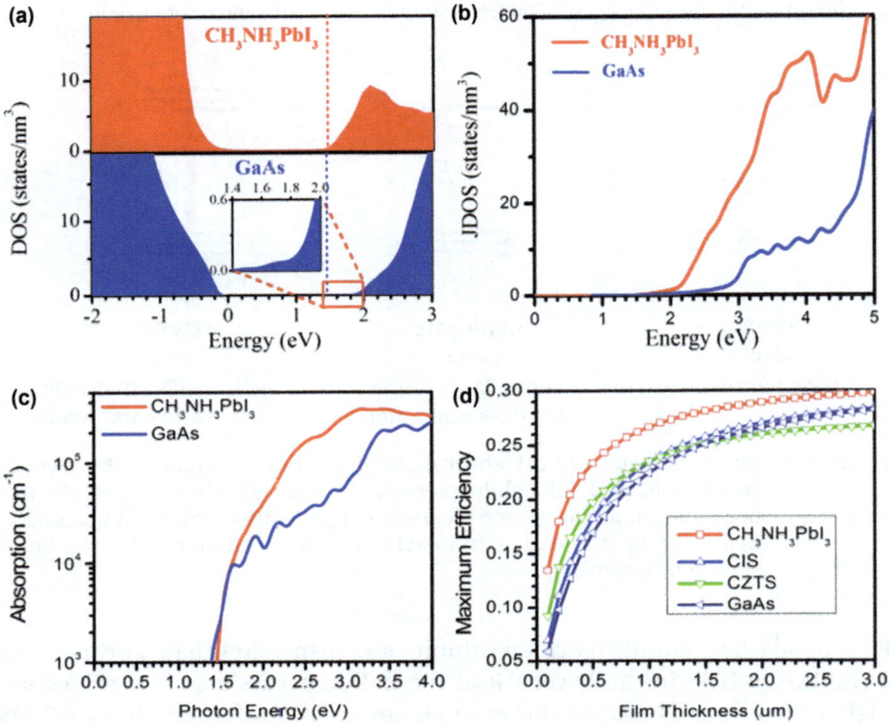

Figure 6.5 (a) DOS of $CH_3NH_3PbI_3$ and GaAs. The VBMs are referred to as zero energy and CBMs are marked as dashed lines. (b) The JDOS of $CH_3NH_3PbI_3$ and GaAs. (c) The optical absorption spectra of $CH_3NH_3PbI_3$ and GaAs. (d) Calculated maximum efficiencies of $CH_3NH_3PbI_3$, CIS, CZTS, and GaAs as a function of film thickness.
Adapted from ref. 18 with permission from John Wiley and Sons © 2014 WILEY-VCH Verlag GmbH & Co. KGaA, Weinheim.

any given thickness. More importantly, halide perovskites are capable of achieving high efficiencies with very thin absorber layers. For example, with a 0.3 μm absorber (similar to the thickness of $CH_3NH_3PbI_{3-x}Cl_x$ in the 15.4% perovskite-based cell[7]), $CH_3NH_3PbI_3$-based cells have a maximum efficiency of 21%, while it is only 13% for the GaAs-based solar cells. The results are consistent with many of today's most common thin-film solar cells (GaAs, CdTe, CIGS, and CZTSS), where the absorbers must be about 2 μm thick to achieve high efficiencies.[57]

6.5 Point Defect Properties

In first- and second-generation solar cells, defect and doping properties are crucial factors affecting the cell performance. The key component is the p–n junction, where the photo-generated electrons and holes are separated and collected with the aid of the internal electric field. The p or n-type absorbers are made from materials with intrinsic defects, or using intentional doping.

On the other hand, intrinsic defects that create deep energy levels in absorbers usually act as Shockley–Read–Hall non-radiative recombination centers and carrier traps, reducing the carrier lifetime, and thus, V_{OC}. A good solar cell absorber must exhibit proper doping and defect properties. For example, the V_{OC} of the next-generation solar cell absorber $Cu_2ZnSn(S,Se)_2$ is much lower than its bandgap, and its record efficiency is around 12%.[57] This could be related to the energetically favorable formation of the dominating defect in $Cu_2ZnSnSe_2$, Cu_{Zn}, which can pin the n-type Fermi energy. It may also be due to the formation of a Sn_{Zn} deep level, which may become a carrier trap and non-radiative recombination center.[58,59]

Several types of defects, including both donor-like and acceptor-like, exist in semiconductors. The relative concentration of these donors and acceptors, which are directly related to defect formation energies, determine the p or n-type conductivity. The formation energy of a defect depends on the chemical potential, which corresponds to environmental factors such as precursors, partial pressure, and temperature. In this way, the experimental conditions determine the formation energy of all the possible defects and thus influence the polar conductivity in the product material. First-principles calculations have been used to study the point defect properties of the α phase of $CH_3NH_3PbI_3$.[33] The chemical potential for equilibrium-growth is as shown in Figure 6.6(top), which indicates that growth of $CH_3NH_3PbI_3$ can occur only in a long and narrow range of chemical potential. Otherwise, secondary phases, such as PbI_2 and CH_3NH_3I (MAI) would also form. Three typical chemical potentials (A, B, and C in Figure 6.6(top)) were chosen to study the formation energy of point defects. All the possible point defects, including three vacancies (V_{MA}, V_{Pb}, V_I), three interstitial (MA_i, Pb_i, I_i), two cation substitutions (MA_{Pb}, Pb_{MA}) and four anti-site substitutions (MA_I, Pb_I, I_{MA}, I_{Pb}), were considered in our calculations. The formation energies of these point defects as a function of the Fermi level position at chemical potentials A, B, and C are shown in Figure 6.6 (bottom left), (bottom center) and (bottom right), respectively. The lowest crossing points between donor and acceptor indicate the positions of Fermi level pinning. The conductivity of $CH_3NH_3PbI_3$ can arise from intrinsically good p-type conduction, but, interestingly, it can also arise from moderate to good n-type conduction when the chemical potential is at A(I-rich/Pb-poor), B (moderate), or C (I-poor/Pb-rich). Such flexible defect properties are distinct from those of conventional thin-film solar cell absorbers, CIGS[60] and CZTSS.[61] This is due to the chemical similarity of the cations in these materials and the presence of Cu antibonding d-orbitals at the VBM, which means that acceptors such as V_{Cu}, Cu_{Ga}, and Cu_{Zn} have very low formation energies. Thus, in equilibrium growth, p-type doping in CIGS and CZTSS is easy, but n-type doping is difficult due to self-compensation.[60] In $CH_3NH_3PbI_3$, the dominant donor MA_i and dominant acceptor V_{Pb} have comparable formation energies, so both p-type and n-type doping are possible in $CH_3NH_3PbI_3$. Our theoretical results on bipolar conductivities are consistent with the experimental findings of both p and n-type $CH_3NH_3PbI_3$ for different growth processes.

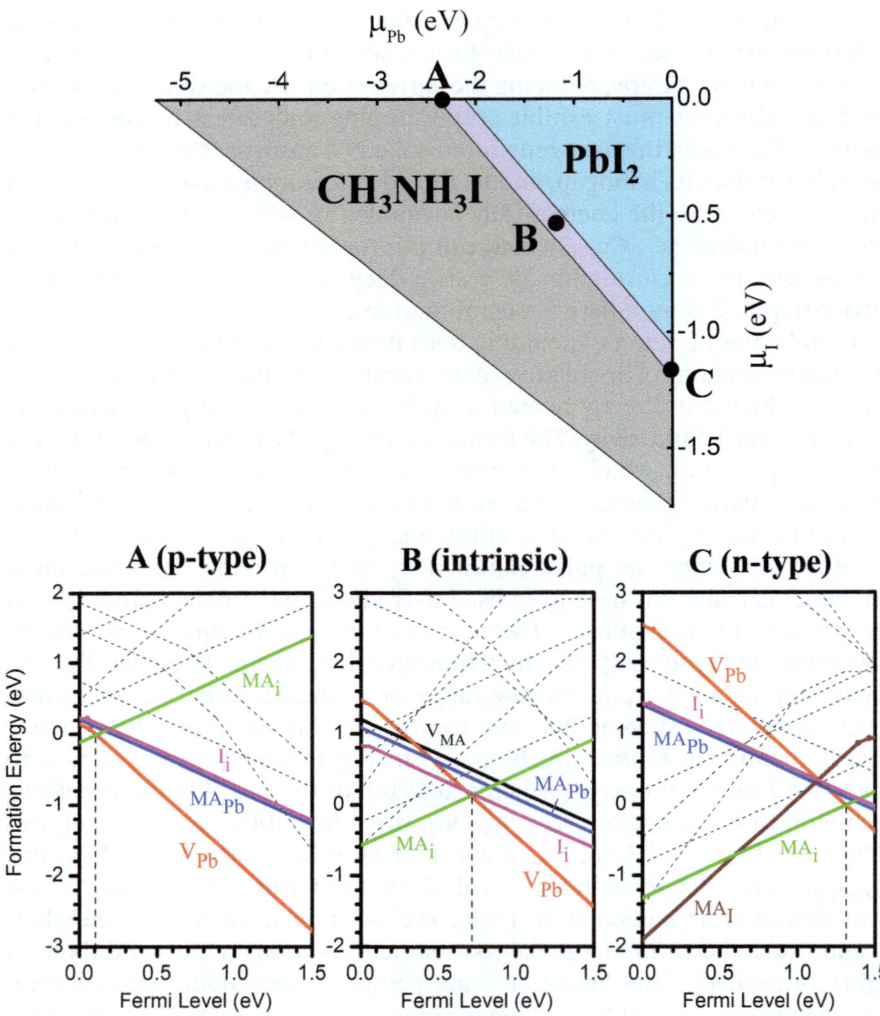

Figure 6.6 The achievable chemical potential for the thermo-equilibrium growth of $CH_3NH_3PbI_3$ (top) and the formation energy of intrinsic point defects in $CH_3NH_3PbI_3$ at chemical potentials A, B, C (bottom). Defects with much higher formation energies are displayed with dashed lines.
Modified with permission from W.-J. Yin, T. Shi and Y. Yan, *Appl. Phys. Lett.*, 2014, **104**, 063903. Copyright (2014), AIP Publishing LLC.[33]

Defect formation energies determine the polar conductivity of a semiconductor, whereas defect transition levels determine the electrical effect of any particular defect. The transition level of a defect is the Fermi level position where the defect can gain or lose electrons and change its charge state. Shallow defect levels, close to band edges, can donate or accept electrons to bulk bands and become effective dopants for absorbers. However, deep level defects near the middle of the bandgap can trap electrons or holes and become SRH non-radiative recombination centers in the absorber.

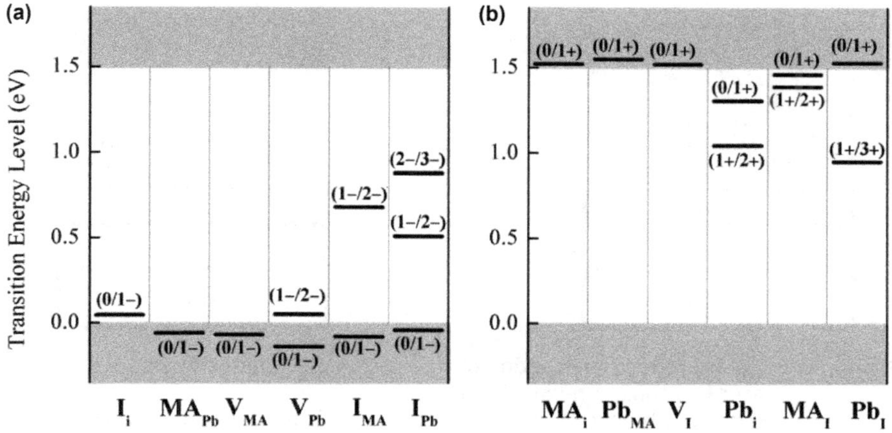

Figure 6.7 The transition energy levels of (a) intrinsic acceptors and (b) intrinsic donors in $CH_3NH_3PbI_3$.
Reprinted with permission from W.-J. Yin, T. Shi and Y. Yan, *Appl. Phys. Lett.*, 2014, **104**, 063903. Copyright (2014), AIP Publishing LLC.[33]

The transition levels of all possible point defects in $CH_3NH_3PbI_3$ are shown in Figure 6.7. The defects with low formation energies, as shown with solid lines in Figure 6.6(bottom), have transition energies lower than 0.05 eV above (below) the VBM (CBM). All defects that create deep levels have high formation energies, as shown with dashed lines in Figure 6.6(bottom). Besides point defects, Kim *et al.*[62] used DFT-GGA to calculate the DOS and partial charge densities of two types of neutral defects in β phase $CH_3NH_3PbI_3$: Schottky defects (equal numbers of positive and negative vacancies) and Frenkel defects (equal numbers of vacancies and interstitials of the same ion). They found that these defect pairs do not create deep gap states. Du[49] used DFT-PBE calculations and similarly found that most of the point defects are shallow in β phase $CH_3NH_3PbI_3$. Alternatively, Du found that for I, a particular configuration may be a deep state, with a role in the electrical properties. So far, it is not clear whether the deep I interstitial state in Du's calculations is derived from different crystal phases or particular defect structures.

Shallow defect properties are closely related to the dual-nature electronic structure of halide perovskites. The shallow dominant acceptors V_{Pb} and MA_{Pb} originate from covalent coupling between the Pb lone-pair s- and I p-orbitals, which increases the VBM so that the acceptors are generally shallower than in cases without strong s–p coupling. The mechanism is similar for the shallow V_{Cu} in CIGS, where the VBM is increased by coupling between the Cu d- and Se p-orbitals.[60] The shallow donors MA_i and V_I are due to high ionicity. MA_i has no covalent bonds to the Pb–I framework and does not create additional gap states. For V_I, our previous study has shown that weak covalence generally leads to shallow anion vacancy levels.[63]

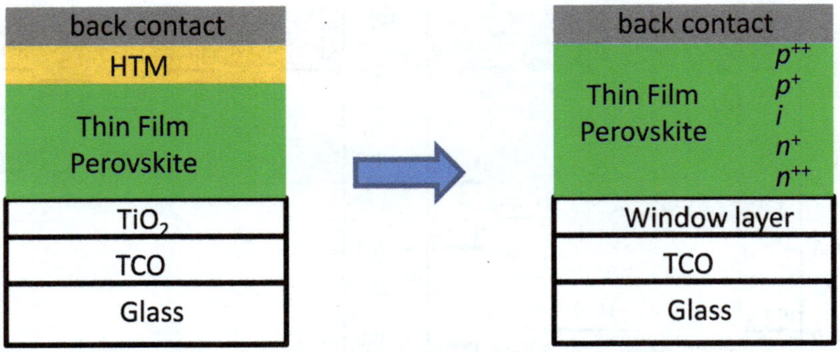

Figure 6.8 The proposed hole-conductor-free perovskite solar cell without an HTM layer by utilizing p-type doping.
Reprinted by permission from ref. 74 with permission from the Royal Society of Chemistry.

The tunable polar conductivity and shallow defect properties may help explain why high-performance perovskite solar cells, with extremely long carrier lifetimes,[19,20] can be produced by a diverse range of growth approaches and a wide variety of solar cell architectures. The superior point defect properties would suggest new architectures for perovskite solar cell devices. The two prevailing architectures are mesoscopic and planar structures. The former mostly uses porous TiO_2 to help electron transport, while the latter does not. Both need a hole transport material (HTM) layer to support hole transport. Organic HTM materials are expensive, and many researchers are seeking inorganic materials as economically viable substitutes,[64–66] even though they generally do not perform as well. However, the HTM layers are not necessary if the absorber can be effectively p-type doped, as proposed in our theory above. In this case, the HTM layer would be replaced by p-type perovskite layers. Electrons and holes could then be separated through an intrinsic p-i-n junction, as depicted in Figure 6.8. Recently, several reports[67–70] have been published on the effective performance of hole-conductor-free perovskite solar cells, demonstrating the potential applications of p-i-n homojunctions in solar cells.

6.6 Grain Boundaries and the Passivating Effect of Cl

Currently, most of the best-performing devices have a perovskite absorber layer thickness around 300 nm. As discussed above, the theoretical maximum efficiency depends strongly on the film thickness. Although halide perovskites have strong optical absorption, a 300 nm $MAPbI_3$ film is not thick enough to absorb sufficient sunlight, and its maximum efficiency for a planar cell was estimated to be about 20.5%.[18] To achieve a cell efficiency of more than 20% experimentally, we need to consider thicker films of up to 1 μm, where the theoretical maximum efficiency is about 26.7%, so that sunlight can be absorbed effectively.[71,72] We note that a recently reported

19.3%-efficient planar cell has an absorber thickness of around 450 nm,[22] which is thicker than previous planar cells with film thickness around 330 nm and efficiencies around 15%–16%.[7] In addition, a thick film is recognized in the photovoltaic field as a way to reduce current leakage and increase shunt resistance, thus increasing the fill factor. This is particularly important for commercialization, in which large-scale manufacturing could be realized by high-throughput methods such as printing, blade coating, or gravure coating.[72] On the other hand, the charge-collection efficiency could be reduced by a thicker film due to electrons and holes traveling longer distances to reach the electrodes, thus having a greater probability of recombination due to defects or GBs. Indeed, the cell efficiency was observed to increase slowly or even decrease when the film thickness was increased,[71,73] which indicates the possible detrimental effect of defects and GBs. However, initial results show that the GBs in $CH_3NH_3PbI_3$ do not introduce deep trap states,[18,74] which is also consistent with experimental electron beam induced current (EBIC) data.[26,28] Therefore, to increase the efficiency of perovskite solar cells beyond 20%, we must understand the fundamental properties of GBs and how they impact the electronic properties in perovskites. Therefore, we studied the structural disorder induced by GBs and its impact on the electronic structures. So far, no atomic structure of $MAPbI_3$ GBs has been reported in experiments. Previous investigations have shown that the electronic structure of GBs in conventional thin-film solar cells[75,76] such as CdTe, $CuInSe_2$, and $Cu_2ZnSnSe_4$, are dominated by the anion–anion and/or cation–cation wrong bonds at the GB plane. For $MAPbI_3$, we chose to study $\Sigma3(111)$ GB and $\Sigma5(310)$ GB, since these two GBs include those wrong bonds. To clearly understand the difference between $MAPbI_3$ and conventional thin-film solar cells, the results for CdTe $\Sigma3(112)$ GB are also provided for comparison. The $\Sigma5(310)$ GB in $MAPbI_3$ is similar to the $\Sigma3(112)$ GB in CdTe in terms of the anion–anion wrong bond, which dominates the formation of defect states within the bandgap.

The relaxed atomic structure around the GB core region and total density of states (DOS) of the $\Sigma3(111)$ and $\Sigma5(310)$ GBs for $MAPbI_3$ and the $\Sigma3(112)$ GB for CdTe are shown in Figure 6.9. It is interesting that $MAPbI_3$ GBs introduce no deep states within the bandgap region. Therefore, the results suggest that GBs in halide perovskites have no strong detrimental electronic effect, which is consistent with previous results.[18,26,28,74] As a comparison, GBs in CdTe introduce deep defect states within the bandgap, as shown in Figure 6.9(f). The partial DOS in Figure 6.9(f) shows that the defect states in CdTe GBs originate from the anion–anion wrong bond (Te1–Te2) and cation–cation wrong bond (Cd1–Cd1′). The electronic structure of *wrong bonds* at $MAPbI_3$ GBs are examined as follows: (i) Pb–Pb *wrong bond*: the Pb–Pb pairs at the core of the $\Sigma3(111)$ [Figure 6.9(a)] and $\Sigma5(310)$ [Figure 6.9(b)] GBs have distances of 3.84 Å and 5.17 Å, respectively, which are too large for Pb^{2+} (1.33 Å) to form strong covalent bonds. The partial DOS of those atoms in Figure 6.9 does not show any defect states; (ii) I–I *wrong bond*: this exists in the $\Sigma5(310)$ GB. Interestingly, a shallow defect state [peak at I3 + I4 partial

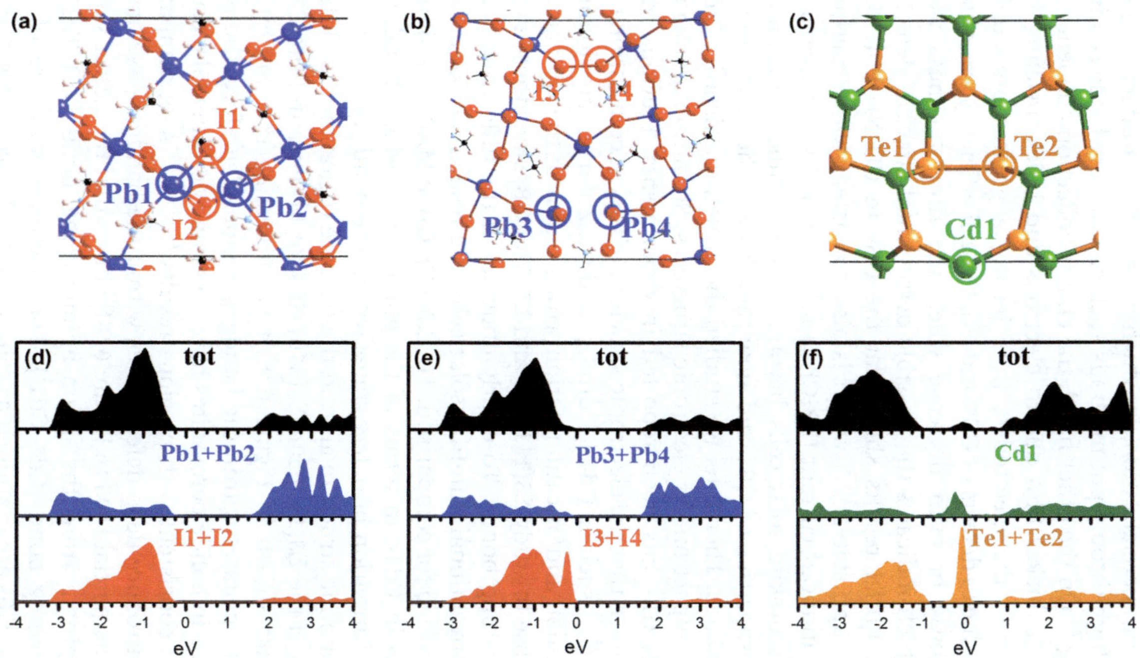

Figure 6.9 (a–c) The atomic structures of the $CH_3NH_3PbI_3$ $\Sigma3(111)$ and $\Sigma5(310)$ GBs, and the CdTe $\Sigma3(112)$ GB after structural relaxation. (d–f) Total DOS of GB models and partial DOS of atoms in GB cores. All the DOS are normalized to per formula unit. Reprinted from ref. 31 with permission from John Wiley and Sons © 2015 WILEY-VCH Verlag GmbH & Co. KGaA, Weinheim.

DOS in Figure 6.9(e)], with a chemical character similar to the deep state in CdTe GBs, is found with its position close to the VBM. Although this state is not very deep, and thus is not as detrimental as that in CdTe, it would increase the hole effective mass or behave like a trap state for holes. These atomic-level findings agree with recent experimental observations from different experimental groups and provide an in-depth understanding of grain size and GB effects on the material properties. Due to the trap state near the VBM, electrons in the $CH_3NH_3PbI_3$ solar cell are extracted more efficiently than holes;[28,77] consequently, holes are found to accumulate in the front of the hole-transport layer,[77] although the effective mass of electrons and holes in pristine $CH_3NH_3PbI_3$ is similar.[18,78] Oga *et al.*[79] detected an extremely shallow trap state with a depth of only 10 meV, and reported that the carrier mobility was greatly impaired by this shallow trap state. They further observed that the conductivity (mainly hole) increases with the increasing grain size, indicating that the trap states (especially for holes) originate from the GBs. D'Innocenzo *et al.*[80] found that the optical gap of $MAPbI_3$ increases with the decreasing grain size, indicating that GBs should not introduce deep gap states to reduce the optical absorption edge, as we predicted. However, the shallow trap states at GBs will reduce the photoluminescence (PL) lifetime with decreasing grain size. A shallow trap state was also confirmed by Shao *et al.*,[81] who observed a small shift of PL peaks before and after PCBM passivation at the GBs, and this level is estimated to be 15 meV away from the band edge.

Before we discuss how to deal with GB-induced trap states, we will discuss why $MAPbI_3$ GBs have no deep gap states. First, the VBM in $MAPbI_3$ is much higher than I 5p, because the strong s–p coupling between Pb 6s- and I 5p-states pushes up the VBM level significantly, as shown in Figure 6.10(b). When an anion–anion wrong bond forms, the anion p–p coupling forms an antibonding ppσ^* state. However, due to the relatively weak p–p coupling (see below) and high VBM, the antibonding ppσ^* state is still below the VBM derived from the strong s–p coupling. On the other hand, in CdTe, without strong s–p coupling, the VBM is not much higher than the Te p-state. When an anion–anion wrong bond is formed, the anion p–p coupling is strong enough to push the antibonding state high above the VBM and act as a deep defect level. The antibonding ppσ^* state could be pushed so high that the electrons can transfer from Te to Cd and form a Cd–Cd bonding state, which will become another defect state within the bandgap. Second, the I–I interaction in $MAPbI_3$ is weaker than the Te–Te interaction in CdTe. The Te–Te bond length in the $\Sigma3(112)$ GB is 2.88 Å, which is much shorter than the I–I bond length (3.72 Å) at the $\Sigma5(310)$ GB in $MAPbI_3$. Considering the ionic size, the Te–Te bond length is 1.39 times the ionic size of Te^{2-}, whereas the I–I bond length is 1.81 times the ionic size of I^{1-}. This indicates that the I–I bonding strength at GBs in $MAPbI_3$ is much weaker than the Te–Te bonding strength at GBs in CdTe. The large I–I bond length in $MAPbI_3$ is a consequence of its extremely large lattice constant due to the large atomic size of the A and B atoms in halide perovskites ABX_3. In perovskites,

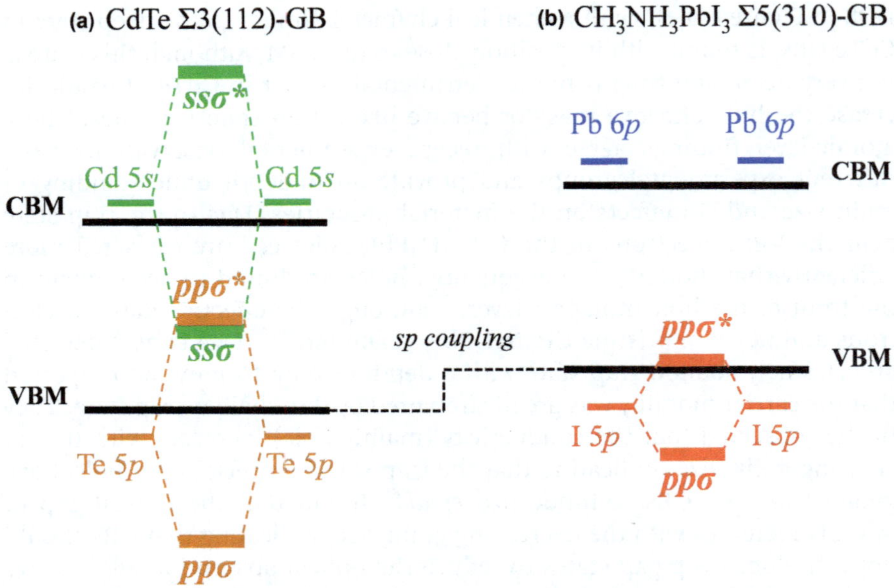

Figure 6.10 Schematic band diagram showing the origin of (a) deep defect states in the CdTe Σ3(112) GB and (b) shallow states in the $CH_3NH_3PbI_3$ Σ5(310) GB. Reprinted from ref. 31 with permission from John Wiley and Sons © 2015 WILEY-VCH Verlag GmbH & Co. KGaA, Weinheim.

A should be much larger than B due to the stability requirement of $(R_A + R_X) / (R_B + R_X) \approx \sqrt{2}$. In Pb-based perovskites, Pb at the B site has a large size and so, Cs or the organic molecule CH_3NH_3 at the A site need to be even larger. The GGA-calculated lattice constant of $MAPbI_3$ is 6.39 Å, as compared to 3.91 Å for the prototypical perovskite $SrTiO_3$. Reduced anion–anion coupling could, in part, explain why large atoms/molecules are adopted for a high-efficiency solar cell absorber.

Another controversial issue is the existence and effect of Cl in $MAPbI_3$ or $MAPbI_{3-x}Cl_x$. Initially, Snaith's group found that mixing some Cl into $MAPbI_3$ could significantly enhance carrier transport, and thus, the cell's efficiency. The atomic ratio of Cl in $MAPbI_3$ was initially thought to be 1/3 ($MAPbI_2Cl$),[9] as in the precursor. However, it was quickly established that the actual Cl concentration in the $MAPbI_{3-x}Cl_x$ material is much smaller than in the precursor.[38,43,82,83] Theoretical calculations show that the temperature of the miscibility gap for $CsPbI_{3-x}Cl_x$ is above 600 K,[84] indicating that it is difficult to mix Cl homogeneously into perovskite iodide when the growth temperature is low. Most recent experiments claim that either negligible or no Cl exists in the final $MAPbI_{3-x}Cl_x$,[14,85] and that Cl may only reside at the interface of $MAPbI_3$ and TiO_2.[86] It has been suggested that the role of Cl is to affect the morphological evolution of $MAPbI_3$ in the chemical process.[85,87,88] However, other recent experimental studies claim to clearly confirm the presence of Cl in the final $MAPbI_3$,[89,90] making the Cl issue more controversial.

Both theoretical and experimental observations suggest that Cl is not likely to be located at the $MAPbI_3$ bulk. We have examined the possibility of Cl being located in the GBs because defect segregation at GBs is a typical phenomenon[75,76,91–93] in thin-film solar cells such as CdTe, $CuInSe_2$, and $Cu_2ZnSnSe_4$. We found that, in the $\Sigma5(310)$ GB, the total energy of Cl_I at I–I wrong bond sites is about 0.2 eV lower than that in the bulk, as shown in Figure 6.11(b), indicating that Cl can spontaneously segregate into the GB region. The fact that Cl favors the I–I wrong bond site at the GBs is due to its smaller size: replacing I by Cl can cause a large strain in the system, thus costing energy; however, it costs less energy at the GB than at the grain interior region. The Cl–I bond length at the GBs increases from 3.72 Å to 3.80 Å, *i.e.*, the bond is weakened. The weakening of the Cl–I bond and the high electronegativity of Cl at the GBs will pull down and thus deactivate the hole trap state. Because many experiments have not detected any Cl in $MAPbI_3$, we can do a straightforward estimate of the amount of Cl necessary to passivate the GBs. Considering the grain size of 200 nm (ranging from 50 nm to 2000 nm in experiments[79,80]), we estimate that a 0.3% atomic ratio of Cl is enough to passivate all the I–I wrong bonds at the GBs. This assumption is based on all the GBs having wrong-bond-derived trap states. Actually, some GBs, such as $\Sigma3(111)$, do not have trap states. As a result, the actual amount of Cl should depend even less on the grain size and the energetic process. This result may explain why a small amount of Cl (perhaps within experimental detectability[27]) can enhance the carrier transportation so much. Evidence for the possible existence of Cl at GBs has been provided by both Kelvin probe force microscopy (KPFM) and EBIC measurements. Edri *et al.*'s experiments detected a small electrostatic barrier ($\sim$40 meV[26]) across the $MAPbI_3$ GBs, whereas no barrier was found for $MAPbI_{3-x}Cl_x$ GBs.[28] A similar effect with O has also been found, as shown in Figure 6.11 (b). However, the difference between O and Cl is that Cl is isovalent with I, whereas O at the I site is an acceptor and makes materials p-type, as shown in Figure 6.11(c), where the calculated Fermi level for O passivation is below the bulk VBM. A recent experiment by Ren *et al.*[94] demonstrated that thermal-assisted oxygen annealing led to a substantial increase in power conversion efficiency. The improvement in device performance was attributed to (i) the reduced localized states at the interface and GBs, and (ii) p-type doping of the hole transport material (HTM). However, the mechanism underlying the observed phenomenon was unclear at that time. Overall, our theoretical results are consistent with this assumption and provide a fundamental understanding of the oxygen passivation mechanism at the $MAPbI_3$ GBs. The only difference suggested by our theory is that oxygen may not only make the HTM more p-type, but may also make $MAPbI_3$ itself p-type,[95] making the p-type interface between $MAPbI_3$ and the HTM layer smoother for hole transportation. Apart from chemical passivation by Cl or O, PCBM/fullerene has also been found to have a passivation effect at the GBs and to eliminate current–voltage (*I*–*V*) hysteresis, although the reason why is still puzzling.[81] Based on our theory, we suggest that the

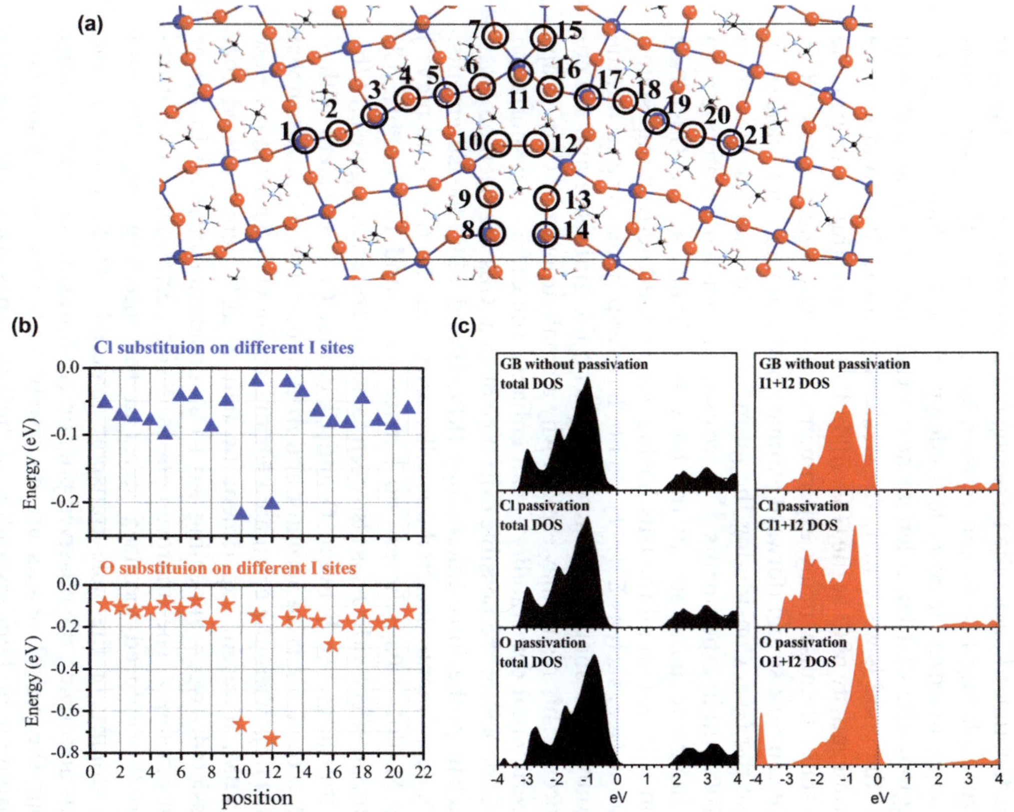

Figure 6.11 (a) Different I positions considered as Cl/O substitution sites in $CH_3NH_3PbI_3$ $\Sigma5(310)$ GB. The model structure is fully relaxed from a snapshot in MD simulation. (b) The relative total energies when Cl/O substitutes different I sites in (a). The asymmetrical behavior is attributed to the random orientations of the CH_3NH_3 molecule. (c) Total DOS and partial DOS at wrong bonds of unpassivated, Cl-passivated and O-passivated $CH_3NH_3PbI_3$ $\Sigma5(310)$ GBs.
Reprinted from ref. 31 with permission from John Wiley and Sons © 2015 WILEY-VCH Verlag GmbH & Co. KGaA, Weinheim.

possible mechanism could be that large fullerene molecules diffusing into the $MAPbI_3$ GBs would stretch the I–I bond length, weaken the wrong bond strength, and thus pulling the trap state below the VBM.

6.7 Conclusions and Outlook

As presented above, first-principles calculations have been proven to be powerful methods for studying the electronic, optical, and defect properties of perovskites, and providing atomistic explanations and guidance for experiments. Theoretical results have demonstrated that halide perovskites exhibit a series of superior electronic and optical properties for solar cell application, such as appropriate bandgap and band alignment, high optical absorption, bipolar carrier conductivity, tunable doping ability, and benign defect properties. There are many other unresolved aspects of perovskite solar cells where theory can help. One is anomalous hysteresis in perovskite solar cells.[96] Frost *et al.*[97] have proposed that the dipole moment of the organic cation could contribute to the hysteresis, although recent experiments suggest that grain sizes and interface physics may play an important role,[98] which requires further investigation. Further, first-principles investigations are needed to optimize the material's properties, and to find new perovskite candidates for high-efficiency, stable solar cells. As previously mentioned, the bandgap of the archetypal perovskite $CH_3NH_3PbI_3$ is larger than the ideal. Band structure engineering of $CH_3NH_3PbI_3$ could be extensively investigated by replacing organic cations, Pb, or I with other choices. Furthermore, the mechanisms of performance degradation are still not well studied, even though long-term stability is a crucial aspect of commercialization. Rapid degradation in moist environments has suggested that water may corrode perovskites. The chemical reactions between H_2O and the perovskite surface need to be studied carefully to unravel the reaction mechanism and to develop methods for stabilizing perovskites. Long-term, stable perovskite solar cells have recently been fabricated in the laboratory *via* chemical composition engineering.[67,99,100] However, the fundamental reasons for the alloy stabilization of the structures require more study. Overall, first-principles calculations will continue to play an important role in the understanding and development of the emerging technology of perovskite solar cells.

References

1. R. E. Cohen, *Nature*, 1992, **358**, 136–138.
2. M. A. Pena and J. L. G. Fierro, *Chem. Rev.*, 2001, **101**, 1981–2017.
3. P. Gao, M. Gratzel and M. K. Nazeeruddin, *Energy Environ. Sci.*, 2014, **7**, 2448–2463.
4. S. A. Bretschneider, J. Weickert, J. A. Dorman and L. Schmidt-Mende, *APL Mater.*, 2014, **2**, 040701.
5. H. J. Snaith, *J. Phys. Chem. Lett.*, 2013, **4**, 3623–3630.

6. N.-G. Park, *J. Phys. Chem. Lett.*, 2013, **4**, 2423–2429.
7. M. Liu, M. B. Johnston and H. J. Snaith, *Nature*, 2013, **501**, 395–398.
8. J. Burschka, N. Pellet, S.-J. Moon, R. Humphry-Baker, P. Gao, M. K. Nazeeruddin and M. Graetzel, *Nature*, 2013, **499**, 316–319.
9. M. M. Lee, J. Teuscher, T. Miyasaka, T. N. Murakami and H. J. Snaith, *Science*, 2012, **338**, 643–647.
10. I. Chung, B. Lee, J. He, R. P. H. Chang and M. G. Kanatzidis, *Nature*, 2012, **485**, 486–U494.
11. A. Kojima, K. Teshima, Y. Shirai and T. Miyasaka, *J. Am. Chem. Soc.*, 2009, **131**, 6050–6051.
12. N.-G. Park, *Mater. Today*, 2015, **18**, 65–72.
13. M. A. Green, A. Ho-Baillie and H. J. Snaith, *Nat. Photonics*, 2014, **8**, 506–514.
14. M. Gratzel, *Nat. Mater.*, 2014, **13**, 838–842.
15. M. D. McGehee, *Nat. Mater.*, 2014, **13**, 845–846.
16. http://www.nrel.gov/ncpv/images/efficiency_chart.jpg.
17. H.-S. Kim, C.-R. Lee, J.-H. Im, K.-B. Lee, T. Moehl, A. Marchioro, S.-J. Moon, R. Humphry-Baker, J.-H. Yum, J. E. Moser, M. Graetzel and N.-G. Park, *Sci. Rep.*, 2012, **2**, 591.
18. W. J. Yin, T. T. Shi and Y. Yan, *Adv. Mater.*, 2014, **26**, 4653–4658.
19. G. C. Xing, N. Mathews, S. Y. Sun, S. S. Lim, Y. M. Lam, M. Gratzel, S. Mhaisalkar and T. C. Sum, *Science*, 2013, **342**, 344–347.
20. S. D. Stranks, G. E. Eperon, G. Grancini, C. Menelaou, M. J. P. Alcocer, T. Leijtens, L. M. Herz, A. Petrozza and H. J. Snaith, *Science*, 2013, **342**, 341–344.
21. N. J. Jeon, J. H. Noh, Y. C. Kim, W. S. Yang, S. Ryu and S. I. Seok, *Nat. Mater.*, 2014, **13**, 897–903.
22. H. P. Zhou, Q. Chen, G. Li, S. Luo, T. B. Song, H. S. Duan, Z. R. Hong, J. B. You, Y. S. Liu and Y. Yang, *Science*, 2014, **345**, 542–546.
23. E. Edri, S. Kirmayer, D. Cahen and G. Hodes, *J. Phys. Chem. Lett.*, 2013, **4**, 897–902.
24. S. Y. Sun, T. Salim, N. Mathews, M. Duchamp, C. Boothroyd, G. C. Xing, T. C. Sum and Y. M. Lam, *Energy Environ. Sci.*, 2014, **7**, 399–407.
25. E. D. Palik, *Handbook of Optical Constants of Solids*, Academic Press, San Diego, CA, 1998.
26. E. Edri, S. Kirmayer, A. Henning, S. Mukhopadhyay, K. Gartsman, Y. Rosenwaks, G. Hodes and D. Cahen, *Nano Lett.*, 2014, **14**, 1000–1004.
27. E. Edri, S. Kirmayer, M. Kulbak, G. Hodes and D. Cahen, *J. Phys. Chem. Lett.*, 2014, **5**, 429–433.
28. E. Edri, S. Kirmayer, S. Mukhopadhyay, K. Gartsman, G. Hodes and D. Cahen, *Nat. Commun.*, 2014, **5**, 3461.
29. S. De Wolf, J. Holovsky, S. J. Moon, P. Loper, B. Niesen, M. Ledinsky, F. J. Haug, J. H. Yum and C. Ballif, *J. Phys. Chem. Lett.*, 2014, **5**, 1035–1039.
30. W.-J. Yin, Y. Yan and S.-H. Wei, *J. Phys. Chem. Lett.*, 2014, **5**, 3625–3631.
31. W.-J. Yin, H. Y. Chen, T. Shi, S.-H. Wei and Y. Yan, *Adv. Elect. Mater.*, 2015, **1**, 1500044.

32. W.-J. Yin, J.-H. Yang, J. Kang, Y. Yan and S.-H. Wei, *J. Mater. Chem. A*, 2015, **A3**, 8926–8942.
33. W.-J. Yin, T. Shi and Y. Yan, *Appl. Phys. Lett.*, 2014, **104**, 063903.
34. J. Miller, *Phys. Today*, 2014, **67**, 13.
35. C. H. Li, X. G. Lu, W. Z. Ding, L. M. Feng, Y. H. Gao and Z. G. Guo, *Acta Crystallogr., Sect. B: Struct. Sci.*, 2008, **64**, 702–707.
36. T. Baikie, Y. N. Fang, J. M. Kadro, M. Schreyer, F. X. Wei, S. G. Mhaisalkar, M. Graetzel and T. J. White, *J. Mater. Chem. A*, 2013, **1**, 5628–5641.
37. C. C. Stoumpos, C. D. Malliakas and M. G. Kanatzidis, *Inorg. Chem.*, 2013, **52**, 9019–9038.
38. J. M. Ball, M. M. Lee, A. Hey and H. J. Snaith, *Energy Environ. Sci.*, 2013, **6**, 1739–1743.
39. I. Chung, J. H. Song, J. Im, J. Androulakis, C. D. Malliakas, H. Li, A. J. Freeman, J. T. Kenney and M. G. Kanatzidis, *J. Am. Chem. Soc.*, 2012, **134**, 8579–8587.
40. D. A. Egger and L. Kronik, *J. Phys. Chem. Lett.*, 2014, **5**, 2728–2733.
41. E. Menendez-Proupin, P. Palacios, P. Wahnon and J. C. Conesa, *Phys. Rev. B: Condens. Matter Mater. Phys.*, 2014, **90**, 045207.
42. Y. Wang, T. Gould, J. F. Dobson, H. M. Zhang, H. G. Yang, X. D. Yao and H. J. Zhao, *Phys. Chem. Chem. Phys.*, 2014, **16**, 1424–1429.
43. S. Colella, E. Mosconi, P. Fedeli, A. Listorti, F. Gazza, F. Orlandi, P. Ferro, T. Besagni, A. Rizzo, G. Calestani, G. Gigli, F. De Angelis and R. Mosca, *Chem. Mater.*, 2013, **25**, 4613–4618.
44. A. Walsh and G. W. Watson, *J. Solid State Chem.*, 2005, **178**, 1422–1428.
45. A. Walsh, D. J. Payne, R. G. Egdell and G. W. Watson, *Chem. Soc. Rev.*, 2011, **40**, 4455–4463.
46. S. H. Wei and A. Zunger, *Phys. Rev. B: Condens. Matter Mater. Phys.*, 1997, **55**, 13605–13610.
47. J. Even, L. Pedesseau, J. M. Jancu and C. Katan, *J. Phys. Chem. Lett.*, 2013, **4**, 2999–3005.
48. L. Lang, J. H. Yang, H. R. Liu, H. J. Xiang and X. G. Gong, *Phys. Lett. A*, 2014, **378**, 290–293.
49. M. H. Du, *J. Mater. Chem. A*, 2014, **2**, 9091–9098.
50. A. Amat, E. Mosconi, E. Ronca, C. Quarti, P. Umari, M. K. Nazeeruddin, M. Gratzel and F. De Angelis, *Nano Lett*, 2014, **14**, 3608–3616.
51. P. Umari, E. Mosconi and F. De Angelis, *Sci. Rep.*, 2014, **4**, 4464.
52. F. Brivio, K. T. Butler, A. Walsh and M. van Schilfgaarde, *Phys. Rev. B: Condens. Matter Mater. Phys.*, 2014, **89**, 155204.
53. H. J. Zhang, C. X. Liu, X. L. Qi, X. Dai, Z. Fang and S. C. Zhang, *Nat. Phys.*, 2009, **5**, 438–442.
54. W. Shockley and H. J. Queisser, *J. Appl. Phys.*, 1961, **32**, 510.
55. L. P. Yu and A. Zunger, *Phys. Rev. Lett.*, 2012, **108**, 068701.
56. L. P. Yu, R. S. Kokenyesi, D. A. Keszler and A. Zunger, *Adv. Energy Mater.*, 2013, **3**, 43–48.

57. M. A. Green, K. Emery, Y. Hishikawa, W. Warta and E. D. Dunlop, *Prog. Photovoltaics*, 2014, **22**, 701–710.
58. A. Polizzotti, I. L. Repins, R. Noufi, S. H. Wei and D. B. Mitzi, *Energy Environ. Sci.*, 2013, **6**, 3171–3182.
59. S. Y. Chen, J. H. Yang, X. G. Gong, A. Walsh and S. H. Wei, *Phys. Rev. B: Condens. Matter Mater. Phys.*, 2010, **81**, 245204.
60. S. B. Zhang, S. H. Wei, A. Zunger and H. Katayama-Yoshida, *Phys. Rev. B: Condens. Matter Mater. Phys.*, 1998, **57**, 9642–9656.
61. S. Y. Chen, X. G. Gong, A. Walsh and S. H. Wei, *Appl. Phys. Lett.*, 2009, **94**, 041903.
62. J. Kim, S. H. Lee, J. H. Lee and K. H. Hong, *J. Phys. Chem. Lett.*, 2014, **5**, 1312–1317.
63. W. J. Yin, S. H. Wei, M. M. Al-Jassim and Y. Yan, *Appl. Phys. Lett.*, 2011, **99**, 142109.
64. A. S. Subbiah, A. Halder, S. Ghosh, N. Mahuli, G. Hodes and S. K. Sarkar, *J. Phys. Chem. Lett.*, 2014, **5**, 1748–1753.
65. P. Qin, S. Tanaka, S. Ito, N. Tetreault, K. Manabe, H. Nishino, M. K. Nazeeruddin and M. Gratzel, *Nat. Commun.*, 2014, **5**, 3834.
66. J. A. Christians, R. C. M. Fung and P. V. Kamat, *J. Am. Chem. Soc.*, 2014, **136**, 758–764.
67. A. Mei, L. Liu, Z. Ku, T. Liu, Y. Rong, M. Xu, M. Hu, J. Chen, Y. Yang, M. Grätzel and H. Han, *Science*, 2014, **345**, 295–298.
68. W. Abu Laban and L. Etgar, *Energy Environ. Sci.*, 2013, **6**, 3249–3253.
69. J. J. Shi, J. Dong, S. T. Lv, Y. Z. Xu, L. F. Zhu, J. Y. Xiao, X. Xu, H. J. Wu, D. M. Li, Y. H. Luo and Q. B. Meng, *Appl. Phys. Lett.*, 2014, **104**, 063901.
70. S. Aharon, B. El Cohen and L. Etgar, *J. Phys. Chem. C*, 2014, **118**, 17160–17165.
71. C. Bi, Y. B. Yuan, Y. J. Fang and J. S. Huang, *Adv. Energy Mater.*, 2014, **2**, 1401616.
72. Z. G. Xiao, Q. F. Dong, C. Bi, Y. C. Shao, Y. B. Yuan and J. S. Huang, *Adv. Mater.*, 2014, **26**, 6503–6509.
73. D. Y. Liu, M. K. Gangishetty and T. L. Kelly, *J. Mater. Chem. A*, 2014, **2**, 19873–19881.
74. W. J. Yin, J. H. Yang, J. Kang, Y. Yan and S.-H. Wei, *J. Mater. Chem. A*, 2015, **3**, 8926–8942.
75. W. J. Yin, Y. L. Wu, R. Noufi, M. Al-Jassim and Y. Yan, *Appl. Phys. Lett.*, 2013, **102**, 193905.
76. W. J. Yin, Y. L. Wu, S. H. Wei, R. Noufi, M. M. Al-Jassim and Y. Yan, *Adv. Energy Mater.*, 2014, **4**, 1300712.
77. V. W. Bergmann, S. A. L. Weber, F. J. Ramos, M. K. Nazeeruddin, M. Gratzel, D. Li, A. L. Domanski, I. Lieberwirth, S. Ahmad and R. Berger, *Nat. Commun.*, 2014, **5**, 5001.
78. G. Giorgi, J. I. Fujisawa, H. Segawa and K. Yamashita, *J. Phys. Chem. Lett.*, 2013, **4**, 4213–4216.
79. H. Oga, A. Saeki, Y. Ogomi, S. Hayase and S. Seki, *J. Am. Chem. Soc.*, 2014, **136**, 13818–13825.

80. V. D'Innocenzo, A. R. S. Kandada, M. De Bastiani, M. Gandini and A. Petrozza, *J. Am. Chem. Soc.*, 2014, **136**, 17730.
81. Y. C. Shao, Z. G. Xiao, C. Bi, Y. B. Yuan and J. S. Huang, *Nat.. Commun.*, 2014, **5**, 5784.
82. G. Grancini, S. Marras, M. Prato, C. Giannini, C. Quarti, F. De Angelis, M. De Bastiani, G. E. Eperon, H. J. Snaith, L. Manna and A. Petrozza, *J. Phys. Chem. Lett.*, 2014, **5**, 3836–3842.
83. J. B. You, Z. R. Hong, Y. Yang, Q. Chen, M. Cai, T. B. Song, C. C. Chen, S. R. Lu, Y. S. Liu, H. P. Zhou and Y. Yang, *Acs Nano*, 2014, **8**, 1674–1680.
84. W. J. Yin, Y. Yan and S. H. Wei, *J. Phys. Chem. Lett.*, 2014, **5**, 3625–3631.
85. M. I. Dar, N. Arora, P. Gap, S. Ahmad, M. Gratzel and M. K. Nazeeruddin, *Nano Lett.*, 2014.
86. E. Mosconi, E. Ronca and F. De Angelis, *J. Phys. Chem. Lett.*, 2014, **5**, 2619–2625.
87. S. T. Williams, F. Zuo, C. C. Chueh, C. Y. Liao, P. W. Liang and A. K. Y. Jen, *Acs Nano*, 2014, **8**, 10640–10654.
88. H. Yu, F. Wang, F. Xie, W. Li, J. Chen and N. Zhao, *Adv. Funct. Mater.*, 2014, **24**, 7102–7108.
89. A. E. Williams, P. J. Holloman, M. J. Carnie, M. L. Davies, D. A. Worsley and T. M. Watson, *J. Mater. Chem. A*, 2014, **2**, 19338–19346.
90. E. L. Unger, A. R. Bowring, C. J. Tassone, V. Pool, A. Gold-Parker, R. Cheacharoen, K. H. Stone, E. T. Hoke, M. F. Toney and M. D. McGehee, *Chem. Mater.*, 2014, **26**, 7158–7165.
91. D. Abou-Ras, S. S. Schmidt, R. Caballero, T. Unold, H. W. Schock, C. T. Koch, B. Schaffer, M. Schaffer, P. P. Choi and O. Cojocaru-Miredin, *Adv. Energy Mater.*, 2012, **2**, 992–998.
92. D. Abou-Ras, B. Schaffer, M. Schaffer, S. S. Schmidt, R. Caballero and T. Unold, *Phys. Rev. Lett.*, 2012, **108**, 075502.
93. C. Li, Y. L. Wu, J. Poplawsky, T. J. Pennycook, N. Paudel, W. J. Yin, S. J. Haigh, M. P. Oxley, A. R. Lupini, M. Al-Jassim, S. J. Pennycook and Y. Yan, *Phys. Rev. Lett.*, 2014, **112**, 156103.
94. Z. W. Ren, A. Ng, Q. Shen, H. C. Gokkaya, J. C. Wang, L. J. Yang, W. K. Yiu, G. X. Bai, A. B. Djurisic, W. W. F. Leung, J. H. Hao, W. K. Chan and C. Surya, *Sci. Rep.*, 2014, **4**, 6752.
95. T. Shi, W.-J. Yin and Y. Yan, *J. Phys. Chem. C*, 2014, **118**, 25350–25354.
96. H. J. Snaith, A. Abate, J. M. Ball, G. E. Eperon, T. Leijtens, N. K. Noel, S. D. Stranks, J. T. W. Wang, K. Wojciechowski and W. Zhang, *J. Phys. Chem. Lett.*, 2014, **5**, 1511–1515.
97. J. M. Frost, K. T. Butler and A. Walsh, *Appl. Mater*, 2014, **2**, 081506.
98. E. L. Unger, E. T. Hoke, C. D. Bailie, W. H. Nguyen, A. R. Bowring, T. Heumuller, M. G Christoforo and M. D McGehee, *Energy Environ. Sci.*, 2014, **7**, 3690–3698.
99. J. H. Noh, S. H. Im, J. H. Heo, T. N. Mandal and S. I. Seok, *Nano Lett.*, 2013, **13**, 1764–1769.
100. N. Pellet, P. Gao, G. Gregori, T. Y. Yang, M. K. Nazeeruddin, J. Maier and M. Gratzel, *Angew. Chem., Int. Ed.*, 2014, **53**, 3151–3157.

CHAPTER 7

Electronic Properties of Metal Halide Perovskites

JACKY EVEN,*[a] LAURENT PEDESSEAU,[a] DANIEL SAPORI,[a] ALAIN ROLLAND,[a] MIKAËL KEPENEKIAN[b] AND CLAUDINE KATAN[b]

[a] Fonctions Optiques pour les Technologies de l'Information, FOTON UMR 6082, CNRS, INSA de Rennes, 35708 Rennes, France; [b] Institut des Sciences Chimiques de Rennes, ISCR UMR 6226, CNRS, Université de Rennes 1, 35042 Rennes, France

*Email: Jacky.Even@insa-rennes.fr

7.1 Introduction to All-Inorganic (AIP) and Hybrid Organic (HOP) Metal Halide Perovskites

Before the perovskite boom in solar-to-electric power conversion,[1–6] layered Hybrid Organic Perovskites (HOP) had already drawn significant interest for their applications in optoelectronics and microelectronics for more than two decades.[7–14] Currently, lead halide based HOP of general formula $CH_3NH_3PbX_3$ (X = I, Br, Cl) are the most popular. However, the presence of lead raises environmental issues, and substitution of Pb has started to be addressed with tin. For both inorganic di-cations ($M = Pb^{2+}$, Sn^{2+}), the crystal structure reveals a three-dimensional perovskite lattice with corner-shared MX_6 octahedra (Figure 7.1a). Replacement of the organic methyl-ammonium (MA: $[CH_3NH_3]^+$) cation has also received attention, including formamidinium (FA: $[HC(NH_2)_2]^+$) and guanidinium (GA: $[C(NH_2)_3]^+$)

RSC Energy and Environment Series No. 16
Unconventional Thin Film Photovoltaics
Edited by Enrico Da Como, Filippo De Angelis, Henry Snaith and Alison Walker

Published by the Royal Society of Chemistry, www.rsc.org

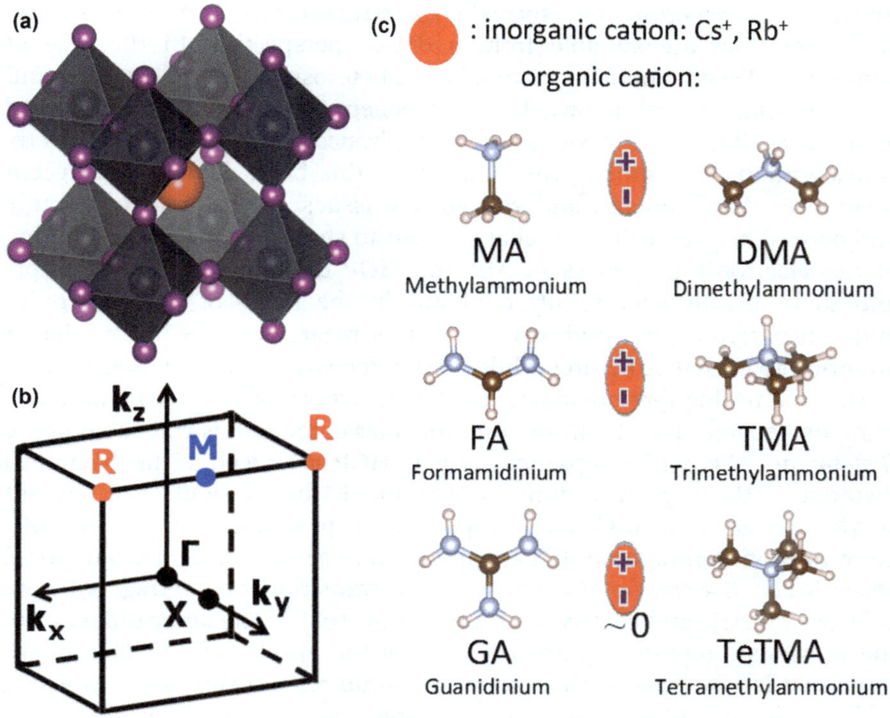

Figure 7.1 (a) Real space 3-D view of the *Pm*-3*m* reference cubic crystal structure of metal halide AIP and HOP of general formulae AMX_3 where A is an organic or inorganic cation sketched by the red ball at the cube center, M = Pb, Sn, Ge and X = I, Br, Cl. (b) Reciprocal space 3-D view showing the first Brillouin Zone (BZ) of the *Pm*-3*m* space group. Points of high symmetry in the cubic BZ are indicated by conventional letters: Γ denotes the origin of the BZ; X is the center of a square face at the BZ boundary, M is a center of a cube edge; and R are the vertices of the cube. Reprinted with permission from ref. 43. Copyright 2014 American Chemical Society. (c) Examples of inorganic and organic cations leading to AIP and HOP, respectively. The position of the organic cation in the cubic phase of HOP crystals is on average at the center of the cube.

cations (Figure 7.1c).[15,16] Analogue all-inorganic perovskites (AIP) also exist within the class of AMX_3 perovskites, with $A = Cs^+$ or Rb^+ (Figure 7.1c).[17–22]

Clearly, the understanding of the physical properties of operational HOP and the design of novel devices with improved performances may greatly benefit from theoretical work. In this context, many computational chemists and physicists have recently taken on the challenge. Prior to 2013, most theoretical studies either focused on layered HOP[23–25] or 3-D AIP.[21,22,26–29] A few papers dealt with 3-D HOP,[23,26,30,31] but overlooked the effect of spin–orbit coupling (SOC). Currently, most of the theoretical work relies on the use of Density Functional Theory (DFT). However, it is difficult to make theoretical predictions or relevant comparisons with physical parameters derived from experiments, especially for structures such as HOP heterostructures, that are suitable from a device perspective. In the case of conventional semiconductor materials and heterostructures (*e.g.* silicon and III–V semiconductors), knowledge often benefits most from semi-empirical approaches based on basic solid-state physics concepts and symmetry considerations.[32–34] Among the chapters of this book dedicated to recent theoretical developments and related key issues in the field of organic and perovskite solar cells, this chapter aims to show how the understanding of the electronic properties of AIP and HOP can benefit from concepts and tools already successfully deployed in the fields of optoelectronics and conventional semiconductors.[34] In this regard, we will pay particular attention to crystal structures while mainly focussing on MA-based HOP.

The room (high)-temperature crystal structure of 3-D HOP and AIP remains debated, but the most commonly accepted one is the cubic phase (*Pm*-3*m*, no. 221), which is pseudo-cubic in HOP as a result of the dynamical disorder of the organic cations in MA- and FA-based HOP.[15,35,36] Recent single crystal X-ray diffraction experiments performed on $CH_3NH_3MI_3$ close to room temperature (RT) support a non-centrosymmetric tetragonal space group (*P*4*mm*, no. 99) and a phase transition to a tetragonal non-centrosymmetric and centred *I*4*cm* phase (no. 108).[15] The latter differs from the previously reported symmetric *I*4/*mcm* (no. 140) phase.[35] This phase transition is associated with a group–subgroup relationship between *P*4*mm* and *I*4*cm* (Figure 7.2),[37,38] similar to the previously proposed *Pm*-3*m*–*I*4/*mcm* phase sequence for $CH_3NH_3PbI_3$ ($T_c = 327.4$ K) and $CH_3NH_3PbBr_3$ ($T_c = 236.9$ K).[35] The low temperature phase of $CH_3NH_3PbX_3$ is expected to belong to the *Pnma* (no. 62) space group with $Z = 4$, 8 and 4 for X = I, Br and Cl, respectively.[36,39,40] We should also mention that the phase transitions in AIP compounds are different; some of these phase transitions have been studied in detail in the past.[18,20] Nevertheless, the *Pm*-3*m* (pseudo)cubic phase plays a central role in the understanding of most of the physical properties of all of these metal halide AIP and HOP, as shown below.[34,41–44]

This chapter is organized as follows. We start with an extensive introduction to the symmetry properties of metal halide perovskites,[33,45,46] which is essential to capture the main physics of this class of semiconductors. The pseudo-cubic phase serves as a reference to investigate 3-D AIP and HOP. It will be shown that many effects, such as strain, phase transitions, relativistic effects, local distortions due to the organic cations in HOP, loss of inversion symmetry, *etc.*, can be considered as perturbations to the pseudo-cubic phase. The importance of spin–orbit coupling is particularly emphasized. Next, starting from the Landau theory, the influence of structural phase transitions in HOP is discussed, with special attention paid to coupled order parameters, Brillouin Zone (BZ) folding, lattice strain and loss of inversion symmetry. The third section addresses the issue of the nature of electronic excitation, namely photoexcited species, charge carriers and Auger effects. The final section offers some concluding reflections.

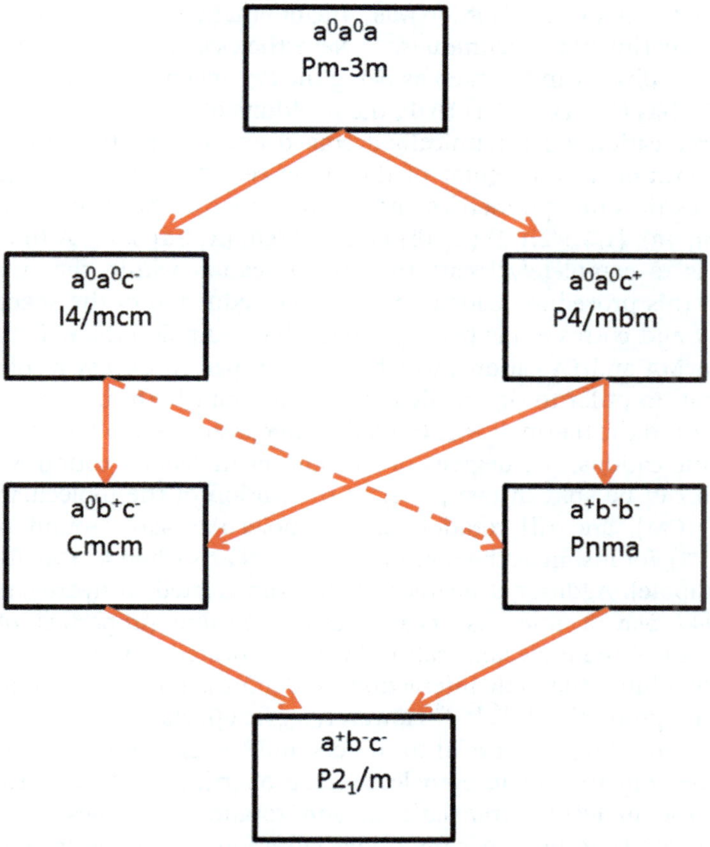

Figure 7.2 Schematic diagram recording the structures for perovskites found by Howard and Stokes (1998). It shows the space group symmetry, along with the Glazer symbol for the tilts. The lines indicate group–subgroup relationships, and a dashed line joining a group with its subgroup means that the corresponding phase transition is, in Landau theory, required to be the first order.
The figure has been reproduced from Howard and Stokes, *Acta Crystallogr., Sect. B: Struct. Sci.*, 2002, **B58**, 565.[37]

7.2 A Pedestrian Guide to Symmetry Properties of the Reference Cubic Structure

7.2.1 General Overview and Lattice Symmetries

Figure 7.1a represents the real space 3-D view of the *Pm-3m* crystal structure of metal halide AMX_3 perovskites, where A can be an inorganic or organic cation (Figure 7.1c), M = Pb, Sn or Ge and X = I, Br or Cl. At high temperatures, the experimental crystal structures of HOP do not reveal strictly ordered and symmetric phases as a consequence of the molecular symmetry of the organic cations, which does not fit the site symmetry of the lattice.

The related dynamical disorder was already analyzed a long time ago by various experimental techniques.[35] Nevertheless, their reference high temperature phase can be taken as being the *Pm*-3*m* cubic phase,[34,41–44] *e.g.*, that of $CsPbX_3$ ($Z = 1$). In 3-D HOP, the positions of the individual atoms of the organic cation are dynamically averaged and occupy the same special Wickoff position at the center of the cubic cell. The *Pm*-3*m* space group severely restricts the symmetries and multiplicities of the species located at (0,0,0) (M, 1a), (1/2,1/2,1/2) (A, 1b) and (1/2,0,0) (X, 3d) sites. A first *ad-hoc* attempt is to completely locate the organic cation within the cubic cell. However, this procedure leads to an artificial reduction of the space group symmetry and corresponds to long-range dipole–dipole interactions in the lattice for MA and FA cations, which are not expected to play a role in the static limit. In order to rigorously take into account both the *Pm*-3*m* lattice symmetry and the thermally-activated disorder associated to the tumbling of the organic cations, the degrees of freedom of molecular cation A located at site 1b can be split in two parts: (i) translation of the molecular center of mass (CM), and (ii) rotation of the molecular axis around position (1/2,1/2,1/2), for instance, the rotation of the C–N axis when A = MA (oriented electric dipole). Additional degrees of freedom related to hydrogen atoms could also play a role. As regards the electronic properties of HOP, replacement of the molecular cation by a Cs^+ located at the molecular CM accurately mimics the ionic interactions and affords reliable band diagrams and related properties.[41,44,47–49] However, steric effects, atomic motion and molecular tumbling are crucial to understand the crystal packing of HOP, their phase transitions and optoelectronic properties.[34,43,44] Besides, separation of the molecular translational and rotational degrees of freedom involves neglecting the linear rotational–vibrational coupling in the lattice, which is allowed by symmetry in the cubic phase (Sections 7.2.2 and 7.2.3). However, at high temperature, these excitations lie in different energy ranges, validating such an approximation. This assumption no longer applies in the low temperature phase, since vibrations and rotations correspond to coupled order parameters that condense simultaneously at the phase transition (Section 7.3.1). This linear coupling is also expected to show itself in the high temperature regime related to low energy polaronic excitations (Section 7.4.4).

The *Pm*-3*m* space group is a symmorphic space group.[33] Its symmetry operations rely on the 48 symmetry operations of the O_h (*m*3*m*) point group and the translational symmetries of the cubic lattice. Among the point group symmetries, inversion plays an important role, yielding either symmetric (g) or antisymmetric (u) irreducible representations (IR) of the symmetry group (Tables 7.1–7.3). Noteworthy, symmetries of sites (M, 1a) and (A, 1b) belong to O_h. Translational symmetry of the lattice must be used to analyze delocalized and collective excitations, such as free carriers, Wannier excitons, phonons and rotational pseudo-spins. It is thus necessary in many cases to switch to a reciprocal space representation of the lattice symmetries.[33] Figure 7.1b shows the reciprocal space 3-D view of the first BZ of the

Pm-3*m* space group, with some points of high symmetry.[43] Reciprocal lattice vectors, $\boldsymbol{K}_m$, are related to the translational periodicity of the reciprocal lattice, thus enabling a representation of the various lattice properties within the first BZ. In order to further address lattice point group symmetries, one has to consider the group of the wave vector $\boldsymbol{k}$, which is the set of space group operations that transform $\boldsymbol{k}$ into itself or an equivalent $\boldsymbol{k}$-vector, $\boldsymbol{k}+\boldsymbol{K}_m$. The star of $\boldsymbol{k}$ is the set of independent wave vectors generated by carrying out all the point group operations on $\boldsymbol{k}$. For a general $\boldsymbol{k}$-vector within the BZ of the *Pm*-3*m* group, the group of $\boldsymbol{k}$ is a subgroup of O_h and the star of $\boldsymbol{k}$ contains several wave vectors. The center of the BZ $\Gamma(0,0,0)$, $R(1/2,1/2,1/2)$, and $M(1/2,1/2,0)$ are special high symmetry $\boldsymbol{k}$-points of the first BZ that play an important role in the physical properties of AIP and HOP.[34,43,44] For Γ and R points, the group of the wave vector is O_h and its star reduces to the vector itself. For this reason, many lattice properties of HOP can be analyzed using the character table of the IR of the O_h point group (Tables 7.1 and 7.2).

In order to account for the site symmetry, either in an isolated octahedron or in the perovskite cubic lattice, it is first necessary to compute its site reducible representation χ_{site}. From the reducible representation χ_{phys}, describing a given on-site atomic or molecular physical parameter (atomic displacement or orbital, molecular pseudospin or static orientation...), one may deduce the corresponding physical representation within an octahedron or a lattice by computing the product $\chi_{site} \otimes \chi_{phys}$. For an isolated

Table 7.1 Character table of the IR of the O_h simple point group[45] and *Pm*-3*m* simple space group[46] at the Γ and R points of the BZ.

O_h	$P_{m\text{-}3m}$	$P_{m\text{-}3m}$	E	$3C_4^2$	$6C_4$	$6C_2$	$8C_3$	i	$3iC_4^2$	$6iC_4$	$6iC_2$	$8iC_3$
A_{1g}	Γ_1^+	R_1^+	1	1	1	1	1	1	1	1	1	1
A_{2g}	Γ_2^+	R_2^+	1	1	−1	−1	1	1	1	−1	−1	1
E_g	Γ_3^+	R_3^+	2	2	0	0	−1	2	2	0	0	−1
T_{1g}	Γ_4^+	R_4^+	3	−1	1	−1	0	3	−1	1	−1	0
T_{2g}	Γ_5^+	R_5^+	3	−1	−1	1	0	3	−1	−1	1	0
A_{1u}	Γ_1^-	R_1^-	1	1	1	1	1	−1	−1	−1	−1	−1
A_{2u}	Γ_2^-	R_2^-	1	1	−1	−1	1	−1	−1	1	1	−1
E_u	Γ_3^-	R_3^-	2	2	0	0	−1	−2	−2	0	0	1
T_{1u}	Γ_4^-	R_4^-	3	−1	1	−1	0	−3	1	−1	1	0
T_{2u}	Γ_5^-	R_5^-	3	−1	−1	1	0	−3	1	1	−1	0

Table 7.2 Character table of the IR of the O_h double point group[45] and *Pm*-3*m* double space group[46] at the Γ and R points of the BZ.

O_h	$P_{m\text{-}3m}$	$P_{m\text{-}3m}$	E	$3C_4^2$	$6C_4$	$6C_2$	$8C_3$	i	$3iC_4^2$	$6iC_4$	$6iC_2$	$8iC_3$
$E_{1/2g}$	Γ_6^+	R_6^+	2	0	$\sqrt{2}$	0	1	2	0	$\sqrt{2}$	0	1
$E_{5/2g}$	Γ_7^+	R_7^+	2	0	$-\sqrt{2}$	0	1	2	0	$-\sqrt{2}$	0	1
$F_{3/2g}$	Γ_8^+	R_8^+	4	0	0	0	−1	4	0	0	0	−1
$E_{1/2u}$	Γ_6^-	R_6^-	2	0	$\sqrt{2}$	0	1	−2	0	$-\sqrt{2}$	0	−1
$E_{5/2u}$	Γ_7^-	R_7^-	2	0	$-\sqrt{2}$	0	1	−2	0	$\sqrt{2}$	0	−1
$F_{3/2u}$	Γ_8^-	R_8^-	4	0	0	0	−1	−4	0	0	0	1

Table 7.3 Character table of the IR of the D_{4h} simple point group[45] and *Pm-3m* simple space group[46] at the *M* point of the BZ.

D_{4h}	$P_{m\text{-}3m}$	E	$2C_2'$	C_2	$2C_4$	$2C_2''$	i	$2iC_2'$	iC_2	$2iC_4$	$2iC_2''$
A_{1g}	M_1^+	1	1	1	1	1	1	1	1	1	1
B_{1g}	M_2^+	1	1	1	−1	−1	1	1	1	−1	−1
A_{2g}	M_3^+	1	−1	1	1	−1	1	−1	1	1	−1
B_{2g}	M_4^+	1	−1	1	−1	1	1	−1	1	−1	1
E_g	M_5^+	2	0	−2	0	0	2	0	−2	0	0
A_{1u}	M_1^-	1	1	1	1	1	−1	−1	−1	−1	−1
B_{1u}	M_2^-	1	1	1	−1	−1	−1	−1	−1	1	1
A_{2u}	M_3^-	1	−1	1	1	−1	−1	1	−1	−1	1
B_{2u}	M_4^-	1	−1	1	−1	1	−1	1	−1	1	−1
E_u	M_5^-	2	0	−2	0	0	−2	0	2	0	0

MX_6 octahedron or a localized excitation around a metal site within the crystal, the local point symmetry is O_h. The central M atom at (0,0,0) is thus described by $\chi_{site} = A_{1g}$. The 6 X atoms at a general position (1/2,0,0) correspond to a $\chi_{site} = A_{1g} + E_g + T_{1u}$. In HOP, considering the 8 nearest molecular neighbors of the central M atom, $\chi_{site} = A_{1g} + T_{2g} + A_{2u} + T_{1u}$.

Next, when translational symmetry is taken into account, one has to compute χ_{site} at multiple points of the BZ. Moreover, among the 6 X atoms surrounding a central M atom, only 3 have to be considered, since the 3 others are equivalent through translations of the cubic lattice. For the same reasons, a single molecular site A, with $\chi_{site} = \Gamma_1^+$ (see Table 7.1 for IR labels), is needed to generate all the molecules within the periodic HOP cubic lattice. At the Γ point, the site symmetries thus correspond to Γ_1^+, $\Gamma_1^+ + \Gamma_3^+$, and Γ_1^+ for M, X and A, respectively. If one reduces the A molecule to an electric dipole, the molecular symmetry would be described by $\chi_{phys} = \Gamma_4^-$, which is incompatible with the lattice symmetry since $\chi_{site} \otimes \chi_{phys} = \Gamma_1^+ \otimes \Gamma_4^- = \Gamma_4^-$ does not contain the totally symmetric representation Γ_1^+. For this reason, molecular orientations are necessarily disordered when taking lattice periodicity into account. It is noteworthy that static molecular configurations remain relevant when analyzing local excitations. At the *R* point, the site symmetries belong to R_1^+, R_4^-, and R_2^- for M, X and A, respectively.

7.2.2 Localized and Collective Vibrations and Strain

In the simplest scenario, vibrations of a molecule or a crystal lattice correspond to degrees of freedom that can be studied in the harmonic approximation of small displacements. For AIP and HOP, it is instructive to start with an isolated MX_6 octahedron, to which the eight CM of the nearest A atoms or molecules (pseudoatoms) are added. The 45 degrees of freedom correspond to 39 vibrational modes, 3 rigid body translations of the overall CM and 3 rigid body rotations of the entire structure. From the inspection of the atomic site symmetries (Section 7.2.1), using the vectorial representation of the O_h group for the displacements $\chi_{phys} = T_{1u}$, the 45 degrees of freedom decompose into: T_{1u} for the central M atom, $A_{1g} + E_g + T_{1g} + T_{2g} + 2T_{1u} + T_{2u}$ for the 6 X atoms and $A_{1g} + E_g + T_{1g} + 2T_{2g} + A_{2u} + E_u + 2T_{1u} + T_{2u}$ for the 8 A atoms or CM. The rigid body translations and rotations of the octahedron correspond respectively to the T_{1u} and T_{2g} IR. The remaining internal vibrations thus decompose into $2A_{1g} + 2E_g + 2T_{1g} + 2T_{2g} + A_{2u} + E_u + 4T_{1u} + 2T_{2u}$. The A_{1g} totally symmetric mode corresponds to the volumetric variation of the dressed octahedron with X and A motions along the M–X bonds and M–A axes, respectively. All the symmetric (g) modes involve motion of X and/or A only.

To describe lattice vibrational properties (phonons) of cubic AIP or HOP, two main differences must be accounted for on the same grounds as above: only 3 halogens and 1 atom (or CM) A shall be considered. For vibrational modes at the Γ point (Figure 7.3),[50] M, A and 3 X yield Γ_4^-, Γ_4^- and $2\Gamma_4^- + \Gamma_5^-$, respectively. The three acoustic modes belong to Γ_4^- whereas the twelve optical phonons correspond to the $3\Gamma_4^- + \Gamma_5^-$ IR. It is thus straightforward to predict that all the optical modes are triply degenerate and none of them is Raman active. Moreover, the Γ_5^- optical phonon mode is neither Raman nor infrared active. The IR decomposition at Γ will be useful to analyze electron–phonon couplings (Section 7.4.2). To determine the representations of the vibrational modes at R, one has to consider different site symmetries. For the optical modes, the M, A and 3 X atoms (or CM) yield R_4^-, R_5^+ and $R_1^+ + R_3^+ + R_4^+ + R_5^+$, respectively. It is noteworthy that neither the metal nor the A atom/molecular CM are involved in the triply

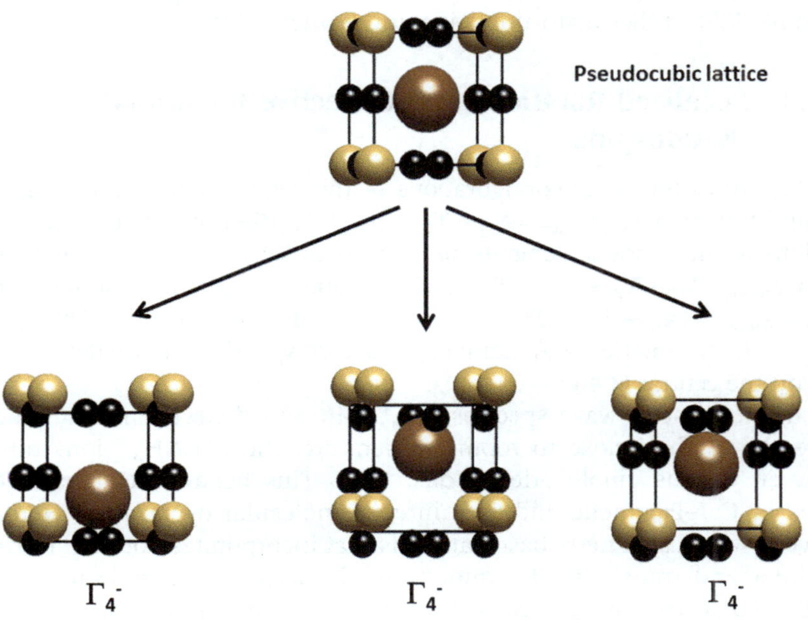

Figure 7.3 Zone center Γ_4^- normal vibrational modes of AMX_3 in its ideal cubic crystal perovskite structure with space group *Pm*-3*m*.

Table 7.4 Some antiferrodistorsive space group changes associated with order parameters for ABX_3 crystal structures.[51] M_3^+ and R_4^+ correspond to octahedra tilts, whereas Γ_3^+, M_2^+, and R_3^+ correspond to secondary parameters and possible Jahn–Teller (JT) distortions with possible origin shift (os).

	M_3^+			R_4^+			Γ_3^+		M_2^+		R_3^+		
Order parameters	q_1	q_2	q_3	q_4	q_5	q_6	q_{tz}	q_{oz}	q_{1JT}	q_{2JT}	q_{4JT}	q_{5JT}	q_{6JT}
I4/mcm	0	0	0	*a*	0	0	*a*	0	0	0	0	0	os
Pnma	0	*a*	0	*b*	0	*b*	$-\boldsymbol{a}/2$	$-\boldsymbol{a}\sqrt{3}/2$	0	*a*	0	0	0
P4/mbm	*a*	0	0	0	0	0	*a*	0	os	0	0	0	0

degenerate R_4^+ phonon modes, but only the X atoms. These modes correspond to rotations of the octahedra. This analysis will also help to describe phase transitions in metal halide AIP and HOP similarly to the approach developed for inorganic perovskite oxides such as $CaTiO_3$ (Section 7.3.1).

The volumetric strain ($q_a = \varepsilon_{xx} + \varepsilon_{yy} + \varepsilon_{zz}$), tetragonal strains ($q_{oz} = \varepsilon_{xx} - \varepsilon_{yy}$ and $q_{tz} = 2\varepsilon_{zz} - \varepsilon_{xx} - \varepsilon_{yy}$) and shear strains ($\varepsilon_{xy}$, ε_{xz}, ε_{yz}) are described by the Γ_1^+, Γ_3^+ and Γ_5^+ IR, respectively. At Γ, and for electron to acoustic phonon coupling, the symmetry properties of the strain tensor components are relevant (Section 7.4.2). In addition, these strain components may couple to the order parameters of certain phase transitions, leading to Jahn–Teller distortions (Γ_3^+). More generally, some of the space group changes associated with order parameters for AIP (*P4/mbm*) and HOP (*I4/mcm* and *Pnma*) are summarized in Table 7.4.[51] M_3^+ and R_4^+ define octahedra tiltings, whereas Γ_3^+, M_2^+, and R_3^+ correspond to secondary parameters and possible Jahn–Teller distortions, including origin shift.

7.2.3 Localized Rotations and Collective Rotational Pseudospins

Starting from the static configurations of the 8 nearest neighbor molecules around a metal having $\chi_{site} = A_{1g} + T_{2g} + A_{2u} + T_{1u}$ (Section 7.2.1), and allowing each molecule to orient along one of the 8 possible $[\pm 1, \pm 1, \pm 1]$ orientations, then $\chi_{phys} = A_{1g} + T_{2g} + A_{2u} + T_{1u}$. The 64 static configurations decompose as $\chi_{site} \otimes \chi_{phys} = 4A_{1g} + 2E_g + 2T_{1g} + 6T_{2g} + 4A_{2u} + 2E_u + 6T_{1u} + 2T_{2u}$. Among the A_{1g} totally symmetric configurations, two correspond to molecules pointing toward the central M atom.

Early millimeter wave spectroscopy, NMR and dielectric measurements have shown that, close to room temperature, the $CH_3NH_3^+$ ions tumble between various dipole orientations.[35,52–54] This behavior can be investigated by Car–Parrinello and path integral molecular dynamics.[55] Here, we proceed with a symmetry-based analysis that incorporates both the intrinsic disorder and translational symmetry of the lattice. The molecular orientations (C–N axis) are described by a discrete pseudospin variable for each molecule that is equal to 1 for a given orientation and 0 for all others. For the whole lattice, occupation numbers may be defined for the orientations. The model thus incorporates in a simple way the thermally-activated disorder of the molecular orientations in a local cubic environment and symmetry. In order to compute the various symmetry equivalent orientations of a dipole in a local cubic environment, we proceed by analyzing the symmetrized combinations of normalized {a,b,c} unit vectors for the O_h group. In that case, the sets of symmetry equivalent vectors can be classified into six scenarios: {1,0,0} (6, A); {1,1,0} (12, B); {1,1,1} (8, C); {a,b,0} (24, D); {a,a,b} (24, E); and {a,b,c} (48, F), where the corresponding multiplicity m and scenario label are given in parentheses. The multiplicity corresponds to the dimension of the point group reducible representation for a given set. The three scenarios (A, B and C) correspond to the simplest sets of unit vectors and have been suggested earlier to rationalize the influence of molecular disorder in phase transitions.[52] In scenarios A, B and C, the molecular dipole points towards a cubic cell facet, an halogen atom, or the center of an octahedron, respectively.[52] However, experimental investigations (calorimetry, X-ray diffraction) do not allow one to differentiate the role played by each scenario in the phase transitions of $CH_3NH_3PbX_3$ (see also Section 7.3.1).

Next, in order to analyze the orientational disorder and its relation to the lattice symmetry, we define a pseudospin that affords a general decomposition among the m possible equivalent orientations. When all molecular configurations are considered, the general pseudospin state for scenario C is an eight-fold vector

$$\left| \begin{matrix} \boldsymbol{N}_1 \\ \boldsymbol{N}_2 \\ \boldsymbol{N}_3 \\ \boldsymbol{N}_4 \\ \boldsymbol{N}_5 \\ \boldsymbol{N}_6 \\ \boldsymbol{N}_7 \\ \boldsymbol{N}_8 \end{matrix} \right.$$

where $\boldsymbol{N}_i$ is the occupation number of one of the eight $[\pm 1, \pm 1, \pm 1]$ orientations. A description of the HOP cubic lattice that includes collective molecular tumbling and translational symmetry is possible within the BZ. At the Γ point, the IR decompositions read $\Gamma_1^+ + \Gamma_3^+ + \Gamma_4^-$ (A), $\Gamma_1^+ + \Gamma_3^+ + \Gamma_5^+ + \Gamma_4^- + \Gamma_5^-$ (B) and $\Gamma_1^+ + \Gamma_5^+ + \Gamma_2^- + \Gamma_4^-$ (C), each IR corresponding to the possible modes of a tumbling scenario. As expected, Γ_1^+ shows up in each decomposition, yielding the same occupation probability for all the orientations $\left(\boldsymbol{N}_i = \frac{\boldsymbol{N}}{\boldsymbol{m}}\right)$ corresponding to the steady state solution. The different possible excitations of the system can be derived from rate equations using a single parameter for the rate of tumbling between nearest neighboring orientations. The deviations of the occupation numbers from the steady state solutions are defined as $\boldsymbol{n}_i$ with $\sum_i \boldsymbol{n}_i = 0$. Allowed symmetrized combinations of these deviations define the possible modes of the system. When computing the representations of the collective molecular tumblings at R, one has to consider phase factors between neighboring cells related to $\chi_{site} = R_2^-$. Symmetry decomposition yields $R_5^+ + R_2^- + R_3^-$, $R_4^+ + R_5^+ + R_2^- + R_3^- + R_4^-$ and $R_1^+ + R_5^+ + R_2^- + R_4^-$ for the IR of scenarios A, B and C, respectively.

7.2.4 Basic Optoelectronic Properties of the Metal Halide Lattice

It is well known that close to the band gap of HOP, optical transitions involve mainly s- and p-orbitals of metal M and p-orbitals of halogen X.[10,44] Leaving aside for a moment SOC, the possible symmetrized linear combinations of atomic orbitals (SLCAO) for an isolated MX_6^{4-} octahedron correspond to the A_{1g}, T_{1u}, $A_{1g} + E_g + T_{1g} + 2T_{1u} + T_{2u}$ IR for the s(M), p(M) and p(X) orbitals, respectively. This shows that two A_{1g} totally symmetric states are hybridized combinations of s(M) and p(X) orbitals and three triply degenerate T_{1u} vectorial states are hybridized combinations of p(M) and p(X) orbitals.

Now, when moving to the description of the HOP cubic lattice, it is useful to describe the energy eigenfunctions as Bloch waves:[34]

$$\psi_{n,k}(\boldsymbol{r}) = e^{i\boldsymbol{k},\boldsymbol{r}} u_{n,k}(\boldsymbol{r}), \tag{7.1}$$

where n is the band index, $\boldsymbol{r}$ the position vector, $\boldsymbol{k}$ the wave vector inside the BZ and $u_{n,k}(\boldsymbol{r})$ the periodic part of the eigenfunction. The corresponding electronic band structures are usually plotted within the BZ (Figure 7.4a). The symmetry of the Bloch functions at various points of the BZ can be analyzed using SLCAO, by analogy with the case of an isolated octahedron, but only 3 X atoms should be considered (Section 7.2.1). At Γ, SLCAO leads to Γ_1^+, Γ_4^- and $2\Gamma_4^- + \Gamma_5^-$ IR for the s(M), p(M) and p(X) orbitals, respectively. This simple analysis shows that totally symmetric Γ_1^+ s–p hybridization is forbidden at Γ, while it is allowed for an isolated octahedron. A similar analysis at R yields R_1^+, R_4^- and $R_1^+ + R_3^+ + R_4^+ + R_5^+$ IR for the s(M), p(M) and p(X) orbitals, respectively. These decompositions help us to understand why the totally symmetric R_1^+ (A_{1g}) corresponds to s(M)–p(X) hybridized atomic orbitals at the top of the valence band (VBM) at point R.[43] Meanwhile, the vectorial representation R_4^- (T_{1u}) relevant for the bottom of the conduction band (CBM) at R involves only hybridization between p(M) orbitals.[44] It turns out that the electronic structure of the reference cubic phase of AIP and HOP (without SOC), has two main differences when compared to GaAs:[34,41] (i) a direct band gap located at the high symmetry point R instead of Γ, and (ii) an inverted band structure: the triply degenerate electronic state is located at CBM instead of VBM. Nevertheless, having this reversal in mind, Bloch functions at R can be labeled by analogy with those of diamond and zinc-blende lattices, namely $|\boldsymbol{X}\rangle$, $|\boldsymbol{Y}\rangle$, $|\boldsymbol{Z}\rangle$ and $|\boldsymbol{S}\rangle$ for the CBM and VBM, respectively.[34,41,47]

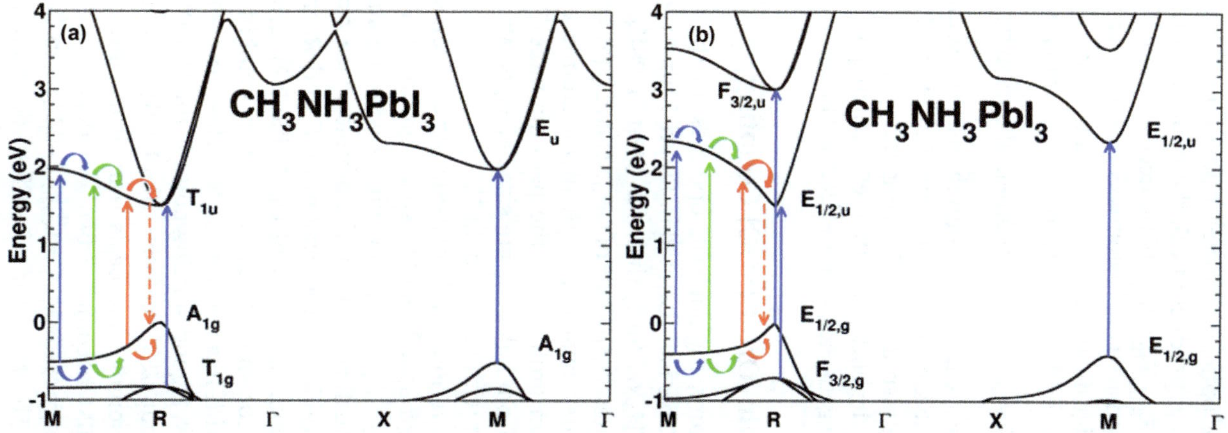

Figure 7.4 Electronic band structure for the high temperature cubic *Pm*-3*m* phase of $CH_3NH_3PbI_3$ without (a) and with (b) SOC at the LDA (Local Density Approximation) level of theory. An upward energy shift of 0.3 eV (a) and 1.4 eV (b) has been applied to match the experimental band gap value at *R*. Labels of irreducible representations (IR) obtained from simple (a) and double (b) group representations are given at *R* and *M* for electronic states close to the band gap. Vertical arrows show various possible optical transitions close to the band gap energy. Optical transitions along the *M*–*R* line generate carriers that can relax easily towards *R*.

Reprinted with permission from ref. 43. Copyright 2014 American Chemical Society.

Dispersion curves close to the critical point R of the BZ can be analyzed within a multiband effective mass model ($\boldsymbol{k}.\mathbf{p}$ method)[32,42] as a function of the $\boldsymbol{k}$-vector. This vector corresponds to the difference with the wave vector at R, $\boldsymbol{k}_R$. Symmetry analysis of the electronic states at both R and M points of the BZ allow identification of optically allowed transitions.[43] The selection rules at R derive from the direct product of CB, the photon operator and VB IR: $R_4^- \otimes T_{1u} \otimes R_1^+ = A_{1g} + E_g + T_{1g} + T_{2g}$ that indeed contains the A_{1g} IR. It should be noted that the opposite conclusion was drawn in early work on $CsSnX_3$, most likely due to significant underestimation of the band-gaps.[56] For each pair of [CB,VB] states, the strength of optical transitions can be derived from: $\frac{2}{m_e}|\langle\psi_{CB}| - i\hbar\nabla|\psi_{VB}\rangle|^2$. The fundamental optical transition can be related to the Kane energy, which is commonly used in semiconductor physics.[32] For $CH_3NH_3PbI_3$, it amounts to 40 and 120 eV at R and M, respectively.[43] But, as will be shown in the following sections, the strength of optical absorption crucially depends on other effects, in particular SOC, which modifies both band degeneracies and effective masses.

7.2.5 Importance of Spin–Orbit (SOC) Coupling

Regardless the crystal structure, a giant SOC operates on the CB of lead-based AIP[28] and HOP.[41,44,47] This is illustrated in Figure 7.4b. Even though it is three times smaller for $CH_3NH_3SnX_3$ than for $CH_3NH_3PbX_3$, it remains sizeable in tin-based perovskites.[42] Analysis of $CH_3NH_3MCl_3$ (M = Pb, Sn, Ge) shows that SOC splittings are consistent with metal atomic energy level tables, with an increase down group-14 of the periodic table.[44] Moreover, SOC interactions stemming from halogens are sizable, especially for iodine-based perovskites, consistent with atomic data.[44]

Applying symmetry considerations while considering electron spin effects (*e.g.* SOC) requires double groups.[45] Double space group IR are deduced from the products of the $D_{1/2} = E_{1/2g}$ electron spin IR with the simple space group IR. Bloch functions transform as IR of the double space group, whose character table is given in Table 7.2 for *Pm*-3*m*. At Γ and R, they can be analyzed using the double point group IR of O_h (Table 7.2). Time reversal symmetry is also important for spin effects, when it is combined (Section 7.2.5) or not (Section 7.3.4) with inversion symmetry close to high symmetry points. More specifically, at R, a good starting point is the $|\boldsymbol{X}\rangle$, $|\boldsymbol{Y}\rangle$, $|\boldsymbol{Z}\rangle$ basis functions of the simple group R_4^- vectorial IR of CBM and the corresponding energy $\boldsymbol{E}^0_{\mathbf{CBM}}$. The spinor state basis transforms into $|\boldsymbol{X}\uparrow\rangle$, $|\boldsymbol{Y}\uparrow\rangle$, $|\boldsymbol{Z}\uparrow\rangle$, $|\boldsymbol{X}\downarrow\rangle$, $|\boldsymbol{Y}\downarrow\rangle$, $|\boldsymbol{Z}\downarrow\rangle$.[43] CBM becomes a two-fold degenerate spin–orbit split-off (SO) state whose energy reads $\boldsymbol{E}_{(\mathbf{so})} = \boldsymbol{E}^0_{\mathbf{CBM}} - 2\Delta_{so}/3$, as SOC leads to an energy splitting $\Delta_{\mathbf{so}}$.[42,43] The remaining four-fold degenerate CB states undergo a positive energy shift: $\boldsymbol{E} = \boldsymbol{E}^0_{\mathbf{CBM}} + \Delta_{\mathbf{so}}/3$. In other words, the simple group R_4^- vectorial representation of the CBM at R splits in a doubly degenerated R_6^- ($E_{1/2u}$) SO state and a fourfold degenerated R_8^- ($F_{3/2u}$) light (LE) and heavy (HE) electron state, (Figure 7.4b). The ground state isotropic optical transition thus stems from the transition between the doubly degenerated R_6^+ VBM and R_6^- CBM states. At R, a series of other transitions are optically allowed, the lowest in energy corresponding to two secondary transitions: $R_8^+ \rightarrow R_6^-$ and $R_6^+ \rightarrow R_8^-$ (Figure 7.4b and Table 7.2).[43] Isotropic optical activity is also found at M, given that for a pseudocubic phase, the star of $\boldsymbol{k}_M$ contains three arms (Section 7.2.1). Besides, SOC leads to a dramatic reduction of oscillator strength that decreases to 17 eV for the fundamental transition at R. In terms of Kane energy, it drops to *ca.* 6 eV. This is about four times smaller than that of GaAs, thus evidencing an additional difference between HOP and conventional semiconductors. In addition, since both VB and CB are now associated with two dispersion parabolas with almost the same effective masses (*i.e.* band curvatures),[57,58] well-balanced carrier transport is obtained for both electrons and holes. Besides, when including SOC in DFT the calculated band gaps of AIP and HOP appear substantially under-estimated.[41,44] This is a well-known limitation that can be cured by including many-body effects using GW self-energy corrections for the band gaps.[41,58–60] This affords more accurate effective masses of the carrier both for conventional semiconductors and metal halide perovskites.

7.3 Phase Transitions

7.3.1 A Landau Theory for HOP with Linear and Linear-quadratic Order Parameter Couplings

The phase transitions introduced in Section 7.1 can be inspected on the basis of Landau theory, already well developed for perovskite materials such $CaTiO_3$, where distortions have been associated with IR of the parent *Pm-3m* space group (Figure 7.2).[37,38] Here, we adapt this model to the case of HOP, with contributions from the organic part.

The symmetry analysis for the *Pm-3m* to *I4/mcm* phase transition shows that the order parameter is related to one of the components of the triply degenerated R_4^+ IR. In $SrTiO_3/CaTiO_3$, this phase transition is attributed to the condensation of one of the triply degenerated octahedron rotations of R_4^+ symmetry (Table 7.4). Symmetry analysis (Section 7.2.2) confirms that such a vibrational mode, mainly related to the inorganic lattice, exists in the high temperature phase of HOP. As condensation of a vibrational mode is usually related to a displacement mechanism, a phonon mode softening should be observed in HOP: (i) at the BZ boundary (R point) by neutron scattering in the high temperature phase, and (ii) at the BZ zone center both by neutron and Raman scattering in the low temperature phase. Symmetry analysis (Section 7.2.3) further shows that the phase transition can also be associated to a mode involving collective tumbling of molecular cations. This is true for scenario B, namely for the C–N axis pointing toward the iodine atoms. It is noteworthy that the R_4^+ IR can also be found for other scenarios with higher indexes than scenarios A–C. It shows that additional order parameters having the same lattice symmetry as the octahedron rotations (R_4^+) are associated with molecular tumblings. From the point of view of Landau theory, these order parameters undergo a linear coupling. Nevertheless, the mechanism associated with these additional order parameters has an order–disorder nature. In other words, even though a linear coupling is predicted between these order parameters in the *Pm*-3*m* phase, the characteristic dynamics appear in different frequency ranges. Detailed derivations of the possible Landau free energy expansions are known.[51] A linear-quadratic coupling is also predicted between the two linearly coupled R_4^+ order parameters and two components of the strain tensor, $\varepsilon_{11}+\varepsilon_{22}+\varepsilon_{33}$ and $2\varepsilon_{33}-\varepsilon_{11}-\varepsilon_{22}$ that correspond to the Γ_1^+ and Γ_3^+ IR, respectively. Such a linear-quadratic coupling is in agreement with available diffraction data recorded on $CH_3NH_3PbI_3$.[61] Below the critical temperature T_c, octahedron rotation was found to vary as $(T_c-T)^{0.25}$ and the tetragonal strain as $(T_c-T)^{0.42}$.[61]

In oxide perovskites, *Pm*-3*m* to *Pnma* and *Pm*-3*m* to *I*4/*mcm* instabilities compete in the same way as in $CH_3NH_3PbI_3$ and $CH_3NH_3PbBr_3$. This instability is related to the simultaneous condensation of R_4^+ and M_3^+ rotational modes of the octahedra. Our symmetry analysis confirms that a M_3^+ vibrational mode, mainly related to the inorganic lattice, exists in the high temperature phase of HOP (Tables 7.3 and 7.4). Additional order parameters with the same symmetry exist for collective molecular tumblings in scenarios A and B. They should exhibit an order–disorder character.

7.3.2 Electronic Properties: Brillouin Zone (BZ) Folding

A rigorous DFT computation including all the atoms in an ordered lattice can be conducted for the low temperature orthorhombic phases (*Pnma*) of $CH_3NH_3PbX_3$ 3-D HOP ($Z=4$ for X = I or Br and $Z=8$ for X = Cl).[36,41,44,48] The electronic band gap remains direct but shows up at Γ, and the number of bands is multiplied by a factor of 4 to 8 ($Z=4$ or 8) with respect to the cubic lattice. At a first sight, the electronic band structure of the *Pnma* phase of $CH_3NH_3PbI_3$ appears very different from that of the reference cubic phase. The tactic of replacing the MA cations by Cs^+ cations located at nitrogen positions (Section 7.2.1) leaves the electronic band diagram close to the band gap unchanged, while it is useful to understand the connection between the two diagrams.[41,47,48] The effect of the *Pm*-3*m* to *Pnma* cell transformation can be explored by considering a $(\vec{a}+\vec{b},\ 2\vec{c},\ \vec{a}-\vec{b})$ supercell of the cubic lattice, while keeping the atoms at the special positions of the *Pm*-3*m* phase (Figure 7.5b, folded-cubic band structure).[34,44] As the cell size undergoes a four-fold increase, it leads to a reduction of the BZ volume by the same amount. Electronic states close to the band gap are folded from *R* of the *Pm*-3*m* BZ to Γ of the *Pnma* supercell BZ. This band folding clarifies the seeming complexity of the band diagram in the low temperature phase. In addition, electronic states from *M* also fold back to Γ and become mixed with *R*-states at the Γ point of the BZ of the supercell.[44] This procedure

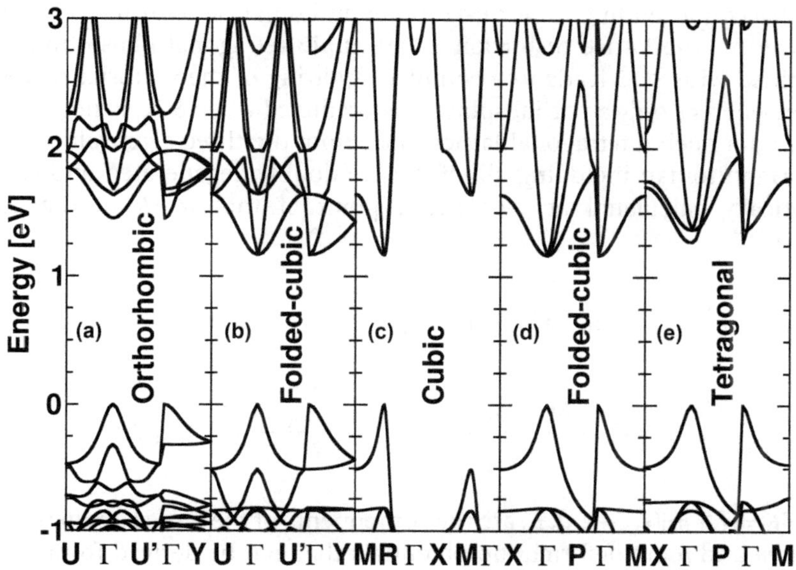

Figure 7.5 Band folding from the pseudo-cubic phase to the low temperature phases of $CH_3NH_3PbI_3$. (a–c) Electronic band dispersion diagram for the *Pm*-3*m* cubic phase of $CH_3NH_3PbI_3$: (c) in its BZ, (b) folded back to the BZ of the *Pnma* orthorhombic phase, and (a) for the real *Pnma* orthorhombic phase of $CH_3NH_3PbI_3$. (d) Electronic band dispersion diagram for the *Pm*-3*m* cubic phase folded back to the BZ of the *I*4/*mcm* tetragonal phase, and (e) for the real *I*4/*mcm* tetragonal phase. The computation was performed at the LDA level of theory and SOC has not been considered.

reveals that, besides electronic band folding, the most important transformation of the diagram is due to atomic displacements.[34,41,44] Thus, it becomes clear that the band gap increase stems from lattice distortions that are mainly related to tilts of the inorganic octahedra and space group symmetry reduction.[62] Similar band folding helps to understand the main features of the electronic band diagram of the *Pm*-3*m* to *I*4/*mcm* cell transformation, or other transformations.

7.3.3 Electronic Properties: Lattice Strain and Octahedron Tilting

Changing the lattice lifts the CBM degeneracy. This is consistent with the group–subgroup relationship between cubic and orthorhombic phases. More generally, the influence of strain (ε_{ij} tensor) and degeneracy lifting can be understood and treated empirically within the $|X\rangle$, $|Y\rangle$, $|Z\rangle$ basis at R by adding a strain perturbation Hamiltonian to the $\boldsymbol{k}.\mathbf{p}$ Hamiltonian.[34,42] The impact of SOC can be empirically rationalized by considering the spinor state basis and adding a perturbation Hamiltonian accounting for SOC.[34] Noteworthy, the lattice distortion at the cubic–tetragonal phase transition combined with SOC leads to a complete splitting of the vectorial representation, whereas only a partial splitting is predicted when SOC is neglected.[42] Indeed, for such a tetragonal lattice distortion, the three eigenvalues of the total Hamiltonian including all effects are doubly degenerated and can be analytically calculated at the critical point of the BZ (R in the cubic phase):[34,42]

$$E_0 = \mathrm{E}^0_{\mathrm{CB}} + (L+m)\varepsilon_{xx} + m\varepsilon_{zz} + \Delta_{\mathrm{so}}/3$$

$$E_{\mp} = E_0 - \frac{\delta E + \Delta_{\mathrm{SO}}}{2} \mp \sqrt{\left(\frac{\delta E - \Delta_{\mathrm{SO}}/3}{2}\right)^2 + 2\Delta^2_{\mathrm{SO}}/9} \tag{7.2}$$

with $\boldsymbol{\delta E} = (\boldsymbol{l} - \boldsymbol{m})(\boldsymbol{\varepsilon_{xx}} - \boldsymbol{\varepsilon_{zz}})$. ε_{ij} are the components of the lattice strain tensors, and l, m and n are deformation potentials that can be derived from experimental data or DFT results. The effect of strain is systematically smaller than that induced by SOC in lead halide compounds. When neglecting SOC ($\Delta_{\mathrm{SO}} = 0$), one of the CBM eigenvalues remains doubly degenerate. Next, close to the *Pm*-3*m*/*I*4/*mcm* phase transition, the strain tensor component should exhibit a quadratic dependence on octahedron rotation. At least for small angles, the band gap should then have a quadratic variation with the rotation angle. DFT simulations evidenced a $[(T_c - T)^{0.25}]^2$ behavior.[62] This needs to be confirmed experimentally.

7.3.4 Electronic Properties: Interplay of SOC and Loss of Inversion Symmetry

Time reversal symmetry yields the general conditions for conjugated spinor states: $\mathrm{E_n}\uparrow(\mathbf{k}) = \mathrm{E_n}\downarrow(-\mathbf{k})$ and $\mathrm{E_n}\downarrow(\mathbf{k}) = \mathrm{E_n}\uparrow(-\mathbf{k})$. Inversion symmetry yields additional conditions: $\mathrm{E_n}\uparrow(\mathbf{k}) = \mathrm{E_n}\uparrow(-\mathbf{k})$ and $\mathrm{E_n}\downarrow(\mathbf{k}) = \mathrm{E_n}\downarrow(-\mathbf{k})$. Combining both symmetries leads to a double spin degeneracy $\mathrm{E_n}\uparrow(\mathbf{k}) = \mathrm{E_n}\downarrow(\mathbf{k})$ across all the dispersion diagrams within the BZ. Indeed, the double group IR at Γ and R is two- or four-dimensional (Table 7.2). When inversion symmetry is lost, the spin degeneracy condition $\mathrm{E_n}\uparrow(\mathbf{k}) = \mathrm{E_n}\downarrow(\mathbf{k})$ can be lost for a general wave vector, except for special high symmetry points.

The loss of inversion symmetry combined with SOC is known to induce a splitting of the spinor bands away from Γ in zinc-blende and würtzite structures.[63] In 3-D HOP, it has been shown that even a small symmetry breaking can lead to a strong spinor splitting, since the SOC effect is giant, as compared to conventional semiconductors.[42] Indeed, the electronic band diagram of the *P*4*mm* phase of $CH_3NH_3MI_3$ close to the critical point A of the BZ (corresponding to R of *Pm*-3*m*) exhibits a splitting of the spinor bands, even though the deviation from the centrosymmetric *Pm*-3*m* phase is small.

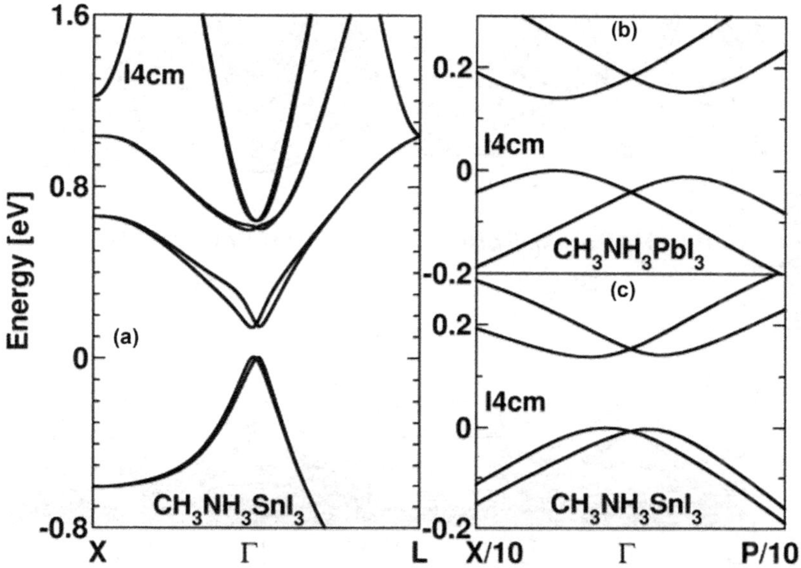

Figure 7.6 (a) Band structure of the *I4cm* phase of $CH_3NH_3SnI_3$ at the LDA level of theory with SOC and zoom close to the critical Γ point showing the Rashba/Dresselhaus-like spinor splitting due to the loss of inversion symmetry for (b) $CH_3NH_3PbI_3$ and (c) $CH_3NH_3SnI_3$.

This effect is more important for the tetragonal *I4cm* phase (Figure 7.6).[42] Such a ***k***-dependent band splitting, also designated as the Rashba/Dresselhaus,[62,64] Dresselhaus,[58] and Rashba[65] effect, still requires experimental evidence in HOP. In addition, we emphasize that the low temperature phase of $CH_3NH_3PbI_3$ is usually refined using a centrosymmetric orthorhombic *Pnma* structure. However, the structure of the low temperature-phase of $CH_3NH_3MI_3$ is still debated: alternative monoclinic, triclinic or non-centrosymmetric *Pna*21 (no. 33) orthorhombic space groups have been suggested, which might also lead to spinor band splitting away from the critical points of BZ.

7.3.5 Electronic Properties: Quantum Confinement

Heterostructures built from 3-D HOP have not yet been explored experimentally, but may afford interesting possibilities for band gap engineering and tuning of optical and transport properties. Figure 7.7 compares the electronic band dispersion diagrams of two short super lattices (SL) of $CH_3NH_3PbI_3/CH_3NH_3PbBr_3$.[34] The stacking axis of both SL corresponds to the long axis **b** of pure $CH_3NH_3PbI_3$ and $CH_3NH_3PbBr_3$ in their *Pnma* phase. Electronic dispersions along Γ–Y are flatter than those of bulk materials, indicative of quantum confinement.[66] Thus, the $CH_3NH_3PbI_3$

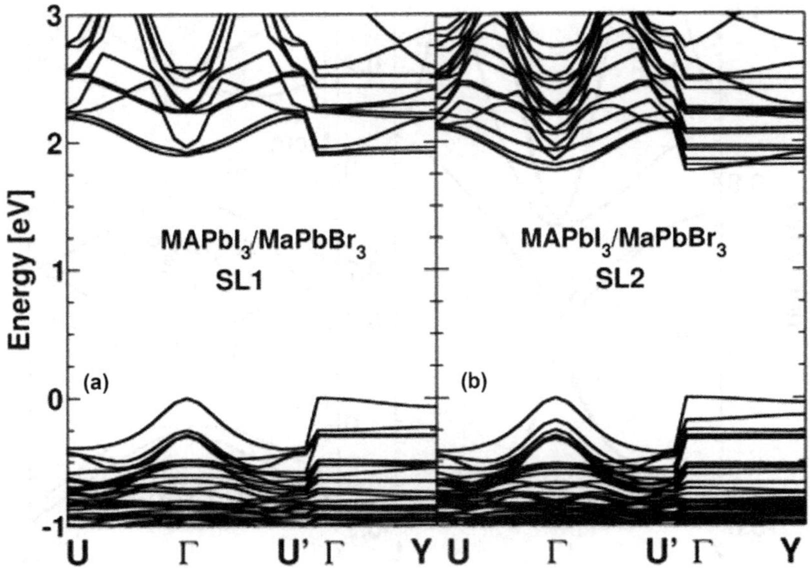

Figure 7.7 Electronic band structure computed at the DFT level with a GGA exchange–correlation functional and without SOC of (a) a SL1 containing a single layer of $CH_3NH_3PbI_3$ and one of $CH_3NH_3PbBr_3$, and (b) SL2 built from two $CH_3NH_3PbI_3$ layers and one layer of $CH_3NH_3PbBr_3$.
Reprinted with permission from ref. 34. Copyright 2015 American Chemical Society.

domain of a $CH_3NH_3PbI_3/CH_3NH_3PbBr_3$ SL can be considered as a quantum well (QW).

This is consistent with the confinement potentials of CB and VB, 0.09 and 0.26 eV, respectively.[67] However, the dispersions along Γ–Y are not completely flat (Figure 7.7) when compared to those of layered HOP.[47,66] This is a clear indication of coupling between the wavefunctions of each QW (so-called SL effect), both in CB and VB, which are delocalized across the $CH_3NH_3PbBr_3$ barrier. This is confirmed by plotting, along the SL axis, the electronic density of the VBM of the short SL built from two $CH_3NH_3PbI_3$ and one $CH_3NH_3PbBr_3$ cells (Figure 7.8).[34] The electronic density is maximum in the $CH_3NH_3PbI_3$ QW, but non-zero in the $CH_3NH_3PbBr_3$ barrier. Moreover, it can be seen that the electronic density variation is based on the same repeated pattern close to lead atoms (Figure 7.8a). This pattern is characteristic of the VBM Bloch functions of both $CH_3NH_3PbI_3$ and $CH_3NH_3PbBr_3$ 3-D bulk materials.[41]

In order to apply the effective mass model to HOP SL, one has to assess the validity of the envelope function approximation (EFA). We start with electronic eigenfunctions of the whole heterostructure given by:[34,66]

$$\psi(\boldsymbol{r}) = \sum_n F_n(\boldsymbol{r})\, U_n(\boldsymbol{r}), \tag{7.3}$$

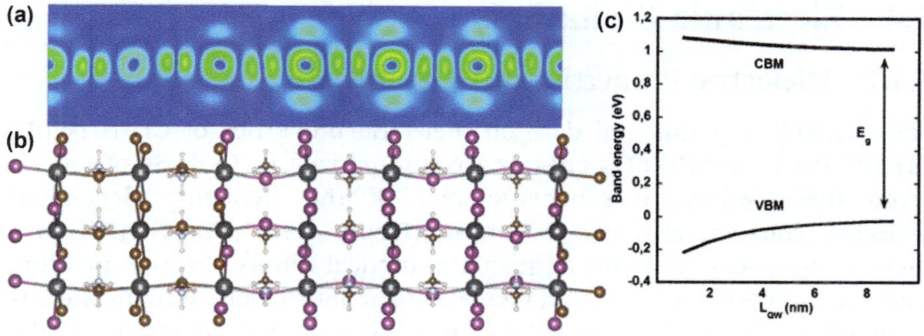

Figure 7.8 (a) Plot of the electronic density at VBM for super-lattice SL2 (Figure 7.7). (b) Overview of the atomic structure. (c) Electron (CBM) and hole (VBM) ground state energies as a function of QW thickness computed with effective masses and confinement potentials.

where $F_n(\boldsymbol{r})$ and $U_n(\boldsymbol{r})$ are expected to be slowly varying and periodic, rapidly oscillating functions, respectively. EFA relies on the assumption that many conventional bulk semiconductors have comparable chemical nature, and Bloch functions of bulk materials at high symmetry points of the BZ differ only slightly. At the VBM or CBM it reads:

$$U_n(\boldsymbol{r}) \approx U^{\mathrm{A}}_{n,0}(\boldsymbol{r}) \approx U^{\mathrm{B}}_{n,0}(\boldsymbol{r}). \tag{7.4}$$

If eqn (7.4) is satisfied, boundary conditions related to the A/B heterostructure, namely $\psi(\boldsymbol{r})$ matching and current conservation, involve the envelope functions $F_n(\boldsymbol{r})$ only. For a simple effective mass model in a 2-D A/B structure having its interface at $z = z_i$:

$$F^{\mathrm{A}}_n(z_i) = F^{\mathrm{B}}_n(z_i) \tag{7.5}$$

$$\frac{1}{m^{\mathrm{A}}}\frac{\mathrm{d}F^{\mathrm{A}}_n(z)}{\mathrm{d}z}\bigg|_{z=z_i} = \frac{1}{m^{\mathrm{B}}}\frac{\mathrm{d}F^{\mathrm{B}}_n(z)}{\mathrm{d}z}\bigg|_{z=z_i}$$

In layered hybrid perovskites, the Bloch functions of A and B partners do not satisfy eqn (7.4).[66] The confinement potentials are correctly defined but cannot be combined with effective mass modelling to predict quantum confinement effects.[66] Thus, DFT remains the best tool to explore theoretically the electronic properties of layered HOP.[47] By contrast, eqn (7.4) is fulfilled in the SL built from A = $CH_3NH_3PbI_3$ and B = $CH_3NH_3PbBr_3$. This suggests that EFA is justified for heterostructures built from 3-D HOP. We note that raw-DFT computation of electronic properties, including quantum confinement and electronic band gap opening, is only qualitative since the effective masses at CBM and VBM are not accurate enough.

7.4 Electronic Excitations

7.4.1 Dielectric Properties of HOP and AIP

The available experimental data on dielectric properties of $CH_3NH_3PbI_3$, $CH_3NH_3PbCl_3$ and $CsPbCl_3$ crystals are summarized in Table S6 of ref. 43. From these values, a schematic plot of the frequency dependent dielectric constant $\varepsilon(\omega)$ can be drawn (Figure 7.9).[43] For $CH_3NH_3PbI_3$, $\varepsilon(\omega) \sim \varepsilon_\infty \approx 5.5$–$6.5$ at high frequencies (optical limit). In the medium frequency range ($\sim$90 GHz), $\varepsilon(\omega)$ increases to about 30, both at room and low temperature. Such an increase can be attributed to polar modes of the inorganic lattice by analogy with $CsPbCl_3$.[43] At room temperature, $\varepsilon(\omega)$ undergoes a further increase, while it is almost temperature independent below the tetragonal to orthorhombic phase transition. This dielectric increment may be attributed to the tumbling of the C–N axis of the MA cations occurring at room temperature in $CH_3NH_3PbX_3$ (Figure 7.9). Recently, a giant dielectric constant (GDC) effect has been reported at very low frequencies where $\varepsilon(\omega)$ amounts to $\sim 10^3$ and $\sim 10^6$ in the dark and under illumination, respectively.[68] This effect has been attributed to photoinduced

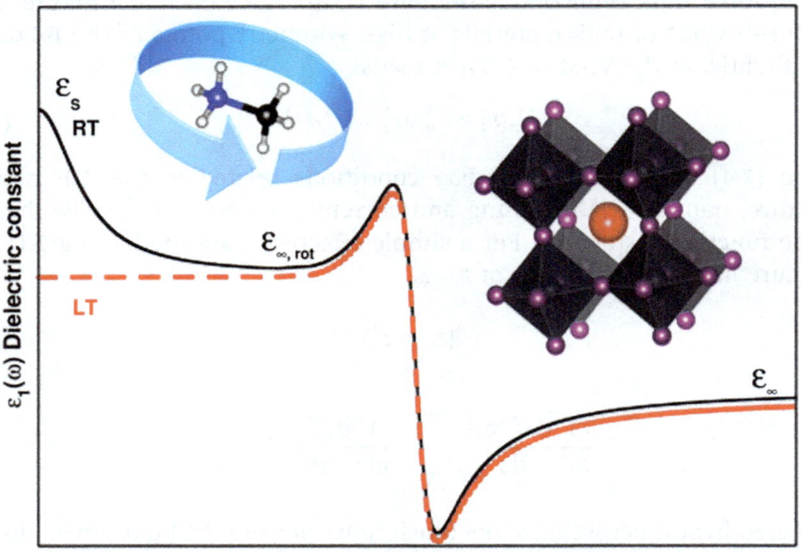

Figure 7.9 Frequency dependence of the dielectric constant of $CH_3NH_3PbX_3$ as derived from available experimental data. The high frequency response is related to vibrational polar phonons stemming from modes of the perovskite lattice (right-hand side). The low frequency region (left-hand side) shows significant differences for the low (LT, red dash line) and room (RT, black line) temperature phases. The static dielectric constant (ε_S) increase at RT is primarily attributed to rotational motion of the $CH_3NH_3^+$ cations.

carriers, assisted by the freedom of rotation of MA, which modify the local unit cell equilibrium and change the polarizability.

7.4.2 From Wannier to Bound Excitons in HOP

In layered HOP, the binding energy of the exciton is very large (200–300 meV) and resonance energies of Wannier-series have been measured and predicted up to the 4s state.[67] Thus, clear exciton signatures can be observed at room temperature both by photoluminescence and absorption.[7–14] Given the large carrier mobility reported for 3-D HOP at room temperature, one may wonder about the nature of photoexcited species: are they free charges or excitons? The possible symmetries of the excitons (without spin-effects) result from the direct product of the IR of the relevant CB state, VB state and envelope function. For the exciton ground state at R with an S type envelope function, the representation yields: $R_4^- \otimes R_1^+ \otimes A_{1g} = T_{1u}$, whereas the crystal ground state IR is A_{1g}. The optical transition from the excitonic state to the crystal ground state is allowed since the IR product of the exciton and the photon operator, $T_{1u} \otimes T_{1u} = A_{1g} + E_g + T_{1g} + T_{2g}$, contains the crystal ground state IR. When considering the spin of the particles but not yet SOC, the electron and hole spin parts combine to form singlet ($S=0$) and triplet ($S=1$) states: $E_{1g} \otimes E_{1g} = \{A_{1g}\} + T_{1g}$. The spatial part combines independently with these two spin states to form excitonic states: $A_{1g} \otimes T_{1u} = T_{1u}$ and $T_{1g} \otimes T_{1u} = A_{1u} + E_u + T_{1u} + T_{2u}$ for the singlet and triplet, respectively. Both are optically active and should be separated by a small exchange energy. These are eigenstates of the exciton without SOC.

However, in lead and tin-based HOP and AIP, SOC is strong and couples triplet and singlet states, both containing the T_{1u} IR. One may instead start from electron and hole spinor states, *i.e.* the monoelectronic eigenstates computed with SOC (Section 7.2.5). The exciton states associated with the top of the VB and the bottom SO state of the CB, are given by: $E_{1/2u} \otimes E_{1/2g} \otimes A_{1g} = A_{1u} + T_{1u}$, while the exciton states associated to the light (LE) and heavy (HE) electron states of the CB yield: $F_{3/2u} \otimes E_{1/2g} \otimes A_{1g} = E_u + T_{1u} + T_{2u}$. The total decomposition of singlet and triplet exciton states is recovered by summing both representations: $A_{1u} + E_u + 2T_{1u} + T_{2u}$. The SO-CB and LE/HE-CB pair states are coupled by the exchange term stemming from the excitonic interaction. The SO-CB pair state yields both optically active (T_{1u}) and inactive (A_{1u}) excitonic states. The latter are related to a triplet spin state.

Excitonic effects can be evaluated numerically using the Bethe–Salpeter Equation (BSE) starting from the monoelectronic states calculated at the DFT level. For the high temperature cubic phases of $CH_3NH_3PbI_3$ and $CH_3NH_3PbCl_3$, enhancement of absorption at the band gap has been evidenced by BSE simulations.[42] However, in the perturbative BSE/DFT approach, screening of the electron–hole interaction due to atomic motion is not taken into account.[43] Such an approach is thus more suited for low temperature phases where these motions are frozen, but calculations are computationally demanding and only resonant states may be addressed accurately.[69,70] This means that, at a high temperature, experimental results on the exciton cannot be fully understood from BSE/DFT calculations, if atomic motions are important.

Nevertheless, exciton screening can also be gauged starting from the experimental absorption spectra of $CH_3NH_3PbI_3$ recorded two decades ago by Ishihara *et al.*[9,43] These data are consistent with exciton quenching (screening and gap-switching) between 159 K and 212 K, in relation to the structural transition at $T_c = 162$ K. A convenient way to simulate exciton screening is to consider a two-particle wave function $\psi(r_e, r_h)$, where $\psi(r_e, r_h)$ is the electron (hole) position. Given that the effective mass approximation works well close to the band gap, the Hamiltonian for a Wannier exciton has two types of solutions: bound pair states and a continuum of pair states having energies larger than the band gap energy. The optical absorption reads:[43]

$$I(\tilde{E})/I_0 = 4\sum_{n=1}^{\infty}\frac{\gamma/n^3}{(\tilde{E}+1/n^2)^2+\gamma^2} + \int_0^{\infty}\frac{d\tilde{E}'}{\pi}\frac{\gamma S(\tilde{E}')\sqrt{\tilde{E}'}}{(\tilde{E}-\tilde{E}')^2+\gamma^2} \quad (7.6)$$

$\tilde{E} = \dfrac{\hbar\omega - E_g}{Ry}$ is the reduced energy, γ a broadening factor that depends on the temperature and $S(\tilde{E}) = \dfrac{2\pi}{\sqrt{\tilde{E}}\left(1 - e^{-2\pi/\sqrt{\tilde{E}}}\right)}$, the absorption enhancement factor of the continuum of pair states.

In the limit of free carriers or a totally screened interaction, the first sum disappears in the optical absorption expression and $S(\tilde{E})$ reduces to 1. Simulated optical absorption spectra using this expression compared to the experimental data yield a reduction of the 1S exciton binding energy for all the spectra above T_c, from *ca.* 15 to 5 meV.[43] Since the value of the exciton binding energy is smaller than kT ($\sim$26 meV) at room temperature, most electron–hole pairs will be ionized yielding free carriers. For $CH_3NH_3PbX_3$ compounds, the exciton resonance below T_c is related to a Wannier-like exciton, already partially screened by polar modes, that becomes completely screened above T_c, yielding almost-free carriers when the orientational motions of the cations are collectively activated by a structural phase transition. This phenomenon is connected with a renormalization of the dielectric constant at high temperature above T_c. The apparent discrepancy between the exciton binding energies deduced at very low temperature (35–70 meV)[71,72] and those estimated below T_c ($\sim$15 meV)[43] for $CH_3NH_3PbX_3$ is most probably related to the two distinct exciton signatures that have recently been observed between 50 K and 100 K.[73] The low temperature exciton is attributed to a bound exciton with a long lifetime with possible triplet enhancement by coupling with the organic moiety, whereas the exciton close to T_c is most probably of Wannier type with a shorter lifetime ($\sim$ns).

7.4.3 Free Carrier Coupling to Phonons and Molecular Rotations

In conventional semiconductors, electron–phonon coupling is known to significantly influence relaxation and broadening mechanisms, carrier transport and optical properties.[33] In III–V semiconductors (GaAs, InP...), one of the basic mechanisms is the Fröhlich interaction between delocalized CBM and VBM states and polar optical phonons (FOP). Often, FOP are at the origin of the formation of localized polarons (Section 7.4.4) and associated to the LO-TO splitting as well as to the dielectric increment between the high and medium frequency range in AIP and HOP (Section 7.4.1). A second mechanism of electron to optical phonon coupling is related to the deformation potential (ODP).[33] ODP can lead to intravalley as well as intervalley carrier scattering and is the only mechanism in group IV non-polar semiconductors (Si, Ge, ...). In IV and III–V conventional semiconductors, it is known that, for symmetry reasons, intravalley ODP vanishes at first order in the CBM.[33] In AIP and HOP, intravalley ODP can also be inferred at first order from group theory for the *Pm*-3*m* space group (Table 7.5): coupling *via* a deformation potential is allowed if the IR of an optical phonon at Γ is included in the decomposition of the square of the electronic IR of CBM and VBM at *R*. For the *Pm*-3*m* space group, it appears that none of the VBM and CBM states can couple to optical phonons *via* a deformation potential, which is a major result in itself.

Turning to the coupling between electrons and acoustic phonons, we first consider the piezoelectric electron–phonon (PZA) coupling that is important in III–V semiconductors, both for CBM and VBM states, while it vanishes in group IV semiconductors for symmetry reasons.[33] In the *Pm*-3*m* phase of AIP and HOP, PZA coupling vanishes for the same reason. Next, electron and acoustic phonons can also interact *via* a deformation potential (ADP). ADP coupling can also be inspected, at first order, from group theory (Table 7.5): coupling *via* a deformation potential is allowed if the IR of a strain component (acoustic phonons) is included in the decomposition of the square of the electronic IR of CBM and VBM at *R*. Moreover, due to SOC, only local volumetric strain is expected to influence ADP. The coupling between

Table 7.5 Possible intravalley deformation couplings (crosses) between CBM and VBM band edge electronic states at the *R* point of the *Pm*-3*m* BZ, and acoustic phonons (ADP, strains), optical phonons (ODP) or pseudospins (PS) at Γ point.

	VBM R_1^+ SOC = 0	VBM R_6^+ SOC ≠ 0	CBM R_4^- SOC = 0	CBM R_6^- SOC ≠ 0
Strain Γ_1^+	X	X	X	X
Strain Γ_3^+	—	—	X	—
Strain Γ_5^+	—	—	X	—
Optical phonons	—	—	—	—
PS_A 1b	X	X	X	X
PS_B 1b	X	X	X	X
PS_C 1b	X	X	X	X

electronic states and acoustic phonons is important for intraband carrier relaxation, whereas the lack of coupling with low energy optical phonons may lead to a phonon bottleneck (Section 7.4.5).

In summary, the coupling between VBM, CBM electronic states and phonons in AIP and HOP is expected to be smaller than that in direct band gap materials like GaAs, when considering that PZA and ODP mechanisms are forbidden. Moreover, SOC limits ADP to local volumetric strains. However, larger FOP could counterbalance this, since AIP and HOP are highly ionic materials. It is worth noting here that the collective tumbling of molecules occurring in HOP affords another coupling mechanism to charge carriers (Table 7.5). VBM and CBM states can couple to all pseudospin modes. Current experimental carrier mobilities reveal little influence of these thermally-activated modes on free carriers. This is probably because the two excitations lie in different energy regimes undergoing weak coupling.

7.4.4 From Free Carriers to Heavy Polarons in HOP

Symmetry analysis of localized electronic states, vibrations, strain and molecular orientations (Sections 7.2.2 to 7.2.5), predicts the formation of dressed quasiparticles, stemming from coupling with acoustic phonons related to local volumetric strain and/or to specific configurations of surrounding molecular cations. This is reminiscent of the concept of the heavy polaron, namely a spatially localized dressed particle (electron or hole) having a large effective mass and a short diffusion length. Such heavy polarons should have dramatic effects on the low frequency dielectric response and on the electric characteristics of the photovoltaic device. In view of the foregoing picture of Wannier excitons screened by concerted molecular motion (Section 7.4.2), one may also expect from symmetry that a strong coupling between a charge carrier and neighboring cations may occur and lead to long-lived trap states. This is typical of the Holstein polaron picture, where the polaron mobility is essentially related to intersite hoping with an effective mass increasing exponentially with the polaron binding energy.[74] We note also that local volumetric strain and local cationic configurations may cooperate to stabilize such a polaronic state. The rotational–vibrational linear coupling is also expected to enhance the formation of low energy polaronic excitations, through the Fröhlich mechanism related to vibrations of the polar lattice (Section 7.4.3).

7.4.5 Many-body and Auger Effects in AIP and HOP

Third-order non-radiative Auger effects may also play a role in AIP and HOP. We may start from free carriers in III–V semiconductors (Figure 7.10a), where an electron in the CB can recombine with a heavy hole (HH) in the VB, either by transferring energy and momenta to a transition within the CB (CCCH or CHCC process) or to a hole jumping from the HH to the SO VB band (CHHS or CHSH process). Since electronic states away from the band gap are involved, full computation requires repeated summations over the

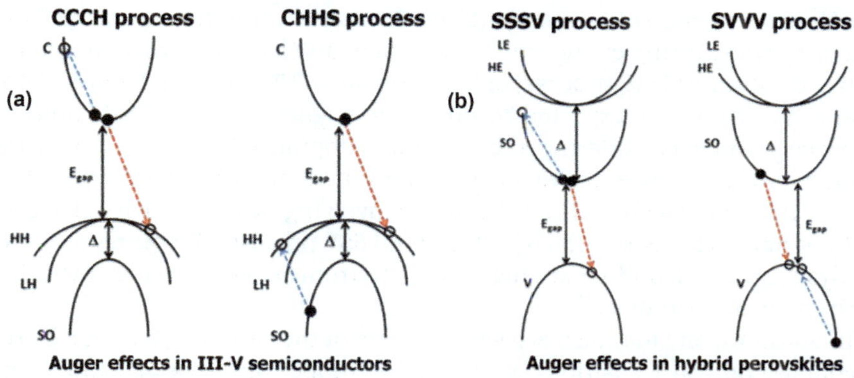

Figure 7.10 Dominant Auger processes in (a) III–V semiconductors and (b) all-inorganic and hybrid perovskites. (a) HH, LH and SO correspond to heavy hole, light hole and hole spin–orbit split-off states of the VB, respectively. (b) HE, LE and SO correspond to heavy electron, light electron and electron spin–orbit split-off states of the CB, respectively. Reprinted with permission from ref. 34. Copyright 2015 American Chemical Society.

particle momenta and careful description of the electronic band structure. The sensitivity to the electronic band gap, effective masses and carrier densities can be estimated from a parabolic approximation of the electronic dispersions.[75–77] In III–V semiconductors, the Auger effect is found to decrease exponentially with the increasing band gap. When SOC is comparable to the band gap, the CHHS process prevails in most cases. However, in n-type III–V semiconductors, the CCCH process is dominant. For large band gap materials like GaAs, where Auger processes are very weak, phonon-assisted Auger processes contribute the most.[75–77]

Now, there are two key differences between HOP (AIP) and III–V semiconductors: (i) reversed band ordering, with SOC on CBM, and (ii) a much larger SOC. Nevertheless, the dependence of the scattering rates on the electronic band gap, effective masses and carrier densities for the two main processes expected in HOP (AIP) can be derived by analogy with III–V semiconductors:[34]

$$R_{\mathrm{Auger,SSSV}} \propto n^2 p \mathrm{e}^{-(E_{\mathrm{T}}-E_{\mathrm{gap}})/kT} \tag{7.7}$$

$$R_{\mathrm{Auger,SVVV}} \propto n p^2 \mathrm{e}^{-(E_{\mathrm{T}}-E_{\mathrm{gap}})/kT} \tag{7.8}$$

where

$$E_{\mathrm{T,SSSV}} = \frac{2m_{\mathrm{s}} + m_{\mathrm{v}}}{m_{\mathrm{s}} + m_{\mathrm{v}}} E_{\mathrm{gap}} \tag{7.9}$$

$$E_{\mathrm{T,SVVV}} = \frac{2m_{\mathrm{v}} + m_{\mathrm{s}}}{m_{\mathrm{s}} + m_{\mathrm{v}}} E_{\mathrm{gap}} \tag{7.10}$$

An electron in the split-off state (S) of CB may recombine with a VB hole, either by transferring energy and momenta to an electronic transition in the split-off CB state (SSSV process) or to a VB hole (SVVV process) (Figure 7.10b). The SVVV process is expected to dominate slightly over the SSSV process, especially when considering the natural p-doping of HOP. However, the overall effect is expected to be small, as it is in large band gap III–V semiconductors (GaAs, InP...). A process involving SOC and heavy electrons (HE) or light electrons (LE) of CB (HSSV or LSSV processes) requires a deeper analysis, since HE and LE are high lying electronic states that may hybridize with molecular orbitals.[41]

In addition, one may also expect that some of the carriers generated by the secondary optical transitions at R may be trapped in the $F_{3/2u}$ (CBM) and $F_{3/2g}$ (VBM) states.[43] Provided that the energy conservation rule is fulfilled, further relaxation from these states may occur by optical phonon-assisted relaxations. Such a "*phonon bottleneck*" is well-known in the case of carrier injection in semiconductor quantum dots. However, in quantum dots, carrier-assisted (Auger relaxation) is the dominant effect in the high injection regime, yielding very fast carrier relaxation. A similar hot hole cooling (intraband process) is consistent with the relaxation after optical excitation reported for $CH_3NH_3PbI_3$.[78]

7.5 Conclusions and Outlook

In this chapter we have summarized our current understanding of the electronic properties and related excitations of metal halide perovskites with particular reference to concepts and tools already deployed extensively in the field of conventional semiconductors and optoelectronics, including intensive use of group theory. This has helped to bring out the most distinctive features that make HOP, and to a less extend AIP, a new and distinct class of semiconductors. First and foremost, in lead halide perovskites, spin–orbit coupling has a dramatic influence not only through its strength but also by alteration of the symmetry selection rules. Next, due to the nature of the atomic orbitals, the band-edge states show a reversed order as compared to conventional semiconductors, and SOC acts on the CB. The presence of organic cations in HOP is another important distinguishing feature. In layered HOP, for instance, it affords quantum and dielectric confinement leading to very robust excitons at room temperature, whereas in 3-D HOP the small cation acts mainly through its rotational degrees of freedom.

Moreover, the properties of AIP and HOP can be understood more clearly considering the cubic *Pm*-3*m* reference structure. It provides the best starting point for modeling, *e.g.* for the design of semi-empirical approaches where effects such as strain and SOC can be included by a perturbative treatment, for symmetry analysis of localized and collective properties as well as for investigating phase transitions. We stress that Brillouin zone folding is a simple yet effective tool to unravel the apparent complexity of HOP and AIP electronic band structures and related properties. This prompts the first issue, the appropriate description and proper understanding require an in-depth knowledge of actual crystallographic phases (space and point groups) and phase transitions, which is still lacking. This would also support theoretical predictions of Rashba and/or Dresselhauss effects in HOP and prompt experimental validation prior to envisaging any application in spintronics.

Among the open questions deserving further work, quite a few stem from symmetry-based predictions. This is especially true considering possible coupling between the various collective or localized degrees of freedom (optical and acoustic phonons, local strain, rotational molecular pseudospins) with electronic excitations (including spin effects) and charge carriers. For example, the role of organic cations in 3-D HOP definitely deserves experimental investigation since they are predicted to screen Wannier excitons at room temperature and at the same time create possibly heavy polarons. We stress that the present study neglects the possible linear rotational–vibrational coupling in the analysis of the high temperature phase. This assumption is expected to fail at low temperature, since rotations and vibrations correspond to coupled order parameters which condense simultaneously at the phase transition critical temperature. Besides, combination of centrosymmetry and direct band gap at the BZ boundary (*R* point of the BZ) leads to vanishing contributions from piezoelectric and deformation potential related couplings between electronic states and phonons. This may explain the nice transport properties of HOP which, according to experimental results, seem to be as good as those of conventional III–V semiconductors.

Acknowledgements

This work was performed using HPC resources from GENCI-CINES/IDRIS grant 2014-c2012096724. The work was supported by the Agence Nationale pour la Recherche (PEROCAI project ANR-10-04), Cellule Energie du CNRS (SOLHYBTRANS Project) and University of Rennes 1 (Action Incitative, Défis Scientifiques Emergents 2015).

References

1. N. G. Park, *J. Phys. Chem. Lett.*, 2013, **4**, 2423–2429.
2. H. J. Snaith, *J. Phys. Chem. Lett.*, 2013, **4**, 3623–3630.
3. H. S. Kim, S. H. Im and N.-G. Park, *J. Phys. Chem. C*, 2014, **118**, 5615–5625.
4. P. Gao, M. Graëtzel and M. K. Nazeeruddin, *Energy Environ. Sci.*, 2014, **7**, 2448–2463.
5. W. Nie, H. Tsai, R. Asadpour, J.-C. Blancon, A. Neukirch, G. Gupta, J. J. Crochet, M. Chhowalla, S. Tretiak, M. A. Alam, H.-L. Wang and A. Mohite, *Science*, 2015, **347**, 522.

6. M. A. Green, A. Ho-Baillie and H. J. Snaith, *Nat. Photonics*, 2014, **8**, 506–514.
7. T. Ishihara, J. Takahashi and T. Goto, *Phys. Rev. B: Condens. Matter Mater. Phys.*, 1990, **42**, 11099–11107.
8. G. C. Papavassiliou, A. P. Patsis, D. J. Lagouvardos and I. B. Koutselas, *Synth. Met.*, 1993, **57**, 3889.
9. T. Ishihara, *J. Lumin.*, 1994, **60 & 61**, 269–274.
10. G. C. Papavassiliou, I. B. Koutselas, A. Terzis and M.-H. Whangbo, *Solid State Commun.*, 1994, **91**, 695–698.
11. D. B. Mitzi, S. Wang, C. A. Feild, C. A. Chess and A. M. Guloy, *Science*, 1995, **267**, 1473.
12. C. R. Kagan, D. B. Mitzi and C. D. Dimitrakopoulos, *Science*, 1999, **286**, 945–947.
13. D. B. Mitzi, C. D. Dimitrakopoulos and L. L. Kosbar, *Chem. Mater.*, 2001, **13**, 3728–3740.
14. G. C. Papavassiliou, G. A. Mousdis, G. Pagona, N. Karousis and M.-S. Vidali, *J. Lumin.*, 2014, **14**, 287–291.
15. C. C. Stoumpos, C. D. Malliakas and M. G. Kanatzidis, *Inorg. Chem.*, 2013, **52**, 9019–9038.
16. G. Giorgi, J.-I. Fujisawa, H. Segawa and K. Yamashita, *J. Phys. Chem. C*, 2015, **119**, 4694.
17. S. Hirotsu, *Phys. Lett. A*, 1972, **41**, 55–56.
18. Y. Fujii, S. Hoshino, Y. Yamada and G. Shirane, *Phys. Rev. B: Solid State*, 1974, **9**(10), 4549.
19. M. Nikl, K. Nitsch, J. Chval, F. Somma, A. R. Phani, S. Santucci, C. Giampaolo, P. Fabeni, G. P. Pazzi and X. Q. Feng, *J. Phys.: Condens. Matter*, 2000, **12**, 1939–1946.
20. D. M. Trots and S. V. Myagkota, *J. Phys. Chem. Solids*, 2008, **69**, 2520.
21. I. Chung, J. H. Song, J. Im, J. Androulakis, C. D. Malliakas, H. Li, A. J. Freeman, J. T. Kenney and M. G. Kanatzidis, *J. Am. Chem. Soc.*, 2012, **134**, 8579–8587.
22. L.-Y. Huang and W. R. L. Lambrecht, *Phys. Rev. B: Condens. Matter Mater. Phys.*, 2013, **88**, 165203.
23. T. Umebayashi, K. Asai, T. Kondo and A. Nakao, *Phys. Rev. B: Condens. Matter Mater. Phys.*, 2003, **67**, 155405–155406.
24. S. Sourisseau, N. Louvain, W. Bi, N. Mercier, D. Rondeau, F. Boucher, J.-Y. Buzaré and C. Legein, *Chem. Mater.*, 2007, **19**, 600–607.
25. J. Even, L. Pedesseau, M.-A. Dupertuis, J.-M. Jancu and C. Katan, *Phys. Rev. B: Condens. Matter Mater. Phys.*, 2012, **86**, 205301.
26. Y. H. Chang, C. H. Park and K. Matsuishi, *J. Korean Phys. Soc.*, 2004, **44**, 889–893.
27. G. Murtaza and I. Ahmad, *Phys. B*, 2011, **406**, 3222–3229.
28. H. Jin, J. Im and A. J. Freeman, *Phys. Rev. B: Condens. Matter Mater. Phys.*, 2012, **86**, 121102(R).
29. H. Ayatullahah, G. Murtaza, S. Muhammad, S. Naeem, M. N. Khalid and A. Manzar, *Acta Phys. Pol.*, 2013, **124**, 102–107.

30. I. Borriello, G. Cantele and D. Ninno, *Phys. Rev. B: Condens. Matter Mater. Phys.*, 2008, **77**, 235214.
31. F. Chiarella, A. Zappettini, F. Licci, I. Borriello, G. Cantele, D. Ninno, A. Cassinese and R. Vaglio, *Phys. Rev. B: Condens. Matter Mater. Phys.*, 2008, **77**, 0451294.
32. S. Chuang, *Physics of Optoelectronic Devices*, ed. J. W. Goodman, Wiley, New York, 1995.
33. P. Y. Yu and M. Cardona, *Fundamentals of Semiconductors*, third edn, Springer, Berlin, Heidelberg, New York, 2005.
34. J. Even, L. Pedesseau, C. Katan, M. Kepenekian, J.-S. Lauret, D. Sapori and E. Deleporte, *J. Phys. Chem. C*, 2015, **119**, 10161–10177.
35. A. Poglitsch and D. Weber, *J. Chem. Phys.*, 1987, **87**, 6373–6378.
36. T. Baikie, Y. Fang, J. M. Kadro, M. Schreyer, F. Wei, S. G. Mhaisalkar, M. Graetzel and T. J. White, *J. Mater. Chem. A*, 2013, **1**, 5628–5641.
37. C. J. Howard and H. T. Stokes, *Acta Crystallogr., Sect. B: Struct. Sci.*, 2002, **58**, 565.
38. H. T. Stokes, E. H. Kisi, D. M. Hatcha and C. J. Howard, *Acta Crystallogr., Sect. B: Struct. Sci.*, 2002, **58**, 934–938.
39. L. Chi, I. Swainson, L. Cranswicka, J.-H. Herb, P. Stephens and O. Knop, *J. Solid State Chem.*, 2005, **178**, 1376–1385.
40. I. P. Swainson, R. P. Hammond, C. Soulliere, O. Knop and W. Massa, *J. Solid State Chem.*, 2003, **176**, 97–104.
41. J. Even, L. Pedesseau, J.-M. Jancu and C. Katan, *J. Phys. Chem. Lett.*, 2013, **4**, 2999–3005.
42. J. Even, L. Pedesseau, J.-M. Jancu and Katan, *Phys. Status Solidi RRL*, 2014, **8**, 31–35.
43. J. Even, L. Pedesseau and C. Katan, *J. Phys. Chem. C*, 2014, **118**, 11566–11572.
44. C. Katan, L. Pedesseau, M. Kepenekian, A. Rolland and J. Even, *J. Mater. Chem. A*, 2015, DOI: 10.1039/c4ta06418f.
45. S. L. Altmann and P. Herzig, *Point-Group Theory Tables*, Clarendon, Oxford, 1994.
46. S. C. Miller and W. F. Love, *Tables of Irreducible Representations of Space Groups and Co-Representations of Magnetic Space Groups*, Preutt, Boulder, 1967.
47. J. Even, L. Pedesseau, M.-A. Dupertuis, J.-M. Jancu and C. Katan, *Phys. Rev. B: Condens. Matter Mater. Phys.*, 2012, **86**, 205301.
48. L. Pedesseau, J.-M. Jancu, A. Rolland, E. Deleporte, C. Katan and J. Even, *Opt. Quantum Electron.*, 2014, **46**, 1225–1232.
49. J. Even, L. Pedesseau and C. Katan, *Proc. SPIE*, 2014, **9140**, 91400Y.
50. G. L. Hua, *J. Phys.: Condens. Matter*, 1991, **3**, 1371.
51. M. A. Carpenter and C. J. Howard, *Acta Crystallogr., Sect. B: Struct. Sci.*, 2009, **65**, 134.
52. N. Onoda-Yamamuro, T. Matsuo and H. Suga, *J. Phys. Chem. Solids*, 1990, **51**(12), 1383.

53. N. Onoda-Yamamuro, T. Matsuo and H. Suga, *J. Phys. Chem. Solids*, 1992, **53**, 935–939.
54. R. E. Wasylishen, O. Knop and J. B. Macdonald, *Solid State Commun.*, 1985, **56**, 581–582.
55. M. A. Carignano, A. Kachmar and J. Hutter, *J. Phys. Chem. C*, 2015, **119**, 8991–8897.
56. I. Lefebvre, P. E. Lippens, M. Lannoo and G. Allan, *Phys. Rev. B: Condens. Matter Mater. Phys.*, 1990, **42**, 9174.
57. V. Giorgi, J.-I. Fujisawa, H. Segawa and K. Yamashita, *J. Phys. Chem. C*, 2014, **118**, 12176–12183.
58. F. Brivio, K. T. Butler, A. Walsh and M. van Schilfgaarde, *Phys. Rev. B: Condens. Matter Mater. Phys.*, 2014, **89**, 155204.
59. P. Umari, E. Mosconi and F. De Angelis, *Sci. Rep.*, 2014, **4**, 4467.
60. E. Menéndez-Proupin, P. Palacios, P. Wahnón and J. C. Conesa, *Phys. Rev. B: Condens. Matter Mater. Phys.*, 2014, **90**, 045207.
61. Y. Kawamura, H. Mashiyama and K. J. Hasebe, *Phys. Soc. Jpn.*, 2002, **71**, 1694.
62. A. Amat, E. Mosconi, E. Ronca, C. Quarti, P. Umari, M. K. Nazeeruddin, M. Grätzel and F. De Angelis, *Nano Lett.*, 2014, **14**, 3608–3616.
63. R. Winkler, *Spin-orbit Coupling Effects in Two-dimensional Electron and Hole Systems*, Springer Tracts in Modern Physics, Springer-Verlag, New York, 2004.
64. C. Quarti, G. Grancini, E. Mosconi, P. Bruno, J. M. Ball, M. M. Lee, H. J. Snaith, A. Petrozza and F. De Angelis, *J. Phys. Chem. Lett.*, 2014, **5**, 279–284.
65. M. Kim, J. Im, A. J. Freeman, J. Ihm and H. Jin, *Proc. Natl. Acad. Sci. U. S. A.*, 2014, **111**, 6900–6904.
66. J. Even, L. Pedesseau and C. Katan, *Chem. Phys. Chem.*, 2014, **15**(17), 3733–3741.
67. K. T. Butler, J. M. Frost and A. Walsh, *Mater. Horiz.*, 2015, **2**, 228–231.
68. E. J. Juarez-Perez, R. S. Sanchez, L. Badia, G. Garcia-Belmonte, Y. S. Kang, I. Mora-Sero and J. Bisquert, *J. Phys. Chem. Lett.*, 2014, **5**, 2390–2394.
69. K. Tanaka, T. Takahashi, T. Kondo, T. Umebayashi, K. Asai and K. Ema, *Phys. Rev. B: Condens. Matter Mater. Phys.*, 2005, **71**, 045312.
70. X. Zhu, H. Su, R. A. Marcus and M. A. Michel-Beyerle, *J. Phys. Chem. Lett.*, 2014, **5**, 3061.
71. M. Hirasawa, T. Ishihara, T. Goto, K. Uchida and N. Miura, *Phys. B*, 1994, **201**, 427.
72. K. Tanaka, T. Takahashi, T. Ban, T. Kondo, K. Uchida and N. Miura, *Solid State Commun.*, 2003, **127**, 619.
73. H.-H. Fang, R. Raissa, M. Abdu-Aguye, S. Adjokatse, G. R. Blake, J. Even and M. A. Loi, *Adv. Funct. Mater.*, 2015, **25**, 2378–2385.
74. F. Ortmann, F. Bechstedt and K. Hannewald, *Phys. Rev. B: Condens. Matter Mater. Phys.*, 2009, **79**, 235206.

75. A. R. Beattie and P. T. Landsberg, *Proc. R. Soc. London, Ser. A*, 1959, **249**, 16–29.
76. G. P. Agrawal and N. K. Dutta, *Long-wavelength Semiconductor Lasers*, Van Nostrand Reinhold, New York, 1986.
77. L. A. Coldren and S. W. Corzine, *Diode Lasers and Photonic Integrated Circuits*, Wiley, New York, 1995.
78. G. Xing, N. Mathews, S. Sun, S. S. Lim, Y. M. Lam, M. Grätzel, S. Mhaisalkar and T. C. Sum, *Science*, 2013, **342**, 344–347.

CHAPTER 8

First Principles Modeling of Perovskite Solar Cells: Interplay of Structural, Electronic and Dynamical Effects

EDOARDO MOSCONI,[a,b] CLAUDIO QUARTI[a,c] AND FILIPPO DE ANGELIS*[a,b]

[a] Computational Laboratory for Hybrid/Organic Photovoltaics (CLHYO), CNR-ISTM, via Elce di Sotto, I-06123, Perugia, Italy; [b] CompuNet, Istituto Italiano di Tecnologia, Via Morego 30, 16163 Genova, Italy; [c] Laboratory for Chemistry of Novel Materials, Universitè de Mons, Place du Park, 20, 7000 Mons, Belgium

*Email: filippo@thch.unipg.it

8.1 Introduction

The field of photovoltaics has experienced in recent years a deep transformation thanks to the exploitation of a new class of solar cells based on mixed organo-halide lead perovskites. These new sun perovskite-based devices have allowed the achievement of amazing photovoltaic efficiencies in a short time, to the point where perovskites have been asserted to be the "*next big thing in photovoltaics*".[1] Kojima *et al.* were the first to employ perovskites for photovoltaic applications in 2009,[2] but the use of lead-halide based

RSC Energy and Environment Series No. 16
Unconventional Thin Film Photovoltaics
Edited by Enrico Da Como, Filippo De Angelis, Henry Snaith and Alison Walker

Published by the Royal Society of Chemistry, www.rsc.org

compounds, such as $KPbI_3$, in this field had already been suggested in the 1980s.[3,4] Following the work of Kojima *et al.*,[2] the attention towards organohalide perovskites has exponentially expanded, sustained by an uncommon increase in their photovoltaic efficiency. Park *et al.*, using the pre-treatment of a TiO_2 surface before perovskite adsorption, were able to improve the power conversion efficiency (PCE) from the 3.8% reported by Kojima *et al.*[2] to 6.5%.[5] At a subsequent time, a further increase in the stability and performance was obtained by employing a solid state hole transporting material (HTM), spiro-OMeTAD, obtaining values of 9.7% and 10.2%, respectively.[6,7] The advance in new structures and the upgraded control over the material's morphology marked a further step forward towards better performing perovskite-based devices, with PCEs of 12.0% reported by Seok and coworkers,[8] 15% by Grätzel and co-workers,[9] and 15.4% by Snaith and co-workers.[10] In particular, this latter result, obtained by returning to a fully planar heterojunction architecture, is of particular technological interest, opening the way for the production of stable and highly reproducible devices. The current PCE record reported in the literature for a perovskite-based solar cell is 19.3%,[11] and a certified NREL efficiency of 20.1% makes this new technology competitive with respect to other long-standing technologies, such as thin-film CdTe devices.[12]

Contemporarily with the remarkable improvements on the techniques for the synthesis of perovskite materials and device architecture design, great attention has been focused on the study of the basic properties of this class of materials. Snaith and collaborators initially demonstrated that methylammonium lead-halides work both as light absorber and electron transporting materials.[7] By utilizing vibrational spectroscopy, the markers of the crystalline order were recognized,[13,14] which highlighted the significance of the orientational transitions of the methylammonium (MA) cation in the $CH_3NH_3PbI_3$ ($MAPbI_3$) perovskite,[15] which is the most studied inorganic cation in perovskite solar cells. By performing time-resolved UV-vis absorption and photoluminescence measurements, it was be possible to reveal the micrometer diffusion length of photo-generated species in $MAPbI_3$ and in its chlorine doped analogue, $CH_3NH_3PbI_{3-x}Cl_x$ ($MAPbI_{3-x}Cl_x$),[16] which demonstrated inherently low charge recombination rates.[17] Further studies by Petrozza and co-workers have demonstrated that free electrons and holes are the main photo-generated species in $MAPbI_{3-x}Cl_x$ under typical photovoltaic working conditions.[18] Hodes *et al.* have further proved the ambipolar response of the photoactive perovskite layer within the solar cell, with carriers generated on the hole transporting and electron transporting perovskite interfaces.[19,20]

8.2 Benchmarking the Computation Toolbox: Sn *vs.* Pb Perovskites

Theoretical and computational analyses and simulations based on first principles approaches, mainly relying on Density Functional Theory (DFT), have acquired a crucial role in our current understanding of hybrid organo-halide perovskites. Standard DFT calculations within the Generalized Gradient Approximation (GGA) are known to simulate geometrical structures for perovskites in good agreement with experimental data.[21–23] On the other hand, a different situation is found for the calculated band gaps of perovskites and, more in general, of various semiconductors, for which underestimation of the theoretical band gap is obtained with DFT-GGA approaches. In particular, this is found for a few oxides of transition metals: an evident example is that of ZnO, for which DFT-GGA underestimates the band gap by *ca.* 2.5 eV. In relation to the present studies, compared to the GLLB-SC functional, the standard PBE functional underestimates the band gap for ABO_3 perovskites, which provides an average deviation with respect to experimental values of 0.5 eV.[23] The use of hybrid DFT functionals, such as B3LYP or HSE06, or post-DFT methods (*e.g.* GW) usually give improved simulated bang gap values with respect to the experimental measurements;[24,25] a combination of these methods was used by Berger *et al.* for the development of a calibrated correction to the band gap calculations for a series of perovskites, obtaining values within 0.2 eV with respect to the experimental data.[25]

For Pb-based perovskites the situation seems quite different, for which simple DFT-GGA calculations have led to band gaps that are, in general, in good agreement with experimental values.[26] For example, calculated band gaps in a range between 1.3–1.6 eV have been reported for $CH_3NH_3PbI_3$,[27,28] comparable to the experimental measured band gap of ∼1.55 eV.[2,7] Also, a 1.68 eV band gap has been calculated for $PbTiO_3$, which compares well with the experimental estimate of 1.70 eV.[21] Most notably, however, such an agreement is not transferred to the – in principle very similar – $CH_3NH_3SnCl_3$ and $CH_3NH_3SnI_3$ perovskites, for which DFT-GGA calculations strongly underestimate the band gaps with respect to experimental values.[22,29] A band gap underestimation is also found for $BaSnO_3$ and $SrSnO_3$, for which, however, DFT-GGA nicely reproduces the difference between the experimental band gaps of the two materials.[30] Similarly, DFT-GGA calculations provide the band gap increase from $CH_3NH_3SnCl_3$ to $CH_3NH_3SnI_3$. What seems to emerge from the analysis of previous computational studies is that DFT-GGA calculations are apparently able to simulate the band gaps of Pb-based perovskites. The unexpected agreement of such calculations with the experimental data[26,27] has underlined the importance of both relativistic and electronic correlation effects. The possibility that such agreement is just fortuitous was proposed by Mosconi *et al.*, who suggested the agreement is possibly induced by an error cancellation that takes place between the electronic correlation and spin–orbit-coupling (SOC) contributions.[26] As a matter of fact, a giant SOC effect was calculated by Even *et al.*, who retrieved the expected underestimate of the band gap evaluated a the SOC-DFT level.[31]

The methylammonium lead iodide ($MAPbI_3$) and its tin-based analogue ($MASnI_3$) perovskites display a tetragonal structure[32] (at different temperatures) although they have different absorption onsets, with $MASnI_3$ and $MAPbI_3$ showing a bang gap of 1.2 and 1.6 eV, respectively.[32,33] Experimental measurements carried out on $CsSnI_3$ and $MASnI_3$ demonstrated that both are able to efficiently transport holes,[32,34] while lead-based compounds, such as $MAPb(I_{1-x}Cl_x)_3$ and $MAPbI_3$, provide higher rates of electron and hole transport, respectively.[7,35] A crucial matter is to understand the reason of the different electronic structure of AMX_3 materials, with M = Sn and Pb, and possibly of mixed Sn/Pb materials, because it would represent a key step towards the perovskite solar cell large-scale uptake. Under this light, a fundamental requirement is to setup computational simulation approaches capable of confidently calculating the band gaps and the optical and electronic properties of these materials, allowing us to design new species and to interpret their properties. As stated above, standard DFT provides reliable structures and stabilities of perovskites,[22,23,36] while it considerably underestimates the band gap of these materials. DFT with asymptotically correction on the functionals partly solve this shortcoming.[23] Many body perturbation theory within the GW approach[37,38] represents the method to improve upon DFT.[25,36]

Contrary to what it is expected, the band gaps calculated at DFT level for $MAPbI_3$ are in surprisingly good agreement, within ± 0.1 eV, with respect to experimental measurements.[26,27] On the other hand, $ASnX_3$ shows a DFT-calculated band gap underestimate of ~ 1 eV.[22,29,36,39] This large difference between the calculated band gaps of the Sn and Pb based perovskites may be associated with the presence of relativistic effects,[26] which are particularly strong for Pb.[40,41] Relativistic effects are usually included in the calculations by the approximated scalar relativistic (SR) and, to a higher level, by spin–orbit coupling (SOC) contributions. A DFT investigation has demonstrated the important role of SOC in $MAPbI_3$, inducing a strong band gap reduction, which is opposite to the GW correction.[31] As a matter of fact, a remarkable band gap underestimation was found,[31] in line with the expected behavior of DFT.

This result underlines the necessity to setup a reliable and efficient computational approach to simulate the electronic structure of both Sn and Pb based perovskites, and possibly of mixed Sn/Pb compounds. The choice of the method involves ideally the inclusion of SOC effects in the GW calculations.[42] Moreover, a very efficient GW implementation is also imperative.[43] Umari *et al.*[44] developed a novel approach to include SOC effects into a computationally efficient GW method. The resulting SOC-GW approach is computationally affordable and allows one to accurately model the electronic, optical and transport properties of $MASnI_3$ and $MAPbI_3$ perovskites, with the possibility to simulate and design new and hypothetical materials.[44]

Geometry optimization of the atomic positions (and cell parameters) of $MAPbI_3$ and $MASnI_3$ have been carried out using a SOC-DFT (SR-DFT) approach starting from a tetragonal *I4cm* structure without any symmetry constraints, with a unit cell containing 48 atoms (four $MAMI_3$ units) and 200 electrons, see the $MAPbI_3$ structure in Figure 8.1.

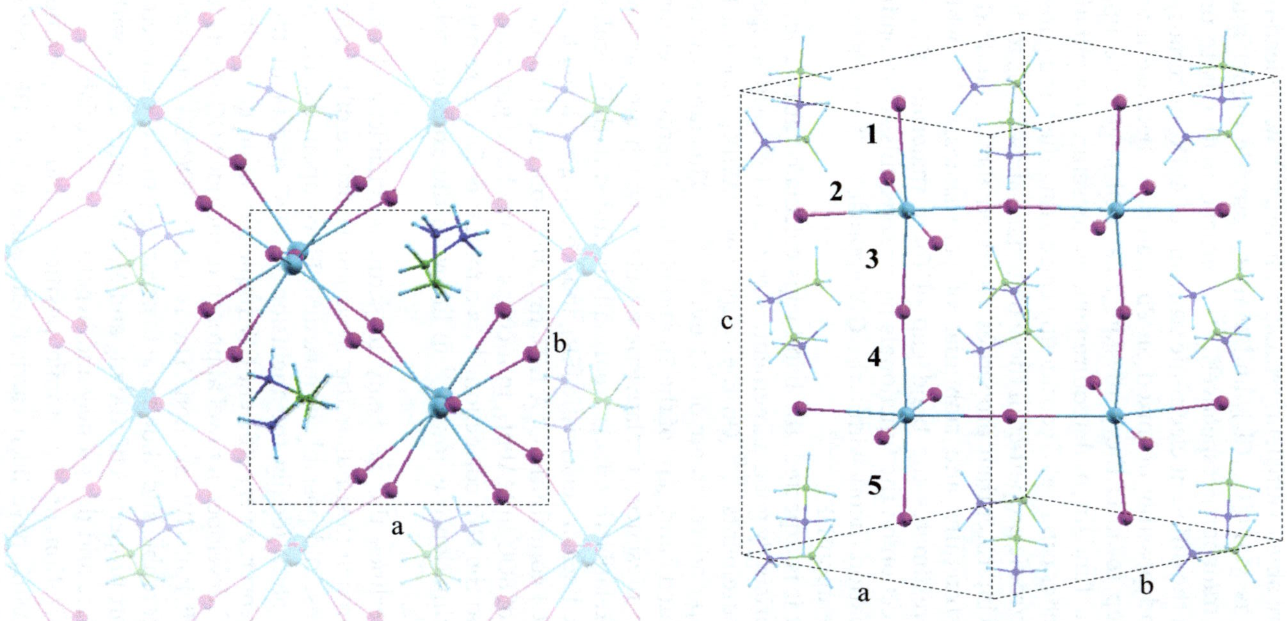

Figure 8.1 SR-DFT optimized structure of $MAPbI_3$ viewed from two different orientations. The unit cell is shown on the left. Calculated (experimental[9,25]) cell parameters for $MAPbI_3$: $a = 8.78$ (8.85–8.86) Å; $c = 12.70$ (12.64–12.66) Å. Calculated (experimental[8,9]) cell parameters for $MASnI_3$ (not shown): $a = 8.71$ (8.76–8.73) Å; $c = 12.46$ (12.43–12.50) Å. Pb = light blue; I = purple; N = blue; C = green.
(Reproduced from ref. 44 with permission from Nature Publishing Group.)

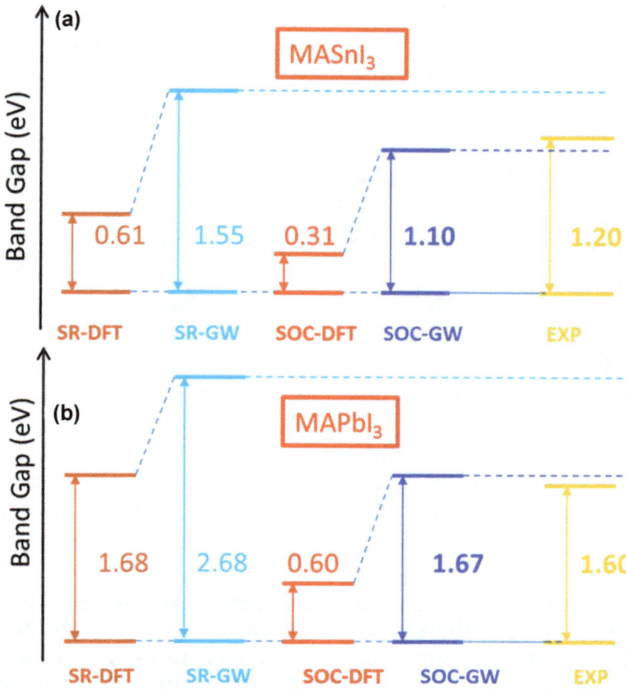

Figure 8.2 Calculated band gaps at various levels of theory for $MASnI_3$ (a) and $MAPbI_3$ (b) perovskites.
(Reproduced from ref. 44 with permission from Nature Publishing Group.)

Figure 8.2 graphically reports the calculated band gap values at different levels of theory. A direct band gap at the Γ point has been found for the investigated species.[22,26] At the scalar relativistic level, SR-DFT shows a band gap of 1.68 and 0.61 eV for $MAPbI_3$ and $MASnI_3$, respectively, comparable to the experimental values of 1.6 and 1.2 eV. Thus, as it can be noticed, the band gap of $MAPbI_3$ is reproduced nicely by SR-DFT, while that of $MASnI_3$ represents an underestimation of 0.6 eV. By looking at the results at the SOC-DFT level, the band gaps of both tin and lead based materials are strongly underestimated, by ~1 eV, in line with the expected behavior of standard GGA-DFT and with previous SOC-DFT results for $MAPbI_3$.[31] However, the calculated band gap at the SOC-DFT level reproduce a qualitatively correct Pb→Sn variation. To overcome this underestimation provided by SOC-DFT, the DFT-calculated band gaps were corrected with SR- and SOC-GW calculations. At the SR-GW level, $MAPbI_3$ shows a ~1 eV overestimate of the band gap (2.67 *vs.* 1.6 eV), while $MASnI_3$ displays only a ~0.3 eV band gap overestimate (1.55 *vs.* 1.2 eV). These results are again showing an unbalanced description of the two simulated materials (Figure 8.2). Finally, only the SOC-GW approach provides a good agreement between the calculated and experimental band gaps (1.67 and 1.10 eV for $MAPbI_3$ and $MASnI_3$, respectively).

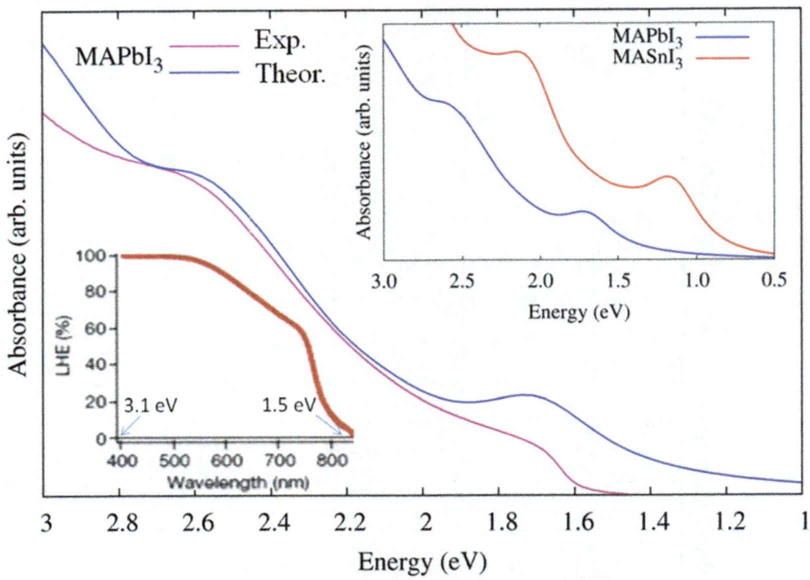

Figure 8.3 Comparison between the experimental UV-vis spectrum of $MAPbI_3$ (red line) and the SOC-GW calculated one (blue line). Notice that the experimental spectrum has been scaled to match the intensity of the calculated one in correspondence with the high energy feature. Top right inset: comparison between the SOC-GW calculated spectra of $MASnI_3$ (red line) and $MAPbI_3$ (blue line). Bottom left inset: LHE for the 15% $MAPbI_3$-based solar cell of ref. 9. The experimental spectrum was recorded at room temperature for $MAPbI_3$ casted on a mesoporous TiO_2 film.
(Reproduced from ref. 44 with permission from Nature Publishing Group.)

Using the electronic structure calculated at the SOC-GW level, we moved on to simulate the optical absorption spectrum of for $MAPbI_3$. In this case, electron–hole interactions have been neglected, and this procedure represents a reasonable approximation to the optical spectra for a semiconductor exhibiting a small band gap.[45] The results are shown in Figure 8.3, in comparison with the experimental data available for $MAPbI_3$.

The calculated data is in good agreement with the experimental UV-vis spectrum. In particular, despite the approximate spectral calculation, the absorption shape of the spectrum at high energies and the signal at ~2.6 eV are nicely reproduced. By comparing the absorption spectrum of $MAPbI_3$ to the absorption spectrum of $MASnI_3$, the latter exhibits, in line with the decreased band gap, a red-shift and a higher intensity (Figure 8.3). Previous calculations reported by Even *et al.*[46] show a 0.6 eV band gap for $CsSnI_3$ by SR-GW. As we will see later, the considerable band gap underestimate is due to the replacement of MA by Cs and the employed cubic symmetry.

In Figure 8.4, the SOC-GW band structures calculated for $MASnI_3$ and $MAPbI_3$ are reported (the bands are derived from SOC-DFT with SOC-GW

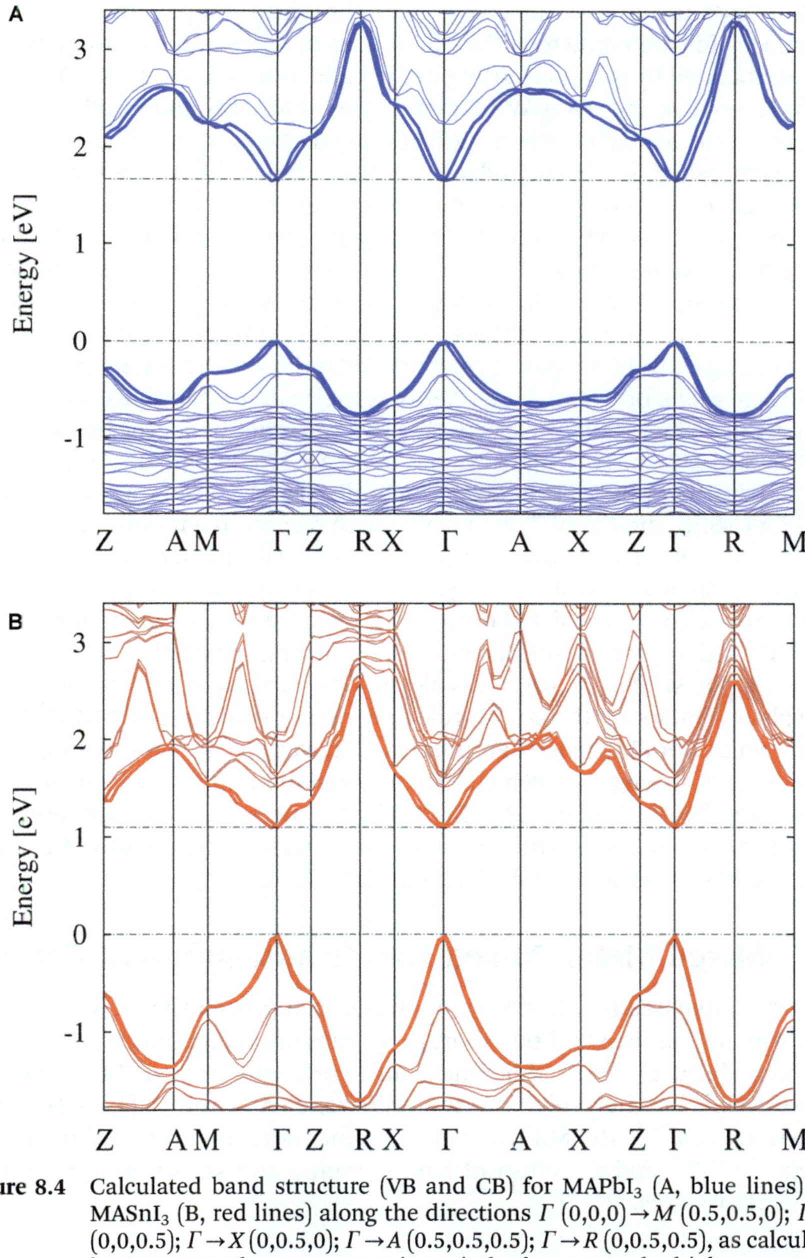

Figure 8.4 Calculated band structure (VB and CB) for $MAPbI_3$ (A, blue lines) and $MASnI_3$ (B, red lines) along the directions Γ (0,0,0)$\rightarrow M$ (0.5,0.5,0); $\Gamma \rightarrow Z$ (0,0,0.5); $\Gamma \rightarrow X$ (0,0.5,0); $\Gamma \rightarrow A$ (0.5,0.5,0.5); $\Gamma \rightarrow R$ (0,0.5,0.5), as calculated by SOC-GW. The energy zero is set in both cases at the highest occupied state.
(Reproduced from ref. 44 with permission from Nature Publishing Group.)

corrected state energies). Apart from the aforementioned band gap variation, inclusion of spin–orbit coupling interactions induces an increased dispersion of the band calculated along the investigated high symmetry directions of the Brillouin zone. The changes in the band shape produce sizable differences for the predicted effective masses of holes and electrons, m_h and m_e, extrapolated by parabolic fitting of the band around the Γ point. As we can see in Figure 8.4, the SOC-GW CB dispersion for $MASnI_3$ and MAPbI3 is quite similar, while the valence band for $MASnI_3$ displays a remarkable stronger dispersion, which is directly reflected in the calculated effective masses.

The simulated carrier reduced masses μ, calculated as $\mu = (m_e \times m_h)/(m_e + m_h)$, can be compared to the experimental measurements for $MAPbI_3$.[47] In particular, values of 0.09, 0.12 and 0.15 m_0 (m_0 is the electron mass) have been reported. As it can been noticed, the calculated maximum, average and minimum SOC-GW μ values obtained for $MAPbI_3$ are 0.17, 0.11 and 0.08 m_0, respectively, in good agreement with the experimental values. SOC-DFT calculations on $CsSnI_3$ and $CsPbI_3$ in their cubic phase have shown 0.12 and 0.04 m_0 values for m_h, respectively.[46] On the other hand, the pseudo-cubic $MAPbI_3$ SOC-DFT has calculated values of 0.29 m_0 for m_e and 0.23 m_0 for m_h.[48] In line with the band gap underestimate discussed above, the values calculated for $CsSnI_3$ and $CsPbI_3$[46] are smaller with respect to our obtained effective masses. As a matter of fact, our results predicted smaller mass values for $MASnI_3$ compared to $MAPbI_3$, in general agreement with the previous calculations and with the predicted decrease of effective masses when introducing SOC-DFT with respect to SR-calculations,[48] and with the smaller m_h value than that of m_e for $MAPbI_3$.[48] SOC-GW calculations have provided improved band gap values, remarking the crucial role of correlation effects and SOC.[44,49] In a recent paper from Menendez Proupin *et al.*, it was also found the importance of the subtle balance between electronic correlation and spin–orbit coupling when using hybrid DFT calculations.[51] Finally, the presence of SOC effects in DFT calculations also introduce some very peculiar features of the band structure known as the Rashba/Dresselhaus effect.[44,50,51]

8.3 Mixed Metal, Mixed Halide and Different Cations

Having established the proper computational approach to nicely reproduce the structural, optical and transport properties of organo-halide perovskites, we now discuss, in a coherent framework, the mixed $MASn_xPb_{1-x}I_3$ perovskites,[44,52] the role of the different A cation ($A = CH_3NH_3^+$, $HC(NH_2)_2^+$, Cs^+),[53] the MA orientation in the inorganic cage and its related effects,[13–15,54,55] and the effect of halide doping and substitution for mixed halide perovskites.[26,56–59]

8.3.1 Mixed Sn/Pb Perovskites

The few investigations that have been focused up to now on the development of new hybrid perovskites as alternatives to the best-performing Pb-based ones, involve mainly analog Sn-based compounds.[32,60,61] The use of mixed Pb/Sn $MASn_xPb_{(1-x)}I_3$ perovskites[62–64] for photovoltaic application has been recently reported, providing a remarkable increment in the extension of the solar spectrum absorption with respect to lead-halide perovskites, moving the absorption onset down near the IR region. These mixed compounds show remarkable efficiencies of 9.8%, and their use is being considered for the reduction of the Pb content in perovskite-based devices.[63] As mentioned by Zuo *et al.*,[63] the exact determination and alignment of the energy levels in $MASn_xPb_{(1-x)}I_3$ perovskite materials is still challenging and the reported values are so far quite scattered,[60–62,64] potentially due to the ease of oxidation of Sn-containing materials.

Under this light, a considerable effort has been invested in the development of a suitable computational approach capable of accurately describing the optical and electronic features of both Sn and Pb compounds.[44,52,65,66] In this spirit, we calculated the electronic properties of mixed Sn/Pb perovskites,[52] exploiting the aforementioned SOC-GW theoretical framework.[44,52] The simulated unit cell contains four $MASn/PbI_3$ units, so we can simulate the pure materials with $x=0$ and $x=1$ and three different Pb:Sn ratios, namely 3:1 ($x=0.25$), 2:2 ($x=0.50$) and 1:3 ($x=0.75$). For the species with $x=0.25$ and $x=0.75$, the position of the minority metal atom is irrelevant; for the system corresponding to $x=0.50$, we have to replace two metal atoms and, for that reason, we have more possible relative positions of the two Pb and Sn atoms in the unit cell. Therefore, all three possible structural combinations were simulated for the case with $x=0.50$, labeled A, B and C in Figure 8.5. Structure A presents the Sn and Pb atoms alternated along the 110 and 001 crystallographic directions, while structures B and C have two consecutive Sn/Pb atoms along the 001 and 110 directions, respectively.

Geometry optimization of the investigated $MASn_xPb_{(1-x)}I_3$ compounds was performed with a scalar relativistic approach employing the $MAPbI_3$ experimental cell parameters[67] shown in Figure 8.5, and by carrying out a variable cell relaxation, see the optimized cell parameters in Table 8.1. For the pure $MASnI_3$ material, we performed the geometry optimization adopting the $MASnI_3$ tetragonal *I4cm* β-phase experimental cell.[32] By carrying out variable cell relaxation, a slight reduction of the cell volume was found with the increase of the Sn percentage. This is in line with X-ray measurements,[64] in which only a slight variation of the X-ray spectra was found between the mixed $MASn_xPb_{(1-x)}I_3$ species compared to the $MAPbI_3$ perovskite. It is noteworthy that the three different configurations explored for $x=0.50$, A, B and C, have the same stability within less than 0.01 eV. This means that, in the mixed Sn/Pb materials, there is no tendency to create separate pure $MAPbI_3$ and $MASnI_3$ domains; on the contrary, these materials are expected to shows very good miscibility. From here on, the A species is referred to as $x=0.5$, corresponding to the fully mixed phase.

To study the optical and electronic features of the series of mixed Sn/Pb species, we carried out additional DFT and GW calculations at SR and SOC levels, see Table 8.2. In particular, the SR-DFT and SR-GW band gap values show an unbalanced description moving from $x=0.00$ to $x=1.00$. On the other hand, SOC-GW and SOC-DFT calculations provided the expected

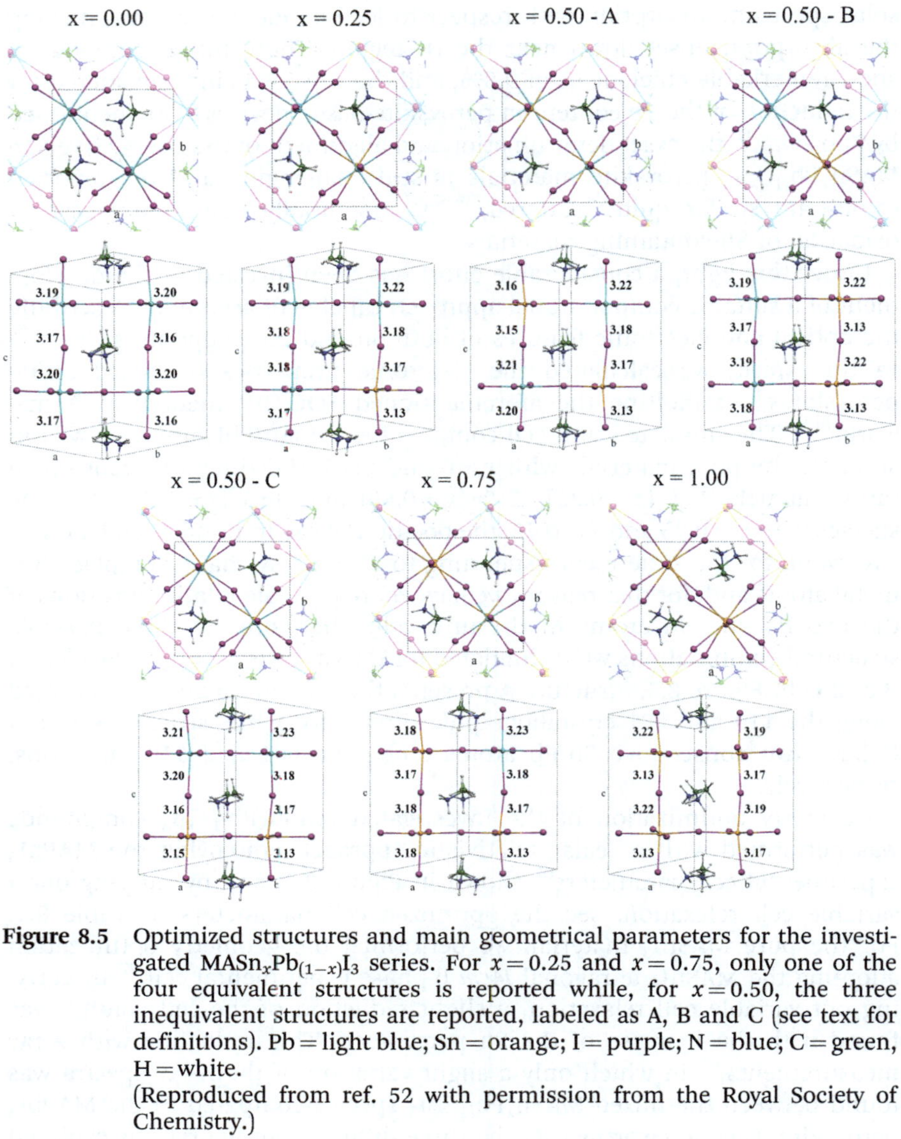

Figure 8.5 Optimized structures and main geometrical parameters for the investigated $MASn_xPb_{(1-x)}I_3$ series. For $x = 0.25$ and $x = 0.75$, only one of the four equivalent structures is reported while, for $x = 0.50$, the three inequivalent structures are reported, labeled as A, B and C (see text for definitions). Pb = light blue; Sn = orange; I = purple; N = blue; C = green, H = white.
(Reproduced from ref. 52 with permission from the Royal Society of Chemistry.)

experimental band gap reduction, in line with what was previously found for the pure Sn- and Pb-based perovskites.[44] Notably, as found previously,[44] only SOC-GW calculations were able to reproduce the experimental band gaps of the investigated materials quantitatively (within ± 0.2 eV), see Table 8.2, with SOC-DFT providing remarkable underestimates. Along with the increase of Sn percentage, we simulated a band gap decrease in line with the experimental measurements reported by Ogomi *et al.*[64] while, for the $x = 1$ case, a slight increase of the band gap was found by Hao *et al.*[62] For each mixed species, a band gap decrease is obtained when using optimized atomic and

Table 8.1 Optimized cell parameters (Å) and cell volume (Å^3) for the investigated mixed $MASn_xPb_{(1-x)}I_3$ perovskites. (Reproduced from ref. 52 with permission from the Royal Society of Chemistry.)

	Sn fraction (*x*)				
Lattice parameters	0.00	0.25	0.50	0.75	1.00
a	8.751	8.742	8.735	8.710	8.681
b	8.735	8.723	8.722	8.697	8.670
c	12.770	12.745	12.678	12.657	12.535
Vol.	976.139	971.889	965.895	958.779	943.438
ΔVol.%	0.0	− 0.4	− 1.0	− 1.8	− 3.3

Table 8.2 Band gap (eV) calculated at different levels of theory for the investigated $MASn_xPb_{(1-x)}I_3$ mixed perovskites. The first entry corresponds to data calculated on optimized geometries employing the experimental cell parameters of $MAPbI_3$. The second entry (in parentheses) refers to optimized atomic and cell parameters. For pure $MASnI_3$, values denoted with * are calculated at the experimental $MASnI_3$ cell parameters. (Reproduced from ref. 52 with permission from the Royal Society of Chemistry.)

	Sn fraction (*x*)				
	0.00	0.25	0.50	0.75	1.00
SR-DFT	1.57	1.27	1.14	0.86	0.66/0.54*
	(1.47)	(1.11)	(0.95)	(0.66)	(0.43)
SR-GW	2.66	2.31	2.15	1.82	1.66/1.55*
	(2.52)	(2.17)	(2.04)	(1.69)	(1.42)
SOC-DFT	0.53	0.39	0.43	0.34	0.39/0.25*
	(0.53)	(0.34)	(0.33)	(0.23)	(0.21)
SOC-GW	1.64	1.37	1.41	1.20	1.31/1.00*
	(1.66)	(1.31)	(1.24)	(1.06)	(0.89)
Exp.[a]	1.51	1.31	1.28	1.23	1.10
Exp.[b]	1.55	1.24	1.17	1.17	1.30

[a]From ref. 64.
[b]From ref. 62.

cell parameters, in agreement with the expected band gap reduction under cell compression conditions.[68]

In order to study the optical features of mixed tin/lead materials, we calculated the SOC-GW absorption spectra for the series of $MASn_xPb_{(1-x)}I_3$, see Figure 8.6. In line with the calculated band gaps, a red shift of the simulated absorption onset was found with the increasing Sn content, in agreement with the experimental data.[64] Most interesting, we also found an increase of the absorption coefficient associated with the increase in Sn percentage, making Sn-based materials a promising alternative to replace the Pb-based perovskites. Moreover, even the calculated absorption intensity of $MASnI_3$ is in line with the experimental measurements published by Snaith and collaborators.[60]

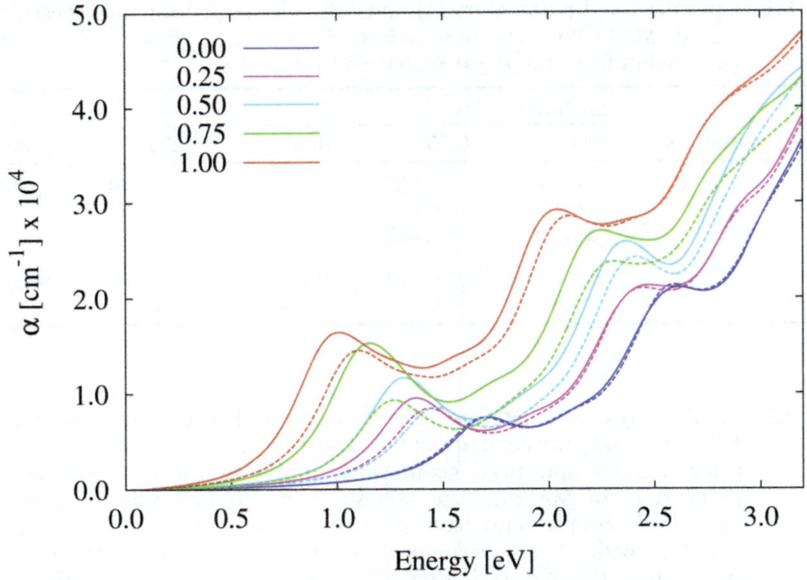

Figure 8.6 Simulated SOC-GW absorption spectra for the $MASn_xPb_{(1-x)}I_3$ series of perovskites. Solid lines refer to optimized atomic and cell parameters; dashed lines for $x = 0.00$–0.75 refer to geometries with $MAPbI_3$ experimental cell parameters, and for $x = 1.00$, it refers to the $MASnI_3$ cell parameters.
(Reproduced from ref. 52 with permission from the Royal Society of Chemistry.)

For the evaluation of the HOMO and LUMO energy levels of mixed metal perovskites and comparison of the VB and CB shifts of the Sn/Pb mixed materials with the experimental data, we chose to align the simulated Densities of States (DOS) calculated at the SOC-GW level with the carbon atom levels of the MA cations, reported in Figure 8.7. Since it is expected that the methyl moieties of the MA cations only interact weakly with the inorganic cavity, it has already been found to be a reasonable approximation to align the energy levels of the Pb and Sn compounds.[44] The DOS evolution by varying x from 0 to 1 is smooth and continuous, in line with the band gap variation, showing a marked rise of the VB edge and a less pronounced rise of the CB edge, which induce the observed reduction in the band gap with the increase in the Sn percentage content in the perovskite materials.

As it can be noticed, the increasing Sn percentage also induces a variation on the shape of the DOS in the proximity to the VB/CB edges, adding strongly hybridized Sn 5s states at the VB top, see Figure 8.7, which leads to a reduction of the hole effective masses, as previously found for $MASnI_3$.[44] Interestingly, while for the $x = 0.75$ species, the CB shape is similar to that of the pure $MASnI_3$ compound, the CB structure for $x = 0.5$ species displays the typical shape of $MAPbI_3$, but shows a decrease in the electron effective masses.

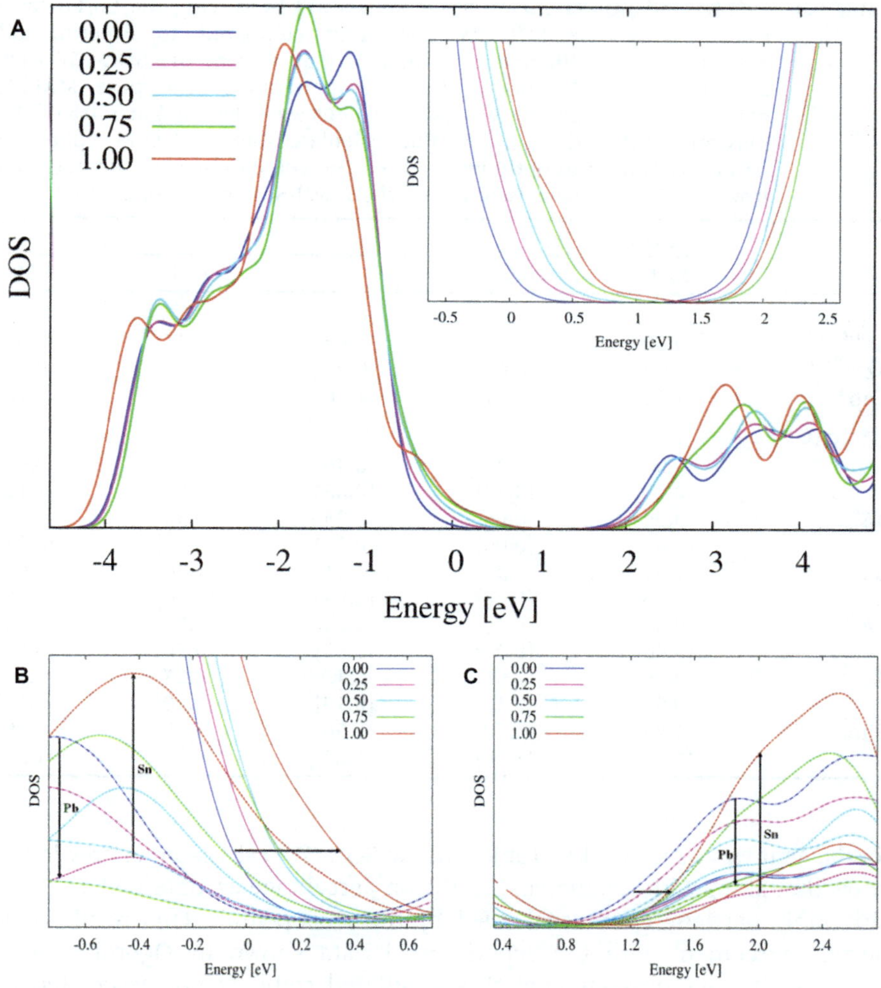

Figure 8.7 (A) DOS calculated with SOC-GW method for $MASn_xPb_{(1-x)}I_3$ by varying x from 0 to 1.00. Values of x from 0.00 to 0.75 refer to the geometry with the $MAPbI_3$ experimental cell and, for $x = 1.00$, they refer to the $MASnI_3$ cell parameters. (B) and (C) SOC-DFT DOS calculated for the investigated series showing contributions from the Pb/Sn atoms (dashed lines) and I atoms (solid lines).
(Reproduced from ref. 52 with permission from the Royal Society of Chemistry.)

Experimental studies by Ogomi *et al.* and Hao *et al.* have shown a marked difference in the variation of the VB and CB energy levels with the relative Sn/Pb ratio. However, Hao *et al.*[62] reported a totally different behavior with a non-monotonic variation of the energy levels when compared to the data by Ogomi *et al.*, in which they found a steady change of the VB and CB edges with the increasing Sn content,[64] obtaining an upshift of the energy for both

Table 8.3 HOMO and LUMO energy levels shifts (eV) by varying the Sn/Pb ratio calculated at the SOC-GW level and compared to experimental measurements.[64] VB and CB shifts evaluated from the SOC-GW DOS are also reported. Energy shifts are referred to the pure $MAPbI_3$ perovskites (set as zero). VB and CB values are calculated at 1.4% of the maximum VB peak of the $MAPbI_3$ calculated SOC-GW DOS. Values in parentheses refer to optimized atomic and cell parameters. (Reproduced from ref. 52 with permission from the Royal Society of Chemistry.)

	Sn content (x)				
	0.00	0.25	0.5	0.75	1.00
Exp.					
ΔE^{HOMO}	0.00	0.27	0.44	0.47	0.66
ΔE^{LUMO}	0.00	0.07	0.21	0.19	0.25
Gap^{EXP}	1.51	1.31	1.28	1.23	1.10
SOC-GW					
ΔE^{HOMO}	0.00	0.34	0.46	0.76	0.99
	(0.00)	(0.35)	(0.65)	(0.93)	(1.11)
ΔE^{LUMO}	0.00	0.08	0.23	0.32	0.35
	(0.00)	(0.00)	(0.23)	(0.32)	(0.33)
$Gap^{H\text{-}L}$	1.64	1.37	1.41	1.20	1.00
	(1.66)	(1.31)	(1.24)	(1.06)	(0.89)
ΔE^{VB}	0.00	0.20	0.35	0.53	0.64
	(0.00)	(0.20)	(0.43)	(0.50)	(0.68)
ΔE^{CB}	0.00	0.08	0.17	0.27	0.19
	(0.00)	(0.10)	(0.19)	(0.20)	(0.18)
$Gap^{VB\text{-}CB}$	1.74	1.62	1.56	1.48	1.29
	(1.87)	(1.77)	(1.63)	(1.57)	(1.37)

levels. As it can be noticed in Table 8.3, the SOC-GW calculated energy levels show that along with the increase in the Sn percentage we obtained a VB up-shift only partially counterbalanced by the CB up-shift. This result is in perfect agreement with the experimental data shown by Ogomi *et al.*[64] Moreover, to demonstrate that this simulated trend is not associated to the different investigated phase *P4mm* reported by Hao *et al*, we also simulated the DOS of pure *P4mm* $MASnI_3$ and compared it to the DOS calculated for the corresponding *I4cm* phase. By doing so, an additional CB up-shift of 0.12 eV was found for the pure $MASnI_3$ *P4mm* phase.

To gain insight into the transport properties of $MASn_xPb_{(1-x)}I_3$ perovskites, the effective carrier masses were calculated using a parabolic band fitting around the band at the Γ point of the Brillouin zone, see Figure 8.8. In particular, by going from $x=0.00$ to $x=1.00$, a gradual inversion of the electron and hole average is found. For Sn-based perovskites, the electron mass was predicted to be greater than the hole mass, while the electron mass was found to be smaller than the hole mass for Pb-based materials; the $x=0.5$ mixed Sn/Pb system exhibits balanced and intermediate hole and electron calculated masses. Moreover, as previously suggested, a SOC k-dependent band splitting (Rashba/Dresselhaus effect)[13,44,46,53,69] was

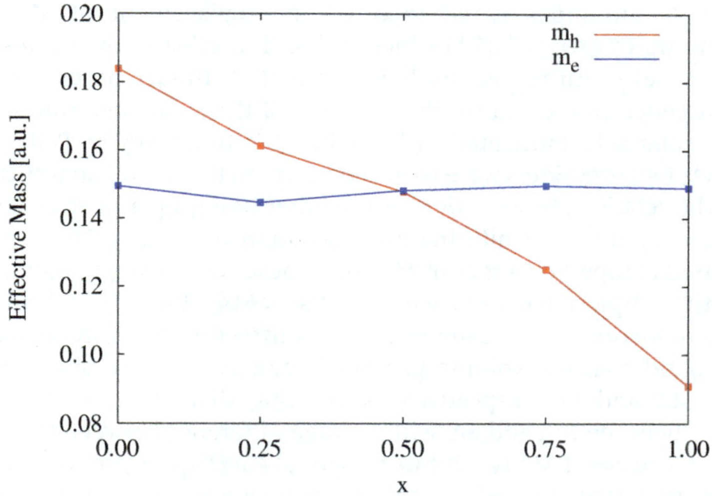

Figure 8.8 Calculated SOC-DFT effective hole and electron masses for the mixed Sn/Pb perovskites.
(Reproduced from ref. 52 with permission from the Royal Society of Chemistry.)

found due to the absence of inversion symmetry in the calculated quasi *I4cm* unit cell, which also slightly changes with the decrease in Pb percentage in the Sn/Pb mixed materials.

In summary, by varying the Sn/Pb content, we found that the 50 : 50 compound exhibits the best promising features, showing perfectly balanced electron and hole transport properties and a red-shifted and increased absorption spectra. At the same time, while perovskite oxidation could become significant above this Sn/Pb ratio, the 50 : 50 compound seems to be sufficiently stable to oxidation, as indicated by the energy level of HOMO placed at 5 eV *vs.* vacuum.

8.3.2 The Role of the A Cation

Band gap modulation can be obtained for a generic ABX_3 perovskite by varying the dimension of the A cation. In particular, the introduction of a large cation is expected to result in a reduction of the band gap in organo-halide perovskites.[22] The Goldschmidt tolerance factor (t) limits the size of the A cation, which for the organo-halide perovskites is defined as:

$$t = \frac{r_A + r_i}{\sqrt{2}(r_{Pb} + r_i)} \tag{8.1}$$

where r_i is the ionic radii. Following eqn (8.1), optimal t values are expected to be between 0.9 and 1 for cubic perovskite structures. If we consider the ionic radii from the crystal structures to be 2.03 and 1.33 Å for I^- and Pb^{2+}, respectively,[70] this gives a possible range of r_A values comprised between

~2.3–2.8 Å. Stoumpos *et al.* reported the replacement of MA by formamidinium, $CH(NH_2)_2^+$ (FA) which induced a red-shift of the absorption onset (~1.5 eV) with respect to that of $MAPbI_3$.[32] In fact, an extended photocurrent generation onset on the red-wing of the solar spectrum has been found in solar cells fabricated with $FAPbI_3$.[71–73] In a study by Pellet *et al.*,[73] mixed MA–FA perovskites were reported in which the band gap was tuned by varying the relative MA/FA ratio. On the other hand, Eperon *et al.* showed a $CsPbI_3$ perovskite[32,71] exhibiting an absorption onset at 1.73 eV, which is blue-shifted compared to that of $MAPbI_3$. These results suggest a role of the size of the employed cation, whereby $Cs^+ < MA < FA$, which leads to an expected reduction of the band gap. X-ray diffraction measurements[32] have indeed found that the volume per $APbI_3$ unit is 222, 248 and 256 Å^3 for $A = Cs^+$, MA and FA, respectively, suggesting that there is only a small difference between MA and FA, while a large difference between Cs^+ and MA is found. Moreover, $FAPbI_3$ exhibits a trigonal *P3m1* space group,[72] similar to a pseudo-cubic structure, while $MAPbI_3$ shows a tetragonal *I4cm* (or *I4/mcm*) space group at room temperature, displaying tilting of the PbI_6 octahedra.[32,74] The similar volumes per AMX_3 unit exhibited by $MAPbI_3$ and $FAPbI_3$ could indicate that the two different perovskites show a comparable structure. However, this is not the case and one might wonder whether the distinctive cell structures (tetragonal *versus* trigonal/pseudo-cubic) could assume a role, beyond the different volume and/or cation size, in determining the optical properties. Finally, the interaction between the cations and the inorganic moiety take place, for the most part, through hydrogen bonding between the acidic hydrogens of MA or FA and the perovskite iodides, which could also affect the material's properties.

With the aim of providing a clear picture of the interplay between size, structure and organic/inorganic hydrogen bond interactions in the determination of the structural and electronic properties of $APbI_3$ perovskites, we carried out a first principles investigation[53] based on state-of-the-art DFT and SOC-GW methods on $APbI_3$ perovskites, with $A = Cs^+$, MA and FA, including the recently reported mixed organic MA–FA materials.[73] We report in Figure 8.9 a comparison between the experimental tetragonal $MAPbI_3$[74] and trigonal $FAPbI_3$,[32] and the corresponding structures calculated by scalar-relativistic (SR) DFT, hereafter 1-MA and P3-MA, obtained adopting experimental cell parameters.[32,67] Analogue structures were found by performing optimization of both the atomic position and lattice parameters, see Table 8.4, providing the calculated cell parameters differ only 1–2% with respect to the experimental values. We refer to the information acquired using the experimental cell parameters to easily compare the different explored structures, but a totally consistent picture is acquired when referring to completely relaxed (nuclear positions and cell parameters) structures (Table 8.4).

It can be noticed that our calculated structures are generally in good agreement,[26] in terms of the arrangement of the PbI_6 octahedra, with the experimental crystal structures. An important difference between $MAPbI_3$

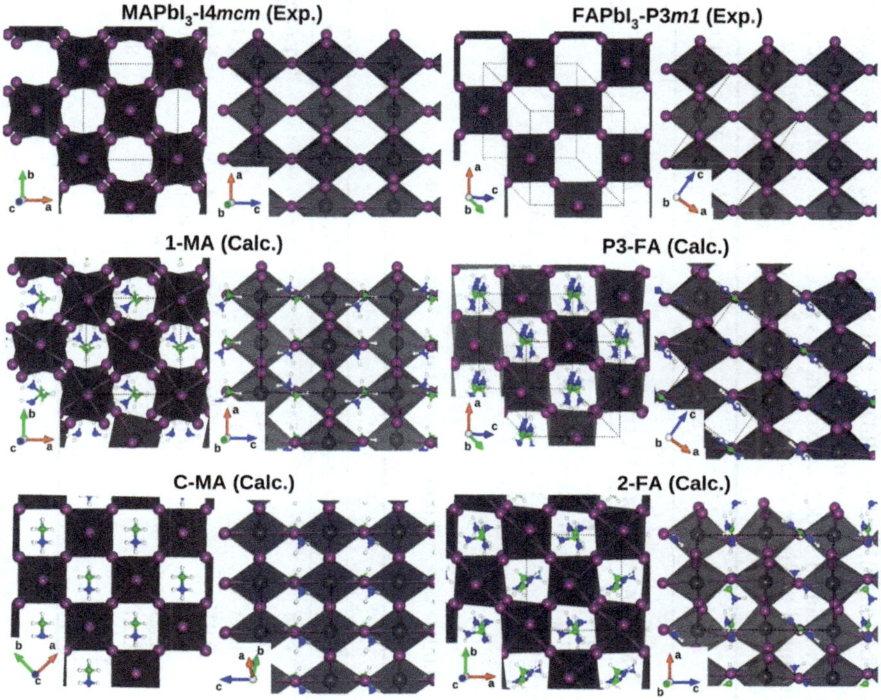

Figure 8.9 Top: Experimental crystal structure of $MAPbI_3$ (from ref. 74) and $FAPbI_3$ (from ref. 8). Bottom: Calculated crystal structures of 1-MA, P3-FA, c-MA and 2-FA, see text for their definitions. The unit cells and crystal axes are also shown in all cases.
(Reproduced with permission from A. Amat, *et al.*, *Nano Lett.*, 2014, **14**, 3608–3616. Copyright (2014) American Chemical Society.[53])

and $FAPbI_3$ found for the 1-MA and P3-FA structures is the notable out-of-phase octahedron tilting that takes place around the *c*-axis in $MAPbI_3$, of $a^0a^0c^-$ type following Glazer notation, while both structures are essentially not distorted along the $MAPbI_3$ tetragonal *c*-axis and the equivalent $FAPbI_3$ direction. The simulated 1-MA structure exhibits an average octahedron tilt angle along the *c*-axis of 22°, which is in line with the value reported by Kawamura *et al.* (21°),[74] and slightly larger than the 16° reported by Stoumpos *et al.*,[32] in which an *I*4*cm* space group was proposed for $MAPbI_3$, different than that reported in ref. 74. As it can be noticed, the tetragonal structures calculated by us, without any symmetry constraints, are characterized by alternating long-short Pb–I bond lengths along the *c*-axis, see 1-MA in Figure 8.9, which can be associated to the ferroelectric *I*4*cm* space group proposed by Stoumpos *et al.*[32] The average of the alternation calculated for the Pb–I bond lengths (3.21 *vs.* 3.16 Å) is in good agreement with the experimental data (3.20 *vs.* 3.13 Å), showing the typical features of ferroelectric distortion.[32] For the sake of comparison, we also calculated the cubic $MAPbI_3$ and $FAPbI_3$ structures (c-MA, Figure 8.9, and c-FA, respectively) and

Table 8.4 Upper panel: calculated band gap (E_g) by SR-DFT, SOC-DFT and their difference (ΔE_g DFT); SOC-GW band gap and experimental data; relative energy per $APbI_3$ unit (eV) ($E_{rel}/APbI_3$) calculated by SR-DFT and SOC-DFT for the investigated systems employing experimental cell parameters, also reported along with the volume per $APbI_3$ unit (vol., in Å^3). Bottom panel: same information for relaxed atomic and cell parameters with calculated structural parameters (Å). The most stable structure for each type corresponds to the zero relative energy. (Reproduced with permission from A. Amat, *et al.*, *Nano Lett.*, 2014, **14**, 3608–3616. Copyright (2014) American Chemical Society.[53])

	$a=b=c=6.33$ Vol. = 254		$a=b=8.86$, $c=12.66$ Å Vol. = 248					$a=b=8.98$, $c=11.01$ Vol. = 256
Exp. cell parameters[32,67]	c-MA	c-FA	1-MA	1-FA[c]	1-MA : FA	2-MA : FA	2-FA	P3-FA
E_g SR-DFT	1.34	1.31	1.66	1.60	1.59	1.58	1.58	1.63
E_g SOC-DFT	0.11	0.09	0.60	0.55	0.53	0.45	0.46	0.45
ΔE_g DFT	1.23	1.22	1.06	1.05	1.06	1.13	1.12	1.18
E_g SR-GW	—	—	2.68	—	—	—	2.65	—
E_g SOC-GW	—	—	1.67	—	—	—	1.48	—
E_g (Exp.)			1.6^a–1.7^b			1.6^c		1.5^c
E_{rel} SR	0.053	0.110	0.000	—	0.006	0.000	0.031	0.000
E_{rel} SOC	0.046	0.110	0.000	—	0.005	0.000	0.035	0.000

	Atomic and cell parameter relaxation						
	c-MA	1-MA	1-FA*	1-MA : FA	2-MA : FA	2-FA	P3-FA
E_g SR-DFT	1.30	1.56	—	1.54	1.54	1.60	1.58
E_g SOC-DFT	0.07	0.57	—	0.54	0.42	0.43	0.39
ΔE_g DFT	1.23	0.99	—	1.00	1.12	1.17	1.19
E_{rel} SR	0.062	0.000	—	0.010	0.000	0.009	0.000
E_{rel} SOC	0.056	0.000	—	0.012	0.000	0.011	0.000
a	6.31	8.78	—	8.85	8.79	8.84	8.88
b	6.31	8.76	—	8.70	8.82	8.85	8.94
c	6.31	12.70	—	12.69	12.64	12.54	10.73
Vol.	251	244	—	244	245	245	250

[a]From ref. 47.
[b]From ref. 76.
[c]Estimated from the optical absorption onset in ref. 10 and 73.

performed the simulation placing FA in the tetragonal structure using the experimental cell parameters of $MAPbI_3$. In particular, we chose to start from two different tetragonal geometries (1-MA and 2-MA) with different cation orientation (head-to-head or head-to-tail), and we obtained 2-FA as the only stable minimum, Figure 8.9. We also simulated a 1-FA* structure, generated by geometry relaxation of the FA cations inserted into the rigid inorganic 1-MA framework. Interestingly, structure 2-FA exhibits a much smaller average of the octahedron tilt angle towards the *c*-axis if compared to 1-MA (7° *vs.* 22°, respectively). Inspired by the recent study on mixed MA–FA cation perovskites by Pellet *et al.*,[73] we also simulated a 2 : 2 MA : FA perovskite employing both the experimental cell parameters of $MAPbI_3$ and then relaxing the cell, see Table 8.4. In this intermediate case, the structures of both type 1 (1-MA : FA) and type 2 (2-MA : FA) could be located. The latter is slightly stabilized (0.02–0.04 eV) over the former, showing a reduced tilt angle around the *c*-axis (6° *vs.* 21°, respectively). Thus, the different cation (MA or FA) – or, in mixed cation systems, their ratio – controls the relative stability of the two structural types, characterized by a different tilt angle.

As we have already suggested above,[26] 1-MA and 2-MA species show a different hydrogen bonding network formed between the acidic MA hydrogens and the perovskite iodine atoms. By searching for structure/property relations, we assessed the number of H–I hydrogen bonds found within the standard bond length/bond angle ranges (2.5–3.5 Å and 0–140°, respectively)[75] for 1-MA, 2-MA, 1-FA* and 2-FA, obtaining estimations of 11, 13, 16 and 18 hydrogen bonds, respectively. This evaluation confirmed our initial theory, and it additionally demonstrated that the FA cation has a higher propensity to form hydrogen bonds. This can be associated with a higher statistical probability of establishing hydrogen bonds due to the larger number of protons of FA respect to MA (4 *vs.* 3), despite MA is possibly a slightly stronger hydrogen bonding donor than FA. As a matter of fact, the ratio of formed hydrogen bonds in the structures containing the FA cation is *ca.* 4/3 that of the corresponding MA cation (*e.g.* for 2-FA and 2-MA structures, the ratio of formed hydrogen bonds 18/13 = 1.38, close to 4/3 = 1.33).

In Table 8.4, we show a summary of the structural, energetic and electronic data calculated for the investigated systems at different levels of theory.

By analyzing the electronic structure of these materials, a direct band gap at the Γ point of the Brillouin zone was found for all the investigated tetragonal structures. However, for the cubic and trigonal species, c-MA/c-FA and P3-FA, a direct band gap was found at the *R* and *A* points, respectively. As previously reported,[26,31,46] SR-DFT simulates band gap values in fortuitous agreement with the experimental data, while SOC-DFT gives underestimate band gap values,[21] which are corrected by the inclusion of electron correlation effects at the SOC-GW level.[44] The qualitative SOC-GW behavior is retrieved by SOC-DFT and it can be employed for comparative purposes. SOC-GW provides calculated band gaps in good agreement with the experimental values,[44] and nicely reproduces the reduction in the band gap

(~0.2 eV) observed moving from the MA to FA cations. Noticeably, the 1-FA* structure, obtained by replacing MA cations with FA in the 1-MA inorganic scaffold, induces a slight decrease of the calculated band gap with respect to that of 1-MA (1.60/0.55 *vs.* 1.66/0.60 eV by SR-DFT/SOC-DFT, respectively), probably induce to the increasing of the network of hydrogen bonding moving from 1-MA to 1-FA* (16 *vs.* 11 hydrogen bonds).

Interestingly, while SR-DFT calculations provide an underestimated band gap reduction going from MA to FA, SOC-DFT, although it underestimates the absolute band gap values, provides the expected band gap reduction moving from 1-MA to P3-FA or 2-FA. A similar behavior is also found by comparing SR- and SOC-GW results. SR delivers only a slight MA → FA band gap reduction, while SOC-GW gives the expected band gap reduction (0.03 *vs.* 0.20 eV). Another interesting aspect is that also the band gap difference between the SR- and SOC-DFT values, which is 1.05–1.06 eV for structures of type 1, increases up to 1.18 eV for the trigonal P3-FA, and to 1.23/1.22 eV for c-MA and c-FA in the cubic phase. Moreover, we obtained the same behavior when we referred to the results obtained by performing atomic and cell relaxation. Finally, the overall picture extracted from our data clearly indicates the presence of a structure-dependent SOC relation, which is enlarged for structures showing a small tilt angle and leading to an effective reduction of the band gap. This has also been demonstrated by focusing on c-MA/c-FA, for which the most evident band gap variation has been found. As a matter of fact, without including SOC effects, $MAPbI_3$ and $FAPbI_3$ give essentially the same band gap values, see Table 8.4, which can be explained by the calculations carried out on the cubic $MAPbI_3$ and $FAPbI_3$ species not providing the correct variation of the band gaps as reported in the experiments.[77] For example, the cubic and the tetragonal species, c-MA and 1-MA, show a band gap difference calculated at the SOC-DFT level of 0.5 eV due to a structure/property response related to SOC. In particular, only when a realistic model is considered, the comparison between the simulated and the experimental bang gap can be meaningful. As it can be noticed, the 2 : 2 MA : FA structure shows a ratio sufficient to stabilize the structure 2-MA : FA, which shows essentially the same tilt angle than that in 2-FA. As a matter of fact, the predicted variation in the structure is in line with the similar electrical conductivity properties (activation energy) that have been observed for the pure $FAPbI_3$ phase and for the ~50 : 50 mixed MA : FA perovskite, see below.[73]

To fully understand the relationship between band gap variation and structure, we used a periodic tetragonal $[CsPbI_3]_4$ system, similar to the 1-MA structure, and we moved on to investigate how the SR- and SOC-DFT electronic structure is modified by changing the octahedron tilting, as defined by the α angle, see Figure 8.10a. This model can be used to clearly understand the effects arising from anisotropy of possible different organic cation orientations, focusing on the inorganic lead-iodide structure. We started from $\alpha = 0$, which represents a pseudo-cubic geometry, displaying an optimized Pb–I distance of 3.23 Å and, as α increased, we decreased the $a = b$ cell

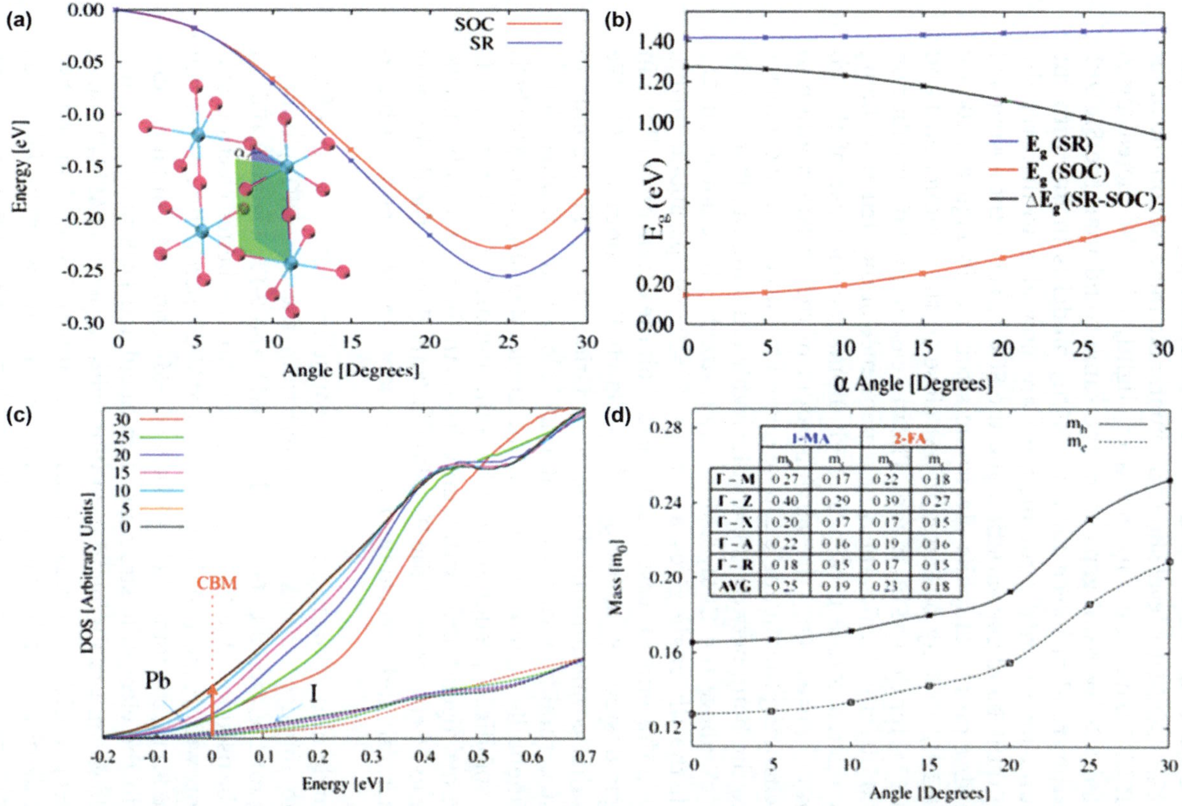

Figure 8.10 (a) Total energy calculated by SR- and SOC-DFT as a function of the α dihedral angle, whose definition is shown in the inset. The zero of the energy is set at the energy of $\alpha = 0$. (b) Variation of the calculated band gap (E_g) by SR- and SOC-DFT and their difference (ΔE_g) for the model $CsPbI_3$ system. (c) Partial lead (solid lines) and iodine (dashed lines) density of states (DOS) around the conduction band minimum as a function of the varying tilt angle α in the $CsPbI_3$ system. (d) Variation of the calculated average effective electron (m_e, dashed line) and hole masses (m_h, solid line) by SOC-DFT as a function of the α titling angle in the $CsPbI_3$ system. The table in the inset shows the SOC-GW calculated effective masses along the high symmetry directions of the tetragonal Brillouin zone for 1-MA and 2-FA. (Reproduced with permission from A. Amat, *et al.*, *Nano Lett.*, 2014, **14**, 3608–3616. Copyright (2014) American Chemical Society.[53])

parameters to keep constant all the Pb–I bond lengths. By carrying out this approach, we found a minimum for $\alpha = 25^\circ$, Figure 8.10a, and this position remains the same at both SR- or SOC-DFT levels of theory, despite a slight destabilization of the distorted structure provided by SOC.

The changes in the bad gap with the variation of the α angle at SR- and SOC-DFT levels are shown in Figure 8.10b. Interestingly, the band gap calculated at the SOC-DFT level shows a remarkable increase; on the other hand, the SR-DFT band gap only increases slightly with the increasing α. The calculated SR-GW band gaps show the same trend than the SR-DFT data. While the trend observed for this $CsPbI_3$ model is in line with the behavior found for the real structures, we are now able to discover the reason behind the band gap variation and the SOC effect in the systems of interest. Moving on to study the electronic structure, in Figure 8.10c we show the SR-DFT Density of States (DOS) close to the conduction band minimum (CBM) for the $CsPbI_3$ model. The increase in the Pb contribution near the CBM directly clarifies the SOC dependence of the structure. Being SOC mainly related to the presence of Pb atoms, an increase of the Pb contribution enhances SOC. The varying extent of the Pb contribution to the DOS by changing the tilt angle can be related to the covalence character associated to the bond between Pb and I, sizably reducing the Pb quantity up to ~0.5 eV above the CBM.[78] Most interestingly, the changes in ΔE_g SR-SOC with the tilt angle gives a perfect ($R^2 = 0.999$) parabolic fitting, indicating that this effect is essentially located on the octahedron tilting in the investigated systems, as proposed by Garcia-Fernandez *et al.*[78] We generalized this covalent/ionic character variation of the Pb–I bonds to clarify the structural behavior with different halogen atoms, where it has been experimentally demonstrated that the cubic $MAPbBr_3$ and $MAPbCl_3$ species are stabilized predominantly by ionic bonds.[67] In line with this analysis, the hydrogen bonds with iodine atoms in the pseudo-cubic phase stabilize the ionic character of the Pb–I bonds. This clarifies the differences in the structural and electronic features between MA- and FA-based perovskites.

The tilt angle also induces a remarkable variation in the average of the electron/hole effective masses, extrapolated from the SOC-DFT parabolic band fitting for the model $CsPbI_3$, and for 1-MA and 2-FA, Figure 8.10d. In particular, the pseudo-cubic $FAPbI_3$ perovskite shows a better electron and hole transport with respect to the tetragonal $MAPbI_3$ perovskite. As we can see in Figure 8.11a, the total SOC-GW DOS of 1-MA and 2-FA exhibit notable differences in the electronic structure. The calculated DOS nicely matches the energy down-shift of VB found in $MAPbI_3$ with respect to $FAPbI_3$[6,72] and, despite a steeper CB/VB increase, provides a similar CB level in the two materials. This difference in the electronic DOS of the two materials leads to sizable differences in the calculated optical absorption spectra, see Figure 8.11b. Both absorption spectra are in perfect agreement with the experimental data. The 2-FA species exhibits a red-shifted absorption onset with respect to that of 1-FA, as expected from the smaller band gap.

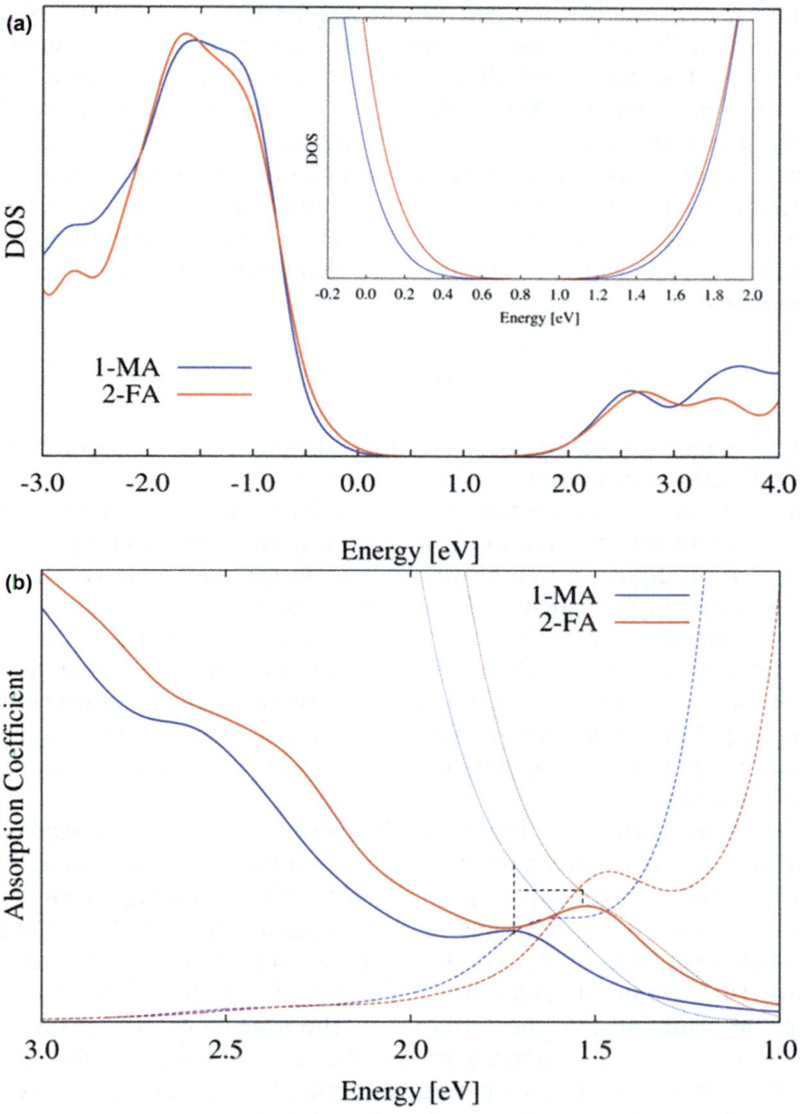

Figure 8.11 (a) Calculated total density of states (DOS, solid lines) by SOC-GW for 1-MA (blue) and 2-FA (red). Inset: zoom around the CBM/VBM regions. (b) Calculated absorption spectra (solid lines) by SOC-GW for 1-MA (blue) and 2-FA (red). Dotted lines are the SOC-GW calculated joint DOS for 1-MA (blue) and 2-FA (red), while dashed lines represent the ratio between the absorption coefficient and the joint DOS for 1-MA (blue) and 2-FA (red), both in arbitrary but consistent units. Black dashed lines mark the absorption maxima for and the related values of the joint DOS.

(Reproduced with permission from A. Amat, *et al.*, *Nano Lett.*, 2014, **14**, 3608–3616. Copyright (2014) American Chemical Society.[53])

Moreover, the part of the spectrum at 2.6–2.4 eV is red-shifted in 2-FA compared to the spectrum of 1-MA, and a less marked shoulder is observed.[71–73] The calculated absorption spectrum of 2-FA shows a higher intensity than that of 1-MA in the whole simulated energy range. As a matter of fact, the red-shifted onset found for 2-FA is directly associated to an increase of the absorption intensity, which induces an improved harvest of sunlight radiation. To clarify the origin of the enlarged spectral intensity found for 2-FA, we refer to the Fermi golden rule interpretation of the spectral intensity, which for a transition from valence (ν) to conduction (c) states reads:

$$W_{\vec{k}} \cong \frac{2\pi}{\hbar} |\langle \nu \,|\, \mathscr{H}' \,|\, c \rangle|^2 \delta[E_c(\vec{k}) - E_\nu(\vec{k}) - \hbar\omega] \tag{8.2}$$

where, for light absorption, H' is the dipole operator, and the second part of eqn (8.2) represents a joint density of states.

From the absorption coefficient and the joint DOS, we can understand what the main contribution to the reduction of the 2-FA band gap between the transition dipole term and the changes in the joint DOS is. The comparison between the calculated absorption spectra and the joint DOS is shown in Figure 8.11b, with the ratio between the spectral intensity and the joint DOS. As it can be seen, the absorption maxima of joint DOS is 30% higher for 1-MA than for 2-FA. Interestingly, these results demonstrate that the enlarged intensity obtained for 2-FA is due to a strong increase in the transition dipole moment generated by the different electronic structures, as discussed above.

Another intriguing property identified with the presence of SOC is the Rashba/Dressalhaus effect, which appears as a k-dependent band splitting in non-centrosymmetric space groups.[20] The effect has a higher impact on the CB than on the VB. The band structures calculated at the SOC-GW level for 1-MA and 2-FA are reported in Figure 8.12, which nicely exhibit a Rashba/Dresselhaus effect on the CB along the $\Gamma \rightarrow M$ direction, which is in the plane orthogonal to the c-axis where the long-short Pb–I distance alternation is located. In particular, the Rashba/Dresselhaus effect can be quantified by calculating the parameter defined as $\alpha_R = 2E_R/k_R$,[79] where E_R is the difference in energy at the Γ point and the effective CB minimum, while k_R is the displacement along a selected direction in the k-point space. In particular, we calculated α_R values of 2.3 and 2.6 eV Å for 1-MA and 2-FA, respectively, at the SOC-GW level. Although these are sizable values, they are 6 times smaller than that for GeTe (16 eV Å), for which a "giant Rashba effect" has been reported.[79] As expected with its higher structure-dependent SOC, 2-FA shows an enlarged Rashba parameter compared to 1-MA. Large values of α_R mean a higher displacement of the CB minimum with respect to Γ that could possibly contribute to reduce carrier recombination, with the consequential increase in the efficiency in perovskite-based devices.

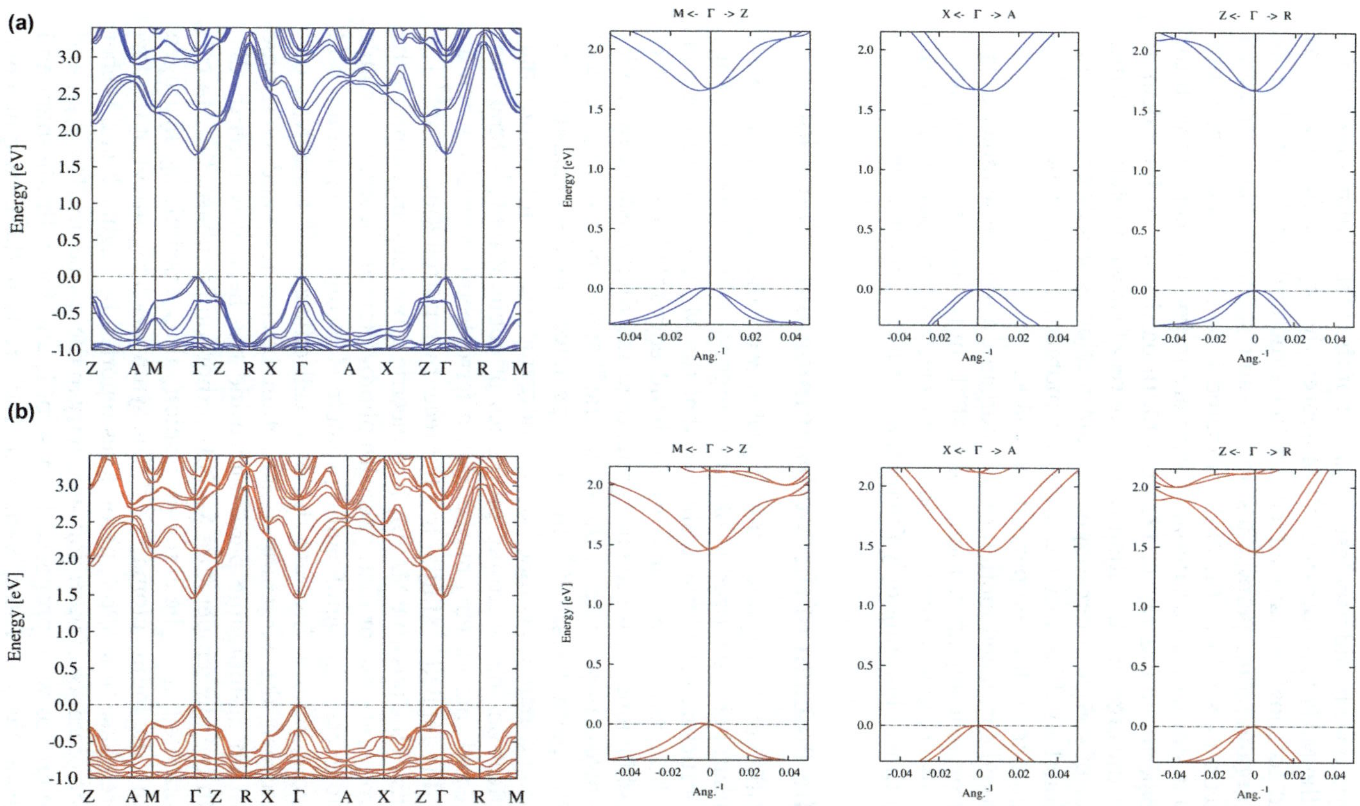

Figure 8.12 Left: SOC-GW calculated band structure for 1-MA (a, blue lines) and 2-FA (b, red lines) along the high symmetry directions of the tetragonal Brillouin zone Γ (0,0,0)$\rightarrow$$M$ (0.5,0.5,0); $\Gamma\rightarrow Z$ (0,0,0.5); $\Gamma\rightarrow X$ (0,0.5,0); $\Gamma\rightarrow A$ (0.5,0.5,0.5); $\Gamma\rightarrow R$ (0,0.5,0.5). The energy zero is set in both cases at the highest occupied state. Right: Zoom of the band dispersion around Γ. (Reproduced with permission from A. Amat, *et al.*, *Nano Lett.*, 2014, **14**, 3608–3616. Copyright (2014) American Chemical Society.[53])

8.3.3 Methylammonium Dynamics in the Perovskite Lattice

Some reports have found very fast dynamics of the methylammonium (MA) cations in the $MAPbI_3$ perovskite.[67,80] Moreover, the organic framework has been found to show relevant effects on the structure[26] and optical properties[81] of organo-halide perovskites. In the same spirit, we have simulated the IR spectra of $MAPbI_{3-x}Cl_x$ and compared it to experimental measurements.[14] Our simulated vibrational spectra are in good agreement with the available experimental data, see Figure 8.13. In particular, we have demonstrated that the organic MA cation can rotate very fast, with rotational times of the order of 4–6 ps. Interestingly, the MA motion is strictly correlated to the vibrational mode associated with the inorganic framework, which is already found at the low temperature of 268 K.

This marked correlated motion of the inorganic and organic moieties is a clear spectroscopic signature found for $MAPbI_3$ in the range 200–300 cm^{-1}. Finally, we also evaluated the variation of the electronic properties along the simulation time, obtaining only fluctuations of a few tenths of eV.

8.3.4 The Problem of the Methylammonium Orientation

The orientation of the organic cation is probably the most complex problem in the structural characterization of hybrid organo-halide perovskites. Early electronic structure calculations showed that the MA cation does not contribute to the energy levels close to the band edges.[27,82] The presence of partially mobile ions within the inorganic cage and their possible dynamics may nevertheless play a role in the structural[26] and electrical properties of organo-halide perovskites.[53,81,83] In this respect, we have demonstrated how the MA cation dynamics, coupled to the inorganic degrees of freedom, may exert an effect on the electronic properties of cubic $MAPbI_3$.[15] Stoumpos *et al.* suggested that $MAPbI_3$ may exhibit a ferroelectric response when an electric field is applied,[32] similar to what recently reported Kutes *et al.*[84] The implications of the possible $MAPbI_3$ ferroelectric properties are currently the focus of strong interest for materials with photovoltaic applications. A stable ferroelectric effect may significantly promote electron/hole separation and may thus contribute to increase the solar cell efficiency.[85] Notably, a correlation between the structure, polarization orientation, and electronic structure of an all-inorganic perovskite oxide has been very recently discussed.[86] In the low temperature $MAPbI_3$ orthorhombic phase, the MA cations are oriented in a head-to-tail fashion, thus maximizing hydrogen bonding with the iodine atoms.[28,87,88] The symmetry assignment of the low temperature phase of $MAPbX_3$ perovskites reported by Poglitsch and Weber excluded the presence of inversion symmetry in these systems. The P_{2221} and P_{na21} space groups were proposed for $MAPbCl_3$ and for both $MAPbBr_3$ and $MAPbI_3$, respectively.[67] Later, Swainson *et al.* revised these assignments, proposing the P_{nma} space group for both $MAPbCl_3$[88] and $MAPbBr_3$,[87] thus excluding possible ferroelectric properties for the low temperature phases.

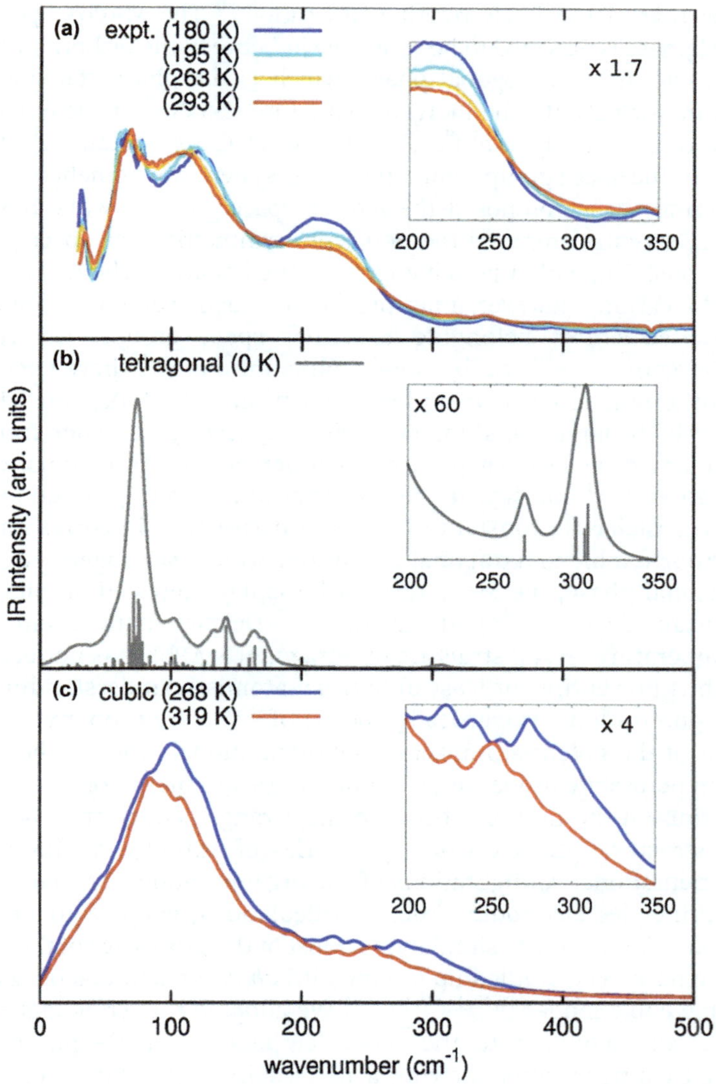

Figure 8.13 Experimental IR spectra of the $MAPbI_{3-x}Cl_x$ perovskite recorded at 180 K, 193 K, 263 K and 293 K; (b) theoretical IR spectrum from the static phonon calculation on the tetragonal structure (proposed in ref. 82 for the $MAPbI_3$ perovskite); (c) theoretical IR spectra from CPMD simulations at 268 K and 319 K. In the insets, we report the detail of the experimental and theoretical spectra in the 200–350 cm^{-1} frequency region, with the respective intensity magnification factor. (Reproduced from ref. 15 with permission from the PCCP Owner Societies.)

Baikie *et al.* have recently confirmed that the $MAPbI_3$ orthorhombic phase belongs to the P_{nma} space group.[28] The high temperature $MAPbX_3$ cubic phases feature MA cations that are dynamically disordered, moving at a rate similar to that of the freely rotating MA cation.[80] The $MAPbX_3$ perovskite tetragonal phase is expected to lie somehow in between the orthorhombic and cubic phases. In the tetragonal phase, which is the phase stable at room temperature for $MAPbI_3$, the inorganic structure shows a characteristic octahedron tilting of the type $a^0a^0c^-$,[32,74] following Glazer notation.[89] The assignment of the space group symmetry of this system was somehow debated: Poglitsch and Weber proposed the *I4/mcm* space group,[67] while Stoumpos *et al.* more recently proposed the non-centrosymmetric *I4cm* space group,[32] which is consistent with a possible ferroelectric behavior. The C_{3v} symmetry of the MA cation, however, may prevent the achievement of genuinely symmetric structures belonging to either space groups. Calorimetric studies by Knop *et al.*[80] and by Onoda-Yamamuro *et al.*[83] have shown that the orthorhombic to tetragonal phase transition in $MAPbX_3$ perovskites is characterized by large transition entropies, suggesting that such transition is accompanied, at least in part, by disorder of the MA cations. Onoda-Yamamuro *et al.*[83] confirmed this observation, detecting a sudden variation of the dielectric constant of $MAPbX_3$ perovskites in correspondence to the orthorhombic to tetragonal transition. Still, it was suggested that, in the tetragonal phase, the MA cations still display preferred orientations[80] and a certain degree of long-range order.[83] Despite many attempts, the room-temperature crystal structure of tetragonal $MAPbI_3$ has not been fully solved, thus preventing, at least in part, an atomistic understanding of its superior photovoltaic properties by means of structure/property relations.

In light of the potential impact of the orientational order of the organic cations on perovskite properties, we show here a summary from a systematic investigation on the interplay between the arrangement of the MA cations and the structural and electronic properties of $MAPbI_3$. We investigated several orientational configurations of the organic cations (see Figure 8.14) by first principles electronic structure calculations, coupled to first principles molecular dynamics simulations. Most of the properties of the $MAPbI_3$ perovskite have been studied up to now with *ab initio* methods on the basis of a simple cubic model.[31,48,49,69,81,90,91] We adopted the tetragonal $MAPbI_3$ unit cell, corresponding to the room temperature stable phase,[92] and considered configurations showing a net resulting orientation of the MA cations, which are likely to present ferroelectric properties, and configurations in which the MA cations show an isotropic distribution. Geometry optimizations at zero temperature were performed to evaluate the relative structural stability and to establish the possible effect of the MA orientation on the electronic properties of the $MAPbI_3$ perovskite. *Ab initio* molecular dynamics simulations were then employed to investigate the dynamics of the MA cations in the tetragonal $MAPbI_3$ perovskite at room temperature with particular emphasis on the possible interconversion among various possible structures under thermal conditions. The impact of the structural features on charge separation and recombination in $MAPbI_3$ thin films was also assessed.

In previous DFT calculations on $MAPbI_3$,[14,26,53,56] we observed that $MAPbI_3$ shows a typical $a^0a^0c^-$ tilting of the octahedra in the tetragonal

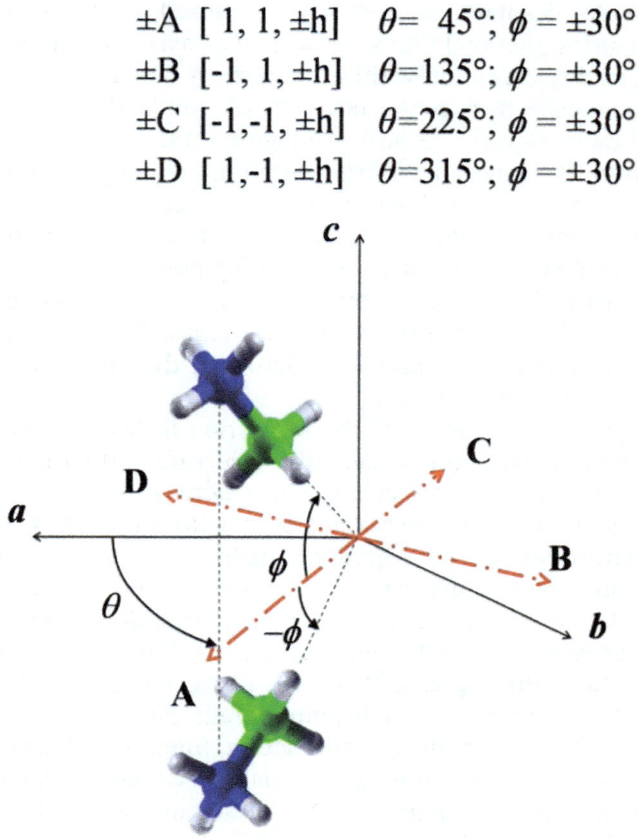

Figure 8.14 Preferred orientations of the organic cation within the perovskite cavity. *a*, *b* and *c* are the crystallographic axes of the perovskite in the tetragonal phase, where the octahedron tilting is along the *c*-axis, *i.e.* along the [001] direction. A, B, C and D represent the projections of the MA cation orientations within the *ab*-plane, as measured by the angle θ. The tilting of the molecule with respect to the *c*-axis is measured by the ϕ angle.
(Reproduced with permission from C. Quarti, *et al.*, *Chem. Mater.*, 2014, **26**, 6557–6569. Copyright (2014) American Chemical Society.[13])

phase. In this type of structure, the MA cations are arranged along specific directions, *i.e.* roughly along the direction connecting the two opposite faces of the pseudo-cubic cell, with a small tilting with respect to the *ab*-plane. This orientation was confirmed also by Filippetti and Mattoni,[93] and it is likely related to the formation of a stronger hydrogen bonding network. The classification of the orientation of the MA cation in the $MAPbI_3$ perovskite is reported in Figure 8.14. With respect to the *ab*-plane, the MA cations can be oriented along the [110], [-110], [-1-10] and [1-10] directions, which correspond the molecules lying in within the *ab*-plane with angles of $\theta = 45°$, 135°, 225° and 315° with respect to the *a*-axis, see Figure 8.14. With respect to the

c-axis, the molecules have two symmetric preferred orientations, $\phi = \pm 30°$, with ϕ measuring the angle formed by the CN axis and the ab-plane. The orientation of the MA cation within the plane is labeled as A, B, C and D, corresponding respectively to $\theta = 45°$, 135°, 225° and 315°, while the ϕ angle is indicated by its sign ($\phi \sim \pm 30°$), see Figure 8.14.

Starting from the typical $a^0a^0c^-$ structure for the inorganic framework, we investigated nine additional guess structures, characterized by either a global preferential orientation or a global isotropic orientation of the MA cations, the former and the latter representing polar and apolar structures, respectively. These investigated structures were labeled as **1–9**, see Table 8.5. In addition, we investigated the structure proposed by Stoumpos *et al.*,[32] which is characterized by a parallel ordering of the MA cations along the c-axis, hereafter **10**, see Table 8.5.

To investigate the dynamics of the MA cation in $MAPbI_3$, we undertook CPMD simulations on two representative structures: (i) structure **1**, which has an isotropic distribution of the MA cations in θ, but shows all the molecules tilted with the same ϕ angle; and (ii) structure **4**, showing an isotropic distribution in both degrees of freedom. The molecular orientation of the MA cations in the investigated structures is summarized in Table 8.5.

DFT calculations within periodic boundary conditions, were carried out with a planewave/pseudopotential method, as implemented in the PW package of Quantum Espresso,[94] on the tetragonal unit cell of $MAPbI_3$, containing 48 atoms and 4 methylammonium cations (see left panel of Figure 8.15). The PBE exchange–correlation functional[95] was used with ultrasoft[96] scalar relativistic pseudopotentials for all atoms. A k-point mesh of $4 \times 4 \times 4$ and two sets of cut-offs, 25–200 Ry and 50–400 Ry, were used, respectively, for the sampling of the first Brillouin zone and the expansion of the wavefunction and density. We performed structural optimization for the

Table 8.5 Orientation of the MA cations in the various structural models investigated and resulting net molecular alignment. The molecular orientation is labeled as indicated in Figure 8.14. (Reproduced with permission from C. Quarti, *et al.*, *Chem. Mater.*, 2014, **26**, 6557–6569. Copyright (2014) American Chemical Society.[13])

	Cation orientation				
Structure	A	B	C	D	Resulting MA dipole direction
1	1−	1−	1−	1−	[001]
2	1−	1−; 1−	1−		[1-10] [001]
3	2−	2−			[010] [001]
4	1+	1−	1−	1+	—
5	2+		2−		—
6	1+; 1−		1+; 1−		—
7	1+; 1−		1+; 1−		—
8	1+; 1−		1+; 1−		—
9	2−		2−		—
10[a]					[001]

[a]In structure **10**, the MA cations are oriented along the c-axis, so this structure does not show any projection of the MA axis on the ab-plane.

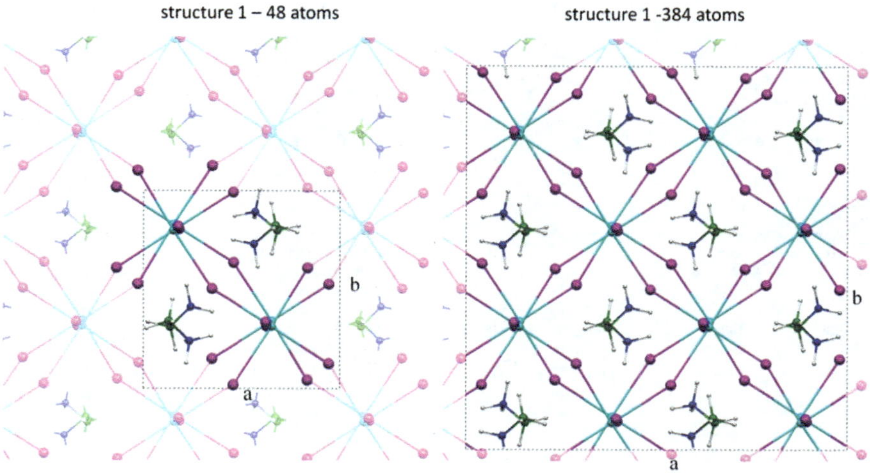

Figure 8.15 Structural models employed in the present work. 48-atom model adopted for the static DFT calculations (left panel) and 384-atom model for the CPMD calculations (right panel). The unit-cell and the crystallographic *a* and *b* axes are highlighted.
(Reproduced with permission from C. Quarti, *et al.*, *Chem. Mater.*, 2014, **26**, 6557–6569. Copyright (2014) American Chemical Society.[13])

cell parameters fixed to the experimental values reported by Poglitsch and Weber[67] and relaxation for both the atomic positions and cell parameters. The electronic structure was also calculated considering spin–orbit coupling (SOC) effects[31] with a 25–200 Ry cut-off on the scalar relativistic (SR) optimized geometries.

CPMD simulations were performed within the Car–Parrinello method employing ultrasoft pseudopotentials,[97,98] as implemented in the Quantum Espresso package.[94] These calculations were carried out on a 2×2×2 supercell of the tetragonal unit (384 atoms and 32 MA cations, right panel of Figure 8.15). The Hamiltonian parameters are the same as those used for the static calculations and in our previous work:[15] the PBE exchange correlation functional, a 25–200 Ry cut-off for the expansion wavefunction–density. CPMD simulations were carried out at the Γ point of the first Brillouin zone. In this case, the cell parameters were fixed to the experimental values.[67] We used an integration time step of 5 au, together with a fictitious electronic mass of 400 au and the real ionic masses, except for the hydrogen atoms, which were replaced by deuterium for computational convenience. CPMD simulations were conducted at ~320 K for a total simulation time of ~12 ps after a few picoseconds of equilibration.

After DFT optimization, all the starting structures roughly preserved the original MA cation orientation, *i.e.* the θ and ϕ angles only slightly varied without undergoing a transition among the various initial orientations, representing the local potential energy surface minima. The optimized geometries of the considered structures are depicted in Figure 8.16. The

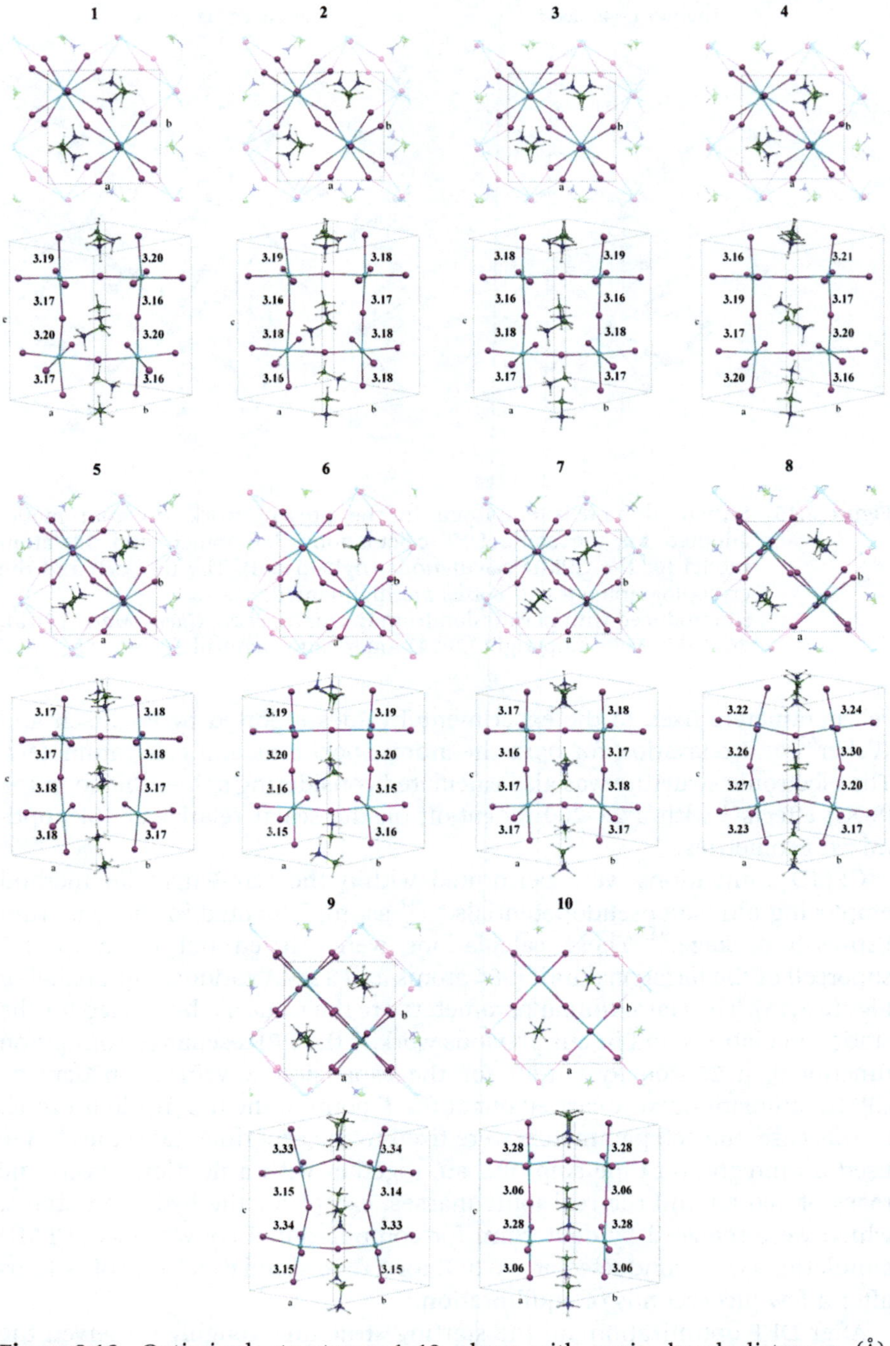

Figure 8.16 Optimized structures **1–10** along with main bond distances (Å). For each structure, we report the view along the [001] and [110] directions.
(Reproduced with permission from C. Quarti, *et al.*, *Chem. Mater.*, 2014, **26**, 6557–6569. Copyright (2014) American Chemical Society.[13])

Table 8.6 Relative energy (ΔE, eV), scalar relativistic (SR) and spin–orbit (SOC) band gaps (E_g, eV) and cell parameters (a, b, c, Å) for all the investigated structures. For each structure, the orientation of the organic cations is reported (MA orient) in relation to the labels of Figure 8.14. Data are presented for calculations at both experimental and optimized cell parameters. (Reproduced with permission from C. Quarti, *et al.*, *Chem. Mater.*, 2014, **26**, 6557–6569. Copyright (2014) American Chemical Society.[13])

		Expt. cell parameters			Optimized cell parameters		
Str.	MA orient.	ΔE	E_g(SR)	E_g(SOC)	a, b, c	Rel. en.	E_g(SR)
1	−D, −A, −C, −B	0.00	1.57	0.53	8.75, 8.74, 12.77	0.02	1.47
2	−B, −A, −C, −B	0.00	1.59	0.57	8.75, 8.73, 12.79	0.00	1.48
3	−B, −A, −A, −B	0.01	1.62	0.60	8.74, 8.73, 12.80	0.00	1.50
4	−B, −C, +D, +A	0.09	1.60	0.53	8.76, 8.74, 12.75	0.11	1.51
5	+A, −C, −C, +A	0.09	1.57	0.51	8.76, 8.72, 12.80	0.08	1.50
6	+A, −A, +D, −D	0.06	1.34	0.30	8.78, 8.77, 12.71	0.08	1.22
7	+C, −A, −C, +A	0.06	1.54	0.48	8.77, 8.77, 12.70	0.11	1.46
8	−C, −A, +A, +C	0.08	1.60	0.46	8.85, 8.86, 12.46	0.11	1.56
9	−A, −C, −C, −A	0.07	1.82	0.80	8.86, 8.81, 12.54	0.10	1.74
10[a]	—	0.47	1.61	0.58	8.85, 8.85, 12.71	0.39	1.62

[a]In structure **10**, the MA cations are aligned along the *c*-axis, so they have no projection on the *ab*-plane.

relative energies, band gaps (all at the Γ point) and structural parameters of the various relaxed structures are summarized in Table 8.6. As previously noted, SR-DFT fortuitously delivers calculated band gap values for lead-halide perovskites in good agreement with the experimental data, due to cancellation of SOC and many-body correlation effects.[26] Introduction of SOC within DFT delivers strongly reduced band gaps,[31] which are properly corrected at the SOC-GW level[44,49] or adopting hybrid functionals.[99] It should also be noted that the effect of SOC is structure-dependent,[53] with SOC-DFT band gaps qualitatively following the results of the more computationally demanding SOC-GW calculations.[44]

As reported in Table 8.6, the isoenergetic structures **1–3** are the most stable among the investigated ones. All other structures, with the exception of **10**, are found within 0.05–0.10 eV above **1–3**, *i.e.* only slightly exceeding kT at room temperature (~0.03 eV). Notably, structures **8** and **9** show tilting of the PbI_6 octahedra along the [110] direction, rather than along the [001] crystallographic direction. Thus, in these two structures, the *c*-axis rotates along the [110] direction. This behavior probably reflects the soft nature of this class of materials, which can accommodate relatively large structural distortions at a small energy cost. The optimized structure **10**, having the MA cations aligned along the *c*-axis, does not show the typical $a^0a^0c^-$ octahedron tilting, Figure 8.16. In addition, the resulting structure is strongly destabilized, lying about 0.5 eV above the most stable **1–3** structures. The results presented above do not change when the cell parameters are allowed to relax along with the atomic positions, Table 8.6, and/or when larger cut-offs are used. This clearly indicates that, in $MAPbI_3$, a tilted orientation of the

MA cations with respect to the *ab*-plane with $\phi \sim 30°$, corresponding to structures **1–3**, is energetically preferred.

Notably, together with a different orientation of the MA cations, the diverse structures show also a different arrangement of the inorganic framework, Figure 8.16. The oriented structures **1**, **2**, **3**, **9** and **10** show an "in-phase" alternation of the apical Pb–I bond lengths that is consistent with the *I*4*cm* space group proposed by Stoumpos *et al.*[32] Such bond length alternation is maximized in structure **10** (~0.2 Å), in relation to the alignment of the MA cations along the *c*-axis, peculiar for this structure. Contrarily, the isotropic structures **4**, **5**, and **8** display an "out-of phase" alternation or no alternation of the apical Pb–I bond lengths, which is consistent with the *I*4/*mcm* space group.[67] This data clearly suggests that the orientation of the MA cation in $MAPbI_3$ strongly influence the inorganic structure, with a ferroelectric-like/antiferroelectric-like structure coherent with *I*4*cm* or *I*4/*mcm* space group symmetry.

Interestingly, the different cation orientations deliver substantially different calculated cell parameters, Table 8.6. In particular, structures **1–7** all provide similar cell parameters, with $a \approx b$ values shorter than the experimental one (8.86 Å) and c values longer than the experimental one (12.66 Å), leading to a c/a ratio of ~1.46, against the ~1.43 experimental ratio. Structures **8** and **9**, instead, show a reverse trend, with elongated $a \approx b$ values and shorter c values, leading to a c/a ratio of ~1.41. This is obviously related to the different octahedron tilt direction of the two types of structures. Notably, the two different tilt directions are found to lie within less than 0.1 eV. Recent calculations have demonstrated that inclusion of van der Waals interactions delivers calculated cell parameters of $MAPbI_3$ in good agreement with experimental values.[51,100–102] Geng *et al.* have pointed out that, even accounting for such interactions, the computed cell parameters depend sizably on the orientation of the organic cations,[99] in line with the presented results.

An interesting finding of our study is that the MA cation orientation has a strong impact on the $MAPbI_3$ electronic properties. As shown in Table 8.6, for all the considered structures except **6** and **9**, the orientation of the MA cations affects the SOC-DFT band gap value within a range of 0.14 eV at fixed cell parameters. SR-DFT band gaps show a reduced variability compared to the SOC-DFT results, due to the aforementioned interplay of structural features and SOC effects.[53] Structure **9** (**6**) shows a sizable band gap increase (decrease) of ~0.2 eV, compared to the more stable structures **1–3**, despite being only 0.07 (0.06) eV higher in energy. These results point out the large variability in band gap values that could be observed if one were able to selectively obtain one type of structure.

The orientation of the MA cations, inducing the structure of the inorganic framework, has relevant implications on the electronic properties of $MAPbI_3$. Considering the oriented structures **1**, **3** and **10**, and the isotropic structure **4**, we computed the SOC-DFT band structure, Figure 8.17. We observed that, despite the general similarity of both the band gaps

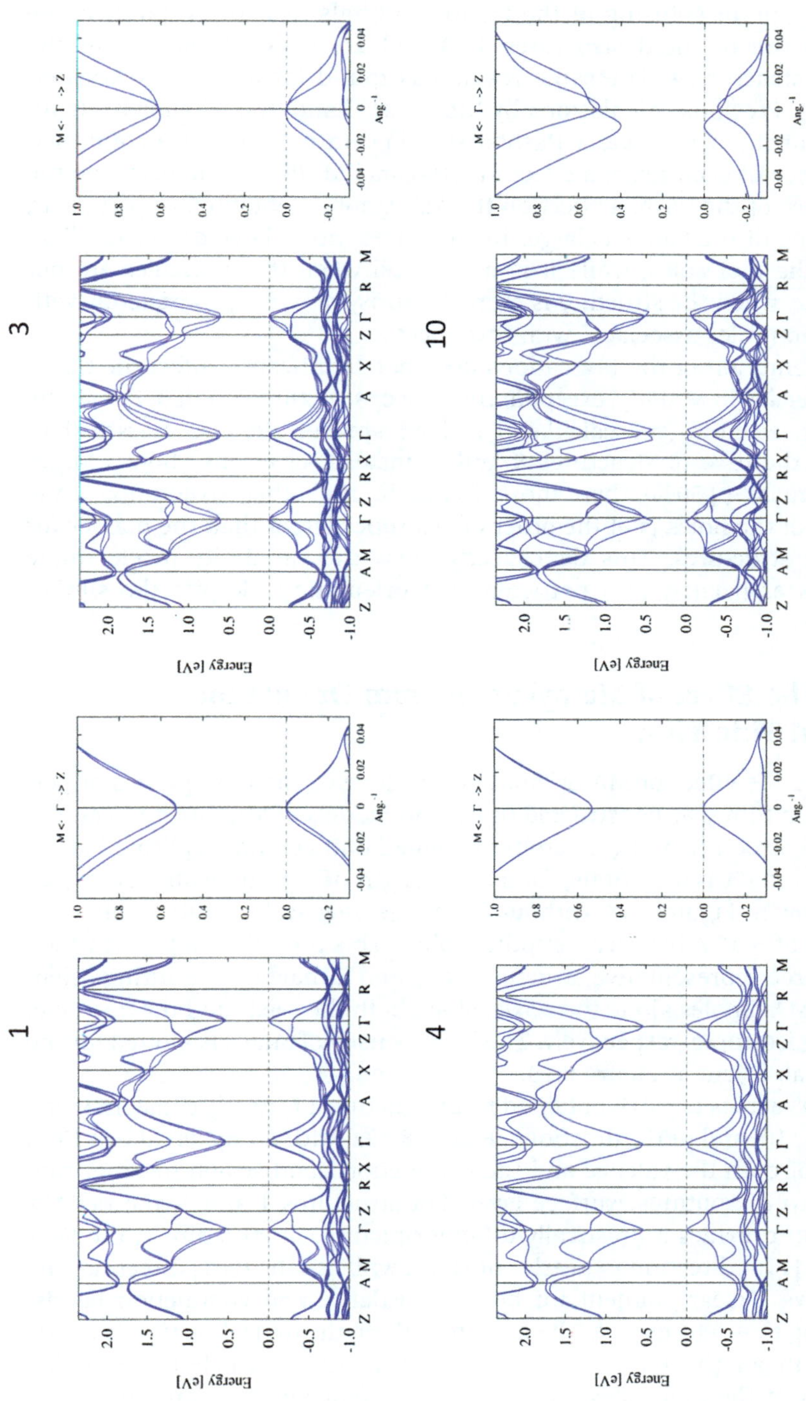

Figure 8.17 SOC-DFT calculated band structures in the first Brillouin zone (left panels) and magnification of the region around the $M \rightarrow \Gamma \rightarrow Z$ direction (right panels) for structures **1**, **3** and **10** (ferroelectric-like) and for structure **4** (antiferroelectric-like). (Reproduced with permission from C. Quarti, *et al.*, *Chem. Mater.*, 2014, **26**, 6557–6569. Copyright (2014) American Chemical Society.[13])

(see Table 8.6) and the band structures, significant differences are observed. A *k*-point splitting of the frontier orbitals near the *Γ* point is observed in the oriented structures **1**, **3** and **10**, which is absent in the isotropic structure **4**. The same result was found for all the investigated structures. Previous calculations by Kim *et al.*[69] and by us[53] already identified this phenomenon, *i.e.* a Rashba splitting. In our work, we confirmed the presence of such Rashba effect in tetragonal $MAPbI_3$, demonstrating the dependence of the Rashba effect on the MA cation orientation. In particular, the splitting of the bands is larger in structures with a large degree of alignment of the molecules, while it is nearly null when the molecules are not aligned, *i.e.* when the structure is more "centrosymmetric", consistently with the selection-rules associated with this effect.

The orientation of the MA cations does not significantly affect the transport properties in $MAPbI_3$, except in the case of structures **8** and **9**. For all the structures, electrons are calculated to have smaller effective masses than holes. In the case of structures **8** and **9**, instead, electrons show a larger effective mass (+50%) than those found in the other structures, while the mass of the holes is of the order of 2–3 times larger than the masses for the other structures. This data clearly show the variability in electronic properties associated to a particular MA orientation, despite the similar energetics.

8.3.5 The Effect of Methylammonium Orientation in Thin Films

To analyze the effect of MA orientation on the electronic properties of perovskite thin films, we constructed finite slab models of $MAPbI_3$ surfaces: the optimized structure of the 48-atom tetragonal unit cell was replicated along the [110] and [001] directions, for a total length of 6 pseudocubic units (see right panels in Figure 8.18) without further geometrical relaxation, allowing a vacuum of ~15 Å between periodic replica. This procedure was performed for the three representative structures **1**, **3** and **4**, having a resulting orientation of MA dipoles along the *c*-axis, along both the *c*-axis and the *ab*-plane, and being isotropic, respectively. The local Density of States is plotted for the various slab layers in Figure 8.18.

The DOS across the slab obtained from structure **4** is nearly constant along both the [110] and [001] directions, Figure 8.18a, with a very small (<0.1 eV) bending of both the valence and the conduction bands when moving from the bulk to the outmost surface layer. For structures **1** and **3**, the net MA orientation provides a drastically different picture. In structure **1**, the DOS along the [110] direction is nearly constant, while in the [001] direction, the DOS shows a clear gradient for both the valence and conduction bands, amounting to ~0.04 eV Å^{-1}. The band gap along the [001] direction is nearly constant in all the layers. The DOS of structure **3**, which has a global alignment of the molecules both within the *ab*-plane and along the *c*-axis, shows a similar gradient of ~0.04 eV Å^{-1} of the valence and conduction

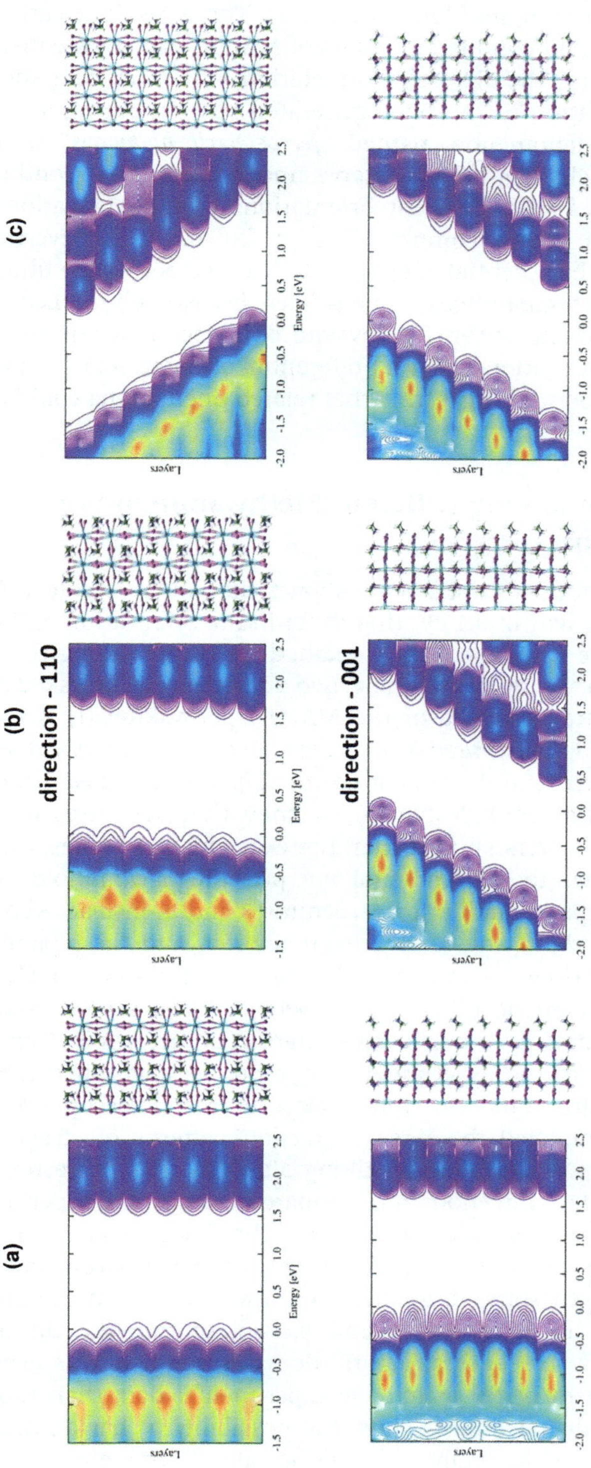

Figure 8.18 Surface plot of the Density Of States of the valence and conduction band for the layer models obtained from (a) structure **4** (no net orientation of the MA cations), (b) structure **1** (net orientation of the MA cations along the [001] direction), and (c) structure **3** (net orientation of the MA cations along the [010] and [001] directions). The upper and the lower panels show the DOS of the finite slabs along the [110] and the [001] direction, respectively. (Reproduced with permission from C. Quarti, *et al.*, *Chem. Mater.*, 2014, **26**, 6557–6569. Copyright (2014) American Chemical Society.[13])

bands along both the [110] and [001] directions. These results clearly indicate that the valence and conduction bands of $MAPbI_3$ can show consistent bending as a function of the MA cation orientation, which can be associated with an easier separation of the photogenerated species during solar cell operation.[103] This phenomenon, named "*ferroelectric highways*" by Frost *et al.*[90] is, however, challenging to observe under operational conditions, since many structures with different orientations of the MA cations are calculated within an energy comparable to a few kT. It is nevertheless interesting to notice the potential of epitaxially-oriented perovskite films as a means to obtain even higher photovoltaic efficiencies. Notably, McLeod *et al.* demonstrated that in the $MAPbI_3$ perovskite obtained by a one-step synthesis protocol, the MA cations show a preferential orientation, which is lost in the two-step synthesis,[104] suggesting that related phenomena could be, in principle, observed.

8.3.6 Conversion among Different Methylammonium Orientations

The results on previous sections have shown that many stable $MAPbI_3$ structures exist, lying within 0.1 eV, that differ in the orientation of the MA cations and/or in the direction of the octahedron tilting (along the [110] direction rather than the [001] one). Now, two questions can be asked about the room temperature structure of the $MAPbI_3$ perovskite: (i) *is there a single crystal of $MAPbI_3$ constituted by just one type of structure, or can two or more structures co-exist?*, and (ii) *how do two different oriented structures interconvert into each other?* In Table 8.6, we show that structures **1**–**7** have the a and b cell parameters shorter than the experimental values, and the c parameter longer than the experimental one, providing a c/a ratio of $\sim$1.46 against the $\sim$1.43 experimental ratio. Structures **8** and **9**, instead, show the opposite trend, with the a and b parameters shorter, and the c parameter longer than the experimental values, and with a c/a ratio of $\sim$1.41. Considering the small energetic difference between the two set of structures and the fact that averaging the cell parameters of structures **1**–**7** and **8**–**9** deliver a value close to the experimental c/a ratio, we envisioned that the two types of structure could possibly coexist in a macroscopic $MAPbI_3$ sample. We thus simulated the X-ray diffraction patterns of structures **1** and **8**, the former having octahedron tilting along the [001] direction and the latter along the [110] direction, and compared them to the experimental data by Stoumpos *et al.*[32] By doing so we found that both structures **1** and **8** reproduce, to some extent, the main features of the experimental XRD patterns, although with some differences. We then calculated the average of the XRD pattern of structures **1** and **8**, representing the diffraction pattern of a system where both structural domains co-exist. This averaged XRD pattern seemed to better match the experimental data than those of the individual structures **1** and **8**. Thus, the possibility of having different crystalline domains, with octahedron tilting along different directions sufficiently large to provide a well-defined XRD signal, seems to be consistent with the experimental data.

To investigate further the co-existence of various different structures, we constructed a new structural model obtained by merging the optimized structures **1** and **8**, stacked along the [001] direction (see Figure 8.19). We optimized this mixed structure both with the experimental cell parameters but allowing them to vary. In both cases, the mixed structure retained the original structural motif with two different tilting in the different model halves. Notably, the calculated cell parameters are now in better agreement with the experimental data, with a c/a ratio of ~1.43 (a = 8.81 Å, b = 8.80 Å, c = 25.27 Å), as found experimentally. The present system was calculated to be 0.16 eV less stable than structure **1** (doubled along c and re-optimized with a k = 4×4×2 point mesh) for the experimental cell parameters (0.22 eV with the optimized parameters). Considering the rather small energetic difference and the fact that the average cell parameters of structures **1–7** and

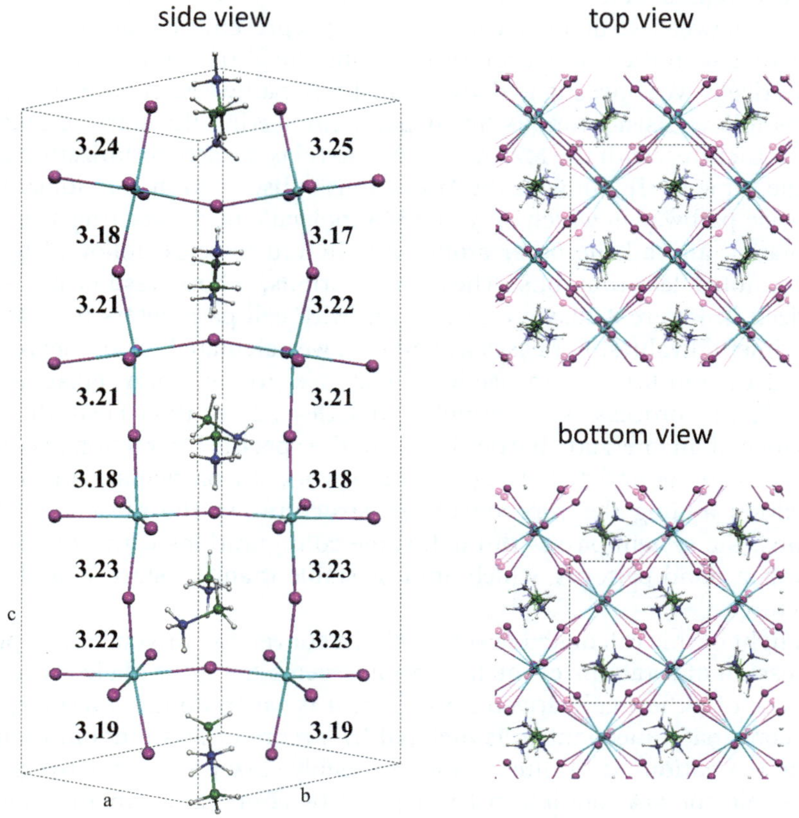

Figure 8.19 Optimized structure of the mixed structure obtained by combining structures **1** and **8**.
(Reproduced with permission from C. Quarti, *et al.*, *Chem. Mater.*, 2014, **26**, 6557–6569. Copyright (2014) American Chemical Society.[13])

8–9 deliver essentially the experimental *c*/*a* ratio, we propose that the $MAPbI_3$ material may possibly be constituted by domains of alternating tilting directions which may dynamically interconvert into each other at room temperature. The picture extracted from our calculations is likely to be consistent with the measurements by Choi *et al.*,[105] who found small crystalline domains composed by roughly one $MAPbI_3$ unit cell.

We now explore the mechanism underlying the transition among the different analyzed structures. Experimental investigations have previously demonstrated that the orientational dynamics of the MA cations takes place in the ps timescale.[67,80] Consistently, we found many possible structures in an energy range of a few kT that are thus accessible at room temperature. The point is, therefore, how a random rotational motion of the MA cations on a ps timescale can be related to the presence of the stable structures that we calculated, and how the transition among these structures can possibly take place.

To clarify these issues, we investigated various interconversion pathways between representative structures. In particular, we considered the conversion between structures **1** and **9**, being representative of polar/apolar structures with the octahedron tilting along the [001] and [110] directions, respectively. We investigated two transition pathways, (i) the inorganic atoms follow a synchronous linear path, connecting the starting and the final structures, with the MA cations free to relax at each optimization step; (ii) the same as (i), but with the MA cations substituted by Cs^+ ions; (iii) a stepwise pathway in which only one MA molecule of the starting structure is rotated, and all the other atoms are allowed to relax, followed by the other molecule's rotations. These calculations, whose results are summarized in Figure 8.20, were carried out with cell parameters fixed to the experimental values. For cases (i) and (iii), we calculated an energy barrier of 0.22 eV and 0.15 eV, respectively. For case (ii), *i.e.* upon replacing MA by Cs^+, the process is essentially barrierless. It is thus clear that the presence of an energetic barrier is due to the specific interactions between the MA cations and the inorganic framework. In particular, during the transition among the two considered structures in the presence of MA cations, the system passes through some configurations with unfavorable hydrogen bond patterns, which are less stable than the starting and final structures.

In light of this calculated lower barrier, a conversion driven by a stepwise succession of rotations of the MA cations seems the most likely process in the case of a thermally driven transition. It is worth noting, however, that this conversion mechanism is dictated by a more or less random rotation of the MA cations, and thus it is a reasonably slow process relative to the time-scale for MA rotation (a few ps).[67,80] By contrast, a synchronous rearrangement of the atoms from one structure to the other is more likely when an external electric field is imposed on the material, decreasing the transition barrier and making the potential felt by the MA cations anisotropic, thus leading to a synchronous rearrangement of the material.

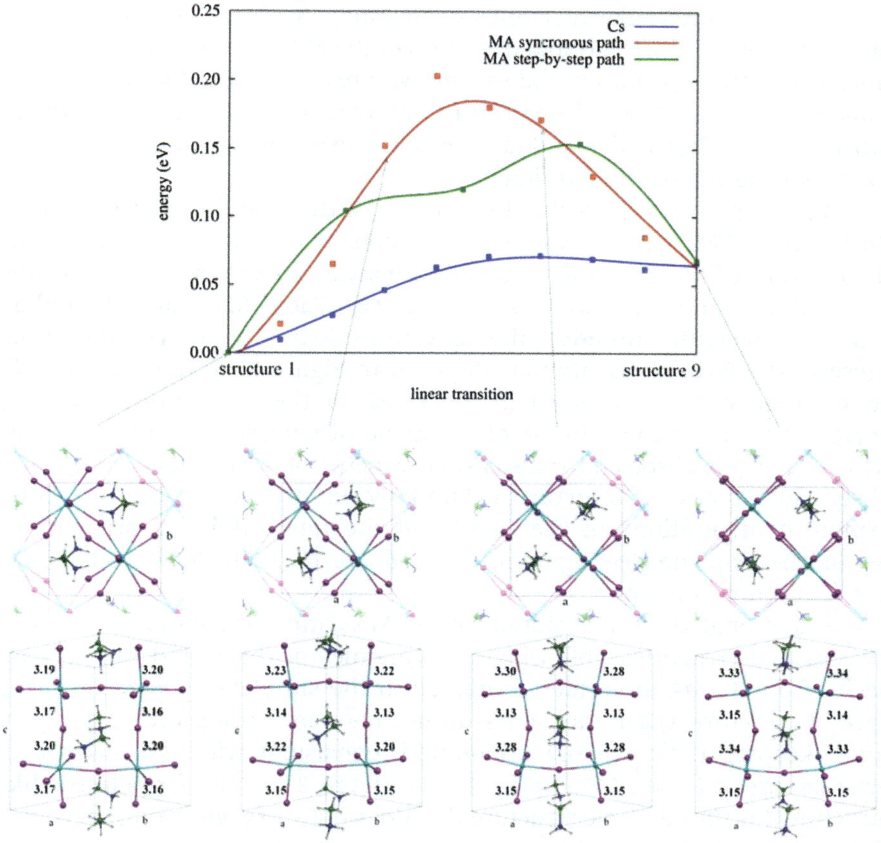

Figure 8.20 Possible mechanisms for the conversion of structure **1** (tilting along the [001] direction) into structure **9** (tilting along the [110] direction). Synchronous transition with Cs cations in place of the MA cations (blue lines); synchronous transition with the MA cation (red line); MA stepwise transition (green line).
(Reproduced with permission from C. Quarti, *et al.*, *Chem. Mater.*, 2014, **26**, 6557–6569. Copyright (2014) American Chemical Society.[13])

In light of the large barriers calculated for such a transition, a slower transition mechanism (relative to the MA rotation time-scale) can be expected. This is consistent with the results from Snaith *et al.*, who observed a hysteresis in the current–potential curve of $MAPbI_3$ that became more marked at a slow potential scanning rate ($\sim$0.001 V s^{-1}).[106] Within this time-scale, enough time is probably given to the material to re-arrange to the new external potential.

8.3.7 Dynamics in Polar and Apolar Structures

To investigate the dynamics of organo-halide perovskites, in particular those of the MA cations, in conditions closer to the operation ones, we carried out

Car–Parrinello molecular dynamics simulations. Such simulations were expected to unravel the rotational dynamics of the MA cations and the possible transition pathways among the various $MAPbI_3$ stable local structures. We thus considered a 2×2×2 supercell for structures **1** and **4**, being representatives of ferroelectric-like and antiferroelectric-like structures, respectively, to investigate the possible differences.

In Figure 8.21, we show the distribution of the ϕ and θ angles sampled during the CPMD simulations for structures **1** and **4**. In both cases, the distribution of the ϕ and θ angles have maxima in correspondence to the orientation of the optimized structures (see Table 8.6) suggesting that, during the dynamics duration, the MA cations do not rotate freely but rather preserve the favored orientations depicted in Figure 8.16. However, significant differences are observed in the width of the distributions and the correspondence of the minima of the angle distribution. A tail in the distribution of ϕ is observed for the case of structure **4**, while this is completely absent in the case of structure **1**. Similarly, in structure **4**, the molecules explore configurations in the middle of the most probable orientations within the *ab*-plane (see Figure 8.14), *i.e.* $\theta = 0$, 180, 270, 360°, while this is not found for structure **1**.

We then analyzed the evolution of the MA cation orientation in the two investigated structures, following the dynamics of the ϕ and θ angles for each of the 32 MA molecules contained in the simulated supercell. In the case of structure **1**, the molecules do not re-orient during the 12 ps dynamics, keeping their original orientation, consistent with the absence of orientations at $\theta = 0°$, 90°, 180°, 270° in Figure 8.22. The ferroelectric-like structure **1** is thus a stable structure that does not show any evolution within the considered time-scale.

For structure **4**, four molecules rotate from their original orientation during the 12 ps dynamics. In Figure 8.22, we have illustrated the evolution of ϕ and θ for one rotating molecule, which rotates from the −B to the +A configuration (see Figure 8.14). During the rotation, this molecule changes both its orientation within the *ab*-plane (*i.e.* θ) and tilting (*i.e.* ϕ) in a concerted fashion. The same behavior has been found for all the four reorienting molecules. Thus, *rotation of the MA cations requires a coupled reorientation within the ab-plane and the concerted modification of the tilting* with respect to it. It is reasonable to associate this observation with the peculiar $a^0a^0c^-$ tilted inorganic framework, which is nearly invariant for an S_4 improper rotation axis. It is also worth noting that the measured timescale of the molecular rotation in $MAPbI_3$ is of the order of a few ps, in agreement with the experimental data available in the literature. We observed that the averaged values of θ and ϕ over all 32 molecules of the model do not deviate significantly from the starting values (0° and 180°); thus, in spite of the greater mobility of the molecules in the present system, we did not observe a transition toward the more stable structure **1** within the investigated time-scale. Nevertheless, structure **4** is far more flexible than ferroelectric-like structure **1**.

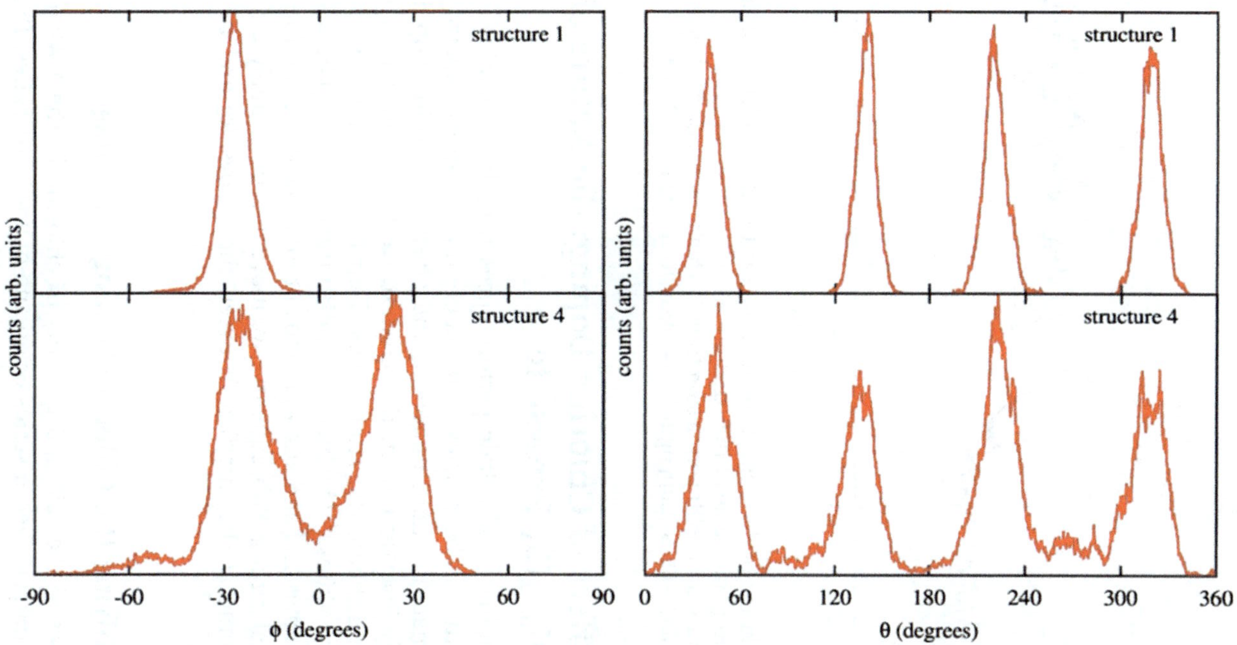

Figure 8.21 Distribution of the ϕ (left) and θ (right) angles, as defined in Figure 8.14, during the CPMD simulation for structures **1** and **4**. (Reproduced with permission from C. Quarti, *et al.*, *Chem. Mater.*, 2014, **26**, 6557–6569. Copyright (2014) American Chemical Society.[13])

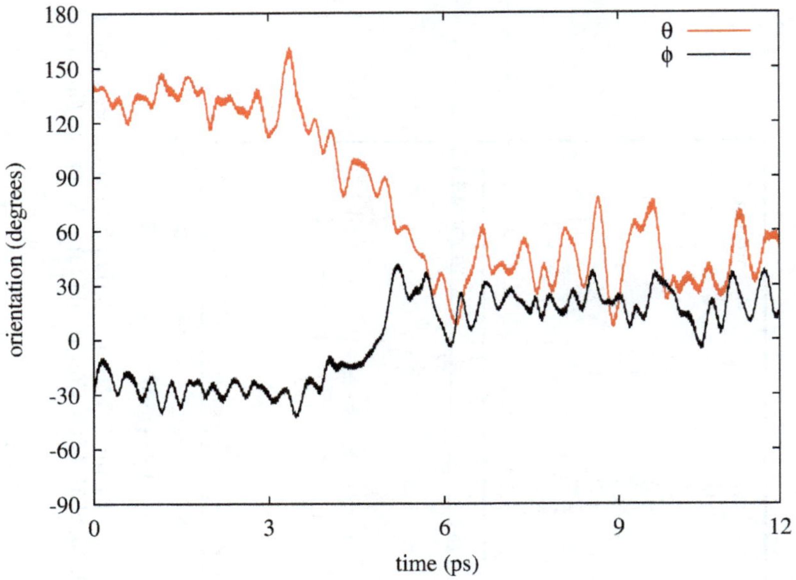

Figure 8.22 Evolution of the θ and ϕ angles from the CPMD simulation of the only rotating molecule of the structure 4.
(Reproduced with permission from C. Quarti, *et al.*, *Chem. Mater.*, 2014, **26**, 6557–6569. Copyright (2014) American Chemical Society.[13])

8.4 The Effect of Chlorine Doping: the Nature of the $MAPbI_{3-x}Cl_x$ Perovskite

Even if outstanding photovoltaic performances have been obtained using $MAPbI_{3-x}Cl_x$, little has been investigated about the real stoichiometry of this mixed halide material, and, moreover, the effect of chloride doping has still been poorly investigated. In this context, we have reported a coupled experimental and theoretical study focused on mixed halide lead methylammonium perovskites,[56] in which we studied the relationship between the I : Cl ratio in the material and attempted to optimize the device performance by compositional tuning. Subsequently, we move now on to analyze the role of chlorine doping at the interface between the perovskite and the TiO_2 semiconductor.[57–59]

8.4.1 Quantifying the Chlorine Doping Percentage

To fully understand the role and quantity of chlorine in the $MAPbI_{3-x}Cl_x$ perovskite we characterized a series of organo-halide perovskites, PS1, PS2 and PS3, synthetized by a process of self-organization. In particular, PS1 was synthesized with a 1 : 1 MAI/PbI_2 ratio; PS2 was produced by dissolving MAI and $PbCl_2$ in a 3 : 1 molar ratio; and PS3 was synthesized using MACl and

PbI_2 in a 1:1 molar ratio. From the XRD analysis, we identified a tetragonal $MAPbI_3$ structure in the PS1 thin films,[2] with cell parameters $a = b = 8.8755(5)$ Å and $c = 12.6735(8)$ Å. While the PS1 film pattern is purely oriented, the PS2 film shows a high degree of orientation. The cell parameters of the PS2 powder were found to be $a = 8.8535(3)$ and $c = 12.6434(4)$ Å, indicating a *ca.* 0.7% reduction of the unit cell volume compared to that of PS1 powder. This difference is small but not negligible, because it is well above the sensitivity of the measurements, and it could be reasonably associated with a certain level of Cl-doping. Thus, $CH_3NH_3PbI_3$ formation is favored when the ratio I:Pb in the precursor is 3:1.

To enhance the incorporation of chlorine, the synthesis of PS3 was carried out using a precursor with a 2:1 I:Pb molar ratio. X-ray analysis of PS3 demonstrated that, in this case, the film is highly oriented, indicating the presence of PbI_2, chlorine-doped $MAPbI_3$ and a further phase that was identified as cubic $MAPbCl_3$.[67] The isolation of the cubic $MAPbCl_3$ phase, besides the Cl-doped $MAPbI_3$ phase, indicates that chloride solubility is quite low in the iodide derivative, in line with the unfavorable formation energies calculated at the DFT level,[26] and that the formation of an extended solid solution is prevented, probably due to the ionic radii difference between Cl^- and I^-. Similar deductions have been made for Sn-based halide perovskites.[107] Thus, contrary to what has been observed in Br/I mixed-halide perovskites, the mixed Cl/I perovskite composition cannot be tuned by varying the Cl:I precursor molar ratio.[108] This conclusion has been further sustained by optical properties measured at room temperature. Indeed, both $MAPbI_3$[5,6,109] and $MAPbI_{3-x}Cl_x$[7] show very similar absorption features. The Tauc plots (Figure 8.23) provide an optical band gap of 1.60 ± 0.01 eV for all the films. The band gap value obtained for $MAPbI_3$ is in good agreement with the experimentally measured values ranging from 1.57[33,108] to 1.63 eV.[110] As for the PS2 and PS3 films, based on results obtained on $MAPbI_{3-x}Br_x$ and $MAPbBr_{3-x}Cl_x$, it was proposed that even in $MAPbI_{3-x}Cl_x$ the optical absorption bands vary linearly with x.[33] By considering this assumption and using a value of 3.11 eV for the band gap of $MAPbCl_3$,[109] we obtained that the measured uncertainty would be of *ca.* 1.3% in both samples.

To model the structural and electronic modifications that may be related to Cl-doping, we carried out periodic DFT calculations considering a large tetragonal cell initially composed of 32 $MAPbI_3$ units. In this way, we could explore Cl doping as small as *ca.* 1% (1 Cl:99 I atoms), but also 0 – which corresponds to the pure $MAPbI_3$–, 4, 8 and 33% doping – which corresponds to a $MAPbI_2Cl$ stoichiometry. Full relaxation of the atomic positions as well as of the lattice parameters was performed. During the variable-cell geometry optimization, we allowed the unconstrained relaxation of the a and b cell parameters to check for possible cell distortions, while the cell angles were fixed to 90°. For the sake of simplicity, we generally considered Cl substitution to take place in the axial octahedral positions[26] while, for $x = 4\%$, we

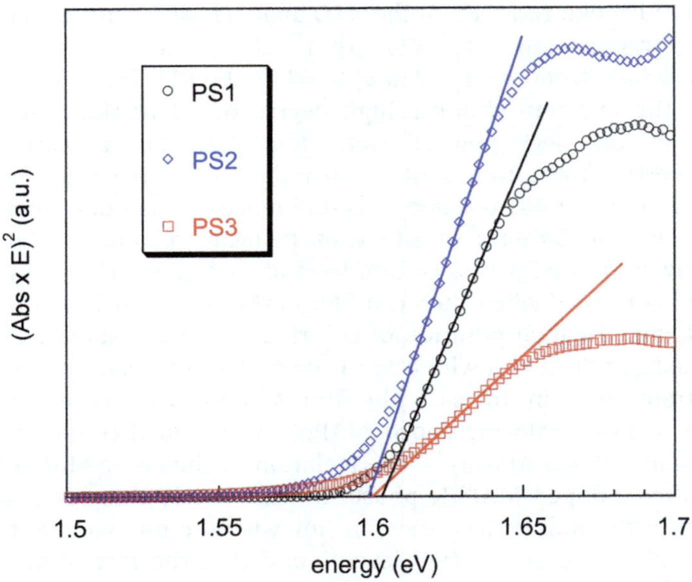

Figure 8.23 Tauc plots of PS1, PS2 and PS3 films.
(Reproduced with permission from S. Colella, *et al.*, *Chem. Mater.*, 2013, **25**, 4613–4618. Copyright (2013) American Chemical Society.[56])

considered both axial and equatorial substitutions. For each doping percentage, we investigated two tetragonal structures (**1** and **2**), generated starting from different orientations of the MA cations. In particular, the **1** and **2** structures were respectively generated using a head-to-head (**1**) and a head-to-tail (**2**) orientation of the MA organic molecules. By performing geometry optimization, while structure **2** maintains a higher order of MA orientation, structure **1** exhibits a strongly disordered MA rearrangement, see Figure 8.24. The structural and electronic calculated data are reported in Table 8.7.

The pure $MAPbI_3$, which corresponds to $x=0$, is more stable as structure **1** with tetragonal symmetry and calculated cell parameters in good agreement with the experimental data ($a=8.76$ Å, very similar to $b=8.75$ Å). As stated before,[26] the SR band gap gives a fortuitous agreement with the expected values. Nevertheless, SR calculated band gap trends, along with Cl doping, are expected to be qualitatively accurate.[26] In particular, we found a continuous reduction in cell volume and a steady band gap increase with the Cl doping values. While **1** is the most stable structure up to $x=8\%$; for $x=33\%$, structure **2** becomes more stable and with an associated decrease in the band gap, as previously reported.[67] Interestingly, a reduction in the cell volume of less than 1% was found the system with $x=4\%$, which is strongly correlated with the experimental data, suggesting a Cl incorporation of less than 4%.

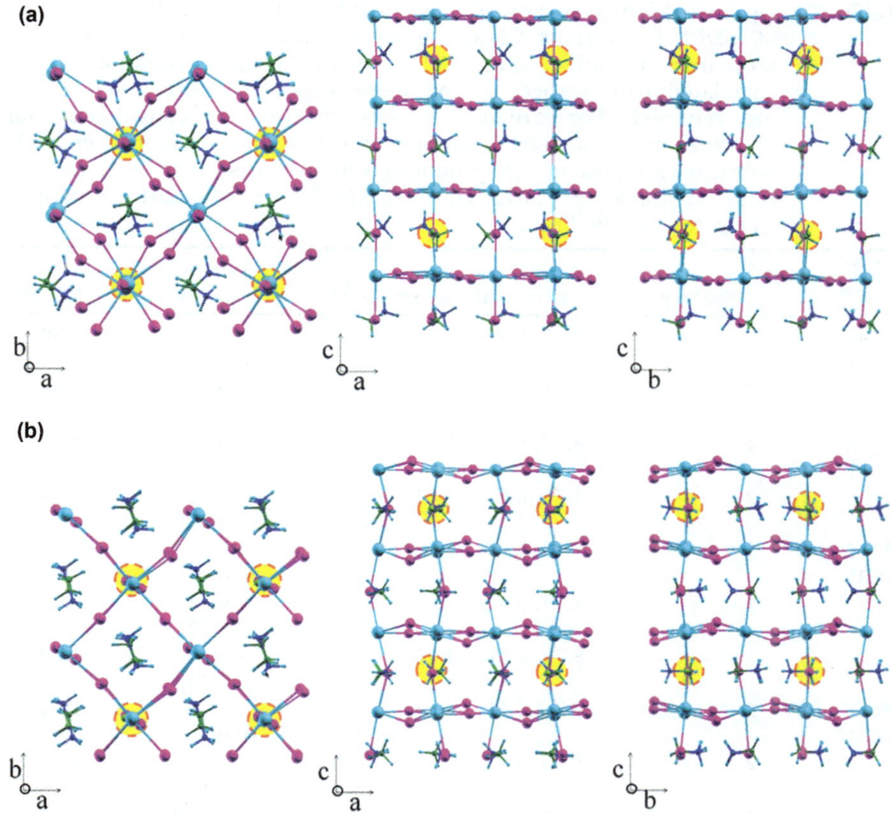

Figure 8.24 Optimized geometries of the 4% chloride-doped $MAPbI_{3-x}Cl_x$, structure **1**ax (a) and **2** (b). Three different geometry orientations are shown. The position of the chloride ions is highlighted by yellow circles; Pb = light blue; I = magenta.
(Reproduced with permission from S. Colella, *et al.*, *Chem. Mater.*, 2013, **25**, 4613–4618. Copyright (2013) American Chemical Society.[56])

8.4.2 Unravelling the Location of Chlorine in Perovskite Thin Films

The most intriguing feature of hybrid perovskites is that they can support both electron and hole transport.[7,35,111] Meso-superstructured and planar heterojunction solar cells have been implemented successfully with $MAPbI_{3-x}Cl_x$,[7,10] on the other hand, nominally identical devices based on the prototype $MAPbI_3$ perovskite have provided lower performances with respect to the Cl-doped materials.[16] This behavior, initially interpreted on the basis of the improved carrier mobility in $MAPbI_{3-x}Cl_x$ compared to that in $MAPbI_3$,[16] was later ascribed to a reduced carrier recombination in $MAPbI_{3-x}Cl_x$[17] and to a different morphology. A possibly related

Table 8.7 Optimized lattice parameters and calculated band gap (EG) by SR and SOC-DFT for various Cl-doping levels. Also the relative energy of structures **1** and **2**, E_{rel} (eV), and the volume variation, ΔV%, calculated with respect to the more stable structure **1** with $x = 0$ are reported. For structure **1** with $x = 4\%$, we considered four chlorides in axial (**1**ax) positions and 2 axial/2 equatorial chloride substitutions (**1**ax-eq). (Reproduced with permission from S. Colella, *et al.*, *Chem. Mater.*, 2013, **25**, 4613–4618. Copyright (2013) American Chemical Society.[56])

Cl : I (%Cl)	Structure	Lattice parameters/2 (Å)	ΔV% (E_{rel} (eV))	EG (eV)
0 : 96 (0.0)	1	8.76, 8.75, 12.74	0.00 (0.00)	1.60
	2	8.71, 9.06, 12.46	+0.68 (+0.04)	1.68
1 : 95 (1.0)	1	8.76, 8.76, 12.71	−0.12 (0.00)	1.61
	2	8.57, 8.95, 12.61	−0.96 (+0.01)	1.71
4 : 92 (4.2)	1ax	8.75, 8.77, 12.63	−0.76 (0.00)	1.64
	1ax-eq	8.72, 8.74, 12.69	−0.91 (+0.02)	1.64
	2	8.77, 9.06, 12.26	−0.25 (+0.03)	1.69
8 : 88 (8.3)	1	8.76, 8.80, 12.53	−1.10 (0.00)	1.74
	2	8.73, 9.12, 11.99	−2.29 (+0.10)	1.66
32 : 64 (33.3)	1	8.91, 8.84, 11.12	−11.49 (+0.02)	1.96
	2	8.87, 9.03, 11.04	−10.43 (0.00)	1.66

phenomenon is the observation of improved charge transport in dye-sensitized solar cells based on $MAPbI_{3-x}Cl_x$ and $MAPbI_{3-x}Br_x$,[56,108] leading to increased photocurrent and/or open circuit voltages compared to $MAPbI_3$. Despite the success of the $MAPbI_{3-x}Cl_x$ mixed halide perovskite, the role of chlorine in the material is still not clear, beyond the phenomenological observation of reduced carrier recombination and/or improved transport. As shown before, combined X-ray diffraction data and computational analysis have indicated only a small Cl-incorporation (~1–4%) into the $MAPbI_3$ lattice,[56] in line with the predicted thermodynamically unfavorable formation of solid $MAPbI_3$/$MAPbCl_3$ solutions.[26,107] Zhao and Zhu also showed very recently the beneficial effect on the photovoltaic properties of adding MACl salt to the standard preparation of $MAPbI_3$, suggesting a peculiar role of chlorine in assisting the perovskite growth, although energy dispersive X-ray (EDX) analysis of the annealed perovskite film did not reveal the presence of chlorine within the technique detection limit (~1%).[112] Similarly, Edri *et al.* obtained a significant enhancement of the photovoltaic performance of the $MAPbBr_{3-x}Cl_x$ perovskite compared to its non-chlorinated analogue.[113] Since no chlorine was detected in the film by EDX, although X-ray photoelectron spectroscopy showed a few % Cl in the material, these authors concluded that chlorine was possibly concentrated at/near the surface.[113]

Apart from the slight change in absorption spectra[18] and a small volume contraction,[56] a peculiar feature of $MAPbI_{3-x}Cl_x$ perovskites is their tendency to grow oriented along the [110] direction.[7–10,28,56,110,112,114–116] On the contrary, the $MAPbI_3$ perovskite shows mainly non-oriented structures, for materials derived from either one or two step syntheses.[8,9,28,110,112,115–117] A possible reason for the differently oriented structures may consist in the perovskite preparation procedure, pointing at either the different solvent (GBL *vs.* DMF/DMSO), the presence of the $PbCl_2$ precursor material, or a combination of these two factors. This observation could be related to the improved photovoltaic performance of $MAPbI_{3-x}Cl_x$ and generically to the aforementioned "chlorine effect", as various works have highlighted the crucial role of perovskite morphology for efficient charge separation and solar cell operation.[26,118] Another interesting feature observed in mixed-halide perovskites is the increased temporal stability of solar cells based on these materials with respect to standard $MAPbI_3$.[108,119]

We have investigated the effect of chlorine on the TiO_2/perovskite interface by state-of-the art first principles electronic structure calculations on extended models.[59] The employed model was made by 3×5×3 perovskite slabs of $(CH_3NH_3)_{60}Pb_{45}I_{150}$ stoichiometry, consistent with the analysis of perovskite surface stoichiometry reported by Mitzi.[120] Starting from this system, we replaced 15 iodine surface atoms by chlorine atoms, Figure 8.25. This choice is justified by the experimental observation that chlorine is basically incorporated into the perovskite lattice only to a very small extent and that it is possibly located at the material surface.[56,113] Since our aim was to study the perovskite/TiO_2 interface, we used throughout the experimental TiO_2 cell parameters ($a = 18.92$ Å, $b = 30.72$ Å) to build the supercells, leaving a 10 Å vacuum along the non-periodic direction orthogonal to the surface. The resulting (110) optimized $MAPbI_3$ slab showed a calculated band gap of 1.96 eV by SR-DFT, which was reduced to 1.32 eV upon inclusion of SOC.[31,32] The chosen set-up provided a lattice mismatch for the (110) surfaces of only +0.36 and +1.92% along the TiO_2 a and b directions, respectively, and of +0.75 and −1.85%, for the pseudo-cubic (001) surfaces. Notably, the tetragonal (001) surface showed a lattice mismatch of −6.40 and −13.52% with the same TiO_2 slab, which would introduce excessive strain into the perovskite sub-system. For the sake of comparison, using the experimental cell parameters for the (110) $MAPbI_3$ surface, the band gap only slightly varied (1.87 eV) with a corresponding 0.4 eV total energy decrease, demonstrating that limited strain is introduced by the small lattice mismatch between the two different materials. The calculated band gap values were higher than those found for the bulk tetragonal phase of the $CH_3NH_3PbI_3$ perovskite calculated at the same level of theory, 1.66 and 0.60 eV by SR- and SOC-DFT,

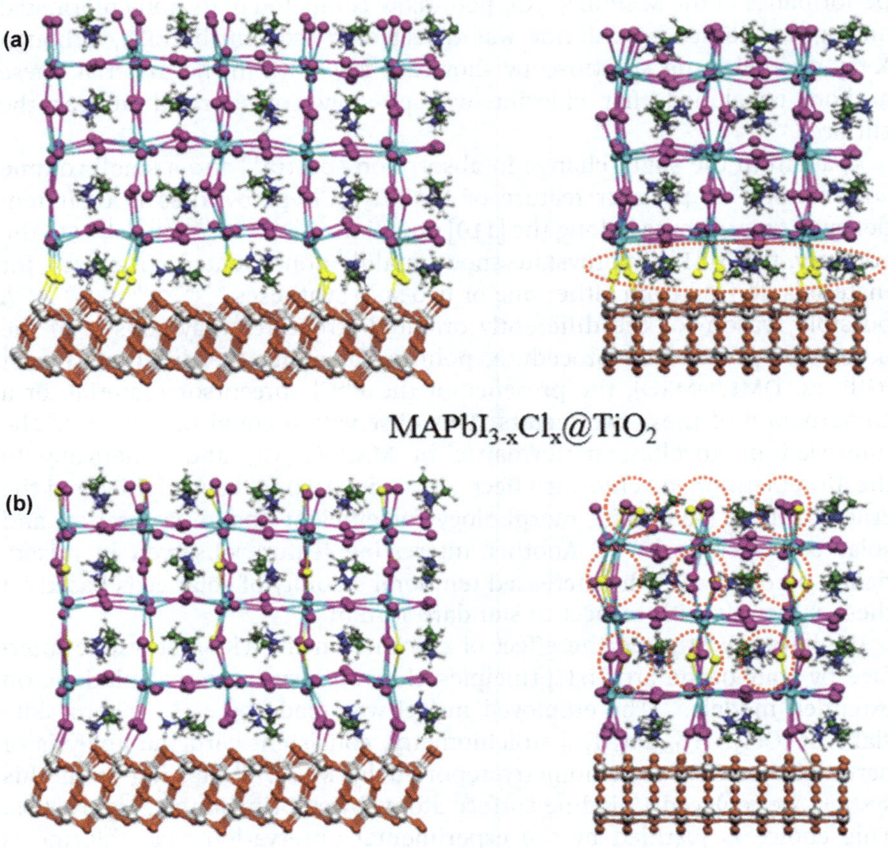

Figure 8.25 Optimized structures of (110) $MAPbI_{3-x}Cl_x$ perovskite slabs interacting with the (101) anatase TiO_2 surface. Notice the different location of chloride atoms (highlighted by red circles) in (a) close to the TiO_2 surface and (b) evenly distributed along the non-periodic direction. Colors: Pb = light blue; I = magenta; Cl = yellow; Ti = light grey; O = red; C = green, N = blue; H = white.
(Reproduced with permission from S. Colella, *et al.*, *J. Phys. Chem. Lett.*, 2014, **5**, 3532–3538. Copyright (2014) American Chemical Society.[57])

respectively,[31] as expected when moving from a periodic to a confined system. The investigated $MAPbI_3$ (001) surface slab showed a larger band gap, 2.37 and 1.50 eV by SR- and SOC-DFT, respectively. Replacement of surface iodine by chlorine atoms does not affect the band gap, since chlorine introduces occupied levels below the valence band edge,[26] see below.

As shown in Figure 8.25, the perovskite models were "deposited" onto a 5×3×2 slab of anatase TiO_2 made by 120 TiO_2 units, exposing the majority of the (101) surface. The perovskite adsorption onto the TiO_2 slab takes place by interaction of the perovskite halide species to under-coordinated titanium sites of the oxide surface.[58] We carried out DFT simulations of the perovskite/TiO_2 interface in order to investigate the energetics of chlorine doping within the perovskite film and at the interface with TiO_2, whilst also assessing its possible effect on the electronic structure at the interface. In particular, we considered the $MAPbI_{3-x}Cl_x$@TiO_2 interface model, in which 15 chlorine atoms were placed at the interface with TiO_2, and another model with 15 chlorine atoms uniformly distributed in the perovskite film along the direction perpendicular to the TiO_2 surface, see Figure 8.25.

Most notably, in line with the unfavorable energetics of chlorine-to-iodine substitution in the bulk material predicted previously,[56] see Section 8.4.1, when displacing the chlorine atoms away from the TiO_2 surface, *i.e.* when dispersing them evenly in the perovskite slab (Figure 8.25b), the total energy was found to increase by 0.7 eV compared to the case when chlorine atoms are all placed on the surface (Figure 8.25a). As a matter of fact, the MAPbI$_{3-x}$Cl$_x$ perovskite binding energy to TiO_2 decreased by 1.3 eV when chloride was displaced away from the TiO_2 surface. These data clearly suggest that chlorine atoms interact preferentially with the TiO_2 surface, rather than being homogeneously distributed in the perovskite film.[56]

To check the reliability of this theoretical prediction, Colella *et al.*[57] performed variable angle XPS measurements on a thin $MAPbI_{3-x}Cl_x$ sample deposited on flat TiO_2 at three different photo-electron take-off angles with respect to the sample surface, *i.e.* 10°, 45° and 70°, with the aim of progressively exploring the perovskite layer depth. All the measurements showed the presence of spectral contributions from perovskite elements (lead, nitrogen, chlorine and iodine) and those from the underneath substrate (titanium and oxygen). In Table 8.8, we summarize the results from the XPS spectra measured in the Cl 2p region at three different p.e. take-off angles. In all the cases, the Cl $2p_{3/2}$ spin–orbit component was centered at 198.4 eV, compatible with the presence of Cl atoms in the form of chloride.[121] The intensity of the Cl signal changed progressively with the p.e. take-off angle of the photo-electrons and, therefore, with the sampling depth. In particular, the intensity of the Cl 2p signal was considerably suppressed to a value below the detection limit at the lowest take-off angle (10°), while it was clearly enhanced (beyond the measurement accuracy) at the highest take-off angle (70°).

As we can see in Table 8.8, in contrast to the cases of the atomic ratios I^N/I^I and I^{Pb}/I^I, which remain constant at the various investigated angles within the statistical uncertainty, the I^{Cl}/I^I atomic ratio y increased significantly

Table 8.8 XPS relative atomic ratio (multiplied by 100) as a function of the p.e. take-off angle. (Reproduced with permission from S. Colella, *et al.*, *J. Phys. Chem. Lett.*, 2014, **5**, 3532–3538. Copyright (2014) American Chemical Society.[57])

	10°	45°	70°
I^{Pb}/I^I	42.2 ± 0.5	41.8 ± 0.5	43.3 ± 0.5
I^N/I^I	18.0 ± 0.5	18.5 ± 0.5	17.5 ± 0.5
I^{Cl}/I^I	<0.6 (noise)	1.2 ± 0.15	2.1 ± 0.1

with the p.e. take-off angle. This trend is an unequivocal indication of an inhomogeneous distribution of the Cl atoms inside the perovskite layer, with most of the chloride species (Cl-rich region) placed in proximity to the TiO_2 surface.

Having established that chlorine atoms are located preferentially at the perovskite/TiO_2 interface, we then studied the effect of interfacial chlorine on the perovskite growth direction and the relative stabilities. Notably, the isolated (110) slab is slightly more stable than the corresponding (001) slab, with a total energy difference of 0.7 and 0.1 eV for the $MAPbI_3$ and $MAPbI_{3-x}Cl_x$ perovskites, respectively. It can be noticed that these surface energies are likely to be affected by the different phase (tetragonal *vs.* pseudo-cubic) simulated here and by the employment of TiO_2 cell parameters, so these data have to be interpreted with care. Upon adsorption of the perovskites onto the TiO_2 slab, see Figure 8.25 for the (110) $MAPbI_{3-x}Cl_x$ case, the calculated energetics changed substantially, with the (110) surface now favored by *ca.* 0.45 eV per surface halide atom against the (001) surface. The (110) surface was consistently calculated to show a higher binding energy to the TiO_2 surface compared to the (001) surface. Moreover, the presence of interfacial chlorine, for both (110) and (001) systems, leads to an increased perovskite binding energy to the TiO_2 surface, calculated as the difference between the TiO_2/perovskite interface and the isolated fragments, of *ca.* 3 eV in terms of total energy difference that is *ca.* 0.2 eV per interacting surface halide, compared to $MAPbI_3$. These results could possibly be related to the observed enhanced temporal stability of devices incorporating mixed halide perovskites, and it points at the perovskite adhesion to the TiO_2 substrate as a possible cause, among other factors, of solar cell deterioration.

In summary, the (110) surface is strongly stabilized through the interaction with TiO_2, with interfacial chlorine atoms leading to an additionally higher binding energy towards TiO_2 in $MAPbI_{3-x}Cl_x$, compared to $MAPbI_3$. For both perovskites, adsorption through the (110) surface induces an enhanced number of shorter X–I bonds compared to the same material adsorbed through the (001) surface, which is consistent with the calculated stabilization of the (110) surface interaction with TiO_2 with respect to the (001) surface arising from better structural matching of the former compared to the latter.

8.4.3 The Electronic Role of Interfacial Chlorine

Based on the structural information described in Sections 8.4.1 and 8.4.2, we now move to analyze the possible electronic effects of interfacial chlorine on the performance of TiO_2-based perovskite solar cells. An important parameter underlying the functioning of the perovskite/TiO_2 interface is the amount of charge transfer occurring in the ground state. This quantity reflects the nature of the interface and the extent of mixing of the electronic states of the interacting components. To quantify the degree of charge transfer across the TiO_2/perovskite interface, we carried out a charge displacement (CD) analysis,[122,38] which allowed us to visualize the displacement of charge occurring from the non-interacting TiO_2 and the perovskite fragments to the interacting heterointerface. The resulting CD curves for (110) $TiO_2/MAPbI_3$ and $TiO_2/MAPbI_{3-x}Cl_x$ interfaces are shown in Figure 8.26, and show a slight increase in the charge donation to TiO_2 in the case of the $MAPbI_{3-x}Cl_x$ perovskite (1 electron *vs.* 0.94 electrons for $MAPbI_3$), along with a strong polarization in both cases, signaled by the strong CD variation across the interfacial perovskite layer region. This induces an increased charge accumulation at the interfacial layer for $MAPbI_{3-x}Cl_x$ compared to $MAPbI_3$, which we quantified as 0.85 *vs.* 0.61 electrons, respectively.

To further investigate the electronic effect of interfacial chloride on the crucial perovskite/TiO_2 interface, we show in Figure 8.27 an isodensity contour plot of the interface Density of States (DOS) for $MAPbI_3$ and interfacial chlorine $MAPbI_{3-x}Cl_x$ perovskite interfaces with TiO_2, calculated including SOC.

The plot in Figure 8.27 basically shows the joint system's integrated DOS as a function of the energy (x axis) and of the distance from the TiO_2 surface (y axis), identified by the dashed horizontal lines. The high DOS red/orange spots in the region associated to the perovskite in Figure 8.27 are associated to the I sp contributions to the valence band (VB), while the onset of the perovskite conduction band (CB, at *ca.* 0.8 eV above the TiO_2 CB, set at 0) is mostly contributed by Pb p states, partly hybridized with I p states.[26] In line with previous calculated electronic structures,[26] chlorine contributes occupied states *ca.* 2 eV below the top of the perovskite VB, see the red circles in Figure 8.27, while it has very weak contributions to the perovskite CB bottom. Despite this indirect effect, the presence of interfacial chloride doping substantially modifies the perovskite CB landscape, providing a strong directional gradient of unoccupied states towards the TiO_2 surface, as indicated by the arrow in Figure 8.27. While a small CB gradient is found also for the $MAPbI_3$ case, due to charge transfer to the TiO_2 surface, this gradient is strongly increased for the $MAPbI_{3-x}Cl_x$ case, inducing the depletion (accumulation) of unoccupied states in the bulk (interface) region, as indicated by the arrow in Figure 8.27. This band-bending towards the TiO_2 surface leads, upon photoexcitation and creation of free carriers, to electrons being strongly funneled towards the TiO_2 semiconductor, while holes are delocalized across the perovskite film and partly pushed in the opposite direction, albeit with a smaller gradient. The overall picture extracted from our results seems consistent with the p-i-n representation of perovskite solar cells derived by Edri *et al.*,[20] and it also accounts for the observation again by Edri *et al.* of the improved electron extraction in $MAPbI_{3-x}Cl_x$ compared to $MAPbI_3$ based solar cells.[113] The asymmetry found in the VB/CB band-bending would, in turn, strongly enhance the charge-collection efficiency at the TiO_2 electron-selective contact, and possibly decrease the electron/hole recombination in the perovskite bulk and, to some extent, the recombination between electrons in

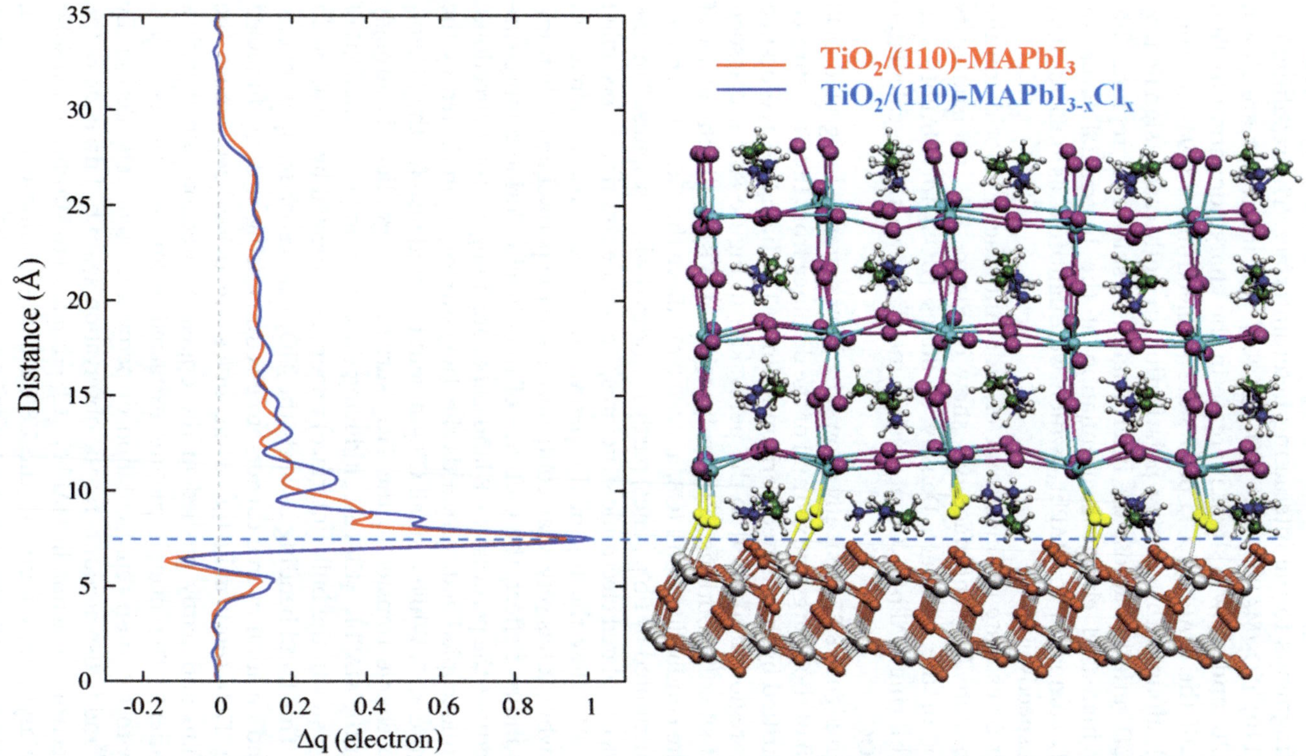

Figure 8.26 Charge displacement analysis of the TiO_2/perovskite interface for the (110) TiO_2/$MAPbI_3$ (red) and TiO_2/$MAPbI_{3-x}Cl_x$ (blue) systems. The strong polarization at the interface between the perovskite and the TiO_2 surfaces is evidenced by the change of sign of the displaced charge (Δq, electrons) at the interface (dashed horizontal line).
(Reproduced with permission from E. Mosconi, *et al.*, *J. Phys. Chem. Lett.*, 2014, 5, 2619–2625. Copyright (2014) American Chemical Society.[59])

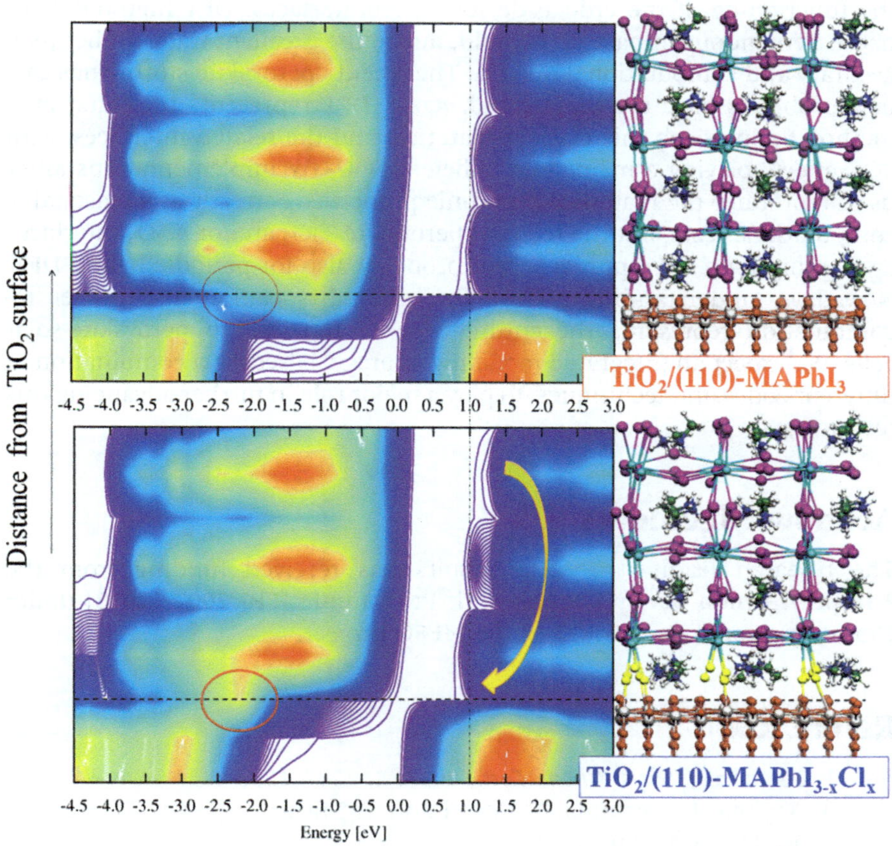

Figure 8.27 Isodensity plot of the integrated Density of States (DOS) as a function of the distance from the TiO_2 surface for $MAPbI_3$ and interfacial chloride $MAPbI_{3-x}Cl_x$ perovskite interfaces with TiO_2. A blue to red color variation indicates an increase of the DOS value. The red circles highlight the energy range where occupied chloride states are found in $MAPbI_{3-x}Cl_x$ and where are lacking in $MAPbI_3$. The arrow in the lower panel indicates the sizable band-bending observed in the $MAPbI_{3-x}Cl_x$ case.
(Reproduced with permission from S. Colella, *et al.*, *J. Phys. Chem. Lett.*, 2014, **5**, 3532–3538. Copyright (2014) American Chemical Society.[57])

the perovskite and holes in the hole transporting material. Our findings can be definitely considered as a platform or launching pad for further studies on the role of chlorine in $MAPbI_{3-x}Cl_x$.

8.5 Conclusions and Outlook

Computational modeling of hybrid/organic PV materials and systems has proved to be an extremely successful way to assist the design and screening of new materials and to provide hitherto inaccessible information related to the functioning of the crucial device heterointerfaces. DFT methods have had a tremendous impact in the field, due to the excellent trade-off between accuracy and computational burden. The recent "perovskite storm" has cast new problems onto the playground. Again DFT represents the "first aid" method to approach this class of materials and the related interfaces, with SOC effects playing a major role. Efficient SOC-GW implementations allow us to reproduce the materials electronic properties correctly at a reasonable computational cost, but the relevant perovskite/TiO_2 and/or Al_2O_3 interfaces are too big for a GW approach, and probably also for periodic hybrid-DFT. Nowadays, understanding exciton formation/dissociation and carrier recombination represents the cutting-edge of research in perovskite-solar cells. We expect an integrated modeling strategy, based on a combination of different codes and techniques, to play a major role in further advancing this technology.

Acknowledgements

The research leading to these results has received funding from the European Union Seventh Framework Programme [FP7/2007%2013] under Grant Agreement No. 604032 of the MESO project.

References

1. J. Bisquert, *J. Phys. Chem. Lett.*, 2013, **4**, 2597–2598.
2. A. Kojima, K. Teshima, Y. Shirai and T. Miyasaka, *J. Am. Chem. Soc.*, 2009, **131**, 6050–6051.
3. A. M. Salau, *Sol. Energy Mater.*, 1980, **2**, 327–332.
4. P. Gao, M. Grätzel and M. K. Nazeeruddin, *Energy Environ. Sci.*, 2014, **7**, 2448–2463.
5. J.-H. Im, C.-R. Lee, J.-W. Lee, S.-W. Park and N.-G. Park, *Nanoscale*, 2011, **3**, 4088–4093.
6. H.-S. Kim, C.-R. Lee, J.-H. Im, K.-B. Lee, T. Moehl, A. Marchioro, S.-J. Moon, R. Humphry-Baker, J.-H. Yum, J. E. Moser, M. Grätzel and N.-G. Park, *Sci. Rep.*, 2012, **2**, 591.
7. M. M. Lee, J. l. Teuscher, T. Miyasaka, T. N. Murakami and H. J. Snaith, *Science*, 2012, **338**, 643–647.
8. J. H. Heo, S. H. Im, J. H. Noh, T. N. Mandal, C.-S. Lim, J. A. Chang, Y. H. Lee, H.-j. Kim, A. Sarkar, K. Nazeeruddin Md, M. Gratzel and S. I. Seok, *Nat. Photonics*, 2013, **7**, 486–491.
9. J. Burschka, N. Pellet, S.-J. Moon, R. Humphry-Baker, P. Gao, M. K. Nazeeruddin and M. Grätzel, *Nature*, 2013, **499**, 316–319.
10. M. Liu, M. B. Johnston and H. J. Snaith, *Nature*, 2013, **501**, 395–398.
11. H. Zhou, Q. Chen, G. Li, S. Luo, T.-b. Song, H.-S. Duan, Z. Hong, J. You, Y. Liu and Y. Yang, *Science*, 2014, **345**, 542–546.

12. M. A. Green, A. Ho-Baillie and H. J. Snaith, *Nat. Photonics*, 2014, **8**, 506–514.
13. C. Quarti, E. Mosconi and F. De Angelis, *Chem. Mater.*, 2014, **26**, 6557–6569.
14. C. Quarti, G. Grancini, E. Mosconi, P. Bruno, J. M. Ball, M. M. Lee, H. J. Snaith, A. Petrozza and F. D. Angelis, *J. Phys. Chem. Lett.*, 2013, **5**, 279–284.
15. E. Mosconi, C. Quarti, T. Ivanovska, G. Ruani and F. De Angelis, *Phys. Chem. Chem. Phys.*, 2014, **16**, 16137–16144.
16. S. D. Stranks, G. E. Eperon, G. Grancini, C. Menelaou, M. J. P. Alcocer, T. Leijtens, L. M. Herz, A. Petrozza and H. J. Snaith, *Science*, 2013, **342**, 341–344.
17. C. Wehrenfennig, G. E. Eperon, M. B. Johnston, H. J. Snaith and L. M. Herz, *Adv. Mater.*, 2014, **26**, 1584–1589.
18. V. D'Innocenzo, G. Grancini, M. J. P. Alcocer, A. R. S. Kandada, S. D. Stranks, M. M. Lee, G. Lanzani, H. J. Snaith and A. Petrozza, *Nat. Commun.*, 2014, **5**, 3586.
19. E. Edri, S. Kirmayer, A. Henning, S. Mukhopadhyay, K. Gartsman, Y. Rosenwaks, G. Hodes and D. Cahen, *Nano Lett.*, 2014, **14**, 1000–1004.
20. E. Edri, S. Kirmayer, S. Mukhopadhyay, K. Gartsman, G. Hodes and D. Cahen, *Nat. Commun.*, 2014, **5**, 3461.
21. H. Lv, H. Gao, Y. Yang and L. Liu, *App. Catal., A*, 2011, **404**, 54–58.
22. I. Borriello, G. Cantele and D. Ninno, *Phys. Rev. B: Condens. Matter Mater. Phys.*, 2008, **77**, 235214.
23. I. E. Castelli, T. Olsen, S. Datta, D. D. Landis, S. Dahl, K. S. Thygesen and K. W. Jacobsen, *Energy Environ. Sci.*, 2012, **5**, 5814–5819.
24. C. Di Valentin, G. Pacchioni and A. Selloni, *Phys. Rev. Lett.*, 2006, **97**, 166803.
25. R. F. Berger and J. B. Neaton, *Phys. Rev. B: Condens. Matter Mater. Phys.*, 2012, **86**, 165211.
26. E. Mosconi, A. Amat, M. K. Nazeeruddin, M. Grätzel and F. De Angelis, *J. Phys. Chem. C*, 2013, **117**, 13902–13913.
27. T. Umebayashi, K. Asai, T. Kondo and A. Nakao, *Phys. Rev. B: Condens. Matter Mater. Phys.*, 2003, **67**, 155405.
28. T. Baikie, Y. Fang, J. M. Kadro, M. Schreyer, F. Wei, S. G. Mhaisalkar, M. Grätzel and T. J. White, *J. Mater. Chem. A*, 2013, **1**, 5628–5641.
29. Y. Takahashi, R. Obara, Z.-Z. Lin, Y. Takahashi, T. Naito, T. Inabe, S. Ishibashi and K. Terakura, *Dalton Trans.*, 2011, **40**, 5563–5568.
30. E. Moreira, J. M. Henriques, D. L. Azevedo, E. W. S. Caetano, V. N. Freire and E. L. Albuquerque, *J. Solid State Chem.*, 2012, **187**, 186–194.
31. J. Even, L. Pedesseau, J.-M. Jancu and C. Katan, *J. Phys. Chem. Lett.*, 2013, **4**, 2999–3005.
32. C. C. Stoumpos, C. D. Malliakas and M. G. Kanatzidis, *Inorg. Chem.*, 2013, **52**, 9019–9038.

33. G. C. Papavassiliou and I. B. Koutselas, *Synth. Met.*, 1995, **71**, 1713–1714.
34. I. Chung, B. Lee, J. He, R. P. H. Chang and M. G. Kanatzidis, *Nature*, 2012, **485**, 486–489.
35. L. Etgar, P. Gao, Z. Xue, Q. Peng, A. K. Chandiran, B. Liu, M. K. Nazeeruddin and M. Grätzel, *J. Am. Chem. Soc.*, 2012, **134**, 17396–17399.
36. F. Chiarella, A. Zappettini, F. Licci, I. Borriello, G. Cantele, D. Ninno, A. Cassinese and R. Vaglio, *Phys. Rev. B: Condens. Matter Mater. Phys.*, 2008, **77**, 045129.
37. L. Hedin, *Phys. Rev.*, 1965, **139**, A796–A823.
38. M. S. Hybertsen and S. G. Louie, *Phys. Rev. B: Condens. Matter Mater. Phys.*, 1986, **34**, 5390–5413.
39. I. Chung, J.-H. Song, J. Im, J. Androulakis, C. D. Malliakas, H. Li, A. J. Freeman, J. T. Kenney and M. G. Kanatzidis, *J. Am. Chem. Soc.*, 2012, **134**, 8579–8587.
40. P. Pyykko, *Chem. Rev.*, 1988, **88**, 563–594.
41. R. Ahuja, A. Blomqvist, P. Larsson, P. Pyykkö and P. Zaleski-Ejgierd, *Phys. Rev. Lett.*, 2011, **106**, 018301.
42. R. Sakuma, C. Friedrich, T. Miyake, S. Blügel and F. Aryasetiawan, *Phys. Rev. B: Condens. Matter Mater. Phys.*, 2011, **84**, 085144.
43. P. Umari, X. Qian, N. Marzari, G. Stenuit, L. Giacomazzi and S. Baroni, *Phys. Status Solidi B*, 2011, **248**, 527–536.
44. P. Umari, E. Mosconi and F. De Angelis, *Sci. Rep.*, 2014, **4**, 4467.
45. G. Onida, L. Reining and A. Rubio, *Rev. Mod. Phys.*, 2002, **74**, 601–659.
46. J. Even, L. Pedesseau, J.-M. Jancu and C. Katan, *Phys. Status Solidi RRL*, 2014, **8**, 31–35.
47. K. Tanaka, T. Takahashi, T. Ban, T. Kondo, K. Uchida and N. Miura, *Solid State Commun.*, 2003, **127**, 619–623.
48. G. Giorgi, J.-I. Fujisawa, H. Segawa and K. Yamashita, *J. Phys. Chem. Lett.*, 2013, **4**, 4213–4216.
49. F. Brivio, K. T. Butler, A. Walsh and M. van Schilfgaarde, *Phys. Rev. B: Condens. Matter Mater. Phys.*, 2014, **89**, 155204.
50. M. R. Filip, C. E. Patrick and F. Giustino, *Phys. Rev. B: Condens. Matter Mater. Phys.*, 2013, **87**, 205125.
51. E. Menéndez-Proupin, P. Palacios, P. Wahnón and J. C. Conesa, *Phys. Rev. B: Condens. Matter Mater. Phys.*, 2014, **90**, 045207.
52. E. Mosconi, P. Umari and F. De Angelis, *J. Mater. Chem. A*, 2015, **3**, 9208–9215.
53. A. Amat, E. Mosconi, E. Ronca, C. Quarti, P. Umari, M. K. Nazeeruddin, M. Grätzel and F. De Angelis, *Nano Lett.*, 2014, **14**, 3608–3616.
54. R. Gottesman, E. Haltzi, L. Gouda, S. Tirosh, Y. Bouhadana, A. Zaban, E. Mosconi and F. De Angelis, *J. Phys. Chem. Lett.*, 2014, **5**, 2662–2669.
55. C. Quarti, E. Mosconi and F. De Angelis, *Phys. Chem. Chem. Phys.*, 2015, **17**, 9394–9409.

56. S. Colella, E. Mosconi, P. Fedeli, A. Listorti, F. Gazza, F. Orlandi, P. Ferro, T. Besagni, A. Rizzo, G. Calestani, G. Gigli, F. De Angelis and R. Mosca, *Chem. Mater.*, 2013, **25**, 4613–4618.
57. S. Colella, E. Mosconi, G. Pellegrino, A. Alberti, V. L. P. Guerra, S. Masi, A. Listorti, A. Rizzo, G. G. Condorelli, F. De Angelis and G. Gigli, *J. Phys. Chem. Lett.*, 2014, **5**, 3532–3538.
58. V. Roiati, E. Mosconi, A. Listorti, S. Colella, G. Gigli and F. De Angelis, *Nano Lett.*, 2014, **14**, 2168–2174.
59. E. Mosconi, E. Ronca and F. D. Angelis, *J. Phys. Chem. Lett.*, 2014, **5**, 2619–2625.
60. N. K. Noel, S. D. Stranks, A. Abate, C. Wehrenfennig, S. Guarnera, A.-A. Haghighirad, A. Sadhanala, G. E. Eperon, S. K. Pathak, M. B. Johnston, A. Petrozza, L. M. Herz and H. J. Snaith, *Energy Environ. Sci.*, 2014, **7**, 3061–3068.
61. F. Hao, C. C. Stoumpos, D. H. Cao, R. P. H. Chang and M. G. Kanatzidis, *Nat. Photonics*, 2014, **8**, 489–494.
62. F. Hao, C. C. Stoumpos, R. P. H. Chang and M. G. Kanatzidis, *J. Am. Chem. Soc.*, 2014, **136**, 8094–8099.
63. F. Zuo, S. T. Williams, P.-W. Liang, C.-C. Chueh, C.-Y. Liao and A. K. Y. Jen, *Adv. Mater.*, 2014, **26**, 6454–6460.
64. Y. Ogomi, A. Morita, S. Tsukamoto, T. Saitho, N. Fujikawa, Q. Shen, T. Toyoda, K. Yoshino, S. S. Pandey, T. Ma and S. Hayase, *J. Phys. Chem. Lett.*, 2014, **5**, 1004–1011.
65. F. De Angelis, *Acc. Chem. Res.*, 2014, **47**, 3349–3360.
66. W.-J. Yin, J.-H. Yang, J. Kang, Y. Yan and S.-H. Wei, *J. Mater. Chem. A*, 2015, **3**, 8926–8942.
67. A. Poglitsch and D. Weber, *J. Chem. Phys.*, 1987, **87**, 6373–6378.
68. I. E. Castelli, J. M. García-Lastra, K. S. Thygesen and K. W. Jacobsen, *APL Mater.*, 2014, **2**, 081514.
69. M. Kim, J. Im, A. J. Freeman, J. Ihm and H. Jin, *Proc. Natl. Acad. Sci. U. S. A.*, 2014, **111**, 6900–6904.
70. R. D. Shannon, *Acta Crystallogr., Sect. A: Cryst. Phys., Diffr., Theor. Gen. Crystallogr.*, 1976, **32**, 751–767.
71. G. E. Eperon, S. D. Stranks, C. Menelaou, M. B. Johnston, L. M. Herz and H. J. Snaith, *Energy Environ. Sci.*, 2014, **7**, 982–988.
72. S. Pang, H. Hu, J. Zhang, S. Lv, Y. Yu, F. Wei, T. Qin, H. Xu, Z. Liu and G. Cui, *Chem. Mater.*, 2014, **26**, 1485–1491.
73. N. Pellet, P. Gao, G. Gregori, T.-Y. Yang, M. K. Nazeeruddin, J. Maier and M. Grätzel, *Angew. Chem., Int. Ed.*, 2014, **53**, 3151–3157.
74. Y. Kawamura, H. Mashiyama and K. Hasebe, *J. Phys. Soc. Jpn.*, 2002, **71**, 1694–1697.
75. E. Arunan, R. Desiraju Gautam, A. Klein Roger, J. Sadlej, S. Scheiner, I. Alkorta, C. Clary David, H. Crabtree Robert, J. Dannenberg Joseph, P. Hobza, G. Kjaergaard Henrik, C. Legon Anthony, B. Mennucci and J. Nesbitt David, Definition of the hydrogen bond (IUPAC Recommendations 2011), *Pure Appl. Chem.*, 2011, **83**, 1637.

76. P. Schulz, E. Edri, S. Kirmayer, G. Hodes, D. Cahen and A. Kahn, *Energy Environ. Sci.*, 2014, **7**, 1377–1381.
77. L. Lang, J.-H. Yang, H.-R. Liu, H. J. Xiang and X. G. Gong, *Phys. Lett. A*, 2014, **378**, 290–293.
78. P. Garcia-Fernandez, J. A. Aramburu, M. T. Barriuso and M. Moreno, *J. Phys. Chem. Lett.*, 2010, **1**, 647–651.
79. D. Di Sante, P. Barone, R. Bertacco and S. Picozzi, *Adv. Mater.*, 2013, **25**, 509–513.
80. R. E. Wasylishen, O. Knop and J. B. Macdonald, *Solid State Commun.*, 1985, **56**, 581–582.
81. F. Brivio, A. B. Walker and A. Walsh, *APL Mater.*, 2013, **1**, 042111.
82. Y. H. Chang, C. H. Park and K. Matsuishi, *J. Korean Phys. Soc.*, 2004, **44**, 889–893.
83. N. Onoda-Yamamuro, T. Matsuo and H. Suga, *J. Phys. Chem. Solids*, 1990, **51**, 1383–1395.
84. Y. Kutes, L. Ye, Y. Zhou, S. Pang, B. D. Huey and N. P. Padture, *J. Phys. Chem. Lett.*, 2014, **5**, 3335–3339.
85. I. Grinberg, D. V. West, M. Torres, G. Gou, D. M. Stein, L. Wu, G. Chen, E. M. Gallo, A. R. Akbashev, P. K. Davies, J. E. Spanier and A. M. Rappe, *Nature*, 2013, **503**, 509–512.
86. F. Wang, I. Grinberg and A. M. Rappe, *Appl. Phys. Lett.*, 2014, **104**, 152903.
87. I. P. Swainson, R. P. Hammond, C. Soullière, O. Knop and W. Massa, *J. Solid State Chem.*, 2003, **176**, 97–104.
88. L. Chi, I. Swainson, L. Cranswick, J.-H. Her, P. Stephens and O. Knop, *J. Solid State Chem.*, 2005, **178**, 1376–1385.
89. A. M. Glazer, *Acta Crystallogr., Sect. A: Cryst. Phys., Diffr., Theor. Gen. Crystallogr.*, 1972, **B28**, 3384–3391.
90. J. M. Frost, K. T. Butler, F. Brivio, C. H. Hendon, M. van Schilfgaarde and A. Walsh, *Nano Lett.*, 2014, **14**, 2584–2590.
91. G. Giorgi, J.-I. Fujisawa, H. Segawa and K. Yamashita, *J. Phys. Chem. C*, 2014, **118**, 12176–12183.
92. R. J. Worhatch, H. Kim, I. P. Swainson, A. L. Yonkeu and S. J. L. Billinge, *Chem. Mater.*, 2008, **20**, 1272–1277.
93. A. Filippetti and A. Mattoni, *Phys. Rev. B: Condens. Matter Mater. Phys.*, 2014, **89**, 125203.
94. P. Giannozzi, S. Baroni, N. Bonini, M. Calandra, R. Car, C. Cavazzoni, D. Ceresoli, L. C. Guido, M. Cococcioni, I. Dabo, A. D. Corso, S. d. Gironcoli, S. Fabris, G. Fratesi, R. Gebauer, U. Gerstmann, C. Gougoussis, A. Kokalj, M. Lazzeri, L. Martin-Samos, N. Marzari, F. Mauri, R. Mazzarello, S. Paolini, A. Pasquarello, L. Paulatto, C. Sbraccia, S. Scandolo, G. Sclauzero, A. P. Seitsonen, A. Smogunov, P. Umari and R. M. Wentzcovitch, *J. Phys.: Condens. Matter*, 2009, **21**, 395502.
95. J. P. Perdew, K. Burke and M. Ernzerhof, *Phys. Rev. Lett.*, 1996, **77**, 3865–3868.

96. D. Vanderbilt, *Phys. Rev. B: Condens. Matter Mater. Phys.*, 1990, **41**, 7892–7895.
97. A. Pasquarello, K. Laasonen, R. Car, C. Lee and D. Vanderbilt, *Phys. Rev. Lett.*, 1992, **69**, 1982–1985.
98. B. Hetényi, F. De Angelis, P. Giannozzi and R. Car, *J. Chem. Phys.*, 2004, **120**, 8632–8637.
99. W. Geng, L. Zhang, Y.-N. Zhang, W.-M. Lau and L.-M. Liu, *J. Phys. Chem. C*, 2014, **118**, 19565–19571.
100. D. A. Egger and L. Kronik, *J. Phys. Chem. Lett.*, 2014, **5**, 2728–2733.
101. Y. Wang, T. Gould, J. F. Dobson, H. Zhang, H. Yang, X. Yao and H. Zhao, *Phys. Chem. Chem. Phys.*, 2014, **16**, 1424–1429.
102. M. M. Szczesniak, R. J. Brenstein, S. M. Cybulski and S. Scheiner, *J. Phys. Chem.*, 1990, **94**, 1781–1788.
103. W. A. Laban and L. Etgar, *Energy Environ. Sci.*, 2013, **6**, 3249–3253.
104. J. A. McLeod, Z. Wu, P. Shen, B. Sun and L. Liu, *J. Phys. Chem. Lett.*, 2014, **5**, 2863–2867.
105. J. J. Choi, X. Yang, Z. M. Norman, S. J. L. Billinge and J. S. Owen, *Nano Lett.*, 2014, **14**, 127–133.
106. H. J. Snaith, A. Abate, J. M. Ball, G. E. Eperon, T. Leijtens, N. K. Noel, S. D. Stranks, J. T.-W. Wang, K. Wojciechowski and W. Zhang, *J. Phys. Chem. Lett.*, 2014, **5**, 1511–1515.
107. K. Yamada, K. Nakada, Y. Takeuchi, K. Nawa and Y. Yamane, *Bull. Chem. Soc. Jpn.*, 2011, **84**, 926–932.
108. J. H. Noh, S. H. Im, J. H. Heo, T. N. Mandal and S. I. Seok, *Nano Lett.*, 2013, **13**, 1764–1769.
109. N. Kitazawa, Y. Watanabe and Y. Nakamura, *J. Mater. Sci.*, 2002, **37**, 3585–3587.
110. J. Qiu, Y. Qiu, K. Yan, M. Zhong, C. Mu, H. Yan and S. Yang, *Nanoscale*, 2013, **5**, 3245–3248.
111. J. M. Ball, M. M. Lee, A. Hey and H. J. Snaith, *Energy Environ. Sci.*, 2013, **6**, 1739–1743.
112. Y. Zhao and K. Zhu, *J. Phys. Chem. C*, 2014, **118**, 9412–9418.
113. E. Edri, S. Kirmayer, M. Kulbak, G. Hodes and D. Cahen, *J. Phys. Chem. Lett.*, 2014, **5**, 429–433.
114. B. Conings, L. Baeten, C. De Dobbelaere, J. D'Haen, J. Manca and H.-G. Boyen, *Adv. Mater.*, 2013, **26**, 2041–2046.
115. H.-B. Kim, H. Choi, J. Jeong, S. Kim, B. Walker, S. Song and J. Y. Kim, *Nanoscale*, 2014, **6**, 6679–6683.
116. Q. Chen, H. Zhou, Z. Hong, S. Luo, H.-S. Duan, H.-H. Wang, Y. Liu, G. Li and Y. Yang, *J. Am. Chem. Soc.*, 2013, **136**, 622–625.
117. P.-W. Liang, C.-Y. Liao, C.-C. Chueh, F. Zuo, S. T. Williams, X.-K. Xin, J. Lin and A. K. Y. Jen, *Adv. Mater.*, 2014, **26**, 3748–3754.
118. G. E. Eperon, V. M. Burlakov, P. Docampo, A. Goriely and H. J. Snaith, *Adv. Funct. Mater.*, 2014, **24**, 151–157.

119. H.-S. Kim, I. Mora-Sero, V. Gonzalez-Pedro, F. Fabregat-Santiago, E. J. Juarez-Perez, N.-G. Park and J. Bisquert, *Nat. Commun.*, 2013, **4**, 2242.
120. D. B. Mitzi, *J. Mater. Chem.*, 2004, **14**, 2355–2365.
121. K. Kishi and S. Ikeda, *J. Phys. Chem.*, 1974, **78**, 107.
122. E. Ronca, M. Pastore, L. Belpassi, F. Tarantelli and F. De Angelis, *Energy Environ. Sci.*, 2013, **6**, 183–193.

CHAPTER 9

Drift Diffusion Modelling of Charge Transport in Photovoltaic Devices

GILES RICHARDSON[a] AND ALISON WALKER*[b]

[a] Mathematical Sciences, University of Southampton, Southampton SO17 1BJ, UK; [b] Department of Physics, University of Bath, Bath BA2 7AY, UK
*Email: a.b.walker@bath.ac.uk

9.1 Introduction

Thin film photovoltaic (PV) devices require a deep understanding of materials and device physics in order not only to improve their power conversion efficiencies but also to achieve rapid energy payback time and sustainability, as discussed by Gong *et al.*[1] for perovskite cells. Currently, extensive trial and error fabrication and characterisation activity is required to optimise PV devices, since in most cases they consist of several layers of different materials so that the number of potential materials combinations is enormous. Multiscale materials models can considerably reduce this time and effort by screening out materials combinations that offer poor device performance and thus make a large impact in how quickly new PV technologies can be developed.

In materials engineering, multiscale models consist of electronic structure calculations at the microscopic level (0.01–0.1 nm) coupled with molecular dynamics models where information about individual atoms is included. These models' output is coarse grained to produce information about groups of atoms and molecules in mesoscale level models (1–100 nm). These

RSC Energy and Environment Series No. 16
Unconventional Thin Film Photovoltaics
Edited by Enrico Da Como, Filippo De Angelis, Henry Snaith and Alison Walker

Published by the Royal Society of Chemistry, www.rsc.org

particles are smeared out into continuously varying densities at the continuum level (1 nm–1 mm) where drift diffusion (DD) models are employed. Mesoscale and DD models predict device characteristics such as current–voltage curves.

Progress has been made in screening perovskite structures using predictions of their microscopic properties such as the electron affinity and ionisation potential *via* advances in quantum chemical computation, *e.g.* De Angelis *et al.*[2], Frost *et al.*,[3] and Mosconi *et al.*[4] Charge transport on the meso/macroscopic scale is as important as microscopic properties, as noted for organic PV by Jackson *et al.*;[5,6] an observation that also holds for perovskite cells. However, there exists no simple and rapid means for *a priori* differentiation between structures through their charge transport characteristics, since predicting charge transport in these cells has proven far more challenging, and so far no quantitatively accurate device models have been published.

A further demand for quantitative charge transport models of perovskite cells comes from the need to explain the observation of hysteresis, where apparent cell performance is influenced by scan rates and preconditioning prior to current–voltage measurements.[7–12] In the presence of hysteresis, measurements of the power conversion efficiency, PCE, and the maximum power point of such cells are meaningless unless the cell history is provided, which is rarely done. As pointed out by Unger *et al.*,[9] an explanation of the origin of hysteresis is crucial to understanding the underlying working principles of lead halide perovskite solar cells, and to determine and improve their efficiency. Beilsten-Edmands *et al.*[13] have used the results of electrical measurements to suggest that the culprit responsible for hysteresis is migration of ions, in the form of vacancies, through the perovskite film. Electron or hole trapping may also contribute to hysteresis, since charges may detrap on a timescale comparable to that associated with the hysteresis. Charge transport models allow us to obtain physical intution on the effect of mobile defects on the operational mechanism of the device and pave the way for the development of cells with improved and reproducible performance.

Whilst drift diffusion (DD) models can be employed for all PV cell technologies, the focus of this chapter is organic and perovskite solar cells, with particular emphasis on the latter. There are only a few published DD models of perovskite cells.[14–17] Of these, only Van Reenen *et al.*[17] have addressed ion migration, in which mobile ions pile up in boundary layers under the influence of the internal field in the cell. The boundary layer width is determined by the density of the mobile defects obtained from *ab initio* theory, $\sim 10^{19}\,\mathrm{cm}^{-3}$,[18] and is approximately 1 nm, much smaller than the perovskite layer width, which is typically 300 nm. Charge carrier density changes of up to 15 orders of magnitude occur across these layers. In ref. 17, Van Reenen *et al.* employ a finite difference based numerical scheme with a resolution of 4 nm and so, they are very far from being able to resolve the boundary layers.

An important parameter for determining whether hysteresis may be caused by mobile ions is the diffusion coefficient of the mobile ions. The reason is that the time required to change the potential profile seen by the mobile charge carriers can be estimated from the time for the defects to diffuse across a Debye layer of 1 nm. This time should be ~0.01 s to be comparable to the time required for a change in a bias voltage of 0.01 V at a scan rate of 1 V s^{-1}, where hysteresis is commonly observed. Eames *et al.*[19] predicted that the fastest defect motion in lead halide perovskites is vacancy-assisted iodide ion diffusion, where the vacancies have a diffusion coefficient $D_+ \sim 10^{-12}$ cm^2 s^{-1}, which is consistent with ions taking 0.1 s to cross the boundary layer. Large values of $D_+ \sim 10^{-4}$ cm^2 s^{-1} and scan rates of 100 kV s^{-1} were used in ref. 17, which excludes the possibility of considering values of D_+ that are compatible with the scan rates at which hysteresis is seen and with the values from microscopic theory.

It is therefore essential to employ a more sophisticated approach than a standard finite difference method to investigate both charge pileup and the time scales associated with mobile defect motion. A first step in setting up a multiscale charge transport model capable of addressing hysteresis was achieved by Richardson *et al.*[20] In this paper, we described a drift diffusion (DD) model that either uses microscopic calculation of D_+ as input, or is sufficiently accurate that we can reverse-engineer by deducing an approximate value of D_+ from comparing our results with experimental data.

This chapter provides an introduction to the drift diffusion approach used to model charge transport in PV devices and to its application to perovskite cells. Since the aim is to make this technique accessible, the chapter offers a 'how to' guide, starting from concepts taken from condensed matter physics and describing some of the tricks of the trade that allow us to look at boundary layers and consider time scales consistent with the view that hysteresis is caused by mobile ions. Mesoscale models for organic PV cells are reviewed elsewhere in this volume by Bartesaghi *et al.*[21] and so only a brief mention of these models is included in this chapter, where we show how they are linked to DD.

Following the introduction, the material is divided into 5 further sections. Section 9.2 briefly reviews the equilibrium statistical mechanics of a semiconductor as essential background material for the non-equilibrium models that we consider in the subsequent 3 sections. Section 9.3 briefly discusses hopping models of charge transport and relates these to the continuum drift diffusion model of the same process. Section 9.4 elaborates on the drift diffusion model introduced in the previous section and shows how it is applied to the particular example of an n–p homojunction solar cell. Furthermore, we show that the current–voltage curve predicted by modelling this device using a drift diffusion model is accurately approximated by a Shockley equivalent circuit. This leads us, in Section 9.5, to discuss Shockley equivalent circuit models of solar cells and their use in inferring device physics from current–voltage data. Finally, Section 9.6 presents results showing how the inclusion of defect-mediated ion motion can explain experimental data on lead halide perovskite solar cells showing significant hysteresis.

9.2 Equilibrium Electron and Hole Distribution in a Semiconductor

We begin by briefly discussing the equilibrium distribution of charge carriers in a conventional (inorganic) semiconductor before going on to consider charge transport processes. More details on the underlying ideas in solid state physics are given in ref. 22. The band diagram of a semiconductor is illustrated in Figure 9.1. Although it is usual to think of the valence and conduction bands as being continuous in energy, they are in fact comprised of a large number of states separated by very small energies. We therefore need to account for the density of states per unit energy in the conduction band, $g_c(E)\delta E$, defined as the number of states per unit volume in a conduction band with energy between E and $E+\delta E$, where δE is a small increment in energy. A similar expression holds for the density of states per unit energy in the valence band, $g_v(E)\delta E$.

Electrons obey a Fermi–Dirac distribution at equilibrium. Thus, the probability $P(E,T)$ of a particular state, with energy E, being occupied at temperature T is given by

$$P(E,T)=\frac{1}{1+\exp((E-E_f)/k_BT)}, \tag{9.1}$$

where E_f is a constant termed the Fermi level, which is chosen so that the sum of the probabilities over all possible states gives the total number of electrons in the system. Assuming we know the density of states. we can now work out the free electron and hole densities,

$$n=\int_{E_c}^{\infty}\frac{g_c(E)}{1+\exp((E-E_f)/k_BT)}\mathrm{d}E,\ p=\int_{-\infty}^{E_v}\frac{g_v(E)}{1+\exp((E-E_f)/k_BT)}\mathrm{d}E. \tag{9.2}$$

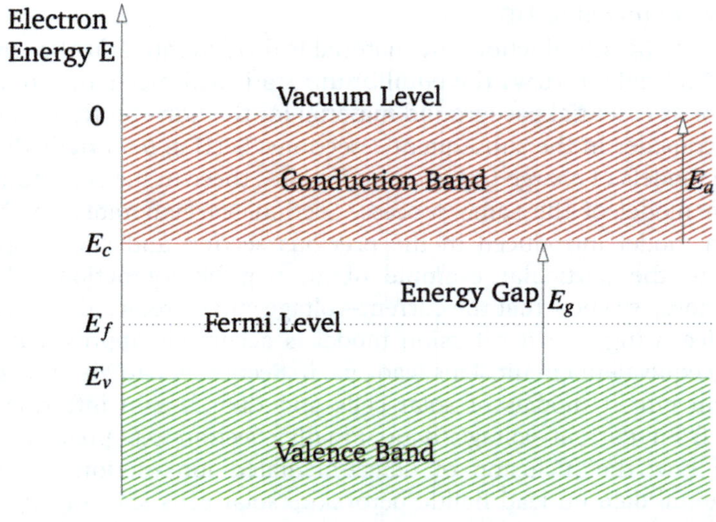

Figure 9.1 The allowed electron energy levels in a semiconductor (at zero electric potential). Here, E_a is the ionisation potential and E_g the gap energy.

Note that the conduction band levels all lie above E_c, while the valence band levels all lie below E_v.

9.2.1 The Boltzmann Approximation

In a typical semiconductor at room temperature, the valence band is very nearly full (*i.e.* almost all the states are occupied), while the conduction band levels are almost all empty (*i.e.* almost all the states are unoccupied). This is a consequence of the energy gap $E_g = E_c - E_v$ being large with respect to $k_B T$ (*i.e.* $E_g \gg k_B T$), and there being almost the right number of electrons to exactly fill the valence states. The latter is equivalent to the following conditions on the Fermi level

$$\frac{E_c - E_f}{k_B T} \gg 1 \quad \text{and} \quad \frac{E_f - E_v}{k_B T} \gg 1. \tag{9.3}$$

Where condition (3) is satisfied, we can accurately approximate p and n by

$$p = N_v(T) \exp\left(-\frac{E_f - E_v}{k_B T}\right) \quad \text{and} \quad n = N_c(T) \exp\left(-\frac{E_c - E_f}{k_B T}\right), \tag{9.4}$$

where the effective conduction band and valence band densities of states are given by

$$\begin{aligned} N_c(T) &= \int_0^\infty g_c(E_c + U) \exp\left(-\frac{U}{k_B T}\right) \mathrm{d}U \quad \text{and} \\ N_v(T) &= \int_0^\infty g_v(E_v - V) \exp\left(-\frac{V}{k_B T}\right) \mathrm{d}V. \end{aligned} \tag{9.5}$$

If, as is usual, the density of states in the vicinity of the conduction band and valence band are approximately parabolic, *i.e.* $g_c(E_c + U) \sim \Upsilon_c U^2$ for $U > 0$ and $g_v(E_v - V) \sim \Upsilon_v V^2$ for $V > 0$ for some constants Υ_c and Υ_v, we can evaluate these integrals explicitly.

9.2.2 The Intrinsic Carrier Density

A consequence of the Boltzmann approximation is that product np is independent of the Fermi level, that is

$$np = n_i^2 = N_c(T) N_v(T) \exp\left(-\frac{E_g}{k_B T}\right) \tag{9.6}$$

and $E_g = E_c - E_v$ is termed the energy gap and n_i the intrinsic carrier density. The reason for doing this is that, at equilibrium in an intrinsic (un-doped) semiconductor, we have $n = p = n_i$.

9.2.3 Doping

Impurities in a semiconducting crystal can lead to dramatic changes in its electrical properties. Foreign atoms in the crystal lattice may preferentially donate electrons to the conduction band (n-type doping) or accept them from the valence band (p-type doping) leading to an imbalance between the number of holes and electrons. In the case of n-type doping, this leads to an immobile net positive charge (the impurity atom loses one of its electrons), while in the case of p-type doping this leads to an immobile net negative charge (the impurity atom gains an electron). If these impurities occur in sufficient quantities, they may increase the number of charge carriers significantly above the intrinsic carrier density. This is a consequence of the net charge in the material having to remain close to zero (charge neutrality). Thus, if N_d denotes the density of donor impurities (fixed positive charges) while N_a denotes the density of acceptor impurities (fixed negative charges), the condition of charge neutrality implies that

$$q(p - n + N_d - N_a) = 0. \tag{9.7a}$$

Hence if there are very few acceptor impurities and a large number of donor impurities with $N_d \gg n_i$, we expect (from eqn. (9.6) and (9.7a)) that

$$n \approx N_d \quad \text{and} \quad p \approx \frac{n_i^2}{N_d}. \tag{9.7b}$$

Such a material is termed n-doped because it has very many more conduction electrons than holes. Alternatively if there are very few donor impurities and a large number of acceptor impurities with $N_a \gg n_i$, we expect from eqn. (9.6) and (9.7a) that

$$p \approx N_a \quad \text{and} \quad n \approx \frac{n_i^2}{N_a}. \tag{9.7c}$$

Such a material is termed p-doped because it has very many more holes than conduction electrons. Doping can be used to enhance the conductivity of a semiconductor because a doped material typically has many more charge carriers than an undoped (or intrinsic) material. It can also be used as a mechanism to separate holes from electrons by p-doping a semiconductor in one region and n-doping it in another. This sets up a potential difference between the two regions that assists separation of holes into the p-doped region and electrons into the n-doped region, and can thus be used to separate solar generated charge carrier pairs in a photovoltaic device.

9.3 Probabilistic and Drift Diffusion Models of Charge Transport

We begin by considering models for particle hopping on a lattice of energy wells and show that, in a certain limit, the probability of finding the particle at a particular time can be approximated by a diffusion equation. We generalise to particle hopping on a lattice with an externally applied potential and show that its probability density satisfies a drift diffusion equation. This analysis is then applied to the motion of charge carriers in a semiconductor. We finish by applying the drift diffusion model of charge carrier transport in a semiconductor to a photovoltaic device formed from a p–n junction, and show that its current–voltage curve is identical to that given by a simple equivalent circuit model of a solar cell.

9.3.1 Rate Equations for Particle Hopping on a Lattice

Consider a particle in one dimensional periodic energy landscape as depicted in Figure 9.2(a). This is characterised by energy wells separated by intervening energy peaks. The particle sitting in one of these wells will, from time to time, have sufficient thermal energy to traverse the energy maximum dividing this well from one of its neighbours. The rate at which this hopping process occurs is given by the Arrhenius equation. The probability of a particle moving from well i to $i+1$ per unit time is $K\exp\left(-\dfrac{\Delta E}{k_{\mathrm{B}}T}\right)$, where k_{B} is Boltzmann's constant, T is absolute temperature, ΔE is the energy difference between two neighbouring wells, and K is a phenomenological rate constant.

9.3.1.1 *The Dynamic Monte Carlo Method*

At this stage we could simulate the motion of the particle simply by getting a computer to throw a dice in order to decide what it does. The simplest way to do this is to pick a small time step Δt with the property that $\Delta t K\exp\left(-\dfrac{\Delta E}{k_{\mathrm{B}}T}\right) \ll 1$ (this means that at any given time step, the chance of

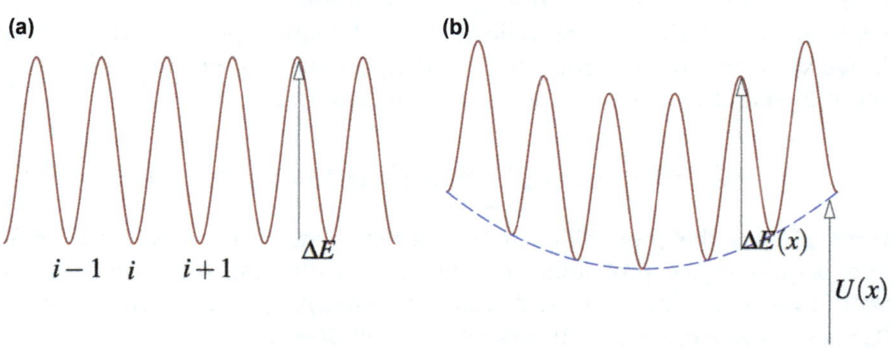

Figure 9.2 The energy landscape of the particle for (a) a lattice with no applied potential and (b) a lattice with applied potential $U(x)$.

the particle moving is small), and then, to pick a random number R evenly distributed between 0 and 1 and implement the procedure:

$$\text{If } 0 \leq R < \Delta t K \exp\left(-\frac{\Delta E}{k_B T}\right) \quad \text{then } \; i(t+\Delta t) = i(t) - 1,$$

$$\text{If } \Delta t K \exp\left(-\frac{\Delta E}{k_B T}\right) \leq R < 2\Delta t K \exp\left(-\frac{\Delta E}{k_B T}\right) \quad \text{then } \; i(t+\Delta t) = i(t) + 1,$$

$$\text{If } 2\Delta t K \exp\left(-\frac{\Delta E}{k_B T}\right) \leq R \leq 1 \quad \text{then } \; i(t+\Delta t) = i(t).$$

This is an easy procedure to implement, but it is numerically inefficient because for most of the time steps nothing happens and the particle remains where it is. A more efficient procedure is to use a random number to determine which event occurs first (in this case there are only two possible events: moving right or left), and then using another randomly generated number to determine the time taken between this event and the last event using the fact that this waiting time obeys a Poisson distribution. The resulting algorithm is termed a Gillespie algorithm after its originator.[23] An example of an application of the Dynamic Monte Carlo method to solar cells is given in ref. 24, and this technique is covered in detail by Bartesaghi *et al.*[21]

9.3.2 A Probabilistic Approach to Particle Hopping

Rather than trying to directly simulate the motion of the particle, it is often sufficient to track the probability of the particle being at any particular lattice point as a function of time. With this in mind, we define $P_i(t)$ as the probability of the particle being in well i at time t, and attempt to write down a series of coupled ordinary differential equations (ODEs) in time for the probabilities $P_i(t)$, considering the evolution of the system over an infinitesimal time interval δt. Suppose that we know what the probability of finding the particle at any of the lattice points at time t (that is, we know $P_i(t)$ for all values i), then the probability that the particle is in the i'th well at time $t+\delta t$ is given approximately by

$$P_i(t+\delta t) = P_i(t)p_{ii}(t) + P_{i+1}(t)p_{i+1i}(t) + P_{i-1}(t)p_{i-1i}(t). \tag{9.8}$$

Here, $p_{ii}(t)$ is the probability of the particle remaining in well i over δt, and $p_{i\pm 1i}(t)$ is the probability of the particle jumping from well $i\pm 1$ to well i over δt. In line with the discussion above, $p_{i\pm 1i} = \delta t\, K \exp[-\Delta E/(k_B T)]$, $p_{ii} = 1 - 2\delta t\, K \exp[-\Delta E/(k_B T)]$. Taking the limit $\delta t \to 0$,

$$\frac{dP_i}{dt} = K \exp\left(-\frac{\Delta E}{k_B T}\right)(P_{i+1}(t) - 2P_i(t) + P_{i-1}(t)). \tag{9.9}$$

9.3.3 Derivation of a Diffusion Equation for Particle Hopping

If there is a very large number of lattice sites – for example, for electron hopping along a line of atoms in a crystal – it is not sensible to track the probability on all lattice sites. Since the probability at neighbouring lattice sites is likely to be very similar, we can seek to replace the probabilities $P_i(t)$ on the lattice sites by a continuum representation of the probability $p(x,t)$, where we suppose x to be a spatial representation of the site's position. Thus, if the lattice sites are evenly spaced a distance δx apart, we can write $x = \delta x$. It is tempting to identify $p(x,t)$ directly with $P_i(t)$ by writing $p(i\delta x,t) = P_i(t)$ but, rather than do this, we instead choose to define $p(x,t)$ as a probability density by writing $p(i\delta x,t) = P_i(t)/\delta x$. Probability densities usually have dimensions of L^{-3} and $p(x,t)$ seems to have dimensions of L^{-1}. However, we note that if we were considering a three-dimensional problem, we would have defined $p(x,y,z,t)$ by $p(i\delta x, j\delta x, k\delta x, t) = P_{ijk}(t)/\delta x^3$, which has the right dimensions and so, given this is a one-dimensional representation of a 3-D problem, the units of $P_i(t)$ are L^{-1}. From eqn (9.9),

$$\frac{\partial p}{\partial t}(x,t) = \left[K\delta x^2 \exp\left(-\frac{\Delta E}{k_B T}\right)\right]\frac{p(x+\delta x,t) - 2p(x,t) + p(x-\delta x,t)}{\delta x^2}. \tag{9.10}$$

If we now make a Taylor expansion of $p(x+\delta x)$ and $p(x-\delta x)$ about x, we find that the leading order in δx, $p(x,t)$ satisfies the diffusion equation:

$$\frac{\partial p}{\partial t} = D\frac{\partial^2 p}{\partial x^2}, \quad D = \left[K\delta x^2 \exp\left(-\frac{\Delta E}{k_B T}\right)\right]. \tag{9.11}$$

9.3.3.1 Interpretation in Terms of Particle Number Density (Concentration)

Thus far, we have assumed only single particle hopping on a lattice but, in reality, we are much more often concerned with a large number of particles on a lattice (*e.g.* free electrons or excitons in an organic semiconductor). This does not present a problem provided that the particles do not interact with each other *via*, for example, Coulomb forces or exclusion effects. We will leave the treatment of interaction *via* Coulomb forces until Section 9.3.4 and note here that exclusion effects only become significant when the number of particles is comparable to the number of lattice sites, so that there is competition between particles for available lattice sites. Hence, provided that particles do not interact *via* a force and their number is much smaller than that of the lattice sites, we can treat them as effectively independent. This means each of the N particles has a probability density $p^{(n)}(x,t)$ (for $n = 1, 2, \ldots, N$) which is, to a good approximation, independent of that for all the other particles. It follows that the particle number density $c(x,t)$ defined by

$$c(x,t) = \sum_{n=1}^{N} p^{(n)}(x,t) \tag{9.12a}$$

obeys the same equation that the probability distribution of each particle obeys, *i.e.* eqn (9.11). Thus,

$$\frac{\partial c}{\partial t} = D\frac{\partial^2 c}{\partial x^2}. \tag{9.12b}$$

9.3.4 Description of Particle Hopping on a Lattice in an Applied Potential

Semiconductors contain relatively large numbers of mobile charged particles (free electrons and holes) which, since they are charged, interact with each other *via* the Coulomb potential. In order to treat this type of scenario, we consider a single particle moving in a potential comprised of a quasi-periodic part (as previously) and a smooth part $U(x)$ (the applied potential); this is illustrated in Figure 9.2(b). In order to make the analysis more general, we allow the energy barrier between the wells (in the absence of the applied potential) ΔE to vary smoothly with x; the implicit assumption being that ΔE varies by a small amount between neighbouring peaks but can vary significantly over many peaks. We denote the energy barrier on passing from well i to well $i+1$ by ΔE. In the presence of the applied potential $U(x)$, the appropriate expressions for these quantities are:

$$\Delta E_{i\to i+1} = U\left(x+\frac{\delta x}{2}\right) - U(x) + \Delta E\left(x+\frac{\delta x}{2}\right), \tag{9.13}$$

$$\Delta E_{i\to i-1} = U\left(x-\frac{\delta x}{2}\right) - U(x) + \Delta E\left(x-\frac{\delta x}{2}\right), \tag{9.14}$$

where, as before $x = i\delta x$. In a similar fashion,

$$\Delta E_{i+1\to i} = U\left(x+\frac{\delta x}{2}\right) - U(x+\delta x) + \Delta E\left(x+\frac{\delta x}{2}\right), \tag{9.15}$$

$$\Delta E_{i-1\to i} = U\left(x-\frac{\delta x}{2}\right) - U(x-\delta x) + \Delta E\left(x-\frac{\delta x}{2}\right). \tag{9.16}$$

In order to obtain a partial differential equation (PDE) for the probability density $p(x,t)$ (as defined by $p_i(t)/\delta x$) we proceed as before and expand in powers of δx (having first substituted for the energy barriers from eqn (9.13)–(9.16)). The calculation is of a similar nature to that used to derive eqn (9.11), and the resulting PDE is:

$$\frac{\partial p}{\partial t} = \frac{\partial}{\partial x}\left(D(x)\left(\frac{\partial p}{\partial x} + \frac{p}{k_B T}\frac{\partial U}{\partial x}\right)\right), \quad D(x) = \left[K\delta x^2 \exp\left(-\frac{\Delta E(x)}{k_B T}\right)\right]. \tag{9.17}$$

Using arguments analogous to those presented at the end of Section 9.3.3, we can show that the same equation is satisfied by the particle number density $c(x,t)$ in three-dimensions for a system in which the number of particles is much smaller than the number of lattice sites:

$$\frac{\partial c}{\partial t} = \nabla \cdot \left(D(x)\left(\nabla c + \frac{c}{k_B T}\nabla U\right)\right). \tag{9.18}$$

9.3.5 Other Processes Giving Rise to Drift Diffusion Models

The above description of particle motion in a periodic energy landscape could easily be adapted to free electrons in the LUMO (lowest unoccupied molecular orbital) of an organic semiconductor (or equally to a hole in the HOMO of an organic semiconductor). However, it does not apply to conduction electrons in an inorganic semiconductor, which are free to move largely unhindered through the crystal lattice. Nevertheless the drift diffusion equation derived above also applies to the motion of electrons in a conduction band and holes in a valence band, except that here the diffusive character of the motion can be attributed to electron scattering in the crystal lattice. The derivation of the drift diffusion equations in an inorganic semiconductor has been outlined by Nelson[24] and Feynman *et al.*[25] This derivation (i) starts from a quantum mechanical description of the wavefunction of an electron in the conduction band of a semiconducting lattice (using Bloch wave functions), (ii) assumes that electrons are in quasi thermal equilibrium at all points in space (*i.e.* scattering of electrons within the crystal is sufficiently fast that the electrons are distributed across energy levels of the conduction band according to a Boltzmann distribution) and (iii), having made this assumption, uses the Boltzmann transport equation to calculate the electron current in terms of the electric field and the spatial distribution of electrons.

9.4 Drift Diffusion Models of Charge Transport in Semiconductors

Charge transport in a semiconductor is characterised by the motion of free electrons (in the conduction band [inorganic] or LUMO [organic]) and holes (in the valence band [inorganic] or HOMO [organic]). Both electrons and holes are charged and therefore interact *via* Coulomb forces. In order to keep track of these interactions, it is usual to define an electric potential ϕ which satisfies Poisson's equation:

$$\nabla \cdot (\varepsilon \nabla \phi) = -\rho(x,t) \tag{9.19}$$

where ε is the permittivity. The charge density ρ is typically approximated in terms of $p(x)$, $n(x)$ and $C_{\text{fix}}(x)$, which are the hole, free electron and the fixed charge densities. $C_{\text{fix}} = N_{\text{d}} - N_{\text{a}}$ is the net doping density which, where needed, can be applied to materials with contributions from charged defects. The relation given by this approximation is

$$\rho(x,t) = q(p(x,t) - n(x,t) + C_{\text{fix}}) \tag{9.20}$$

where q is the elementary charge. In making this approximation, we have smeared out the effects of the individual point charges by replacing the delta-functions in the charge density (considering each charge as a point charge) with a smooth function $q(p(x,t) - n(x,t) + C_{\text{fix}})$ through finding the total contribution from the point charges in a volume $(\delta x)^3$ where δx is much smaller than the distance over which ϕ varies significantly. This is a sensible approach to adopt where opposite charges do not tend to pair up, and indeed it works well for semiconductors. However, there are systems, such as electrolytes, where this approximation is not particularly good and alternative approaches have to be adopted.

Assuming that there is very little charge recombination or generation, the resulting equations for n, p and ϕ can be found by combining eqn (9.19) with (9.20), writing $c = p$ and $U = q\phi$ in eqn (9.18), and writing $c = n$ and $U = -q\phi$ in eqn (9.18), to obtain the set of coupled PDEs

$$\frac{\partial p}{\partial t} = \nabla \cdot \left(D_p \left(\nabla p + \frac{q}{k_{\text{B}}T} p \nabla \phi \right) \right), \tag{9.21}$$

$$\frac{\partial n}{\partial t} = \nabla \cdot \left(D_n \left(\nabla n - \frac{q}{k_{\text{B}}T} n \nabla \phi \right) \right), \tag{9.22}$$

$$\nabla \cdot (\varepsilon \nabla \phi) = q(n - p - C_{\text{fix}}). \tag{9.23}$$

Here D_p and D_n are the diffusion coefficients of holes and mobile electrons, respectively.

There are 3 PDEs for the 3 variables, p, n and ϕ, and we therefore have the right number of equations for these unknowns. However, there are effects missing from the physics. Firstly, we have entirely neglected charge generation and recombination, which are key processes if we want to model a solar cell. Secondly, we have implicitly assumed that we are considering only a single material when we may want to investigate devices with junctions between a number of different semiconducting materials.

9.4.1 Currents, Fluxes and Carrier Concentration

In order to aid the physical interpretation of their solutions, it is helpful to recast the drift diffusion equations for free electrons and holes in terms of alternative variables (*e.g.* quasi-Fermi levels). We begin by rewriting eqn (9.18) (or alternatively eqn (9.21) and (9.22)) in terms of conservation laws for the carrier concentrations. On writing $c=p$ in eqn (9.18), and subsequently writing $c=n$ in eqn (9.18), and re-arranging the resulting equations, we find

$$\frac{\partial p}{\partial t}+\nabla.(p\nu_p)=0 \quad \text{where } \nu_p=-\frac{D_p}{k_\mathrm{B}T}\nabla\left(k_\mathrm{B}T\ln\left[\frac{p}{c_p^*}\right]+U_p\right), \tag{9.24}$$

$$\frac{\partial n}{\partial t}+\nabla.(n\nu_n)=0 \quad \text{where } \nu_n=-\frac{D_n}{k_\mathrm{B}T}\nabla\left(k_\mathrm{B}T\ln\left[\frac{n}{c_n^*}\right]+U_n\right). \tag{9.25}$$

Here, it is not immediately clear what the reference concentrations $c_p{}^*$ and $c_n{}^*$ should be, and it does not matter, provided we only consider a single homogenous material in which these concentrations are constant so that their gradients are zero. In fact, it turns out (for sound statistical mechanical reasons) that we should identify $c_p{}^*$ and $c_n{}^*$ with the effective densities of states in the valence and conduction bands, respectively (see eqn (9.5)) so that $c_p{}^*=N_\mathrm{v}$ and $c_p{}^*=N_c$.

We may identify eqn (9.24) and (9.25) as conservation equations for p and n, respectively, in which ν_p and ν_n represent the average velocities of holes and electrons respectively. Eqn (9.24) and (9.25) for the average velocities can then be re-expressed in terms of thermodynamic quantities as

$$\nu_p=-\frac{D_p}{k_\mathrm{B}T}\nabla\mu_p \quad \text{where } \mu_p=k_\mathrm{B}T\ln\left(\frac{p}{N_v}\right)+U_p, \tag{9.26}$$

$$\nu_n=-\frac{D_n}{k_\mathrm{B}T}\nabla\mu_n \quad \text{where } \mu_n=k_\mathrm{B}T\ln\left(\frac{n}{N_c}\right)+U_n. \tag{9.27}$$

Here μ_p and μ_n are the electrochemical potentials of holes and free electrons, respectively. The hole current density j_p (*i.e.* that portion of the current carried by holes) and the electron current density j_n are related to the average carrrier velocities by $j_p=qp\nu_p$ and $j_n=-qn\nu_n$, from which it follows that

$$j_p=M_p p\nabla E_\mathrm{Fp} \text{ and } j_n=M_n n\nabla E_\mathrm{Fn}. \tag{9.28}$$

Here, the electron and hole mobilities (M_n and M_p, respectively) and the electron and hole quasi Fermi levels (E_Fn and E_Fp respectively) are given by

$$M_n=\frac{qD_n}{k_\mathrm{B}T}, \quad M_p=\frac{qD_p}{k_\mathrm{B}T}, \quad E_\mathrm{Fn}=\mu_n, \quad E_\mathrm{Fp}=-\mu_p. \tag{9.29}$$

9.4.1.1 Electron and Hole Potentials and Quasi-Fermi Levels

We now need to specify the electron and hole potentials (U_n and U_p, respectively) more precisely. We begin by considering the mobile electrons. These electrons lie close to the lower edge of the conduction band and so, in the absence of an electric field (*i.e.* zero electric potential), their energy is $-E_a$, where E_a is the electron affinity, *i.e.* the energy required to remove an electron from the conduction band edge to the vacuum level (see Figure 9.1). The electrostatic part of the potential is $-q\phi$. Adding these two contributions together gives:

$$U_n(x) = -E_a(x) - q\phi(x). \tag{9.30}$$

The holes lie close to the upper edge of the valence band and, in order to calculate their potential, we calculate the potential of an electron lying there and note that the potential of a hole (*i.e.* the lack of an electron) is just minus that of the electron it replaces. It follows that

$$U_p(x) = (E_a(x) + E_g(x)) + q\phi(x), \tag{9.31}$$

where E_g is the energy gap between the valence and the conduction band, and $-(E_g + E_a)$ is the energy of an electron at the upper edge of the valence band at zero electric potential (see Figure 9.1). It is also common to write the electron and hole potentials in terms of the electron energies (including the effect of the electric potential) at the band edges:

$$U_n(x) = E_c(x) \quad \text{and} \quad U_p(x) = -E_v(x). \tag{9.32}$$

Here, E_c and E_v are the energies at the conduction and valence band edges, respectively. It now becomes clear why E_{Fn} and E_{Fp} are termed quasi-Fermi levels because, if we take our definitions of the quasi-Fermi levels in eqn (9.29), substitute for the chemical potentials μ_p and μ_n from eqn (9.26) and (9.27), and rewrite the resulting expression in terms of p and n using eqn (9.32), we obtain

$$p = N_v \exp\left(-\frac{E_{Fp} - E_v}{k_B T}\right) \quad \text{and} \quad n = N_c \exp\left(-\frac{E_c - E_{Fn}}{k_B T}\right). \tag{9.33}$$

These expressions look almost identical to eqn (9.4), the formulae for the electron and hole densities at equilibrium, except that the Fermi level in the equation for n has been replaced by the quasi-Fermi level E_{Fn} and the Fermi level in the equation for p has been replaced by the quasi-Fermi level E_{Fp}. However, the formulae in eqn (9.33) do not help, because we cannot use them to determine the carrier concentrations n and p, and it is much better to think of the quasi-Fermi levels as being determined by the carrier concentrations rather than *vice versa*. With this in mind, we use the definitions of the quasi-Fermi levels in eqn (9.26), (9.27) and (9.29), together with those of the electron and hole potentials to write:

$$E_{\mathrm{Fn}} = k_{\mathrm{B}}T\ln\left(\frac{n}{N_{\mathrm{c}}}\right) - E_{\mathrm{a}} - q\phi,\ E_{\mathrm{Fp}} = -\,k_{\mathrm{B}}T\ln\left(\frac{p}{N_{\mathrm{v}}}\right) - E_{\mathrm{a}} - E_{\mathrm{g}} - q\phi. \quad (9.34)$$

9.4.2 Carrier Generation and Recombination

Thinking back on our treatment of a semiconductor at equilibrium, it becomes apparent that we have forgotten something. Our equations for the conservation of electrons Eqn (9.24) and (9.25) allow only for transport of electrons within a band, but not for the interchange of electrons between the conduction and valence bands. Yet, when we discussed the equilibrium problem, we assumed that there was thermal equilibrium between the electrons in the two bands and so, we clearly need to take into account the transfer of electrons between bands, even if this effect might be small with respect to other processes. Otherwise, we cannot expect the solution of our model to settle down eventually to equilibrium. Eqn (9.24) and (9.25) are conservation equations for holes and free electrons, respectively. Moving an electron from the valence band to the conduction band results in the creation of a hole in the valence band and a free electron in the conduction band. Similarly, a free electron can only move down from the conduction band into the valence band if there is hole for it to fill, and so this process therefore results in the annihilation of a free electron and a hole. In other words, the motion of electrons between bands results in the creation (or annihilation) of an equal number of holes and free electrons. It follows that the only sensible way to alter eqn (9.24) and (9.25) is by adding the same term to the right-hand side of both these equations so that:

$$\frac{\partial p}{\partial t} + \nabla\cdot(p\,\nu_p) = G_{\mathrm{therm}},\quad \frac{\partial n}{\partial t} + \nabla\cdot(n\,\nu_n) = G_{\mathrm{therm}}. \quad (9.35)$$

Here, it is to be understood that G_{therm} is the volumetric rate of generation of holes and free electrons, which can be either positive (if the net rate of electrons jumping from the conduction to the valence band is positive) or negative (if the net rate of electrons jumping from the conduction to the valence band is negative). Furthermore, we require that, if the semiconductor is isolated and there are no spatial gradients so that $\nu_n = \nu_p = 0$, the solution to eqn (9.35) settles to the thermodynamic equilibrium given in eqn (9.6) for long times (*i.e.* $np \rightarrow n_i^2$ as $t \rightarrow +\infty$). If we think first about the recombination of free electrons in the conduction band with holes in the valence band, since both species are rare, we might reasonably expect that the rate limiting step is getting the free electron close enough to the hole so that they can annihilate. The rate at which these two species come within a critical reaction radius per unit volume is clearly proportional to the product of their concentrations np. Putting these facts together, we end up with a simple (and correct) model for the thermal generation rate

$$G_{\text{therm}} = \gamma(n_i^2 - np) \tag{9.36}$$

for some material dependent rate constant γ.

9.4.2.1 Solar Generation

Central to the use of semiconductors in solar cells is their ability to absorb light and turn the absorbed energy into electron–hole pairs. Clearly, such pairs can only be generated directly if the absorbed photon has more energy than the band gap, which is one of the reasons that the choice of material with the right band gap is so important. If the band gap is large, not many solar photons are turned into free-electron hole pairs, but the electron hole pairs that are generated have a large potential energy, corresponding to a high open circuit voltage V_{oc} and a small short-circuit current J_{sc}. On the other hand, if the band gap is small, more solar photons will generate electron hole pairs, but these will have low potential energy corresponding to a high J_{sc} and low V_{oc}.

The generation of charge pairs in inorganic semiconductors typically occurs directly throughout the bulk of the semiconductor. However, charge pair generation in organic materials is typically a two stage process. Photons are absorbed in the bulk of the semiconductor to produced an exciton, which can be thought of as a coulombically bound electro–hole pair. The exciton binding energy is much greater than the thermal energies because the dielectric constant for organic materials is ~3–4, much smaller than the dielectric constant for inorganic semiconductors, typically ~13. This exciton can diffuse through the material, but it has a relatively short lifetime before it releases its energy as heat. Such excitonic materials can only be used in photovoltaics if they form an interface with another material at which it is energetically favourable for an exciton to dissociate into a free-electron in one material and a hole in the other.

As charge generation in organic PV is discussed elsewhere in this volume,[21] we will only consider the direct form of generation, which is appropriate for perovskites since their dielectric constant is 24. At the most basic level, we ought to be able to treat such carrier generation in a similar fashion to thermal generation, by adding a term $G_{sol}(x)$ to the right-hand side of eqn (9.35). Of course, the exact details of the size of G_{sol} and its spatial variation are extremely important and depend on the band gap, the radiation spectrum and how much radiation has been absorbed in other parts of the cell before reaching the point x. However, these are not the focus of this discussion and so, for simplicity, we will assume G_{sol} is a constant (which can only really be justified in a very thin cell that absorbs only a small fraction of the incident light). It is worth noting that the solar generation rate by sufficiently energetic photons is intimately linked to the rate constant of bimolecular recombination γ. This is because solar generation and bimolecular recombination are essentially the same processes, but in reverse. For more information on how these rates should be calculated, see ref. 26. Assuming that the recombination and generation rates are independent of each other can lead to nonsensical predictions such as devices that produce more power than they absorb through solar radiation.

9.4.2.2 *Recombination* via *Trapped States*

Thus far, we have assumed that eletron–hole pairs only recombine *via* bimolecular recombination, in which an electron from the conduction band drops directly into a hole in the valence band. This is represented by the second (negative) term on the right-hand side of eqn (9.36) (*i.e.* $-\gamma np$). As noted above, bimolecular recombination is the reverse process to solar generation, since the potential energy lost by the conduction band electron can be emitted as a photon or as phonons. Indeed, bimolecular recombination is responsible for light emission in LEDs (light emitting diodes), which are devices that turn electrical energy into light energy. However, other recombination mechanisms are possible and, indeed, turn out to be more significant than bimolecular recombination in many photovoltaic devices. These alternative forms of recombination usually occur when an electron from the conduction band becomes trapped in an energy state lying in the forbidden energy gap. Such states are termed trap states and are generally associated with imperfections in the semiconductor. The most common model for recombination *via* trapped states is the Shockley–Read–Hall (SRH) model that states that the rate of recombination R_{trap} has the form:

$$R_{\text{trap}} = \frac{np - n_i^2}{\tau_1 n + \tau_2 p + \tau_3 n_i}, \tag{9.37}$$

where τ_1, τ_2 and τ_3 are timescales associated with the trapping process. Other more exotic recombination processes are discussed in ref. 26 and 27.

9.4.3 The Full Equations

We are now in a position to write down the full equations for a semiconductor, which may be doped, and which absorbs light to generate charge carrier pairs. These are given by eqn (9.28) and the generalisation of eqn (9.35) to include solar generation G_{sol} and recombination *via* trapped states, R_{trap}.

$$\frac{\partial p}{\partial t} + \frac{1}{q}\nabla.(j_p) = \gamma(n_i^2 - np) + G_{\text{sol}}(x) - R_{\text{trap}} \quad \text{where } j_p = M_p p \nabla E_{\text{Fp}}, \tag{9.38}$$

$$\frac{\partial n}{\partial t} - \frac{1}{q}\nabla.(j_n) = \gamma(n_i^2 - np) + G_{\text{sol}}(x) - R_{\text{trap}} \quad \text{where } j_n = M_n n \nabla E_{\text{Fn}}, \tag{9.39}$$

$$\nabla \cdot (\varepsilon \nabla \phi) = q(n - p - C_{\text{fix}}), \tag{9.40}$$

where the quasi-Fermi levels are defined in eqn (9.34).

9.4.3.1 Boundary Conditions

Boundary conditions on these equations need to be specified where the semiconductor forms a junction with a current collectors. The latter are usually metallic or have metallic properties. In most devices, it is usual to assume that the metallic contact and the semiconductor are in thermal equilibrium. Even this assumption still yields rather complicated boundary conditions which typically result in a solution with a thin layer in which the potential changes rapidly. However, in practice, the contacts are usually chosen so that they do not unduly influence the behaviour of the device. Where this is the case, a sensible set of boundary conditions for the interface between a doped semiconductor and a metallic contact consists of imposing continuity of the potential at the contacts, requiring that the minority carrier current flux across the boundary is zero and ensuring that the carrier charge density is equal to the doping charge density, that is,

$$J_{\min}\cdot n|_{\partial\Omega}=0,\ \ \phi|_{\partial\Omega}=V,\ \ n-p|_{\partial\Omega}=C_{\text{fix}}. \tag{9.41}$$

Here V is the potential of the contact, n is the unit normal to the boundary and $J_{\min}$ is the current flux of the minority carrier.

9.4.4 A Simple One-dimensional Inorganic Solar Cell: The n–p Homojunction

Here, we will consider a solar cell formed from a planar inorganic semiconductor sandwiched between two contacts (see Figure 9.3) at $x=-L$ and $x=L$, and look to calculate its current–voltage curve. In order to separate solar generated charges, the semiconductor is n-doped in $-L<x<0$ and p-doped in $0<x<L$. At equilibrium, as illustrated in Figure 9.3, this results in band bending and a potential difference between the two sides of the cell that acts to drive apart the electrons and holes created by solar generation. In order to simplify the problem, we consider a cell with a high degree of symmetry. In particular, we assume that the doping in the two

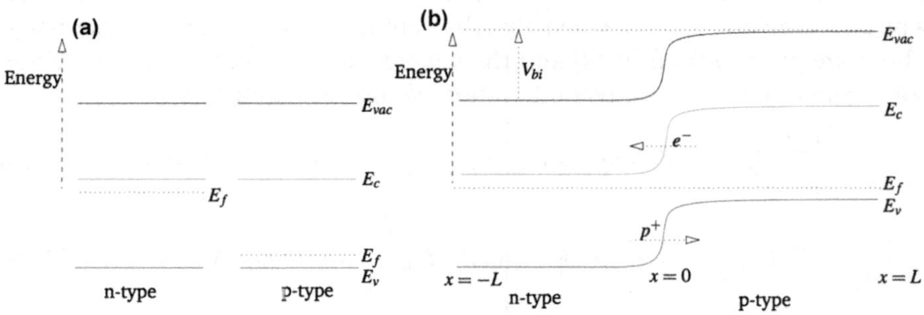

Figure 9.3 (a) Energy levels in the isolated n-type and p-type materials. (b) Band bending and the built-in voltage V_{bi} when two materials form an n–p junction.

halves of the cell is equal and opposite (such that $C_{\text{fix}}=qN_{\text{dop}}$ in $-L<x<0$ and $C_{\text{fix}}=-qN_{\text{dop}}$ in $0<x<L$) and that the mobilities of holes and electrons are equal. We shall also assume (rather unrealistically) that recombination is entirely bimolecular (so that $R_{\text{trap}}=0$). With these assumptions, the governing eqn (9.38)–(9.40) may be written as:

$$\frac{\partial j_p}{\partial x}=q(\gamma(n_i^2-np)+G_{\text{sol}}),\quad j_p=-qD\left(\frac{\partial p}{\partial x}+\frac{p}{V_T}\frac{\partial \phi}{\partial x}\right), \tag{9.42}$$

$$\frac{\partial j_n}{\partial x}=-q(\gamma(n_i^2-np)+G_{\text{sol}}),\quad j_n=qD\left(\frac{\partial n}{\partial x}-\frac{n}{V_T}\frac{\partial \phi}{\partial x}\right), \tag{9.43}$$

$$\frac{\partial^2 \phi}{\partial x^2}=\begin{cases}\frac{q}{\varepsilon}(n-p-N_{\text{dop}}) & \text{in } -L<x<0\\ \frac{q}{\varepsilon}(n-p+N_{\text{dop}}) & \text{in } 0<x<L\end{cases}. \tag{9.44}$$

Here, we have chosen to rewrite the mobility M in terms of the diffusion coefficient using the relationship $D=k_BTM/q$ and have defined $V_T=k_BT/q$. The quantity V_T, termed the thermal voltage, plays an important role in semiconductor physics, and at room temperature $V_T\approx 25$ mV. Here we use the boundary conditions of eqn (9.41) at both contacts. On the left contact ($x=-L$) the minority carriers are holes, while on the right contact they are electrons. We choose to split the voltage difference V between the two contacts into two parts, the built-in voltage V_{bi} and the applied voltage V_{ap} by writing $V=V_{\text{bi}}-V_{\text{ap}}$. The boundary conditions then become:

$$j_n|_{x=-L}=0,\qquad \phi|_{x=-L}=\frac{V_{\text{bi}}-V_{\text{ap}}}{2},\qquad n-p|_{x=-L}=N_{\text{dop}}, \tag{9.45}$$

$$j_p|_{x=L}=0,\qquad \phi|_{x=L}=\frac{V_{\text{ap}}-V_{\text{bi}}}{2},\qquad p-n|_{x=-L}=N_{\text{dop}}. \tag{9.46}$$

9.4.4.1 *Rescaling*

We now have the option of trying to solve this model computationally for physically realistic parameters. However, this does not really allow us to develop much physical intuition about the problem and it turns out that a much better way of proceeding is to try and solve the model asymptotically, *i.e.* approximately, making use of the fact that certain dimensionless parameters (these are sometimes termed dimensionless groups) are either very small or very large in all practical devices. As an example of this approach, let us consider the ratio of the Debye length L_D (*i.e.* the characteristic length scale defined by eqn (9.46)) to the half-width of the cell. Here, on assuming that the characteristic voltage of the problem is V_T and the characteristic carrier density is N_{dop} we find:

$$L_D = \left(\frac{\varepsilon V_T}{qN_{\text{dop}}}\right)^{1/2}.$$

Given $\varepsilon \sim 4\varepsilon_0$ and $N_{\text{dop}} \sim 10^{22}$–$10^{25}$ m^{-3} gives L_D in the range 10^{-9}–10^{-8} m. Comparing this to a typical half-width of a cell, which might be 10^{-6}–10^{-5} m, we see that the dimensionless parameter $\lambda = L_D/L$ is very small: $\lambda = O(10^{-4}) - O(10^{-2})$.

In order to identify all the dimensionless parameters in the problem, we rescale it in such a way as to leave a dimensionless problem characterised by a minimal set of dimensionless parameters. This process is called non-dimensionalisation and, providing that we choose to scale our variables with sensible values, leads to considerable insight into the physics of the problem.

9.4.4.2 Non-dimensionalisation

We rescale the variables in the problem as follows

$$\phi = V_T\bar{\phi}, \quad n = N_{\text{dop}}\bar{n}, \quad p = N_{\text{dop}}\bar{p}, \quad G_{\text{sol}} = G_{\text{1-sun}}\bar{G}, \quad j_n = qLG_{\text{1-sun}},$$
$$x = L\bar{x}, \quad V_{\text{bi}} = V_T\bar{V}_{\text{bi}}, \quad V_{\text{ap}} = V_T\bar{V}_{\text{ap}}, \quad j_p = qLG_{\text{1-sun}}$$

The charge pair generation term is scaled by the generation that would be expected under illumination of one sun $G_{\text{1-sun}}$. On substituting these rescaled variables into eqn (9.42)–(9.46), we obtain the following system on dimensionless equations and boundary conditions:

$$\frac{\partial \bar{j}_p}{\partial \bar{x}} = \Theta(N_i^2 - \bar{n}\bar{p}) + \bar{G}, \quad \bar{j}_p = -\kappa\left(\frac{\partial \bar{p}}{\partial \bar{x}} + \bar{p}\frac{\partial \bar{\phi}}{\partial \bar{x}}\right), \tag{9.47}$$

$$\frac{\partial \bar{j}_n}{\partial \bar{x}} = -\Theta(N_i^2 - \bar{n}\bar{p}) - \bar{G}, \quad \bar{j}_p = \kappa\left(\frac{\partial \bar{n}}{\partial \bar{x}} - \bar{n}\frac{\partial \bar{\phi}}{\partial \bar{x}}\right), \tag{9.48}$$

$$\frac{\partial^2 \bar{\phi}}{\partial \bar{x}^2} = \begin{cases} \dfrac{1}{\lambda^2}(\bar{n} - \bar{p} - 1) & \text{in} \quad -1 < \bar{x} < 0 \\ \dfrac{1}{\lambda^2}(\bar{n} - \bar{p} + 1) & \text{in} \quad 0 < \bar{x} < 1 \end{cases}, \tag{9.49}$$

$$\bar{j}_p|_{\bar{x}=-1} = 0, \quad \bar{\phi}|_{\bar{x}=-1} = \frac{\bar{V}_{\text{bi}} - \bar{V}_{\text{ap}}}{2}, \quad \bar{n} - \bar{p}|_{\bar{x}=-1} = 1, \tag{9.50}$$

$$\bar{j}_n|_{\bar{x}=1} = 0, \quad \bar{\phi}|_{\bar{x}=1} = \frac{\bar{V}_{\text{ap}} - \bar{V}_{\text{bi}}}{2}, \quad \bar{p} - \bar{n}|_{\bar{x}=1} = 1, \tag{9.51}$$

where the dimensionless parameters or groups in the problem are

$$N_i = \frac{n_i}{N_{\rm dop}}, \quad \lambda = \frac{1}{L}\left(\frac{\varepsilon V_T}{qN_{\rm dop}}\right)^{1/2}, \quad \Theta = \left(\frac{\gamma N_{\rm dop}^2}{G_{\text{1-sun}}}\right), \quad \kappa = \left(\frac{DN_{\rm dop}}{G_{\text{1-sun}}L^2}\right). \tag{9.52}$$

9.4.4.3 *Dimensionless Parameters and Their Size*

We have already seen that $\lambda = L_{\rm D}/L$ gives the ratio of the Debye length to the half-width of the cell. The next parameter we consider is N_i, which gives the ratio of the intrinsic carrier density to the doping density. The intrinsic carrier density in silicon at room temperature is roughly 10^{16} m^{-3}, and so taking $N_{\rm dop} \sim 10^{20}$–10^{22} m^{-3} (as above) leads to an estimate of $N_i = O(10^{-4})$ – $O(10^{-8})$. Θ gives the ratio of the square of the typical charge densities obtained by balancing solar generation with bimolecular recombination to the square of the doping density. κ gives the ratio of the typical current density produced by a potential difference across the cell of size $O(V_T)$ to the current density produced by charge generation at one sun. Thus if $\kappa \gg 1$, this suggests the current produced by illumination of the solar cell at one sun can be transported away by a very small potential diffference across the cell or, to put it another way, that the ohmic resistance is negligible. It is nearly always the case that $\kappa \gg 1$, indeed if it were not, the cell would be very inefficient.

9.4.4.4 *Simplifying the Problem Using its Symmetry*

The problem is symmetric about $\bar{x} = 0$ as can be seen by making the substitution

$$\bar{x} = -\bar{X}, \quad \bar{p} = \bar{N}, \quad \bar{n} = \bar{P}, \quad \bar{\phi} = -\bar{\Phi}, \quad \bar{j}_n = \bar{j}_p, \quad \bar{j}_p = \bar{j}_n,$$

in eqn (9.47)–(9.51). If this substitution is made, we retrieve exactly the same problem but with the variables replaced as follows:

$$\bar{x} \to \bar{X}, \quad \bar{n} \to \bar{N}, \quad \bar{p} \to \bar{P}, \quad \bar{\phi} = -\bar{\Phi}, \quad \bar{j}_n \to \bar{j}_n, \quad \bar{j}_p \to \bar{j}_p,$$

It is thus apparent that the solution to eqn (9.47)–(9.51) has the symmetry

$$\bar{p}(-\bar{x}) = \bar{n}(\bar{x}), \quad \bar{n}(-\bar{x}) = \bar{p}(-\bar{x}), \quad \bar{\phi}(-\bar{x}) = -\bar{\phi}(\bar{x}), \quad \bar{j}_p(-\bar{x}) = \bar{j}_n(\bar{x}), \quad \bar{j}_n(-\bar{x}) = \bar{j}_p(\bar{x}). \tag{9.53}$$

We need therefore only to solve in $0 < \bar{x} < 1$, where we impose the boundary conditions

$$\bar{p}|_{\bar{x}=0} = \bar{n}|_{\bar{x}=0}, \quad \bar{\phi}|_{\bar{x}=0} = 0, \quad \bar{j}_p|_{\bar{x}=0} = \bar{j}_n|_{\bar{x}=0}.$$

The problem we need to solve in $0 < \bar{x} < 1$ can thus be stated as

$$\frac{\partial \bar{j}_p}{\partial \bar{x}} = \Theta(N_i^2 - \bar{n}\bar{p}) + \bar{G}, \quad \bar{j}_p = -\kappa\left(\frac{\partial \bar{p}}{\partial \bar{x}} + \bar{p}\frac{\partial \bar{\phi}}{\partial \bar{x}}\right), \tag{9.54}$$

$$\frac{\partial \bar{j}_n}{\partial \bar{x}} = -\Theta(N_i^2 - \bar{n}\bar{p}) - \bar{G}, \quad \bar{j}_p = \kappa\left(\frac{\partial \bar{n}}{\partial \bar{x}} - \bar{n}\frac{\partial \bar{\phi}}{\partial \bar{x}}\right), \tag{9.55}$$

$$\frac{\partial^2 \bar{\phi}}{\partial \bar{x}^2} = \frac{1}{\lambda^2}(\bar{n} - \bar{p} + 1), \tag{9.56}$$

$$\bar{p}|_{\bar{x}=0} = \bar{n}|_{\bar{x}=0}, \quad \bar{\phi}|_{\bar{x}=0} = 0, \quad \bar{j}_p|_{\bar{x}=0} = \bar{j}_n|_{\bar{x}=0}, \tag{9.57}$$

$$\bar{p} - \bar{n}|_{\bar{x}=1} = 1, \quad \bar{\phi}|_{\bar{x}=1} = \frac{\bar{V}_{\text{ap}} - \bar{V}_{\text{bi}}}{2}, \quad \bar{j}_n|_{\bar{x}=1} = 0, \tag{9.58}$$

and we can determine the solution in $-1 < \bar{x} < 0$ from the symmetry conditions (eqn (9.53)).

9.4.4.5 Asymptotic Solution to the Dimensionless Model in the Limit $\lambda \ll 1$ and $\kappa \gg 1$

We now seek to solve the problem approximately by making use of the fact that $\kappa \gg 1$ and $\lambda \ll 1$. We start by noting that, since $\kappa \gg 1$, the right-hand sides of eqn (9.56) and (9.57) must both be very small, *i.e.* the following approximate equations must hold:

$$\frac{\partial \bar{p}}{\partial \bar{x}} \sim -\bar{p}\frac{\partial \bar{\phi}}{\partial \bar{x}}, \quad \frac{\partial \bar{n}}{\partial \bar{x}} \sim \bar{n}\frac{\partial \bar{\phi}}{\partial \bar{x}}$$

It is not hard to show that these have solutions

$$\bar{p} = A\exp(-\phi) \quad \text{and} \quad \bar{n} = B\exp(\phi) \tag{9.59}$$

for some arbitrary constants A and B. Applying the conditions in eqn (9.57) on $\bar{x} = 0$ implies that $B = A$, while application of the conditions in eqn (9.58) on $\bar{x} = 1$ implies that

$$A = \frac{1}{\exp\left(\frac{\bar{V}_{\text{bi}} - \bar{V}_{\text{ap}}}{2}\right) - \exp\left(-\frac{\bar{V}_{\text{bi}} - \bar{V}_{\text{ap}}}{2}\right)} = \frac{1}{2\sinh\left(\frac{\bar{V}_{\text{bi}} - \bar{V}_{\text{ap}}}{2}\right)},$$

so that eqn (9.59) can now be rewritten as:

$$\bar{p} = \frac{\exp(-\bar{\phi})}{2\sinh\left(\frac{\bar{V}_{\text{bi}} - \bar{V}_{\text{ap}}}{2}\right)}, \quad \bar{n} = \frac{\exp(\bar{\phi})}{2\sinh\left(\frac{\bar{V}_{\text{bi}} - \bar{V}_{\text{ap}}}{2}\right)}. \tag{9.60}$$

Substituting for $\bar{p}$ and $\bar{n}$ in eqn (9.56), we find the following Poisson–Boltzmann equation for $\bar{\phi}$:

$$\frac{\partial^2 \bar{\phi}}{\partial \bar{x}^2} = \frac{1}{\lambda^2}\left(\frac{\sinh(\bar{\phi})}{\sinh\left(\frac{\bar{V}_{\text{bi}} - \bar{V}_{\text{ap}}}{2}\right)} + 1\right), \tag{9.61}$$

with boundary conditions:

$$\bar{\phi}\big|_{\bar{x}=0}=0 \quad \text{and} \quad \bar{\phi}\big|_{\bar{x}=1}=-\frac{\overline{V_{\text{bi}}}-\overline{V_{\text{ap}}}}{2}. \tag{9.62}$$

9.4.4.6 *The Small λ Limit*

Given $\lambda \ll 1$, the term in the brackets on the right-hand side of eqn (9.61) must be very small, *i.e.*

$$\frac{\partial^2\bar{\phi}}{\partial\bar{x}^2}=\frac{1}{\lambda^2}\left[\frac{\sinh(\bar{\phi})}{\sinh\left(\dfrac{\overline{V_{\text{bi}}}-\overline{V_{\text{ap}}}}{2}\right)}+1\right]$$

The only solution to this approximate equation is

$$\bar{\phi}\sim-\frac{\overline{V_{\text{bi}}}-\overline{V_{\text{ap}}}}{2} \tag{9.63}$$

which obviously satisfies the second of the boundary conditions in eqn (9.62), but not the first. The problem here is that we cannot expect to neglect the highest derivative in an ODE and still satisfy its boundary conditions, so that our naive approximation in eqn (9.63) fails. Problems in which we are tempted to neglect the highest derivative, but get into trouble by doing so, are called singular perturbation problems. And, as we shall see, their solutions often have boundary layers in which the solution varies rapidly. Indeed, as we shall show, this is precisely what happens here and

$$\bar{\phi}\sim-\frac{\overline{V_{\text{bi}}}-\overline{V_{\text{ap}}}}{2} \quad \text{for} \quad \lambda \ll x<1. \tag{9.64}$$

That is, the approximation only fails in a narrow region of size $O(\lambda)$ around $\bar{x}=0$ (this is illustrated in Figure 9.4).

If the solution is indeed as we have postulated in Figure 9.4, then we ought to be able to investigate the rapid transition from $\bar{\phi}=0$ at $\bar{x}=0$ to the approximate solution of eqn (9.64) for $\bar{x}\gg\lambda$, by rescaling $\bar{x}$ with λ by writing $\bar{x}=\lambda z$

On making this substitution into eqn (9.61) and (9.62), we get

$$\frac{\partial^2\bar{\phi}}{\partial z^2}=\left(\frac{\sinh(\bar{\phi})}{\sinh\left(\dfrac{\bar{V}_{\text{bi}}-\bar{V}_{\text{ap}}}{2}\right)}+1\right) \quad \text{with} \quad \bar{\phi}\big|_{z=0}=0, \tag{9.65}$$

And, if it is indeed to tend towards the approximate solution of eqn (9.64), for large z, we must impose the far-field condition

$$\bar{\phi}\rightarrow-\frac{\bar{V}_{\text{bi}}-\bar{V}_{\text{ap}}}{2} \quad \text{as} \quad z\rightarrow+\infty. \tag{9.66}$$

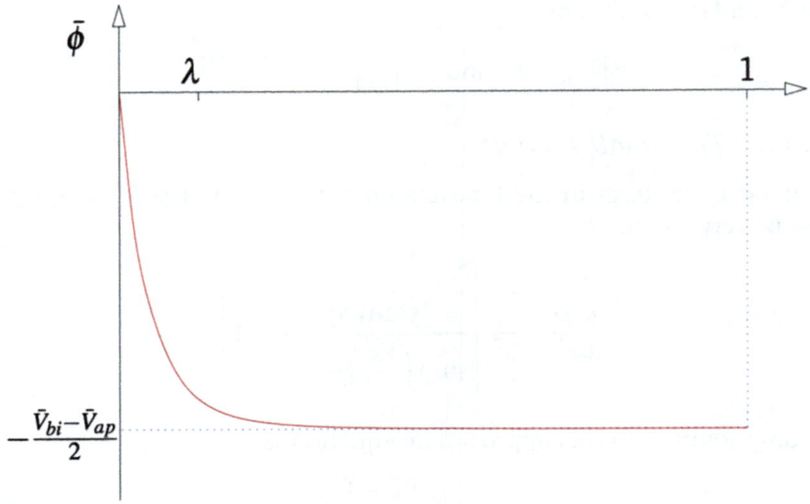

Figure 9.4 Illustration of the solution of eqn (9.61) and (9.62) for $\bar{\phi}$ with $\lambda \ll 1$.

Provided one believes that there is a solution to eqn (9.65) satisfying the far-field condition in eqn (9.66), it is now fairly easy to find a good approximation to the current–voltage curve for the device. In fact, it turns out that, by using a phase-plane analysis, it is relatively straightforward to show that such a solution exists, provided that $\bar{V}_{\text{bi}} - \bar{V}_{\text{ap}} > 0$.

As $\bar{n}$ and $\bar{p}$ are known nearly everywhere in $0 < \bar{x} < 1$, except very close to $\bar{x} = 0$, we can get a very good approximation to the amount of recombination occurring, and hence work out a current–voltage curve. Since $\bar{\phi} \sim (\overline{V_{\text{ap}}} - \overline{V_{\text{bi}}})/2$ in almost all of $0 < \bar{x} < 1$, it follows from eqn (9.60) that

$$\bar{n} \sim \frac{\exp\left(\dfrac{\overline{V_{\text{bi}}} - \overline{V_{\text{ap}}}}{2}\right)}{2\sinh\left(\dfrac{\overline{V_{\text{bi}}} - \overline{V_{\text{ap}}}}{2}\right)}, \quad \bar{p} \sim \frac{\exp\left(\dfrac{\overline{V_{\text{bi}}} - \overline{V_{\text{ap}}}}{2}\right)}{2\sinh\left(\dfrac{\overline{V_{\text{bi}}} - \overline{V_{\text{ap}}}}{2}\right)} \tag{9.67}$$

9.4.4.7 *The Built-in Voltage* $\bar{V}_{bi}$

When there is zero solar generation $\bar{G} \equiv 0$ and zero applied voltage $\bar{V}_{\text{ap}} = 0$, and the device is allowed to reach equilibrium, $\bar{j}_n = \bar{j}_p = 0$. These assumptions combined with eqn (9.54) and (9.55) imply that $\bar{n}\bar{p} = N_i^2$. Subsitution of the values of $\bar{n}$ and $\bar{p}$, as determined in eqn (9.69) with $\bar{V}_{\text{ap}} = 0$ into this relation gives the equation for $\bar{V}_{\text{bi}}$:

$$\sinh^2\left(\frac{\bar{V}_{\text{bi}}}{2}\right) = \frac{1}{4N_i^2}.$$

N_i, the ratio of the intrinsic carrier density to the doping density, is very small, so $\bar{V}_{\text{bi}}$ is large and given (approximately) by

$$\overline{V_{\text{bi}}} \sim 2\ln\left(\frac{1}{N_i}\right), \quad \text{so} \quad V_{\text{bi}} = 2V_T \ln\left(\frac{N_{\text{dop}}}{n_i}\right) \tag{9.68}$$

which is the result we would get from looking at the band bending and lining up the Fermi levels.

We now look at the case where the generation term $\bar{G}$ (which is assumed constant) and the applied voltage $\bar{V}_{ap}$ are both non-zero, and attempt to calculate the total curren t density $\bar{j}=\bar{j}_n+\bar{j}_p$ as a function of the applied voltage. The equations and boundary conditions for the current densities become on substituting for $\bar{n}$ and $\bar{p}$ from eqn (9.67) in eqn (9.54), (9.55), (9.57) and (9.58),

$$\frac{\partial \bar{j}_p}{\partial \bar{x}} = \Theta\left(N_i^2 - \frac{1}{4\sinh^2\left(\frac{\bar{V}_{bi}-\bar{V}_{ap}}{2}\right)}\right) + \bar{G}, \tag{9.69}$$

$$\frac{\partial \bar{j}_n}{\partial \bar{x}} = -\Theta\left(N_i^2 - \frac{1}{4\sinh^2\left(\frac{\bar{V}_{bi}-\bar{V}_{ap}}{2}\right)}\right) - \bar{G}, \tag{9.70}$$

$$\bar{j}_p|_{\bar{x}=0} = \bar{j}_n|_{\bar{x}=0}, \quad \bar{j}_n|_{\bar{x}=0} = 0. \tag{9.71}$$

Notably, the terms on the right-hand side of eqn (9.69) and (9.70) are constant, so that they are strightforward to integrate. In particular,

$$[\bar{j}_p - \bar{j}_n]_{\bar{x}=0}^1 = 2\left(\bar{G} - \Theta\left(\frac{1}{4\sinh^2\left(\frac{\bar{V}_{bi}-\bar{V}_{ap}}{2}\right)} - N_i^2\right)\right)$$

so that, since $\bar{j}_p - \bar{j}_n|_{\bar{x}=0} = 0$ from eqn (9.57) and $\bar{j}_n|_{\bar{x}=0} = 0$ from eqn (9.58), it follows that the dimensionless current density $\bar{j} = \bar{j}_n = \bar{j}_p$ is given by

$$\bar{j} = \bar{j}_p|_{\bar{x}=1} = 2\left(\bar{G} - \Theta\left(\frac{1}{4\sinh^2\left(\frac{\bar{V}_{bi}-\bar{V}_{ap}}{2}\right)} - N_i^2\right)\right).$$

Furthermore, since $\bar{V}_{bi} - \bar{V}_{ap}$ is typically large (*i.e.* in dimensional terms many thermal voltages), we can approximate $4((\bar{V}_{bi} - \bar{V}_{ap})/2) \sim \exp(\bar{V}_{bi} - \bar{V}_{ap})$ so that, on substituting for V_{bi} from eqn (9.70), we obtain

$$\bar{j} \sim 2\left[\bar{G} - N_i^2\Theta(\exp\{\overline{V_{ap}}\} - 1)\right] \tag{9.72}$$

Rewriting this in terms of the original dimensional variables, using the definitions of the rescaled variables above eqn (9.47) and (9.53), we find

$$j = 2qL\left(G - \gamma n_i^2\left(\exp\left(\frac{V_{ap}}{V_T}\right) - 1\right)\right). \tag{9.73}$$

As we shall show in Section 9.5, this corresponds precisely to the current–voltage relation of a current source strength $2qLG$ in parallel with an ideal diode (ideality factor 1) with reverse saturation current density $2qL\gamma n_i^2$.

9.5 Shockley Equivalent Circuit Models of Photovoltaic Devices

Drift diffusion models of charge transport, although they are capable of accurately describing the behaviour of photovoltaics, have the disadvantage of being complicated and hard to solve. Indeed, numerical solutions of these models in the right parameter regimes are frequently very difficult to obtain with any degree of accuracy. When comparing modelling with experimental data, it is often easier to make use of phenomenological equivalent circuit models. At their very simplest, these consist of a current source term representing the current generated by solar radiation in parallel with a diode (see Figure 9.5(a)) which represents the effects of charge recombination. Indeed, as we shall show, the current–voltage curve that we obtained from our approximate analysis of a drift diffusion model of an n–p homojunction (eqn (9.73)) is precisely of this form.

9.5.1 The Current–Voltage Curve of a Diode

The current I through a diode is related to the voltage V across it by the relation

$$I = I_s\left[\exp\left(\frac{V}{N_{id}V_T}\right) - 1\right], \tag{9.74}$$

where I_s and N_{id} are termed the reverse saturation current and the ideality factor respectively. Since this relation is asymmetric, the diode (as a device) has directionality as illustrated in Figure 9.6.

9.5.1.1 *The Simple Shockley Equivalent Circuit (Figure 9.5(a))*

We now use the diode current–voltage curve, defined in eqn (9.74), in the simple equivalent circuit in Figure 9.5(a). Note that the diode in Figure 9.5(a)

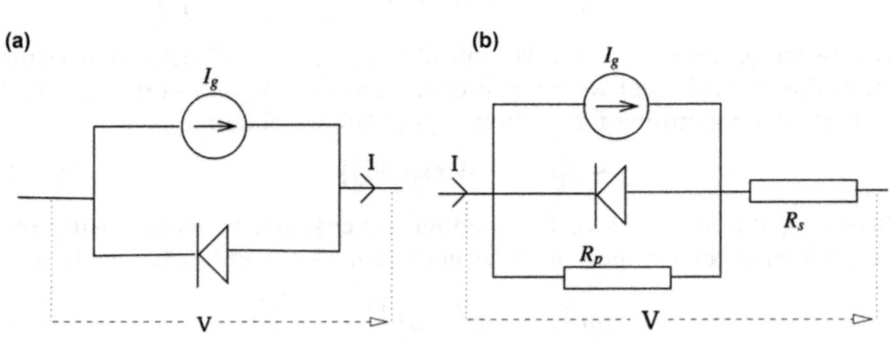

Figure 9.5 Shockley equivalent circuits for a solar cell. (a) Simple circuit for an n–p diode, (b) the generic Shockley equivalent circuit.

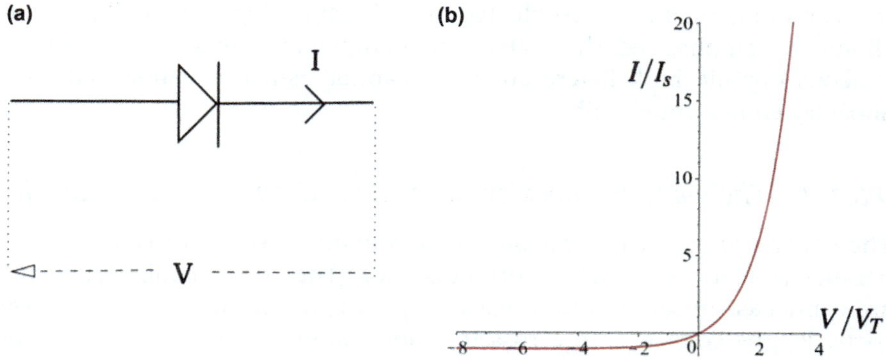

Figure 9.6 (a) Illustration of the direction of the current flow I through, and voltage V across, a diode. (b) The current–voltage curve for a diode with ideality $N_{\rm id} = 1$.

points in the opposite direction to that in Figure 9.6. The current flowing through the diode, from left to right, is thus

$$I_{\rm diode} = -I_{\rm s}\left[\exp\left(\frac{V}{N_{\rm id}V_T}\right) - 1\right].$$

Adding this current to the current $I_{\rm g}$ from the current source gives the total current I flowing through the circuit

$$I = I_{\rm g} - I_{\rm s}\left[\exp\left(\frac{V}{N_{\rm id}V_T}\right) - 1\right]. \tag{9.75}$$

Note that this current–voltage curve has exactly the same form as the current–voltage curve we predicted from the n-p homojunction in Section 9.4.4, *i.e.* eqn (9.73), provided we identify

$$I_{\rm g} = 2AqLG, \quad I_{\rm g} = 2Aq\gamma n_i^2, \quad N_{\rm id} = 1.$$

Here, A is the area of the homojunction so that $I = jA$. Kirchartz *et al.*[28] have shown how to link $N_{\rm id}$ to recombination rates. Bimolecular recombination gives $N_{\rm id} = 1$, SRH recombination gives $N_{\rm id} = 2$ and Auger recombination, $N_{\rm id} = 2/3$.

9.5.1.2 Other Devices

Whilst our analysis of the n–p homojunction in Section 9.4.4 shows that it is appropriate to fit current–voltage data for such a device to an equivalent circuit model, it is not obvious that this approach is appropriate for other devices. Roughly speaking, if the device has planar material and doping interfaces that run parallel to the contacts, an equivalent circuit description will provide a consistent approximation of its behaviour. In ref. 29, the analysis of a planar organic bulk heterojunction results in a current–voltage curve that is well approximated by a Shockley equivalent circuit, whilst in ref. 31 a similar analysis is conducted for a planar (tri-layer) perovskite cell. It should be emphasised that cells with non-planar geometry (*e.g.* heterogeneous organic bulk heterojunctions) cannot usually be modelled by a Shockley equivalent circuit.

9.5.1.3 The Generic Shockley Equivalent Circuit (Figure 9.5(b))

The generic Shockley equivalent circuit contains a shunt resistance R_p in parallel with a diode and a current source. The three parallel elements are connected in series with a resistor R_s. By splitting the voltage into two parts: V_1 (the voltage drop across the diode, shunt resistance and current source) and V_2 the voltage drop across R_s, we obtain the following set of equations

$$V = V_1 + V_2,$$

$$I_{\text{diode}} = -I_s\left[\exp\left(\frac{V_1}{N_{\text{id}} V_T}\right) - 1\right] \quad \text{current through diode,}$$

$$I_{\text{shunt}} = -\frac{V_1}{R_p} \quad \text{current through shunt resistor,}$$

$$I = I_g + I_{\text{shunt}} + I_{\text{diode}} \quad \text{conservation of current (Kirchoff),}$$

$$V_2 = -IR_s \quad \text{current through series resistor.}$$

It is straightforward to combine these expressions to obtain a single implicit expression for the current–voltage curve:

$$I = I_g - I_s\left[\exp\left(\frac{V + IR_s}{N_{\text{id}} V_T}\right) - 1\right] - \frac{V + IR_s}{R_p}. \tag{9.76}$$

9.5.1.4 The Physical Basis for Series and Shunt Resistances

We have already seen that charge recombination in an n–p homojunction leads to diode-like behaviour. So what processes give rise to the shunt resistance and series resistance used to model real photovoltaic devices? The series resistance R_s can arise as a result of the resistance of the contacts or because of the resistance of the semiconductor itself. The latter effect can be fairly readily incorporated into the analysis that we conducted in Section 9.4.4. The shunt resistance is, in some ways, more interesting because it usually arises from shorting between the contacts. This might be because there are regions where one of the doped layers is absent, and the other spans the entire width of the cell, or it may be because dendrites have grown directly from one contact to the other. In either case, a low value of R_p can be attributed to some fault in the manufacture of the cell. This is therefore clearly something to look out for when fitting current–voltage curves to an equivalent circuit.

9.5.1.5 Use of Equivalent Circuits Models to Fit Real Data

Here, we show a couple of examples of fits to real current–voltage data. In Figure 9.7, we show fitted data taken from Potscavage *et al.*[30] for current–voltage curves measured from three different designs of bilayer heterojunction organic solar cells. In the cases of the data plotted with triangles and squares, the ideality factor $N_{id} \approx 2$ corresponds to SRH recombination, while in the case of data plotted with circles, $N_{id} \approx 1.7$.

In Figure 9.8, we show a comparison of an experimental current–voltage curve obtained from a perovskite cell and the fit using a simple equivalent circuit (taken from ref. 31). Here the ideality factor is $N_{id} \approx 3$. Whilst this high ideality factor might be attributable to Auger recombination, we believe it is more likely related to the hysteretic behaviour of perovskite cells, discussed in Section 9.6.

9.6 Inclusion of Defect-mediated Ion Motion

As noted in the introduction, diffusion of mobile ions aided by defects, such as vacancy assisted iodide ion diffusion,[19] is a likely explanation for hysteresis in perovskite cells. The vacancies behave in a similar way to mobile ions in an electrolyte and create an electrical double layer. A discussion

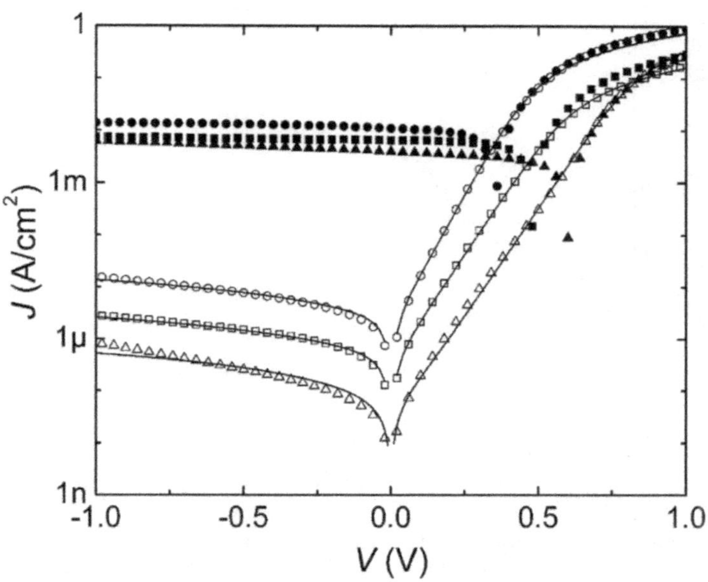

Figure 9.7 Fitting of an equivalent circuit model to current–voltage data from bilayer heterojunction organic solar cells. Filled symbols represent cells under illumination, open symbols cells in the dark. Notice the current has been plotted on a log scale.
Reprinted with permission from W. J. Potscavage, S. Yoo and B. Kippelen, *Appl. Phys. Lett.*, 2008, **93**, 193308. Copyright 2008, AIP Publishing LLC.[30]

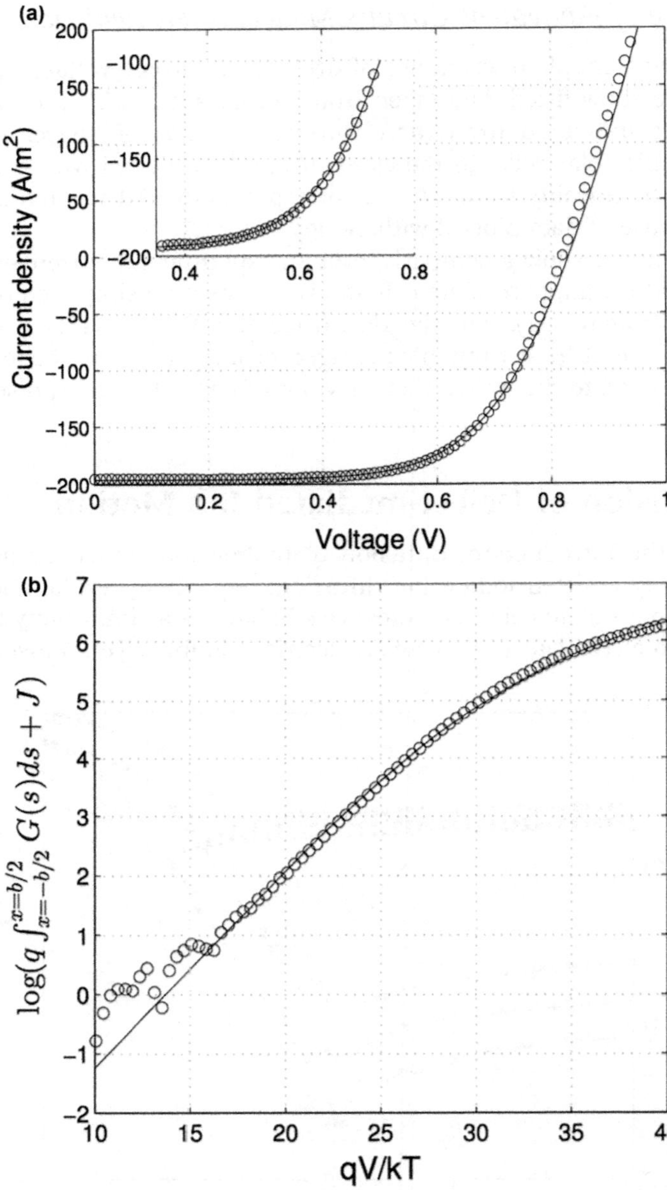

Figure 9.8 Fitting of an equivalent circuit model to current–voltage data from a perovskite cell prepared by Giles Eperon. From the data it is inferred that the ideality factor $N_{id} \approx 3$. For more details see ref. 31. Note that the current has been defined in an opposite sense to that in the text – to agree with the text the figure should be flipped about the V-axis.
Reprinted from J. M. Foster, H. J. Snaith, T. Leijtens and G. Richardson, *SIAM J. Appl. Math.*, 2014, **74**, 1935. [31]

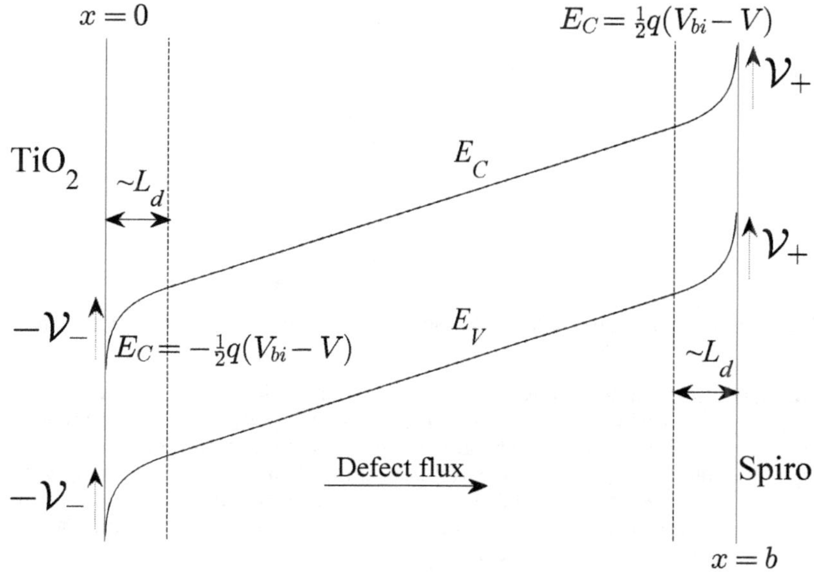

Figure 9.9 Schematic representation of the conduction band edge profile $E_C(x)$ and valence band profile $E_V(x)$. We assume the applied bias V is zero under short circuit conditions and is positive on the spiro side. The positions of the Debye layers adjacent to the interfaces with the blocking layers are indicated. The defect flux direction assumes positively charged defects. Reproduced from ref. 20 with permission from the Royal Society of Chemistry.[20]

of the electrostatic interactions in a double layer formed by mobile ions in an electrolyte is provided in the textbook by Bockris and Reddy,[32] based on the ideas of Helmholtz in the 1850s. For perovskite cells, the double layer is created by positively charged vacancies accumulated in a region of the order of the Debye layer width on one side of the perovskite, and by a region where vacancy concentration falls below its equilibrium value with a net negative charge. This double layer causes band bending at the edges of the perovskite layer, illustrated in Figure 9.9.

Our model couples electrons, holes and defect mediated ion motion suggesting that slow moving ions are indeed the origin of the hysteresis. The method developed for solving the stiff equations in the drift diffusion model is widely applicable to other double layer problems occurring in electrochemical applications such as the evolution of transmembrane potentials in living cells.

Figure 9.10, panel (a) shows how $E_C(x)$ evolves as the bias V is scanned at a constant rate due to vacancy motion, and panel (b) shows how the vacancy concentration profile changes with time.

To circumvent the difficulties of a DD model that allows for the narrow boundary layers where defects pile up described in the introduction, an approximate approach is adopted based on asymptotic expansions. Matched asymptotic expansions similar to those described in ref. 31 have been used

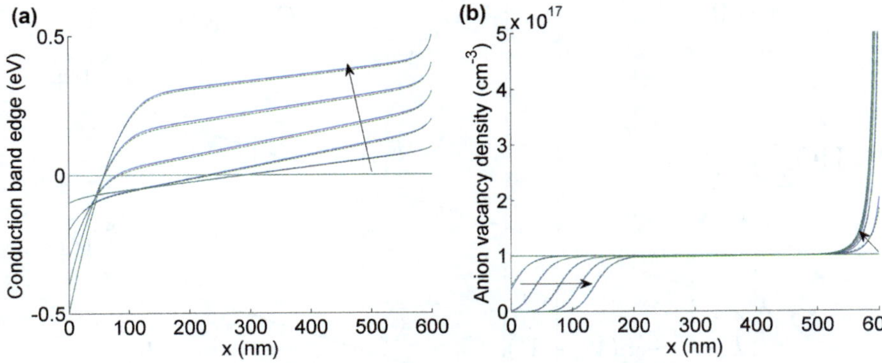

Figure 9.10 Comparison between the numerical (solid blue) and the asymptotic (dashed green) solutions to $E_C(x)$ (panel a) and I^- vacancy concentration profiles (panel b). The arrow indicates the direction of increasing time t as the applied voltage $V(t)$ is scanned from 1 V to 0 V taking values at $V = 0.8$, 0.6, 0.4, 0.2, 0 V. At $t = 0$, $E_C(x) = 0$ for all x. We assume an equilibrium vacancy concentration $N_0 = 1 \times 10^{17}$ cm^{-3} and a vacancy diffusion coefficient of 1×10^{-11} cm^2 s^{-1}.
Reproduced from ref. 20 with permission from the Royal Society of Chemistry.[20]

to derive a capacitance relation between the charge Q (per unit area) in the boundary layers and the potential drop across either layer as a function of time. The field in the perovskite layer between the boundary layers was obtained with the boundary layer potential drop used as inputs in a 1-D DD model for electron and hole densities (n and p) within the perovskite. The generation rate used in the model is the Beer–Lambert law while charge recombination was assumed to obey Shockley–Read–Hall theory.

The numerical solution to the DD model was performed using the package Chebfun.[33] This scheme is an example of a spectral method. Here, a grid spacing is not fixed, unlike in the finite difference approach. Our method makes use of adaptive procedures that aim to find the required number of collocation points automatically so as to represent each function to roughly machine precision. The solution to the DD model yields electron and hole current densities j_n and j_p, respectively, which are used to obtain an expression for the total current density $J(t)$. When plotted against the applied voltage $V(t)$ this yields the hysteresis curves.

Figure 9.11(a) shows a series of snapshots of the solution at evenly spaced time intervals in the scans down from $V = 1$ V to $V = 0$ and back up from $V = 0$ up to $V = 1$ V, respectively. Features of the current–voltage curves at varying scan rates can be explained in terms of a time varying internal electric field across the perovskite layer. We have compared the predictions of our model with experimental measurements on a cell *chosen* to demonstrate high levels of hysteresis in Figure 9.11(b). The agreement between these results and the model solutions is very good. Deviation between the experimental and theoretical results is most significant for the fastest scan rate 1 V s^{-1} and when the applied voltage is increased from 0 V to 1 V. The

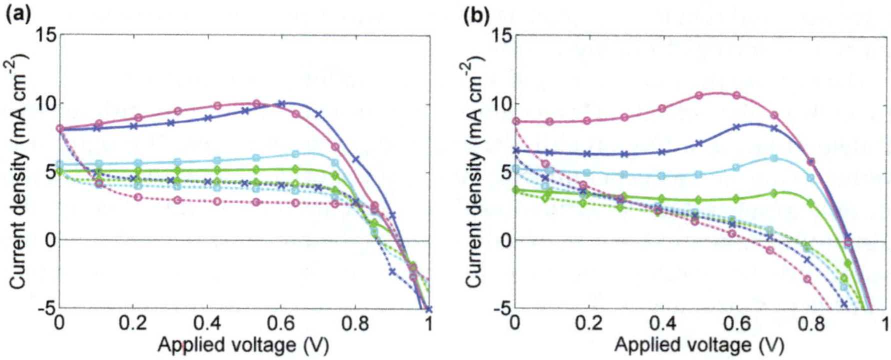

Figure 9.11 (a) Calculated current–voltage curves without allowing for preconditioning; (b) measured current–voltage curves for cell 1 after 2 previous scan cycles and preconditioning for 5 s at 1.2 V. Solid lines show the 1.2 V–0 V scan; broken lines show the 0 V–1.2 V scan. Scan rates are 1 V s^{-1} (magenta, circles), 500 mV s^{-1} (blue, crosses), 250 mV s^{-1} (cyan, squares), 100 mV s^{-1} (green, diamonds).
Reproduced from ref. 30 with permission from the Royal Society of Chemistry.[20]

deviation can be understood from the need to model cell preconditioning accurately.

The moving ion hypothesis provides a strong explanation for the influence of the scan rate. For slower scan rates, the ions/vacancies have more time to respond to the change in the applied voltage than for faster scans; the ionic charge thus screens the bulk electric field more effectively on slower scans than on fast ones, and thus the field in the bulk of the perovskite layer remains relatively small on such slow scans and results in a limited hysteresis. For intermediate scan rates the hysteresis is the strongest and, at the other extreme, if the scans are sufficiently fast, the ions do not have a chance to build up in the Debye layers, and so there is very limited rescreening and consequently little hysteresis (and increased short circuit current).

9.7 Conclusions and Outlook

This chapter has shown the importance of charge transport modelling in perovskite cells, noting that, to date, only a handful of papers has been published on this topic. We have indicated how a DD model forms the basis of a multiscale model, provided that the DD model can handle the varying length and time scales required. Much of the chapter covers the ideas behind DD modelling for solar cells and the mathematical tricks such as rescaling and asymptotical expansions. We have shown how equivalent circuit parameters are obtained from the model, allowing interpretation of the current–voltage characteristics. The final section shows that our DD model can work at the length and time scales needed to consider the influence of mobile ions on the current–voltage characteristics, and that we have achieved remarkably good agreement with experimental measurements on cells showing strong hysteresis.

Having laid the foundations of a modelling effort, we will refine the model to look at the effects of preconditioning and to model explicitly charge transport layers sitting on either side of the perovskite layer. Through these refinements, and a deeper understanding of the origins of hysteresis, we will be able to suggest device structures that reduce hysteresis and so reduce the amount of trial and error involved in finding better cells. We can therefore speed up the development of perovskite cells with improved and reproducible performance.

Acknowledgements

We would like to thank Jamie Foster, Nicola Courtier and Simon O'Kane for many useful discussions and for reading this chapter. The research leading to these results has received funding from the European Union Seventh Framework Programme [FP7/2007-2013] under grant agreement 316494 and from the UK Physical and Engineering Research Council grant EP/J017361/1.

References

1. J. Gong, S. B. Darling and F. You, *Energy Environ. Sci.*, 2015, **8**, 1953.
2. F. De Angelis, *chapter in this volume*, 2015.
3. J. M. Frost, K. T. Butler, F. Brivio, C. H. Hendon, M. van Schilfgaarde and A. Walsh, *Nano Lett.*, 2014, **14**, 2584.
4. E. Mosconi, A. Amat, M. K. Nazeeruddin, M. Grätzel and F. De Angelis, *J. Phys. Chem. C*, 2013, **117**, 13902.
5. N. E. Jackson, B. M. Savoie, L. X. Chen and M. A. Ratner, *J. Phys. Chem. Lett.*, 2015, **6**, 1018.
6. N. E. Jackson, B. M. Savoie, T. J. Marks, L. X. Chen and M. A. Ratner, *J. Phys. Chem. Lett.*, 2015, **6**, 77.
7. H. J. Snaith, A. Abate, J. M. Ball, G. E. Eperon, T. Leijtens, N. K. Noel, S. D. Stranks, J. T. W. Wang, K. Wojciechowski and W. Zhang, *J. Phys. Chem. Lett.*, 2014, **5**, 1511–1515.
8. R. S. Sanchez, V. Gonzalez-Pedro, J. W. Lee, N. G. Park, Y. S. Kang, I. Mora-Sero and J. Bisquert, *J. Phys. Chem. Lett.*, 2014, **5**, 2357–2363.
9. E. L. Unger, E. T. Hoke, C. D. Bailie, W. H. Nguyen, A. R. Bowring, T. Heumuller, M. G. Christoforo and M. D. McGehee, *Energy Environ. Sci.*, 2014, **7**, 3690–3698.
10. W. Tress, N. Marinova, T. Moehl, S. M. Zakeeruddin, M. K. Nazeeruddin and M. Grätzel, *Energy Environ. Sci.*, 2015, **8**, 995–1004.
11. T. C. Sum, S. Chen, G. Xing, X. Liu and B. Wu, *Nanotechnology*, 2015, **26**, 342001.
12. X. Li, X. Wang, W. Zhang, Y. Wu, F. Gao and J. Fang, *Org. Electron.*, 2015, **18**, 107.

13. J. Beilsten-Edmands, G. E. Eperon, R. D. Johnson, H. J. Snaith and P. G. Radaelli, *Appl. Phys. Lett.*, 2015, **106**.
14. T. Minemoto and M. Murata, *J Appl. Phys.*, 2014, **116**, 054505.
15. F. Liu, J. Zhu, J. Wei, Y Li, M. Lv, S. Yang, B. Zhang, J. Yao and S. Dai, *Appl. Phys. Lett.*, 2014, **104**, 253508.
16. X. He, K. Jin, C. Ge, C. Wang, H. Lu and G. Yang, *Europhys. Lett.*, 2013, **102**, 37007.
17. S. Van Reenen, M. Kemerink and H. J. Snaith, *J. Phys. Chem. Lett.*, 2015, **6**, 77.
18. A. Walsh, D. O. Scanlon, S. Chen, X. G. Gong and S.-H. Wei, *Angew. Chem.*, 2015, **127**, 1791.
19. C. Eames, J. M. Frost, P. R. F. Barnes, B. C. O'Regan, A. Walsh and M. S. Islam, *Nat. Commun.*, 2015, **6**, 7497.
20. G. Richardson, S. E. J. O'Kane, R. G. Niemann, T. A. Peltola, J. M. Foster, P. J. Cameron and A. B. Walker, *Energy Environ. Sci.*, 2016, DOI: 10.1039/C5EE02740C.
21. D. Bartesaghi, N. J. van der Kaap and L. J. A. Koster, *3D simulations of organic solar cells, chapter in this volume*, 2015.
22. J. Nelson, *The Physics of Solar Cells'*, Imperial College Press, London, 2003, ch. 4.
23. D. T. Gillespie, *J. Phys. Chem.*, 1977, **81**, 2340.
24. A. B. Walker, *Topics in Current Chemistry*, Springer-Verlag, Vienna, 2013.
25. R. P. Feynman, R. B. Leighton and M. Sands, *Lectures on Physics*, Addison Wesley, New York, 2003, vol. III, ch. 13–14.
26. T. Kirchartz, B. E. Pieters, J. Kirkpatrick, U. Rau and J. Nelson, *Phys. Rev. B: Condens. Matter Mater. Phys.*, 2011, **83**, 115209.
27. A. Schenk and U. Krumbein, *J. Appl. Phys.*, 1995, **78**, 3185.
28. T. Kirchartz, F. Deledalle, P. S. Tuladhar, J. R. Durrant and J. Nelson, *J. Phys. Chem. Lett.*, 2013, **4**, 2371.
29. J. M. Foster, J. Kirkpatrick and G. Richardson, *J. Appl. Phys.*, 2013, **114**, 104501.
30. W. J. Potscavage, S. Yoo and B. Kippelen, *Appl. Phys. Lett.*, 2008, **93**, 193308.
31. J. M. Foster, H. J. Snaith, T. Leijtens and G. Richardson, *SIAM J. Appl. Math.*, 2014, **74**, 1935.
32. J. O'M Bockris, A. K. N Reddy and M. Gamboa-Aldeco, *Modern Electrochemistry 2A*, Springer, US, New York, 2000.
33. T. A. Driscoll, N. Hale and L. N. Trefethen, *Chebfun Guide*, Panufty Publications, Oxford, 2014.

CHAPTER 10

Small Molecule Organic Solar Cells

JOHANNES WIDMER, CHRISTIAN KÖRNER AND KARL LEO*

Institut für Angewandte Photophysik, Technische Universität Dresden, George-Bähr-Straße 1, 01069 Dresden, Germany
*Email: karl.leo@iapp.de

10.1 Introduction

Renewable energies are currently being investigated intensively due to limited fossil fuel resources and climate change. Among the technologies available, photovoltaics (PV) has by far the largest potential for terrestrial use. Although the standard technology based on crystalline silicon has made impressive progress in terms of efficiency and cost, thin-film alternatives are the subject of intensive investigation. Among those, organic PV devices show great promise since they are almost completely based on carbon and can be manufactured in very efficient roll-to-roll techniques at room temperature. Furthermore, organic solar cells are flexible, light-weight, and potentially transparent, which opens up completely new applications, *e.g.*, in building-integrated PV and in transportation. Serious research on organic PV devices started in 1986 with the realization of the first reasonably efficient thin film cell by Tang.[1] The key innovation was the heterojunction employing a donor and an acceptor material for exciton dissociation, which was later extended to bulk heterojunctions with mixed donor and acceptor.[2,3] Since then, the efficiency and lifetime of organic PV have increased continuously. The current state of the art (which means photovoltaic devices >1 cm^2 and certification by an independent lab) is represented by a 13.2% efficiency cell by

RSC Energy and Environment Series No. 16
Unconventional Thin Film Photovoltaics
Edited by Enrico Da Como, Filippo De Angelis, Henry Snaith and Alison Walker

Published by the Royal Society of Chemistry, www.rsc.org

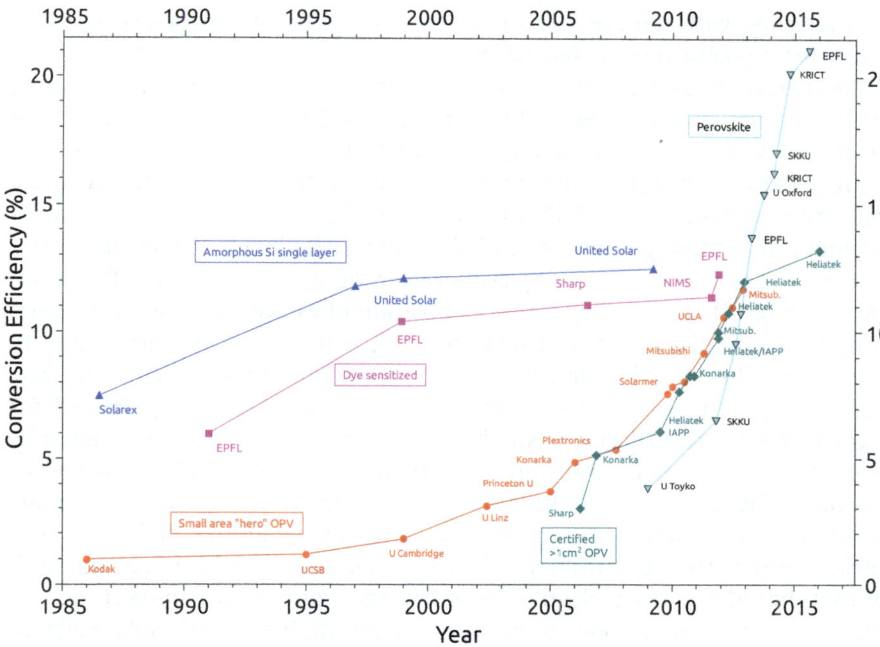

Figure 10.1 Development of record efficiency *vs.* time for organic solar cells and some other thin-film technologies.[7]

Heliatek.[4] Figure 10.1 shows the development of record efficiencies over time. It is obvious that the growth in efficiency of organic PV was rather slow for some time. However, in the past few years, the growth in efficiency has accelerated rapidly due to an improved understanding of the working principles and the discovery of many new materials. However, a new phase of stagnation can be observed since the end of 2012. Recent estimates indicate that single junction cells should be able to reach about 15% and tandem cells about 20% efficiency.[5,6] With the current record value of 13.2%, the performance of organic solar cells is comparable to that of amorphous silicon and dye-sensitized solar cells.

It should be mentioned that solar cells based on organic–inorganic perovskites, which are discussed elsewhere in this volume, have recently achieved spectacular advances and have displayed the most rapid growth in efficiencies ever observed in the field of PV. Within a few years of development of the first cells, efficiencies of more than 20% have been achieved, putting perovskite cells alongside the best inorganic thin film PV systems such as CIGS and CdTe. Despite the rapid progress, there are still several open questions. For instance, the only perovskite material class that has so far attained these excellent efficiencies is based on lead, which is a critical issue in terms of acceptance as a "green" PV technology. Attempts to replace lead have been largely unsuccessful so far. Also, these organic–inorganic perovskites suffer from low thermal stability, and it remains to be seen whether they will achieve the lifetimes needed for an extensive outdoor usage (for a recent review, see ref. 8).

The field of organic PV can be separated in different subfields, depending on the molecular structures and technology used to deposit the films, with the majority of researchers using solution-based technologies, mostly spin coating. These techniques have the advantage that they can be scaled to large-area printing production. Solution-based coating can be done either with polymers[9] or with oligomers (so-called small molecules).[10,11] The alternative approach is to prepare layers by evaporation of small molecules. The main advantage of this technology is that deposition of multilayer systems is more straightforward than with wet technologies, where problems arise due to dissolution of underlying layers.[12] The evaporation approach helps in basic research since it is possible to design samples specifically optimized for characterization techniques. Also, the experience with the only commercially successful organic electronics product so far, the OLED display, has shown that the ability to deposit multi-layers in a dry solvent-free process has many advantages for a stable mass production process, which has led to a 100% market share for this technology.[13,14] Nevertheless, it is highly likely that both solution-based and dry technologies will coexist, at least for some time.

In this chapter, we review our recent work on small-molecule solar cells. In particular, we address the p-i-n-structure, which allows solar cell layer stacks to be designed with almost unlimited freedom with regards to optical design and choice of electrodes. The doping of organic materials by co-evaporation of dopant molecules is reviewed, and the materials relevant for doped layers in solar cells are discussed. Furthermore, the p-i-n structure is ideally suited to realize multi-junction devices since the doped layers allow stacking of several junctions without further design restrictions. The chapter is closed by a discussion of optimized solar cells with new thiophene-based absorbers that have been the basis for the current state of the art cells discussed above.

10.2 Device Structures

In this section, we discuss some of the basic structures for organic PV. In particular, we introduce the p-i-n structure and explain the motivation behind its use. In the subsequent discussion, we describe the realization of such structures by controlled doping and its influence on parameters such as the built-in voltage of organic solar cells.

10.2.1 The p-i-n Structure

For decades, the silicon solar cell has been the archetype of photovoltaic devices. Its functioning is based on a p–n junction that provides a built-in field separating the photo-generated electrons and holes (Figure 10.2 top). This picture was extended by the pioneering work of Würfel, who pointed out that a solar cell does not necessarily need to contain a p–n junction.[15] Instead, he introduced a generalized solar cell model that describes the solar

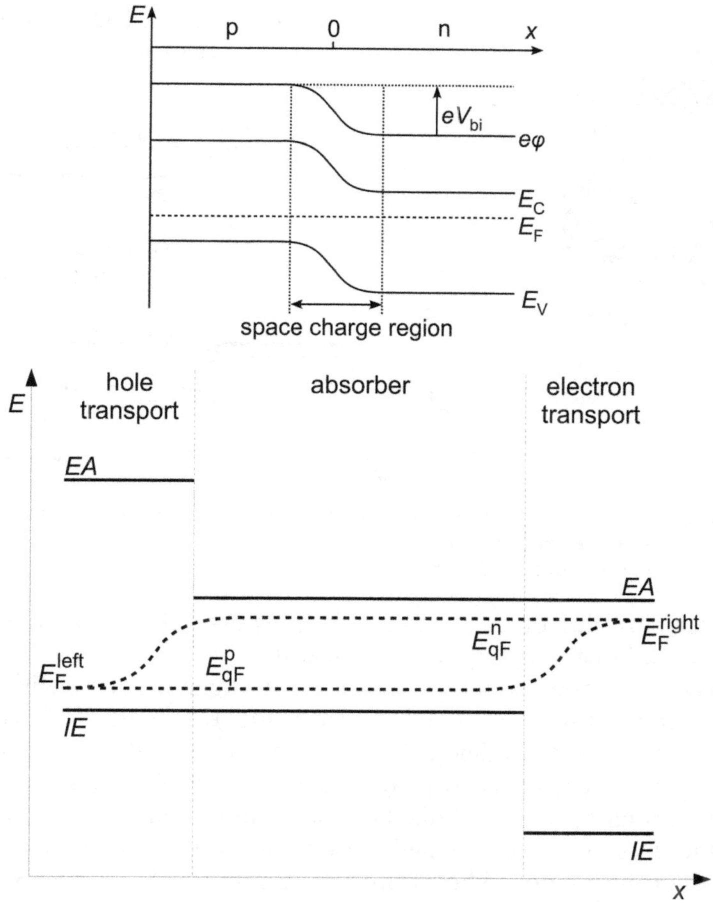

Figure 10.2 Top: classical solar cell scheme based on a p–n homojunction. (Reproduced with permission from ref. 18.) Bottom: Würfel concept of solar cells.[19] (Based on ref. 15.)

cell as a thermodynamic engine where an absorber layer is sandwiched between two selective outer contact layers, which allow only electrons to pass on one side and only holes on the other. In this picture (see Figure 10.2 bottom), doped layers and a built-in voltage are not needed. The open circuit voltage is given by the splitting between the electron and hole quasi-Fermi-levels E_F^{left} and E_F^{right}. If the Fermi levels are only controlled by the density of carriers in the absorber, carriers are transported preferentially by diffusion to the contacts. At the maximum power point, when the cells deliver a current as well as a voltage, the electric field is directed against the carrier transport. Thus, there must be large carrier concentrations in the centre of the cell, which leads to high recombination. We have proven in recent publications,[16,17] that the lack of a sufficient built-in potential leads to strongly S-shaped current–voltage curves with low fill factors, which results in much reduced power delivery (see Section 10.2.2 for a more detailed discussion). We conclude that, at least for the organic solar cells discussed

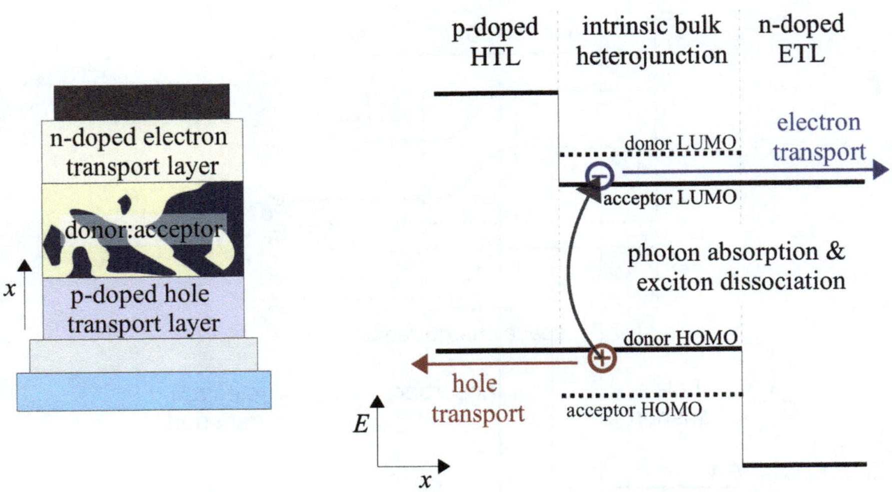

Figure 10.3 Scheme of a p-i-n-solar cell. Left: layer sequence. Right: simplified energy level diagram.
(Figure adapted from ref. 19.)

here, a certain value of the built-in voltage is essential to guarantee good currents for all operating regimes of our cells.

We have developed a p-i-n concept[20] for organic solar cells[21,22] that combines the advantages of the classical and the Würfel cell concepts. Figure 10.3 shows schematically the layer sequence and the energy diagram. The undoped ("intrinsic") active layer (which can be either a planar donor–acceptor arrangement or a bulk heterojunction containing a mixture of donor and acceptor) is embedded between an n-doped electron transport layer (n-ETL) and a p-doped hole transport layer (p-HTL). The energy levels of the ETL and HTL are chosen in such a way that the lowest unoccupied molecular orbital (LUMO) level of the absorber system continues without a barrier into the transport layer. For the holes, the highest occupied molecular orbital (HOMO) is arranged accordingly. Due to the significantly larger energy gap of ETL and HTL, the barriers for the "wrong" carrier are quite high so that the idealized picture of the Würfel model assuming selective "membranes" that are only transparent for one type of charge carrier is met rather well.

The p-i-n structure now combines several advantages:

- Due to the good energetic alignment and the high built-in voltage, carrier transport to the contacts is enhanced.
- The large band gaps of the transport layers make them transparent in the absorption region of the solar cells, thus, they can serve as "window" layers that can be used to optimize the thin-film optics of the cell.
- The highly-doped transport layers form a well-conducting contact to the electrodes, even if the work function of the electrode is not optimized.
- Finally, when p-i-n structures are stacked on top of each other, the p–n junction formed between them serves as a highly efficient and electrically,

as well as optically, almost lossless recombination contact, without the need for further measures such as metallic nanoparticle recombination centres.[23]

Finally, it should be mentioned that the p-i-n structure is probably used more frequently than is generally realized. For instance, many polymer solar cells are built on PEDOT:PSS, a highly p-type doped organic polymer. Furthermore, they only work well if a thin film of a lithium or caesium compound is deposited between the absorbing polymer and the metal cathode. It is well known that these materials decompose when the metal cathode is evaporated on top, with the Li or Cs acting as n-dopants, creating an n-type layer between the polymer and the electrode.

10.2.2 The Built-in Voltage

10.2.2.1 Origin of the Built-in Voltage in Organic Solar Cells

The high conductivities and well-defined Fermi levels E_F of the doped layers in a p-i-n type solar cell bring an important benefit for the intrinsic absorber layers. The built-in field F_{BI} is spatially confined to the electrically less conductive intrinsic layers with lower charge density. A major part of the gradient in the vacuum level, originating from the energy level alignment of the various semiconductor materials used in a device, drops over these layers. This enhanced built-in field supports exciton splitting at the donor–acceptor interface[24–27] as well as the selective extraction of photo-generated charges.[28] This makes the output current j less dependent on the external voltage V_{ext} and, thus, improves the fill factor FF of the solar cell.[29] In materials where intrinsic exciton splitting plays a significant role, F_{BI} can even enhance the photo-current by field-enhanced splitting of excitons.[25,30,31]

The evolution of the built-in field for an ideal p-i-n type solar cell is illustrated in Figure 10.4, which compares the theoretical "flat-band" case of the materials without contact with the external voltage-free (short-circuit, unilluminated) equilibrium case of the materials in contact. It is shown that the difference of the Fermi levels ΔE_F of the doped transport layers in the "flat band" case turns into the built-in voltage V_{BI} (also named built-in potential) in the short-circuit case. As a consequence, V_{BI} can be directly controlled by adjusting the Fermi levels E_F of the doped transport layers, *e.g.*, by means of varying the doping concentration, as will be discussed below.

An additional voltage V_{ext} at the device contacts can either be applied externally, or it may be caused by illuminating the device in a non-short-circuit situation, or both. In this case, the current flowing in the device is generally $j \neq 0$ and the Fermi level is not spatially constant, with the Fermi level difference at the contacts equalling the external voltage. The spatial distribution of the electric field F_{ext} caused by V_{ext} in the device is not spatially uniform. Especially in the highly conductive doped transport layers with high charge carrier densities, the voltage drop is very small compared to

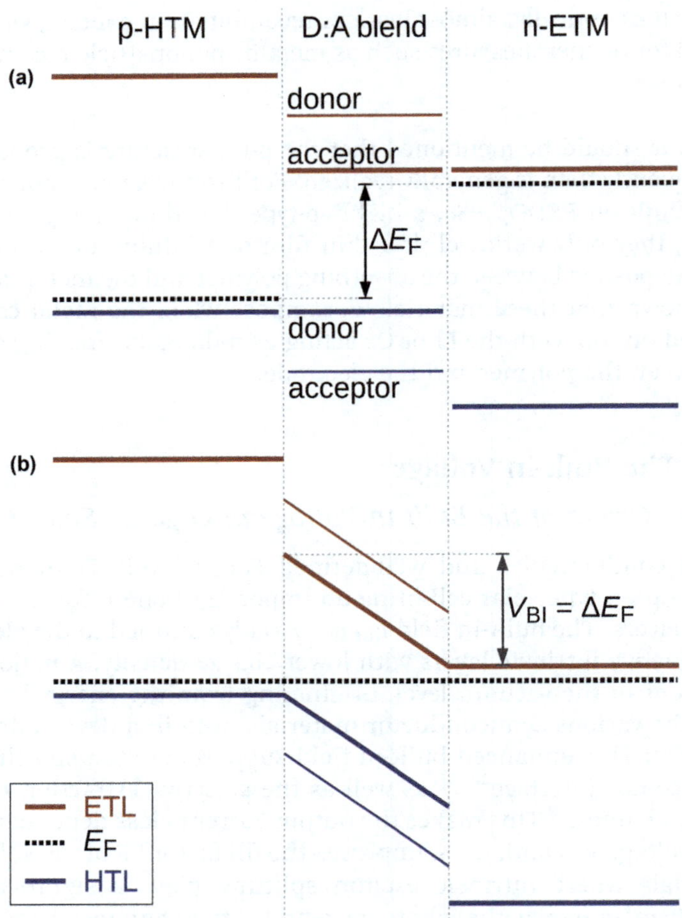

Figure 10.4 Origin of the built-in voltage V_{BI} in an ideal p-i-n type solar cell comprising a p-doped hole transport material (p-HTM, left), an intrinsic donor:acceptor (D:A) system* (middle), and an n-doped electron transport material (n-ETM, right). The materials are characterized by their transport gap between the electron transport level (ETL; red) and their hole transport level (HTL; blue), as well as their Fermi level (dotted dark grey). The D:A system is characterized by the transport gap between the electron transport level of the acceptor and the hole transport level of the donor. (a) In the "flat-band" case of the materials without contact, the difference between the Fermi levels of the doped layers is denoted as ΔE_F. (b) With the materials in contact in a voltage-free case, *i.e.* a common Fermi level, the voltage dropping over the D:A system equals ΔE_F and is denoted as the built-in voltage V_{BI}. * representing a blend, planar, or hybrid planar-blend heterojunction.

the less conductive intrinsic donor–acceptor (D:A) system. Therefore, it is a good approximation to neglect the voltage drop in the doped layers and to attribute the external voltage drop completely to the intrinsic D:A blend.

In consequence, the actual electric field in the blend arises from the internal voltage V_{int} equalling the sum of V_{BI} and V_{ext}.

$$V_{int} = V_{BI} + V_{ext} \tag{10.1}$$

10.2.2.2 Non-ideal Devices

In real devices, the alignment of the hole transport levels of donor and hole transport layer – as well as the alignment of the electron transport levels of acceptor and electron transport layer – may not be perfect.[32] However, these additional energy steps are typically small compared to V_{BI} and only lead to minor quantitative changes, leaving the qualitative origin and nature of V_{BI} unchanged. As a further simplification, Figure 10.4 shows the energy levels in the donor:acceptor (D:A) blend linearly decreasing, indicating a spatially constant F_{BI} inside the blend. However, it should be kept in mind that in a more detailed view obtained, for example, from drift-diffusion calculations, the spatial distribution of F_{BI} might be non-uniform, leading to a curved $V(x)$ profile.[33,34] The main reason for this bending is charge carrier diffusion from doped layers or other regions with a high charge carrier density into intrinsic layers with an originally lower charge density. However, the magnitude of V_{BI} is insensitive to this simplification, so that the preceding discussion remains valid. On this basis, we shall discuss the built-in voltage rather than the built-in field, and refer to the spatial distribution of the built-in field only where necessary for an adequate discussion. Unless stated otherwise, the spatially constant distribution of F_{BI} is assumed as an approximation, and F_{BI} can be related to V_{BI} *via*

$$F_{BI} \approx \frac{V_{BI}}{d} \tag{10.2}$$

with d being the effective thickness of the intrinsic D:A system.

In the case of larger energy steps, *e.g.* injection barriers $\gg k_B T$ between photo-active and transport layers, V_{BI} might be reduced by the transport levels of the doped layers. In this context, it is instructive to make a clear differentiation between photo-current and photo-voltage generation. While V_{BI} supports the extraction of charges, it is not primarily required for the build-up of a photo-voltage. This statement is readily demonstrated by observing that V_{OC} can well exceed V_{BI}, though in this case the FF is simultaneously reduced by an S-shape of the j–V curve.[16,35]

10.2.2.3 Bulk, Planar, and Hybrid Heterojunction

The discussion up to now has been for the case of a bulk heterojunction (BHJ), where the D:A blend was treated as one effective material with one effective dielectric constant. In the case of planar heterojunction (PHJ) or D:A systems with more than two layers, *e.g.* a hybrid planar bulk heterojunction consisting of a D:A blend layer sandwiched between neat D and A layers, the above discussion about V_{BI} is qualitatively equally valid. In this case, the D:A blend is replaced by a sequence of adjacent intrinsic layers representing the D:A system, and the precise partial V_{BI} in each of these layers depends on the thicknesses of the layers, the dielectric constants of the used materials, and the relative alignment of the energy levels in the stack.

10.2.2.4 Beyond p-i-n

The above discussion of V_{BI} in a p-i-n type device is equally valid for n-i-p type devices[16,22] and other, not purely organic, p-i-n type devices such as hybrid perovskite solar cells with energetically aligned p- and n-doped transport layers.[36,37] In contrast, generalization to other organic or thin-film solar cell architectures, such as m-i-p (m stands for metal in direct contact to the intrinsic D:A system), p-i-m (as often used for polymer devices), or m-i-m stack designs, needs further concretization. In an m-i-m stack, V_{BI} is defined by the work function difference ΔW of the used contact materials[38,39] and further details of the organic–metal interface.[40] In a simplified m-i-m picture with only one layer of an effective semiconductor system between conducting contacts with given work functions, V_{OC} is limited by ΔW.[41] This is in contrast to p-i-n devices, where V_{OC} can exceed ΔW of the metal contacts, due to the electric de-coupling by the doped layers.[33] In hybrid cases with doped and undoped contacts, the interplay of both effects has to be considered accordingly. Often, V_{BI} can validly be approximated as the difference between the Fermi levels adjacent to the least conductive layer(s). In many of these cases, however, the approximation of a spatially uniform F_{BI}, as assumed in eqn (10.2), is not a good assumption any more, due to charge carrier diffusion and consequent energy level deformation and other interface effects near undoped contacts.[42]

10.2.2.5 Measuring the Built-in Voltage

The effective local electric field F in a molecular material can be measured in an electroabsorption (EA) measurement by observing the Stark effect, which influences the absorption coefficient, α, of a molecule in the presence of an electric field.[43–46] Due to the superposition of the mesoscopic local field F on the microscopic energy landscape within one molecule, the molecule and its orbitals are deformed and, thus, the absorption is slightly shifted. In a material of randomly oriented molecules, this effect leads to an absorption change at particular wavelengths that is proportional to the square of the local mesoscopic electric field. This dependence can be exploited to measure the local field, since application of a small external AC signal, with amplitude ΔV and frequency ω, modulates the local field and induces a change in the absorption coefficient, $\Delta\alpha$, at the frequency ω, given approximately by

$$\Delta\alpha \underset{\sim}{\propto} \Delta V \cdot \frac{\partial F^2}{\partial V} \propto \Delta V \cdot F \tag{10.3}$$

The first proportionality in eqn (10.3) is a good approximation for small amplitudes ΔV.

An example of the measurement of the spectrally-resolved change in absorption at various external DC bias voltages is shown in Figure 10.5. The right-hand side proportionality in eqn (10.3) implies that the measurable change in absorption $\Delta\alpha$ is only present if the local field F is non-zero, *i.e.* even if the proportionality constants are unknown and quantitative evaluation of the strength of $\Delta\alpha$ is not possible, the case $F=0$ can be well distinguished experimentally. This observation has enabled the development of a measurement protocol for the quantitative determination of V_{BI}.[29,48] If, when a zero external DC voltage, $V_{ext}=0$ is applied to a device, a modulated response $\Delta\alpha$ is found in a range of the absorption spectrum that can clearly be attributed to the active layer, it is known that it is proportional to F_{BI},

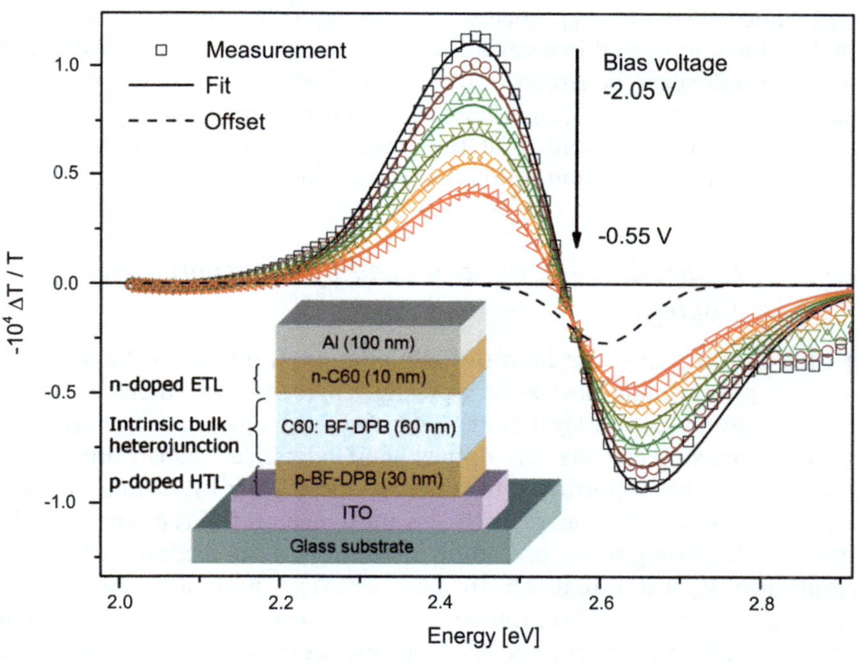

Figure 10.5 Electroabsorption measurement of an organic p-i-n diode (inset: device stack, illumination from bottom) with a transparent bottom electrode and a reflecting metal top electrode. The measurement is performed in reflection geometry measuring the relative change in transmission $\Delta T/T$ after the light is reflected at the top electrode. The main signal arises from C_{60}, because the other materials are transparent in the respective wavelength range. As a deviation from the ideal behaviour, a quantitative fit (solid lines) reveals a small offset signal (dashed line), which can be attributed to non-isotropic molecules.[47] This offset leads to a deviation from zero signal at the isosbestic point, which would be expected in an undisturbed spectrum.
(Figure adapted from ref. 17 under Creative Commons License (CC-by) – see original publication for further details.)

though the direction and value of F_{BI} are still unknown. In the next step, an additional external DC voltage V_{ext} is applied to the sample, which increases or decreases the local field and consequently $\Delta\alpha$, depending on whether the internal and external fields are aligned parallel or anti-parallel, yielding the direction of F_{BI}. A systematic variation of V_{ext}, finally, leads to a linear dependence $\Delta\alpha(V_{ext})$ with its zero-crossing at the point where F_{ext} exactly compensates F_{BI}. Provided that the assumption that well-conducting transport layers have a negligible voltage drop is valid in p-i-n devices as discussed above, the V_{ext} variation can now be directly related to V_{BI}.

$$\text{When } \Delta\alpha(V_{ext}) \approx 0, \quad \text{then } V_{BI} - V_{ext}. \tag{10.4}$$

In practical cases, this direct detection of V_{BI} is often hampered by the presence of additional signals affecting the absorption spectrum. Especially at V_{ext} values close to $-V_{BI}$, injected charges typically induce an additional signal that is manifested in a deviation from the linearity of $\Delta\alpha$ as a function of V_{ext}. This effect can be circumvented by considering only values of V_{ext} in a reduced range where $\Delta\alpha(V_{ext})$ is linear and linearly extrapolating from this region to $\Delta\alpha = 0$. The validity of the resulting V_{BI} can be confirmed by checking that the evaluation yields the same value at different wavelengths.

10.2.2.6 Controlling the Built-in Field by Transport Layer Doping

For direct control over the built-in field, eqn (10.2) offers two handles, V_{BI} and d. In photovoltaic devices, d typically represents the thickness of the intrinsic donor–acceptor layer system, which is well optimized in a trade-off between opposing trends of optical and electrical constraints.[28,49,50] Consequently, the opportunity to tune F_{BI} specifically by means of V_{BI} is the more attractive alternative, with less influence on the remaining device properties. According to the preceding discussion in the section 10.2.2.4, the magnitude of V_{BI} is dominated by the energetic separation of the Fermi levels of the doped electron and hole transport layers. These Fermi levels, E_F, depend on the respective transport levels of the transport materials as well as on the charge carrier concentration in them. However, variation of the transport material can lead to undesired effects like a strong S-shape in the j–V characteristic, which drastically reduces the FF.[51] However, the charge density – and consequently E_F in a given material – can be controlled directly by means of the doping concentration.[52,53]

Compared to thickness variation, variation of the doping concentration has a much weaker influence on the device performance through secondary effects, such as the FF, due to a change in conductivity of the doped layers and through the short-circuit current density j_{SC} due to parasitic absorption by the dopant molecules. In view of these rather weak influences on the power conversion efficiency (PCE), varying the doping concentration is considered a suitable handle to control V_{BI} and investigate its influence on the solar cell performance.

An experimental confirmation of the correlation between the doping concentration and V_{BI} is given by Siebert-Henze *et al.*[17,29,48] in a series of p-i-n diodes based on a donor–acceptor heterojunction. They comprise C_{60} as acceptor with a measurable EA signal in the visible spectral range, and donors that are transparent in the respective range. These devices are used as suitable model systems for solar cells, following the donor–acceptor approach and working as weakly blue-absorbing solar cells. Compared to solar cells with active materials that absorb strongly in the visible range, these cells have the advantage of yielding reliably-evaluable EA spectra in the experimentally well-accessible visible range. The measurement is performed on samples with the stacking shown in Figure 10.5 (inset), which are illuminated through a transparent bottom contact made of indium tin oxide (ITO), and have a reflecting metal top contact. The electron transport layer consists of the fullerene C_{60} n-doped (n-dopant: $W_2(hpp)_4$ (tetrakis(1,3,4,6,7,8-hexahydro-2*H*-pyrimido[1,2-*a*]pyrimidinato) ditungsten(II))[52]) with a doping concentration c_n varying between 0.5 wt% (weight percent) and 32 wt%. The hole transport layer consists of BF-DPB (*N,N′*-((diphenyl-*N,N′*-bis)9,9,-dimethyl-fluoren-2-yl)-benzidine[54]) p-doped (p-dopant: F6-TCNNQ (2,2′-(perfluoronaphthalene-2,6-diylidene) dimalononitrile)[55]) with a doping concentration c_p varying between 1 wt% and 17.5 wt%. These variations of the doping concentration are related to the corresponding E_F values by a separate set of UPS measurements on equally doped layers, showing that the variation of c_n leads to a Fermi level shift in the electron transport layer of 0.29 eV and the variation of c_p to a Fermi level shift in the hole transport layer of 0.52 eV. Figure 10.6 (top) shows the common increase of the two quantities ΔE_F and V_{BI}, confirming the influence of ΔE_F on V_{BI}. Quantitative deviations from the unity relation might be attributed to non-complete level bending in the doped layers[56] and similar parasitic effects in non-ideal devices.

To exclude effects from the D:A system itself as origin for the V_{BI} variation, the general invariance of the energetic structure of the active material towards the doping concentration in the transport layers is shown separately by characterizing the effective donor–acceptor gap E_g^{eff}. In a temperature dependent V_{OC} measurement of solar cells, the extrapolation of a linear $V_{OC}(T)$ characteristic towards 0 K yields V_0, which can be directly identified with E_g^{eff}/e.[57] We performed a measurement series of solar cells based on a bulk heterojunction of zinc-phthalocyanine and C_{60} (stack sequence: glass/ITO/1 nm p-dopant F6-TCNNQ/20 nm BF-DPB p-doped with F6-TCNNQ/30 nm blend of zinc-phthalocyanine and C_{60} 1 : 1 by volume/10 nm C_{60}/20 nm C_{60} n-doped with $W_2(hpp)_4$/100 nm Al) and with varying c_n and c_p values. V_0 stays constant throughout the whole series, as shown in Figure 10.7. This observation leads to the conclusion that the doping concentration does not have a substantial influence on the effective gap of the photo-active material system.

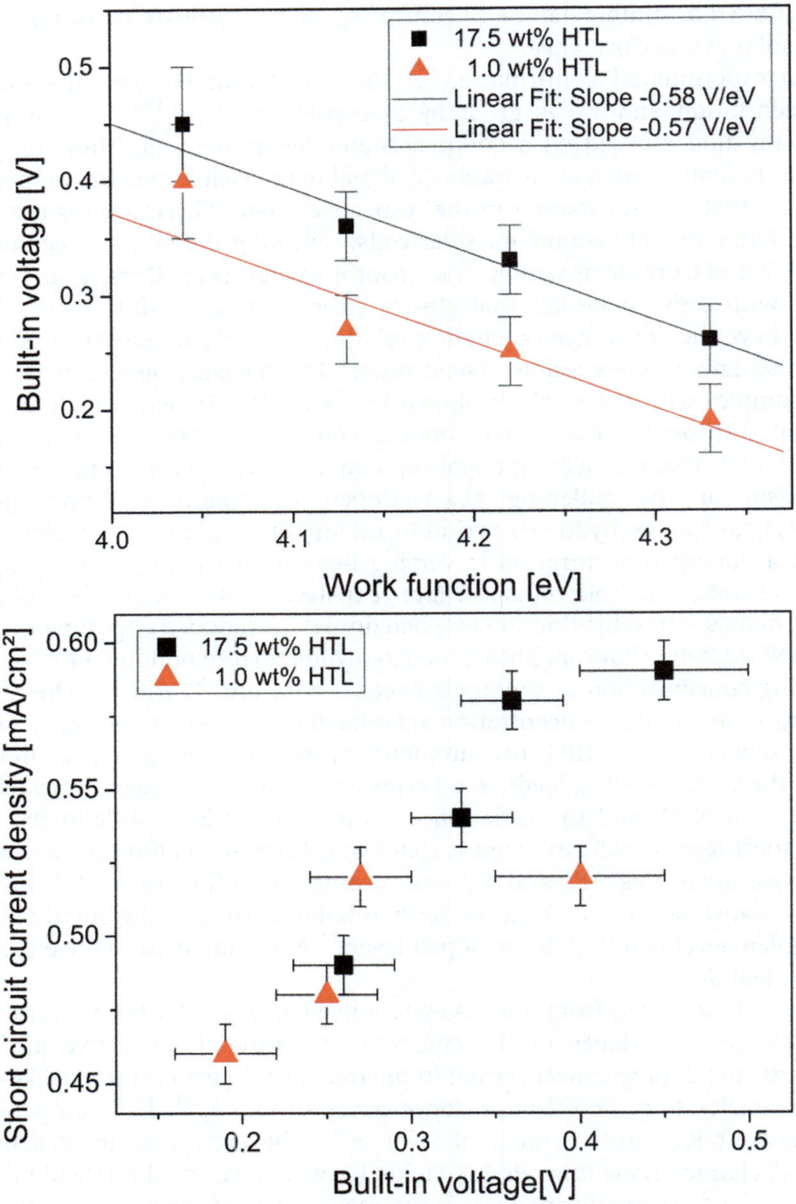

Figure 10.6 Top: the built-in voltage V_{BI} as a function of the work function of the electron transport layer, which is varied by means of changing the doping concentration (see Figure 10.5 for the device stack). Two measurement series are made, one with $c_p = 17.5$ wt% (black squares) and one with $c_p = 1.0$ wt% (red triangles). The lines show linear fits through the respective sets of data points. Bottom: short-circuit current density of the same devices, which all comprise the same absorber layers, plotted as a function of V_{BI}. A clear positive trend can be observed, demonstrating the positive influence of V_{BI} on charge generation and extraction. (Figures taken from ref. 17 under Creative Commons License (CC-by) – see original publication for further details.)

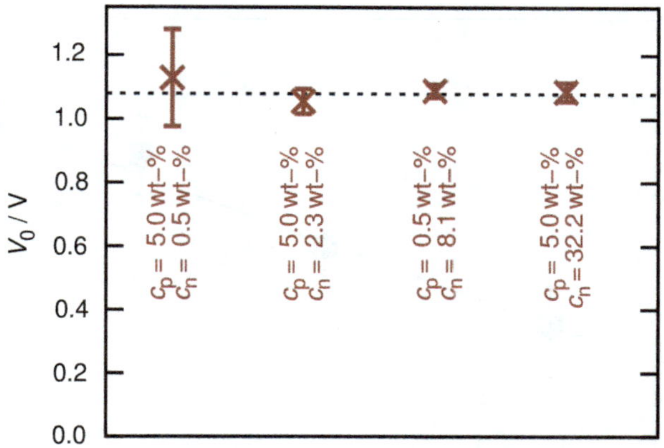

Figure 10.7 The effective energy gap in terms of V_0, the extrapolated low temperature limit of V_{OC}, in solar cells with the same active material and device stack but varying doping concentrations, *i.e.* varying V_{BI}. All V_0 values, independently of V_{BI}, are equal within the measurement accuracy and coincide with the previously published value of 1.08 V[57] for this material system (dotted line). The error bars reflect the statistical accuracy of the extrapolation.

In conclusion, the substantial increase of V_{BI} with c_p and c_n, as shown in Figure 10.6 by Siebert-Henze *et al.* in connection with the data in Figure 10.7, clearly demonstrates that the built-in field can be controlled by transport level doping, which is further confirmed by the improvement of FF and j_{SC} in solar cell characteristics, see Figure 10.6 (bottom).[22,29] For the design of highly efficient organic solar cells, this finding adds an additional rule for optimization: increasing the doping concentration in the transport layers improves not only their conductivity but also the supportive built-in field in the active layers. Thus, an optimized trade-off is to be found between the typically occurring parasitic absorption by the dopant molecules impairing j_{SC} through a reduced light intensity in the active layers, and the enhancement in j_{SC} and FF through the improvement of the conductivity and built-in voltage.

10.2.3 Doped Transport Layers

For the optimization and design of devices, doped transport layers (as used for example in a p-i-n device stack) offer a number of valuable possibilities, which are outlined below.

10.2.3.1 Freedom of Metal Selection

As shown in Figure 10.8, the built-in field is only weakly dependent on the contact metal used. It is observed that although the metal work function (WF) varies between $WF^{Al} = 4.2$ eV for aluminium and $WF^{Au} = 5.3$ eV for

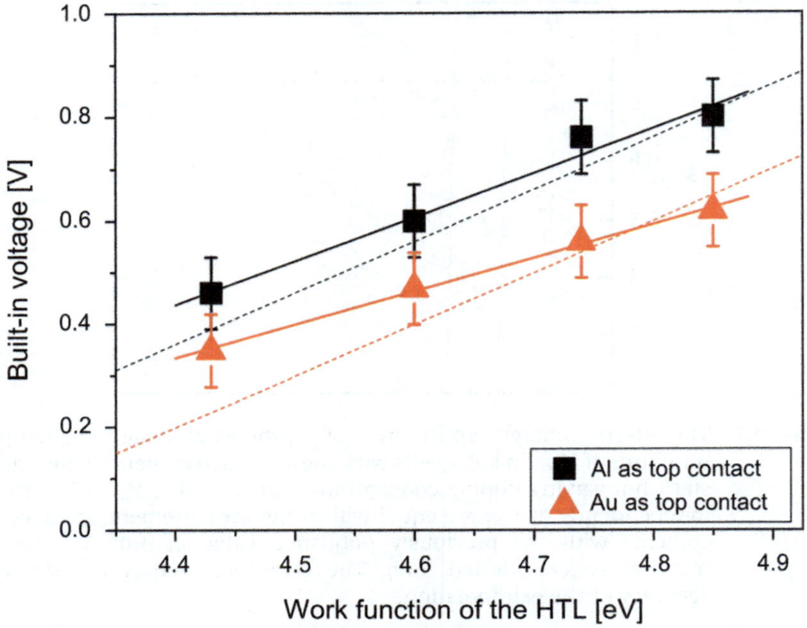

Figure 10.8 The built-in voltage V_{BI} as a function of the work function of the hole transport layer, which is varied by means of changing the doping concentration. Two measurement series of PHJ diodes are investigated, one with an aluminium (black squares) and one with a gold (red triangles) top contact. Solid lines show a linear fit through the respective set of data points, dashed lines show a slope-one relation as comparison. (Figure taken from ref. 29). The devices are designed analogously to the devices shown in Figure 10.5, but with MeO-TPD (*N*,*N*,*N*′,*N*′-tetrakis(4-methoxyphenyl)-benzidine) used both as donor and hole transport material. For details, see ref. 29. Reprinted from Organic Electronic, 15, E. Siebert-Henze *et al.*, Electroabsorption studies of organic p-i-n solar cells: Increase of the built-in voltage by higher doping concentration in the hole transport layer, 563–568, Copyright (2014), with permission from Elsevier.

gold by more than 1 eV, V_{BI} shifts by less than 0.2 eV. This comparison demonstrates that the energetics of the photo-active heterojunction of a solar cell is largely decoupled from the contact metal WF by the doped transport layer. Further confirmation is provided by the observation that V_{OC} also stays approximately constant when the metal contacts are changed. For example, by comparing V_{OC} data from the samples with moderate doping concentrations from Figure 10.8, it can be seen that V_{OC} has a constant value of 0.533(7) V in all four samples, in spite of the large shift of the metal WFs. This independence allows us to optimize the contact metal independently of its WF, with the focus on optical properties, device stability, and processing constraints. This allows the use of technologically highly interesting concepts like transparent conducting metal contacts for ITO replacement and for semi-transparent solar cells.[58,59]

10.2.3.2 Optical Cavity Optimization

The thickness of a typical organic thin-film solar cell is in the range of the coherence length of sunlight. That is, in a device with a reflecting back contact and substantially incomplete absorption in one pass of the light through the absorber layers, the interference between incident and reflected light governs the optical field distribution. The reflected light – which has a reduced intensity – leads to a partially standing wave with the first maximum of the field amplitude at a wavelength-dependent position with an optical pathway of $\lambda/4$ from the reflecting contact. Since also the transparent contact (*e.g.* ITO) typically has a refractive index differing from the organic materials, this interface offers another – though weaker – partially reflecting surface leading to an optical cavity with wavelength-selective field enhancement (this effect is often intentionally enhanced in the design of organic light emitting diodes (OLEDs)).[60] These effects can be detrimental if not properly tuned, but on the other hand they can substantially improve device performance if carefully optimized to optically match the requirements of efficient sunlight absorption. This optimization can be achieved directly by fine adjustment of the thickness of the transport layers.[61,62] This approach is of particular importance in the context of high performance tandem and multi-junction devices, as will be discussed below.

10.2.3.3 All-organic Recombination Contact

Doped transport layers also allow implementation of an all-organic conversion contact to convert a hole current into an electron current and *vice versa*.[63] This kind of conversion contact is required in stacked tandem devices, where two solar cells are connected in series.[49] Such a conversion contact can also be realized by means of a thin metal layer or a layer of metal nanoparticles.[64,65] However, the additional metal interlayer is avoided in the all-organic concept, where two p-i-n type solar cells ("sub-cells") are stacked on top of each other, and the highly p-doped hole transport layer of one sub-cell is in direct contact with the highly n-doped electron transport layer of the other sub-cell.[63] Even in single junction cells, this type of recombination contact can have advantages if, for example, one transport layer is less transparent or less conductive, its thickness can be minimized by partially substituting it by a complementary transport layer,[66] leading to such structures as n-i-p-n solar cells.

10.2.3.4 Advanced Cell Concepts

Advanced solar cell concepts include several refinements of the standard cell design discussed up to now. They include ternary blends[67–69] with more than one donor or more than one acceptor in a bulk heterojunction, and cascade solar cells, similarly combining more than two absorber materials in a nominally planar device structure.[70,71] An interesting approach that bridges photovoltaic technologies based on different material classes is the use of the p-i-n concept with doped organic transport layers for perovskite solar cells[36] as discussed elsewhere in this volume, making them even compatible with organic multi-junction stacks like those discussed at the end of the following section.

10.3 High-efficiency Devices

10.3.1 No Need for Ultra-high Mobilities?

In order to obtain high power conversion efficiencies, several prerequisites must be fulfilled. The absorption of the dye must be strong, charge transport must be efficient, and energy losses should be minimal. Inefficient charge transport is often regarded as the major drawback for typical small molecules used in organic solar cells, as they exhibit rather low charge carrier mobilities in the range of 10^{-5} to 10^{-4} $cm^2\ V^{-1}\ s^{-1}$. Still, those molecules can achieve very high FFs, as long as the absorber layers are thin. This condition is feasible if the absorption is strong enough to harvest most of the incoming light.

Organic solar cells based on small molecules (SMOSCs) have proven to be able to compete with the highest FFs obtained in the field. One example is the molecule class of diindenoperylenes (DIP). With the core molecule itself, FFs above 74% were shown for PHJ devices with 50 nm of DIP as the donor and 80 nm of C_{60} as the acceptor.[72] Such high FF values have been attributed to the strongly crystalline growth of DIP supporting charge transport and leading to high hole mobilities of 5×10^{-2} $cm^2\ V^{-1}\ s^{-1}$.[73] However, we have demonstrated that macroscopic crystallinity is not necessarily a precondition for high FFs in organic solar cells. Schuenemann *et al.*[74] compared the DIP core molecule to a phenyl-substituted derivative (Ph4-DIP). The four phenyl groups rotate out of the molecular plane, leading to strong disruption of the crystalline growth in thin films. The mobility measured in organic field-effect transistors (OFETs) drops by one order of magnitude compared to that obtained with DIP (from 5×10^{-4} $cm^2\ V^{-1}\ s^{-1}$ to 5×10^{-5} $cm^2\ V^{-1}\ s^{-1}$), and X-ray diffraction (XRD) as well as atomic force microscopy (AFM) investigations confirmed that the layers are rather amorphous. Nevertheless, the use of this molecule in a PHJ device using 15 nm of donor and 25 nm of C_{60} as the acceptor achieved a FF of 76%. Similar values were obtained for a derivative with additional propyl chains[75] or another six-membered ring as terminal groups.[76] For the latter derivative, a remarkably high PCE of 5.2% was demonstrated in a PHJ device. Besides the DIP derivatives, dicyanovinyl-substituted oligothiophene materials (DCV*n*T) have recently impressively demonstrated their potential for high efficiencies (exemplary molecular structures are shown in Figure 10.9).[62,77] Despite its similarly low OFET hole mobility of 7×10^{-5} $cm^2\ V^{-1}\ s^{-1}$, DCV5T-Me[33] achieves FFs of up to 73% in PHJ solar cells giving a PCE of 4.8% using only 6 nm of donor and 15 nm of C_{60} as the acceptor.[78]

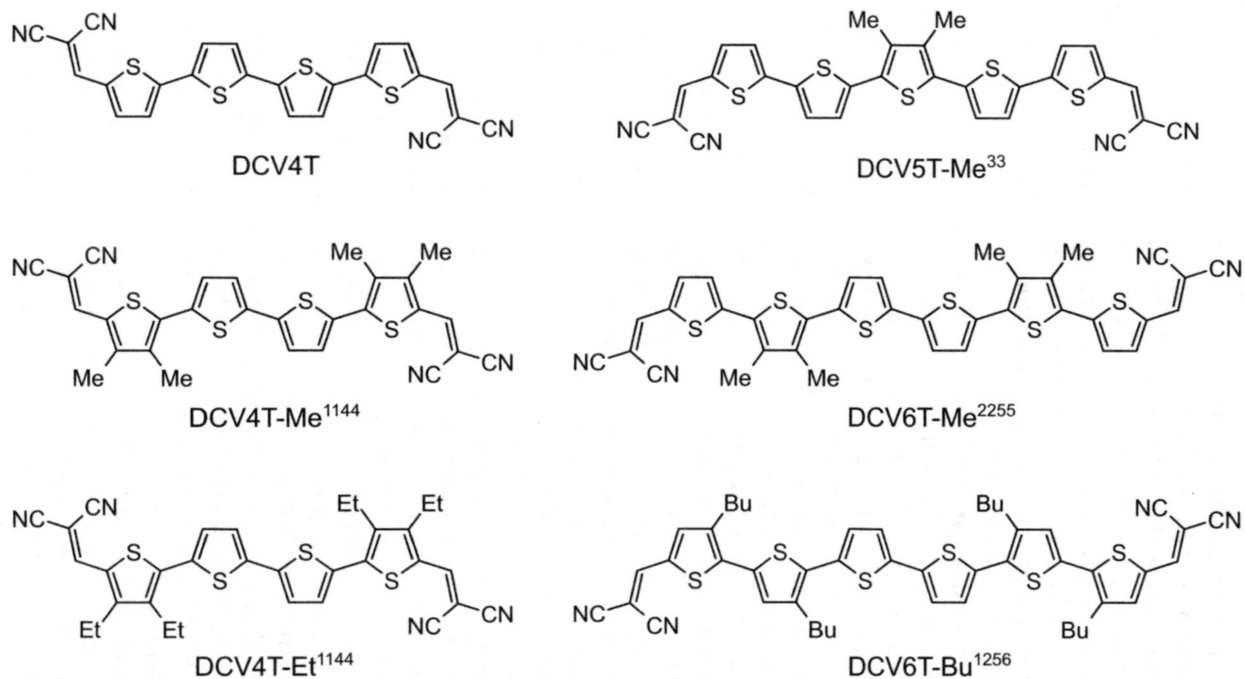

Figure 10.9 Molecular structures of some DCV*n*T materials. The name denotes the length of the oligothiophene chain, followed by the type of alkyl side chains and to which rings they are attached to.

However, in most cases the overall efficiencies of PHJ devices do not exceed 3–4% due to the limitation of the layer thickness by the low exciton diffusion length and, consequently, low overall absorption. Therefore, in order to reach higher efficiencies, the step from a simple PHJ to a device containing a donor:acceptor mixed phase BHJ is inevitable, but usually comes at the cost of a lower FF. The main reason for the decrease in FF is the disadvantageous morphology of the blend. The donor and acceptor phases are strongly intermixed such that the molecular stacking within the phases is disturbed, resulting in suppressed crystallization and large positional and energetic disorder, which hinders effective charge transport. Furthermore, the charge density increases, and recombination of photo-generated charge carriers is enhanced (see *e.g.* ref. 79 and 80).

In contrast to solution processing of solar cells, vacuum deposition offers fewer control options for the morphology. The most important parameters are the evaporation rate[81,82] and the temperature of the substrate during layer deposition.[70] Further possibilities are the choice of the sub-layer,[83] the substrate angle,[84] co-evaporation of additives,[85] and post-processing measures like thermal annealing,[86] solvent annealing,[78] or high pressure.[87,88] In the past, substrate heating turned out to be an effective means both to increase the phase separation of the blend and at the same time force the molecules within their respective phase to crystallize.[89,90] We have successfully applied this technique and have studied the consequences in more detail, *e.g.* for ZnPc[80,91] and a DCV6T derivative.[92] In these cases, substrate temperatures between 90 °C and 140 °C during deposition turned out to be optimal for the morphology. We found that heating led to increased donor/acceptor phase separation, higher mobilities of holes and electrons, higher crystallinity, a positive effect on the free charge carrier generation efficiency, and reduced bimolecular recombination. For the DCV6T derivative, Wynands *et al.*[92] used photoluminescence (PL) quenching and XRD measurements to demonstrate that the increased phase separation is accompanied by a more crystalline morphology of the donor phase. Using a DCV6T-Bu derivative (see Figure 10.9), the PCE could be increased in this way from 2.1% to 4.9% for a substrate temperature of 90 °C.[93] For a similar DCV5T derivative, Ziehlke *et al.*[94] used photo-induced absorption (PIA) spectroscopy and impedance spectroscopy to show that substrate heating increases the lifetime of photo-generated charge carriers in the blend, which can again be attributed to a stronger phase-separated mixture of donor and acceptor.

However, substrate heating is not beneficial for the performance of SMOSCs in all cases. The optimum deposition temperature depends on the respective donor–acceptor combination. Recently, we investigated a DCV4T derivative and observed a continuous decrease of the PCE upon deposition of the photo-active blend layer on a heated substrate even at moderate temperatures.[95] AFM measurements revealed a strong tendency of these molecules to form big clusters on the surface, which can be attributed to the lack of alkyl side chains. The observed decrease in efficiency is mainly due to reduced absorption, which can have two possible reasons. The first is re-orientation of the molecules, from lying flat to standing upright, which can reduce the absorption due to the mismatch between the transition dipole moment along the molecular backbone and the electric field vector of the incoming light under normal incidence.[72] However, XRD measurements as well as angular dependent photo-current measurements could not finally confirm this explanation. The second possibility is based on a simple geometric argument. Agglomeration of the molecules in big clusters leads to a lower surface coverage and an island growth to thicknesses larger than the absorption length $1/\alpha$, and therefore a higher overall transmission. In general, those observations confirm the expectation that very compact molecules already tend to form crystalline films without substrate heating. Other examples are *e.g.* DCV3T derivatives[96] or also the well-known materials DIP[97] or pentacene.[81,98]

Finally, substrate heating has been applied to a system employing DCV5T-Me[33] as the donor material[72] in combination with C_{60} as the acceptor. For a substrate temperature of 90 °C, the PCE was increased from 5.0% to 7.1% compared to deposition on an unheated substrate, a change that is mainly due to an increase of the FF from 50% to 68% (stack sequence: glass/ITO/5 nm bis-HFl-NTCDI (*N*,*N*-bis(fluoren-2-yl)-naphthalenetetracarboxylic diimide)[99] n-doped with 7 wt% $Cr_2(hpp)_4$ (tetrakis(1,3,4,6,7,8-hexahydro-2*H*-pyrimido[1,2-*a*]pyrimidinato)dichromium(II))[52]/15 nm C_{60}/30 nm blend of DCV5T-Me[33] and C_{60} 2 : 1 by volume/5 nm BPAPF (9,9-bis[4-(*N*,*N*-bis-biphenyl-4-yl-amino)phenyl]-9*H*-fluorene[121]/40 nm BPAPF p-doped with 10 wt% NDP9/1 nm NDP9/100 nm Al). Again, the higher crystallinity of the heated film is mainly responsible for improved extraction of charge carriers, yielding higher FFs. With more thorough optimization, we were able to increase the efficiency even further to a certified efficiency value of 8.3%,[62] which is among the highest values achieved to date for single-junction vacuum-deposited SMOSCs.

In summary, small molecule based organic solar cells are – in spite of their low charge carrier mobilities – not necessarily transport limited as long as the photo-active layers are thin. The above-mentioned efficiency of 8.3% (with FFs well above 60%) was achieved with a material with only average mobility, resulting in an optimum blend layer thickness of no more than 40 nm. Nevertheless, the substrate heating method shows that, especially for blend layers – which are a key concept to achieve very high efficiencies[8] – an improvement of the charge transport properties does usually increase the performance of those devices. Recently, so-called energy cascade devices with a new device concept showed efficiencies of 8.4%.[70] In these devices, only pure materials are stacked on top of each other, and so drawbacks from mixed films are avoided. Thus, one of the goals for material optimization is to further increase charge carrier mobilities to reach FFs beyond 75%, in spite of the fact that thick and fully-absorbing blend layers are used. Beyond charge extraction, an improvement of charge transport is also beneficial for charge carrier generation, which will be discussed below.

10.3.2 Charge Generation at the DCV*n*T:C_{60} Interface

Before addressing the correlation between charge carrier generation yield and the charge carrier mobility, we review the basic properties of the materials used and the important processes governing the photo-generation process.

The class of DCV*n*T materials introduced above has proven to be a highly suitable model system to study structure–property relationships and to obtain design rules for high-efficiency OPV devices. One of the advantages is that the HOMO and LUMO energy levels are separately determined by the electron-poor oligothiophene unit (donor part) and the electron-rich dicyanovinyl units (acceptor parts of the molecule), respectively. By changing only the backbone length, it is possible to investigate the effect of the HOMO position on the charge carrier generation yield[100] or, furthermore, the effect of the increased conformational degrees of freedom on the charge transport properties.[101] Non-aromatic alkyl side chains attached to the backbone affect the molecular packing in pure films[96,102] as well as in blends with the acceptor C_{60},[50] thus determining charge transport and recombination properties, while leaving the energy levels of the frontier orbitals virtually undisturbed.

The shift of the HOMO level brought about by extending the oligothiophene backbone and the concomitant red-shift of the absorption lead to a fundamental difference in the charge transfer properties between DCV3T and longer derivatives ($n = 4$–6). In general, for a charge transfer between donor and acceptor species to be efficient, a driving force (*i.e.* an energy difference) is necessary to overcome the exciton binding energy. This driving force is usually provided by the energy of the resulting charge transfer (CT) state at the interface between donor and acceptor being lower in energy than the singlet exciton states of both donor and acceptor. An analysis of the energy levels of DCV3T and C_{60} reveals that the CT state energy is rather high and very similar to the absorption onset of C_{60} (we use 6.0 eV for the ionization potential of DCV3T[96] and 4.0 eV for the electron affinity of C_{60},[103] giving a CT state energy of approximately 1.7 eV assuming a Coulomb binding energy of 0.3 eV in the CT state). Furthermore, due to a spectral overlap between the emission of DCV3T and the absorption of C_{60}, the desired charge transfer from an exciton on DCV3T to the CT state is replaced by an excitonic energy transfer to C_{60} (see Figure 10.10). This mechanism was demonstrated by the observation of efficient quenching of the PL of DCV3T and sensitized emission from C_{60} in such blended films.[104,105] For this reason, the combination of DCV3T and C_{60} is not appropriate for solar cell applications (functional solar cells have been demonstrated with DCV3T as the acceptor in combination with ZnPc or DIP as donors[106,107]).

The extension of the donor part beyond three thiophene units replaces the excitonic energy transfer by the desired charge transfer from the donor molecule to the acceptor molecule due to the shift of the CT energy below the singlet energy level of C_{60}. Experimentally, this process is demonstrated by equally efficient PL quenching close to 100%, but without any sensitized

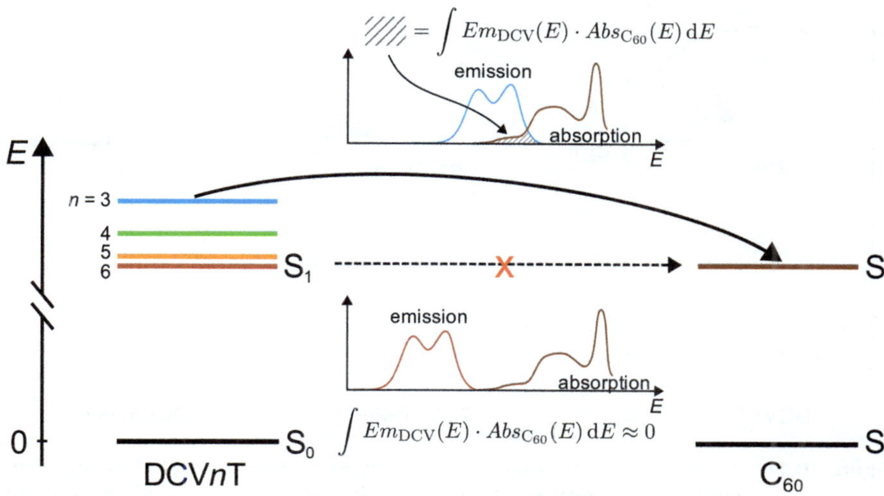

Figure 10.10 Sketch of the energy levels of DCV*n*T donors and the acceptor C_{60}. Due to the overlap between DCV3T emission and C_{60} absorption, the undesired energy transfer to C_{60} is feasible, preventing the generation of free charge carriers. The lowering of S_1 for longer derivatives suppresses this process.

C_{60} emission.[100,104] Nevertheless, rather low photo-generation yields are observed for derivatives with $n = 4$. Detailed photo-physical studies using near-steady state photo-induced absorption (PIA) spectroscopy revealed that, instead of the CT state, the triplet state of those oligothiophene molecules is the lowest energy state in this system (see Figure 10.11).[100,108] As a consequence, back transfer from the CT state at the donor–acceptor interface leads to a high population of the donor triplet state, acting as a loss mechanism in solar cells.

In contrast to DCV4Ts, DCV5Ts and DCV6Ts deliver high photo-current densities and high quantum efficiencies as well as high fill factors.[62,77,92,93,109] Interestingly, however, a strong triplet population is observed in these materials similar to that in DCV4T derivatives. The fact that high charge generation yields are observed nevertheless can be explained by a reverse process such as the thermal regeneration of the CT state from triplet excitons. The final yield of photo-generated charge carriers depends sensitively on the different rates of processes occurring at the donor–acceptor interface, for example the initial charge transfer, recombination from the CT to the donor triplet state, regeneration of the CT state from triplet excitons, and the final separation of CT states into free charge carriers. The latter process is discussed below.

10.3.3 Correlating Charge Carrier Generation Yield with Charge Carrier Mobility

The final separation of CT states into free charge carriers is still under investigation. Previous studies have discussed the influence of the excess

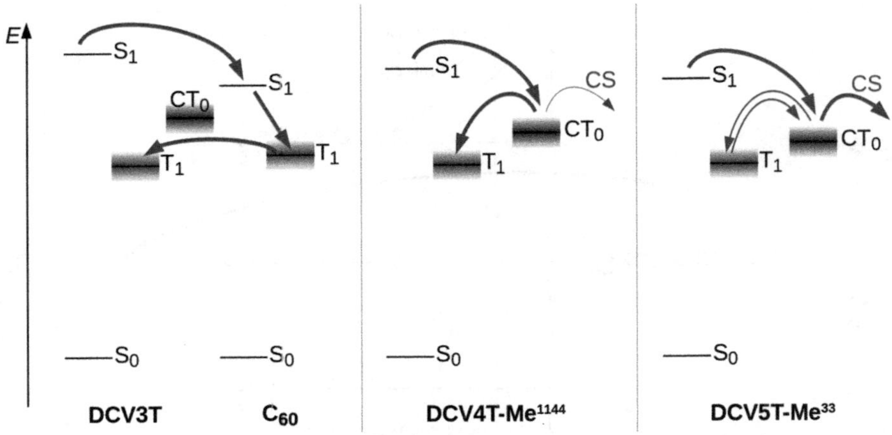

Figure 10.11 Schematic energy diagram of three exemplary DCV*n*T derivatives in combination with C_{60} as acceptor. Initial absorption and consequent relaxation in the DCV*n*T is always considered to happen from the ground state S_0 to the relaxed excitonic state S_1. In all cases, the energy of the relaxed triplet state T_1 is lower than the energy of the relaxed charge transfer state CT_0. The differences between the energy levels vary, leading to different processes occurring with high probability, as indicated by the arrow strengths. In DCV3T, inter-system crossing $S_1 \rightarrow T_1$ occurs on C_{60}, followed by a back-transfer to the DCV3T T_1 state. The CT state is not populated efficiently. In the molecules with longer thiophene backbones, the CT state is populated and charge separation (CS) can occur.

energy of the initial exciton on the separation yield (hot process)[111] *versus* a cold process, indicated by efficient charge generation, upon direct excitation of the CT state.[112] Besides energetic factors, it has been argued that the morphology of the thin film – especially directly at the heterojunction – might affect the delocalization of charge carriers at the interface, thereby facilitating the dissociation of the CT state.[113]

Within the group of DCV*n*T molecules, the growth of crystals and thin films is determined mainly by the side chains attached to the molecule backbone. In a comparison of several DCV4T derivatives, the closest and most interconnected molecular packing in single crystals was observed for the methylated derivative, accompanied by the highest hole mobility measured in OFET devices.[102] As the energy levels hardly change for different side chains, the influence of the morphology on the charge carrier separation yield can be studied nicely in this system. In order to access the differences in charge carrier generation yield for these derivatives, we performed temperature dependent PIA measurements on blend layers with the acceptor C_{60}. In PIA measurements, spectral features of donor cations (holes on the donor) can be identified unambiguously, and the respective generation rates and lifetimes can be extracted.[108] In temperature dependent measurements, we observed that the generation rate increases for higher temperatures, indicating a temperature-activated charge-generation process (see Figure 10.12a). The corresponding activation energies, E_a, determined with a simple Arrhenius-like equation lie in the range of 40 to 100 meV (see Figure 10.12b).[110] We associate these activation energies with the temperature dependence of the charge carrier mobility. This connection between charge carrier generation and the mobility can be understood by considering that the charge carriers have to escape their mutual Coulomb binding to be separated. The macroscopic parameter that characterizes the movement of charge carriers is the charge carrier mobility. Although the microscopic mobility at the interface region between donor and acceptor may differ from the macroscopic average over large distances, the same trends are expected for both, *i.e.* a higher charge carrier mobility corresponds to more efficient charge carrier movement induced, for example, by closer packing, higher crystallinity or lower disorder. In this way, charge separation is enhanced by high charge carrier mobility and hindered by low mobility. Figure 10.12b shows that the activation energy for three different DCV4T derivatives and a DCV6T derivative decreases with the increasing hole mobility of the donor material. We therefore attribute the observed thermal activation energy of charge carrier generation E_a to the energetic disorder in the system, which is in turn correlated with charge carrier mobility. The first link is supported by CELIV (charge extraction by linearly increasing voltage) measurements for the DCV6T compound that give a value of 109 meV for the energetic disorder, which is similar to the activation energy determined with PIA (see Figure 10.12b).[110,85] Similar disorder values were also found in charge transport simulations with simulated morphologies based on measured crystal structures.[101]

As mentioned above, DCV4T derivatives show rather low EQE values, which is due to recombination to the donor triplet state. However, the activation energy for charge generation in this case is much lower than that for the DCV6T compound, which exhibits higher EQE values in solar cells. In the Arrhenius-like equation in Figure 10.12a, this difference is accounted for by a scaling parameter g_1, which leads to a slightly higher generation rate for the sexithiophene compared to the quaterthiophene compound, despite of its higher activation energy. As an illustration, Figure 10.12a also shows simulated curves where the activation energies are swapped. The theoretically best combination of these parameters would be a high g_1 and a low activation energy (dash-dotted line), *i.e.* a high-mobility, low disorder DCV6T compound. Therefore, in the context of organic solar cell devices, the best results are expected for a DCV5T or DCV6T molecule – with appropriate energy levels to reduce triplet recombination – paired with a high hole mobility (or low energetic disorder).

We conclude that – apart from charge extraction – charge carrier mobility also plays a role in charge generation. In this context, a hole mobility of the order of $10^{-4}\,cm^2\,V^{-1}\,s^{-1}$ already leads to activation energies close to the thermal energy at room temperature accompanied by highly efficient exciton splitting. A further increase of the charge carrier mobility is probably not required for highly efficient charge generation.

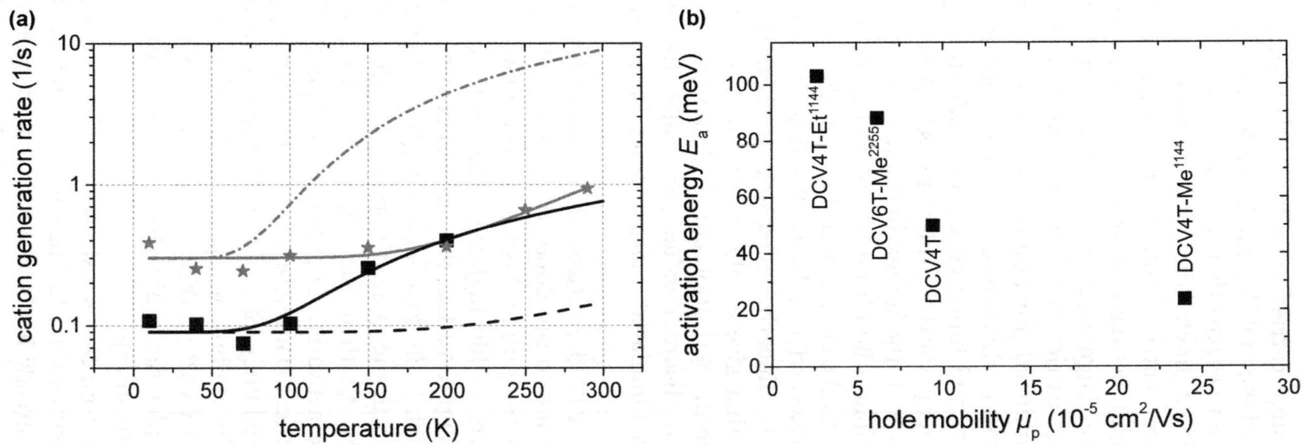

Figure 10.12 (a) Temperature dependence of the cation generation rate in blends of DCV4T-Me:C_{60} (black squares) and DCV6T-Me:C_{60} (grey stars). The fit (lines) is performed according to an Arrhenius-like behaviour for the generation rate g: $g = g_0 + g_1\exp(-E_a/k_BT)$. The dashed and dash-dotted lines are simulated curves where the activation energies of the two materials were swapped. (b) Correlation between the activation energies determined from temperature dependent cation generation rates like in (a) and the OFET hole mobilities of the donor for four different oligothiophene donor materials. The chemical structures of the materials are depicted in Figure 10.9 (the data was first published in ref. 110).

10.3.4 Multi-junction Cells

The conversion of sunlight into electric energy is subject to fundamental thermodynamic limitations of the PCE. For a single solar cell, this limit is described by the Shockley–Queisser limit.[114] This limit can be increased substantially by combining several solar cells in a multi-junction device that harvests complementary parts of the solar spectrum,[115] pushing towards a theoretical thermodynamic efficiency limit of 85%.[15]

In organic photovoltaics, the arrangement of several solar cells in one multi-junction device is a technologically attractive option since thin-film devices can be directly processed on top of each other, efficiently connected by transparent conversion contacts (see also the section on the all organic recombination contact).[62,116–119] Especially for vacuum-deposited devices, several multi-layer stacks can be combined directly with virtually no technological constraints regarding the choice of materials or the design of the sub-cells. In the resulting series connection, the sub-cells add up in terms of their voltages, while the total current density is the same in all sub-cells.[119] This concept is particularly beneficial because ohmic losses (a persistent challenge in organic semiconductors, as well as for the transparent electrodes that are required) are proportional to the current density but independent of the voltage.

The optical design of organic thin-film devices allows optimization of the electromagnetic field distribution for targeted light absorption in the desired parts of the device *cf.* section 10.2.3.2. An example of optical optimization of a tandem device is shown in Figure 10.13. It illustrates a typical scenario where the use of multi-junction devices is beneficial. Some organic absorber materials have a limited charge carrier mobility,[119] leading to the need for very thin absorber layers to ensure a high FF. If such an absorber layer is too thin to absorb a large part of the incident light, it can be complemented in a tandem solar cell approach by a second solar cell based on the same absorber materials, *i.e.* a homo-tandem solar cell. The crucial role of optical design in thin-film tandem solar cells is illustrated by devices of this kind fabricated by Schueppel *et al.*, who achieved a 2-fold increase of the photo-current and the PCE simply by varying the thickness of a non-absorbing spacer layer, reaching an efficiency of 3.8% in an optimized device based on ZnPc (zinc-phthalocyanine) as the donor and C_{60} as the acceptor.[120] Using the modern high-efficiency absorber DCV5T-Me (2,2′-((3″,4″-dimethyl-[2,2′:5′,2″:5″,2‴:5‴,2⁗-quinquethiophene]-5,5⁗-diyl)bis(methanylylidene))-dimalononitrile)[8] as a donor, Meerheim *et al.* demonstrated that careful optimization of a homo-tandem solar cell can achieve an efficiency of 9.6%, which is an improvement of 14% compared to the highest value of 8.3% PCE for a single cell of the same material.[62] The reason why the efficiency is so far from being doubled is, on the one hand, the competing absorption in both sub-cells in those spectral ranges where the materials are strongly absorbing and, on the other hand, the requirement to match the current density and the fill factor in both series-connected sub-cells. For both sub-cells, the absorbing

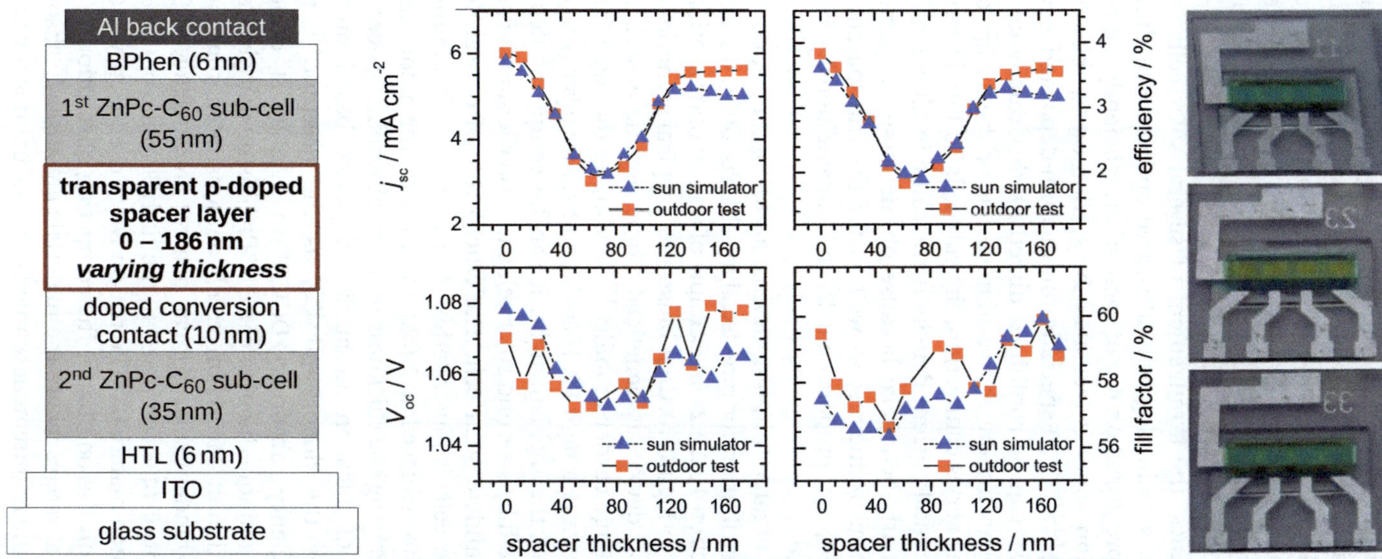

Figure 10.13 Homo-tandem device with varying thickness of the transparent spacer layer between the sub-cells. The device stack is shown on the left-hand side (for details, see ref. 120), and the key figures of the device performance are shown in the graphs in the centre as functions of the spacer thickness. The photographs on the right-hand side show the devices with 0 nm, 74 nm, and 124 nm spacer thickness. The solar cells are the four approximately square-shaped areas in a row in the middle of the substrate. The middle picture shows the lightest-coloured device, illustrating the reduced absorption due to placement of the 2nd sub-cell in the minimum of the interference pattern.
(Adapted with permission from R. Schueppel *et al.*, *J. Appl. Phys.*, 2010, **107**, 44503. Copyright (2010), AIP Publishing LLC[120]).

layers must be thinner compared to the optimized single cell in order to achieve the same short-circuit current density in both sub-cells, and the resulting overall current density is decreased substantially as a consequence. The same paper showed that a homo-triple-cell made by stacking three DCV5T-Me[8]:C_{60} cells on top of each other even outperforms the tandem cell and achieves an efficiency of 9.7%.

As an alternative to combining sub-cells with the same absorber materials, the absorption range of a tandem solar cell can be enhanced by combining sub-cells consisting of different absorber materials with complementary absorption ranges. This is usually realized by using different donor materials in the sub-cells covering different parts of the green to red part of the solar spectrum. On the other hand, C_{60}, one of the most popular acceptor materials, has a weaker absorption than many donor materials and can be employed in all sub-cells to harvest the high-energy part of the incident light efficiently. Using this concept, Riede *et al.*[122] achieved 6.1% PCE with a non-disclosed red-absorbing donor in 2010. In the same year, Cheyns *et al.*[123] achieved 5.2% PCE with a fully disclosed device structure and, one year later, Riede *et al.*[66] achieved 6.1% again, this time with disclosed absorber materials. In 2012, Uhrich *et al.*[124] reported certified PCE values of 10.7% with proprietary materials, and recently, a PCE of 10.0% was reported with fully disclosed materials for a hetero-tandem device comprising one sub-cell of DTDCTB (2-((7-(5-(dip-tolylamino)thiophen-2-yl)benzo[*c*][1,2,5]thiadiazol-4-yl)methylene)malononitrile) and C_{60} complemented by a second sub-cell of DBP (dibenzo{[*f,f'*]-4,4′,7,7′-tetraphenyl}diindeno[1,2,3-*cd*:1′,2′,3′-lm]perylene, see ref. 125) and C_{70}.[65] In the same study, the authors also combined the homo- and hetero-tandem approaches using two identical DBP:C_{70} sub-cells in a triple-junction device with DTDCTB:C_{60}. In this way, the short-circuit current density was reduced by 26% while the voltage improved by 50%, yielding an overall efficiency of 11.1%.

One of the major challenges in the design of multi-junction devices is the matching of FF and j_{SC} of all sub-cells. Due to the serial connection, the sub-cell with the lowest FF dominates the FF of the whole device, and the same applies for j_{SC}. Since the sub-cells within a multi-junction device are exposed to a different illumination spectrum and spatial field distribution than in the same cell in a single-junction device, the design of a multi-junction device generally requires careful simulation and optimization. It is especially noteworthy how well careful optoelectronic simulations are able to reproduce and even predict the dependence of the achieved current density and the EQE for varying layer thicknesses.[61,62,120] Furthermore, proper characterization of multi-junction devices is challenging because illumination by a typical solar simulator is neither sufficient to exactly meet the required AM1.5 illumination spectrum, nor it is possible to make a simple mismatch correction by an illumination pre-factor, as is usually done for single-junction devices.[126] For this reason, careful characterization with differing illumination spectra is required in order to obtain the correct PCE of a multi-junction device at standard reporting conditions.[127]

Current commercial work by Heliatek demonstrates that the efficient roll-to-roll production of organic tandem solar cells on flexible substrates is possible. Thanks to the usage of small molecule organic semiconductors in a vacuum deposition process, the efficiency difference between batch-to-batch lab cells on glass and roll-to-roll "solar films" is low, leading to a cell efficiency of above 7% on a 30 cm wide module connecting 24 tandem cells in series.

In summary, tandem and multi-junction cells have two major advantages: on the one hand, they can harvest a larger part of the solar spectrum compared to single-junction cells; on the other hand, the output current density is not larger and in many cases even smaller than in single-junction cells, *i.e.* the efficiencies of the sub-cells add up *via* the voltage, and the ohmic losses, which are proportional to the current density, are kept low. In this way, multi-junction cells combine the advantages of organic semiconductor processing technology with possibilities to compensate the more problematic peculiarities of organic dyes, such as their limited conductivity and narrow absorption ranges, to realize solar cells with strongly enhanced power conversion efficiencies, with Heliatek currently having achieved 12.3% PCE.[4]

10.4 Conclusions

In this chapter, we have reviewed our recent work on evaporated small-molecule solar cells, placing it in the context of other closely related research. The solar cells are based on the p-i-n-structure, which uses intrinsic absorber layers embedded between n-doped and p-doped transport layers. This design leads to stable devices and high built-in voltages, and it allows for easy optical optimization of the cells as well as the straightforward realization of multi-junction devices. We have shown that both the p- and n-doping of the transport layers can be controlled very well and that optimization of the energy levels can substantially improve the performance.

Furthermore, we have discussed the systematic study of a class of thiophene donor absorbers that have shown excellent photovoltaic properties. These properties can be explained by the suitable packing scheme of the molecules, which leads to a highly absorbing orientation with good transport properties, high currents and fill factors. The voltages achieved are also among the highest reported relative to the absorption gap. However, the performance of these materials depends in a very subtle manner on the detailed molecular structure and the preparation conditions. Inadequate transport in the bulk heterojunction remains as the main problem since it limits the absorber layer thickness and thus the external quantum efficiency. Also V_{OC}, although excellent compared to most other organic material systems, might be improved further to reach voltage losses as small as those in the best inorganic semiconductors.

Finally, we discuss multi-junction cells, which are realized by stacking several p-i-n cells on top of each other. By careful optimization, it is possible to increase the efficiency well beyond the single junction case. Nevertheless, the achieved gain is still limited by the lack of efficient absorber materials in the near infrared.

References

1. C. W. Tang, *Appl. Phys. Lett.*, 1986, **48**, 183–185.
2. M. Hiramoto, H. Fujiwara and M. Yokoyama, *Appl. Phys. Lett.*, 1991, **58**, 1062.
3. N. S. Sariciftci, L. Smilowitz, A. J. Heeger and F. Wudl, *Science*, 1992, **258**, 1474–1476.
4. Heliatek GmbH, www.heliatek.com, accessed 15th March 2016.
5. L. J. A. Koster, S. E. Shaheen and J. C. Hummelen, *Adv. Energy Mater.*, 2012, **2**, 1246–1253.
6. M. C. Scherer and N. S. Sariciftci, *Prog. Polym. Sci.*, 2013, **38**, 1929–1940.
7. Organic semiconductors world, www.orgworld.de, accessed December 3rd, 2015.
8. G. Niu, X. Guo and L. Wang, *J. Mater. Chem. A*, 2015, **3**, 8970–8980.
9. I. Etxebarria, J. Ajuria and R. Pacios, *J. Photonics Energy*, 2015, **5**, 57214.
10. X. Chen, X. Liu, M. a. Burgers, Y. Huang and G. C. Bazan, *Angew. Chem., Int. Ed.*, 2014, **53**, 14378–14381.
11. Y. Huang, E. J. Kramer, A. J. Heeger and G. C. Bazan, *Chem. Rev.*, 2014, **114**, 7006–7043.
12. A. Buckley, *Woodhead Publ. Ser. Electron. Opt. Mater.*, 2013, 36.
13. J. W. Hamer, *U.S. OLED Lighting Manufacturing Status and Trends*, OLEDWorks, 2014, http://apps1.eere.energy.gov/buildings/publications/pdfs/ssl/hamer_USMfgOLED_sandiego2014.pdf.
14. E. Fujimoto, H. Daiku, K. Kamikawa, E. Fujimoto and Y. Matsumoto, *SID Symp. Dig. Tech. Pap.*, 2010, **41**, 695.
15. P. Würfel and U. Würfel, *Physics of Solar Cells: From Basic Principles to Advanced Concepts*, Wiley VCH, Weinheim, 2010.
16. C. Uhrich, D. Wynands, S. Olthof, M. K. Riede, K. Leo, S. Sonntag, B. Maennig and M. Pfeiffer, *J. Appl. Phys.*, 2008, **104**, 43107.
17. E. Siebert-Henze, V. G. Lyssenko, J. Fischer, M. Tietze, R. Brueckner, M. Schwarze, K. Vandewal, D. Ray, M. Riede and K. Leo, *AIP Adv.*, 2014, **4**, 47134.
18. C. Falkenberg, PhD thesis, IAPP, TU Dresden, 2011.
19. J. Widmer, PhD thesis, IAPP, TU Dresden, 2014.
20. J. Huang, M. Pfeiffer, A. Werner, J. Blochwitz, S. Liu and K. Leo, *Appl. Phys. Lett.*, 2002, **80**, 139–141.
21. J. Drechsel, B. Maennig, F. Kozlowski, D. Gebeyehu, A. Werner, M. Koch, K. Leo and M. Pfeiffer, *Thin Solid Films*, 2004, **451–452**, 515–517.
22. B. Maennig, J. Drechsel, D. Gebeyehu, P. Simon, F. Kozlowski, A. Werner, F. Li, S. Grundmann, S. Sonntag, M. Koch, K. Leo, M. Pfeiffer, H. Hoppe,

D. Meissner, N. S. Sariciftci, I. Riedel, V. Dyakonov and J. Parisi, *Appl. Phys. A: Mater. Sci. Process.*, 2004, **79**, 1–14.
23. R. Timmreck, S. Olthof, K. Leo and M. K. Riede, *J. Appl. Phys.*, 2010, **108**, 33108.
24. S. Albrecht, W. Schindler, J. Kurpiers, J. Kniepert, J. C. Blakesley, I. Dumsch, S. Allard, K. Fostiropoulos, U. Scherf and D. Neher, *J. Phys. Chem. Lett.*, 2012, **3**, 640–645.
25. L. Onsager, *Phys. Rev.*, 1938, **54**, 554–557.
26. C. L. Braun, *J. Chem. Phys.*, 1984, **80**, 4157–4161.
27. A. Petersen, A. Ojala, T. Kirchartz, T. Wagner, F. Würthner and U. Rau, *Phys. Rev. B: Condens. Matter Mater. Phys.*, 2012, 85.
28. W. Tress, K. Leo and M. Riede, *Phys. Rev. B: Condens. Matter Mater. Phys.*, 2012, **85**, 155201.
29. E. Siebert-Henze, V. G. Lyssenko, J. Fischer, M. Tietze, R. Brueckner, T. Menke, K. Leo and M. Riede, *Org. Electron.*, 2014, **15**, 563–568.
30. R. R. Chance, *J. Chem. Phys.*, 1973, **59**, 2269.
31. D. Ray, M. Furno, E. Siebert-Henze, K. Leo and M. Riede, *Phys. Rev. B: Condens. Matter Mater. Phys.*, 2011, 84.
32. C. Falkenberg, S. Olthof, R. Rieger, M. Baumgarten, K. Muellen, K. Leo and M. Riede, *Sol. Energy Mater. Sol. Cells*, 2011, **95**, 927–932.
33. W. Tress, K. Leo and M. Riede, *Adv. Funct. Mater.*, 2011, **21**, 2140–2149.
34. U. Würfel, D. Neher, A. Spies and S. Albrecht, *Nat. Commun.*, 2015, **6**, 6951.
35. W. Tress, A. Petrich, M. Hummert, M. Hein, K. Leo and M. Riede, *Appl. Phys. Lett.*, 2011, **98**, 63301.
36. L. E. Polander, P. Pahner, M. Schwarze, M. Saalfrank, C. Koerner and K. Leo, *APL Mater.*, 2014, **2**, 81503.
37. O. Malinkiewicz, A. Yella, Y. H. Lee, G. M. M. Espallargas, M. Graetzel, M. K. Nazeeruddin and H. J. Bolink, *Nat. Photonics*, 2014, **8**, 128–132.
38. C. Deibel and V. Dyakonov, *Rep. Prog. Phys.*, 2010, **73**, 96401.
39. F. Bloom, W. Wagemans, M. Kemerink and B. Koopmans, *Phys. Rev. Lett.*, 2007, **99**, 1–4.
40. O. J. Sandberg, M. Nyman and R. Österbacka, *Phys. Rev. Appl.*, 2014, **1**, 24003.
41. D. Rauh, A. Wagenpfahl, C. Deibel and V. Dyakonov, *Appl. Phys. Lett.*, 2011, **98**, 133301.
42. H. Ishii, K. Sugiyama, E. Ito and K. Seki, *Adv. Mater.*, 1999, **11**, 605–625.
43. J. Stark, *Nature*, 1913, **92**, 401.
44. J. Stark, *Ann. Phys.*, 1914, **348**, 965–1047.
45. G. U. Bublitz and S. G. Boxer, *Annu. Rev. Phys. Chem.*, 1997, **48**, 213–242.
46. W. Stampor, *Chem. Phys.*, 2004, **305**, 77–84.
47. P. A. Lane, J. Rostalski, C. Giebeler, S. J. Martin, D. D. C. Bradley and D. Meissner, *Sol. Energy Mater. Sol. Cells*, 2000, **63**, 3–13.
48. E. Siebert-Henze, PhD thesis, IAPP, TU Dresden, 2014.
49. A. Hadipour, B. de Boer and P. W. M. Blom, *Adv. Funct. Mater.*, 2008, **18**, 169–181.

50. C. Koerner, PhD thesis, IAPP, TU Dresden, 2013.
51. W. Tress, *J. Photonics Energy*, 2011, **1**, 11114.
52. T. Menke, D. Ray, J. Meiss, K. Leo and M. Riede, *Appl. Phys. Lett.*, 2012, **100**, 93304.
53. M. Tietze, L. Burtone, M. Riede, B. Lüssem and K. Leo, *Phys. Rev. B: Condens. Matter Mater. Phys.*, 2012, **86**, 35320.
54. S.-H. Hwang, Y. K. Kim, Y. Kwak, C.-H. Lee, J. Lee and S. Kim, *Synth. Met.*, 2009, **159**, 2578–2583.
55. P. K. Koech, A. B. Padmaperuma, L. Wang, J. S. Swensen, E. Polikarpov, J. T. Darsell, J. E. Rainbolt and D. J. Gaspar, *Chem. Mater.*, 2010, **22**, 3926–3932.
56. S. Olthof, R. Meerheim, M. Schober and K. Leo, *Phys. Rev. B: Condens. Matter Mater. Phys.*, 2009, **79**, 245308.
57. J. Widmer, M. Tietze, K. Leo and M. Riede, *Adv. Funct. Mater.*, 2013, **23**, 5814.
58. S. Schubert, J. Meiss, L. Müller-Meskamp and K. Leo, *Adv. Energy Mater.*, 2013, **3**, 438–443.
59. S. Schubert, PhD thesis, IAPP, TU Dresden, 2015.
60. M. Furno, R. Meerheim, S. Hofmann, B. Lüssem and K. Leo, *Phys. Rev. B: Condens. Matter Mater. Phys.*, 2012, **85**, 115205.
61. C. Falkenberg, K. Leo and M. K. Riede, *J. Appl. Phys.*, 2011, **110**, 124509.
62. R. Meerheim, C. Körner and K. Leo, *Appl. Phys. Lett.*, 2014, **105**, 63306.
63. R. Timmreck, S. Olthof, K. Leo and M. Riede, *J. Appl. Phys.*, 2010, **108**, 33108.
64. M. Hiramoto, M. Suezaki and M. Yokoyama, *Chem. Lett.*, 1990, **19**, 327–330.
65. X. Che, X. Xiao, J. D. Zimmerman, D. Fan and S. R. Forrest, *Adv. Energy Mater.*, 2014, **4**, 1400568.
66. M. Riede, C. Uhrich, J. Widmer, R. Timmreck, D. Wynands, G. Schwartz, W. M. Gnehr, D. Hildebrandt, A. Weiss, J. Hwang, S. Sundarraj, P. Erk, M. Pfeiffer and K. Leo, *Adv. Funct. Mater.*, 2011, **21**, 3019–3028.
67. Y. Kim, M. Shin, H. Kim, Y. Ha and C.-S. Ha, *J. Phys. D: Appl. Phys.*, 2008, **41**, 225101.
68. J. A. Mikroyannidis, D. V. Tsagkournos, S. S. Sharma, A. Kumar, Y. K. Vijay and G. D. Sharma, *Sol. Energy Mater. Sol. Cells*, 2010, **94**, 2318–2327.
69. J. Min, T. Ameri, R. Gresser, M. Lorenz-Rothe, D. Baran, A. Troeger, V. Sgobba, K. Leo, M. Riede, D. M. Guldi and C. J. Brabec, *ACS Appl. Mater. Interfaces*, 2013, **5**, 5609–5616.
70. K. Cnops, B. P. Rand, D. Cheyns, B. Verreet, M. A. Empl and P. Heremans, *Nat. Commun.*, 2014, **5**, 3406.
71. K. Cnops, B. P. Rand, D. Cheyns and P. Heremans, *Appl. Phys. Lett.*, 2012, **101**, 143301.
72. J. Wagner, M. Gruber, A. Hinderhofer, A. Wilke, B. Bröker, J. Frisch, P. Amsalem, A. Vollmer, A. Opitz, N. Koch, F. Schreiber and W. Brütting, *Adv. Funct. Mater.*, 2010, **20**, 4295–4303.

73. M. Horlet, M. Kraus, W. Brütting and A. Opitz, *Appl. Phys. Lett.*, 2011, **98**, 233304.
74. C. Schuenemann, A. Petrich, R. Schulze, D. Wynands, J. Meiss, M. P. Hein, J. Jankowski, C. Elschner, J. Alex, M. Hummert, K.-J. Eichhorn, K. Leo and M. Riede, *Org. Electron.*, 2013, **14**, 1704–1714.
75. J. Meiss, M. Hermenau, W. Tress, C. Schuenemann, F. Selzer, M. Hummert, J. Alex, G. Lackner, K. Leo and M. Riede, *Phys. Rev. B: Condens. Matter Mater. Phys.*, 2011, **83**, 165305.
76. M. Hirade and C. Adachi, *Appl. Phys. Lett.*, 2011, **99**, 153302.
77. R. Fitzner, E. Mena-Osteritz, A. Mishra, G. Schulz, E. Reinold, M. Weil, C. Koerner, H. Ziehlke, C. Elschner, K. Leo, M. Riede, M. Pfeiffer, C. Uhrich and P. Bäuerle, *J. Am. Chem. Soc.*, 2012, **134**, 11064–11067.
78. B. A. Gregg, *J. Phys. Chem.*, 1996, **100**, 852–859.
79. J. Kniepert, I. Lange, J. Heidbrink, J. Kurpiers, T. J. K. Brenner, L. J. A. Koster and D. Neher, *J. Phys. Chem. C*, 2015, **119**, 8310–8320.
80. A. F. Bartelt, C. Strothkämper, W. Schindler, K. Fostiropoulos and R. Eichberger, *Appl. Phys. Lett.*, 2011, **99**, 143304.
81. A. A. Günther, J. Widmer, D. Kasemann and K. Leo, *Appl. Phys. Lett.*, 2015, **106**, 233301.
82. W. Tress, K. Leo and M. Riede, *Sol. Energy Mater. Sol. Cells*, 2011, **95**, 2981–2986.
83. C. Schünemann, C. Elschner, A. A. Levin, M. Levichkova, K. Leo and M. Riede, *Thin Solid Films*, 2011, **519**, 3939–3945.
84. N. Li and S. R. Forrest, *Appl. Phys. Lett.*, 2009, **95**, 123309.
85. T. Kaji, M. Zhang, S. Nakao, K. Iketaki, K. Yokoyama, C. W. Tang and M. Hiramoto, *Adv. Mater.*, 2011, **23**, 3320–3325.
86. U. Mayerhöffer, K. Deing, K. Gruß, H. Braunschweig, K. Meerholz and F. Würthner, *Angew. Chem., Int. Ed.*, 2009, **48**, 8776–8779.
87. C. F. Shih, K. T. Hung, J. W. Wu, C. Y. Hsiao and W. M. Li, *Appl. Phys. Lett.*, 2009, **94**, 143505.
88. Y. Matsushima, Esaki and C. Adachi, *Appl. Phys. Lett.*, 2014, **105**, 243301.
89. T. M. K. Suemori, T. Miyata, M. Hiramoto and M. Yokoyama, *Jpn. J. Appl. Phys.*, 2004, **43**, L1014–L1016.
90. M. Vogel, J. Strotmann, B. Johnev, M. C. Lux-Steiner and K. Fostiropoulos, *Thin Solid Films*, 2006, **511-512**, 367–370.
91. S. Pfuetzner, C. Mickel, J. Jankowski, M. Hein, J. Meiss, C. Schuenemann, C. Elschner, A. a. Levin, B. Rellinghaus, K. Leo and M. Riede, *Org. Electron.*, 2011, **12**, 435–441.
92. D. Wynands, M. Levichkova, M. Riede, M. Pfeiffer, P. Baeuerle, R. Rentenberger, P. Denner and K. Leo, *J. Appl. Phys.*, 2010, **107**, 14517.
93. D. Wynands, M. Levichkova, K. Leo, C. Uhrich, G. Schwartz, D. Hildebrandt, M. Pfeiffer and M. Riede, *Appl. Phys. Lett.*, 2010, **97**, 73503.
94. H. Ziehlke, L. Burtone, C. Koerner, R. Fitzner, E. Reinold, P. Bäuerle, K. Leo and M. Riede, *Org. Electron.*, 2011, **12**, 2258–2267.

95. C. Koerner, C. Elschner, N. C. Miller, R. Fitzner, F. Selzer, E. Reinold, P. Bäuerle, M. F. Toney, M. D. McGehee, K. Leo and M. Riede, *Org. Electron.*, 2012, **13**, 623–631.
96. H. Ziehlke, R. Fitzner, C. Koerner, R. Gresser, E. Reinold, P. Bäuerle, K. Leo and M. K. Riede, *J. Phys. Chem. A*, 2011, **115**, 8437–8446.
97. U. Hörmann, J. Kraus, M. Gruber, C. Schuhmair, T. Linderl, S. Grob, S. Kapfinger, K. Klein, M. Stutzman, H. Krenner and W. Brütting, *Phys. Rev. B: Condens. Matter Mater. Phys.*, 2013, **88**, 235307.
98. W.-S. Hu, Y.-T. Tao, Y.-F. Chen and C.-S. Chang, *Appl. Phys. Lett.*, 2008, **93**, 53304.
99. C. Falkenberg, Unpubl. Work, 2016.
100. R. Schueppel, K. Schmidt, C. Uhrich, K. Schulze, D. Wynands, J. Brédas, E. Brier, E. Reinold, H.-B. Bu, P. Baeuerle, B. Maennig, M. Pfeiffer and K. Leo, *Phys. Rev. B: Condens. Matter Mater. Phys.*, 2008, **77**, 85311.
101. M. Schrader, C. Körner, C. Elschner and D. Andrienko, *J. Mater. Chem.*, 2012, **22**, 22258.
102. R. Fitzner, C. Elschner, M. Weil, C. Uhrich, C. Koerner, M. Riede, K. Leo, M. Pfeiffer, E. Reinold, E. Mena-Osteritz and P. Bäuerle, *Adv. Mater.*, 2012, **24**, 675–680.
103. H. Yoshida, *MRS Proc.*, 2013, **1493**, 3–4.
104. R. Schueppel, C. Uhrich, M. Pfeiffer, K. Leo, E. Brier, E. Reinold and P. Baeuerle, *ChemPhysChem*, 2007, **8**, 1497–1503.
105. R. Schüppel, PhD thesis, IAPP, TU Dresden, 2007.
106. C. Uhrich, R. Schueppel, A. Petrich, M. Pfeiffer, K. Leo, E. Brier, P. Kilickiran and P. Baeuerle, *Adv. Funct. Mater.*, 2007, **17**, 2991–2999.
107. R. Scholz, R. Luschtinetz, G. Seifert, T. Jägeler-Hoheisel, C. Koerner, K. Leo and M. Rapacioli, *J. Phys.: Condens. Matter*, 2013, **25**, 473201.
108. C. Koerner, H. Ziehlke, R. Gresser, R. Fitzner, E. Reinold, P. Bäuerle, K. Leo and M. Riede, *J. Phys. Chem. C*, 2012, **116**, 25097–25105.
109. R. Fitzner, E. Reinold, A. Mishra, E. Mena-Osteritz, H. Ziehlke, C. Koerner, K. Leo, M. Riede, M. Weil, O. Tsaryova, A. Weiß, C. Uhrich, M. Pfeiffer and P. Bäuerle, *Adv. Funct. Mater.*, 2011, **21**, 897–910.
110. C. Koerner, M. P. Hein, V. Kažukauskas, A. Sakavičius, V. Janonis, R. Fitzner, P. Bäuerle, K. Leo and M. Riede, *ChemPhysChem*, 2014, **15**, 1049–1055.
111. G. Grancini, M. Maiuri, D. Fazzi, A. Petrozza, H.-J. Egelhaaf, D. Brida, G. Cerullo and G. Lanzani, *Nat. Mater.*, 2012, **12**, 29–33.
112. K. Vandewal, S. Albrecht, E. T. Hoke, K. R. Graham, J. Widmer, J. D. Douglas, M. Schubert, W. R. Mateker, J. T. Bloking, G. F. Burkhard, A. Sellinger, J. M. J. Fréchet, A. Amassian, M. K. Riede, M. D. McGehee, D. Neher and A. Salleo, *Nat. Mater.*, 2013, **13**, 63–68.
113. R. D. Pensack, C. Guo, K. Vakhshouri, E. D. Gomez and J. B. Asbury, *J. Phys. Chem. C*, 2012, **116**, 4824–4831.
114. W. Shockley and H. J. Queisser, *J. Appl. Phys.*, 1961, **32**, 510–519.
115. A. De Vos, *J. Phys. D: Appl. Phys.*, 1980, **13**, 839–846.

116. M. Hiramoto, M. Suezaki and M. Yokoyama, *Chem. Lett.*, 1990, **19**, 327–330.
117. M. Riede, T. Mueller, W. Tress, R. Schueppel and K. Leo, *Nanotechnology*, 2008, **19**, 424001.
118. R. Timmreck, J. Meiss, A. Merten, R. Schueppel, M. K. Riede and K. Leo, in Proceedings of the 24th European Photovoltaic Solar Energy Conference, Hamburg, Germany, 2009.
119. A. Colsmann, J. Junge, C. Kayser and U. Lemmer, *Appl. Phys. Lett.*, 2006, **89**, 203506.
120. R. Schueppel, R. Timmreck, N. Allinger, T. Mueller, M. Furno, C. Uhrich, K. Leo and M. Riede, *J. Appl. Phys.*, 2010, **107**, 44503.
121. Y.-L. Tung, S.-W. Lee, Y. Chi, Y.-T. Tao, C.-H. Chien, Y.-M. Cheng, P.-T. Chou, S.-M. Peng and C.-S. Liu, *J. Mater. Chem.*, 2005, **15**, 460.
122. M. Riede, C. Uhrich, R. Timmreck, J. Widmer, D. Wynands, M. Levichkova, M. Furno, G. Schwartz, W. Gnehr, M. Pfeiffer and K. Leo, in 2010 35th IEEE Photovoltaic Specialists Conference, Institute of Electrical and Electronics Engineers, 2010, pp. 513–517.
123. D. Cheyns, B. P. Rand and P. Heremans, *Appl. Phys. Lett.*, 2010, **97**, 122–124.
124. C. L. Uhrich, R. Meerheim, T. Mueller, F. Lindner, K. Walzer, B. Maennig, A. Weiss and M. Pfeiffer, *Proc. SPIE*, 2012, **8477**, 847705.
125. L.-Y. Lin, Y.-H. Chen, Z.-Y. Huang, H.-W. Lin, S.-H. Chou, F. Lin, C.-W. Chen, Y.-H. Liu and K.-T. Wong, *J. Am. Chem. Soc.*, 2011, **133**, 15822–15825.
126. V. Shrotriya, G. Li, Y. Yao, T. Moriarty, K. Emery and Y. Yang, *Adv. Funct. Mater.*, 2006, **16**, 2016–2023.
127. R. Timmreck, T. Meyer, J. Gilot, H. Seifert, T. Mueller, A. Furlan, M. M. Wienk, D. Wynands, J. Hohl-Ebinger, W. Warta, R. a. J. Janssen, M. Riede and K. Leo, *Nat. Photonics*, 2015, **9**, 478–479.

CHAPTER 11

Modeling Organic Solar Cells: What are the Challenges Ahead?

JÉRÔME CORNIL,[a] FRÉDÉRIC CASTET[b] AND DAVID BELJONNE*[a]

[a] Laboratory for Chemistry of Novel Materials, University of Mons, Place du Parc 20, BE-7000 Mons, Belgium; [b] Institut des Sciences Moléculaires, UMR CNRS 5255, Université de Bordeaux, Cours de la Libération 351, FR-33405 Talence, France
*Email: David.Beljonne@umons.ac.be

11.1 Introduction

The search for new routes towards cheap and efficient production of electricity based on sunlight conversion (*i.e.*, the photovoltaic effect) has triggered the development of novel technologies involving organic semiconductors (organic photovoltaics)[1] or hybrid inorganic/organic devices such as dye-sensitized solar cells[2] and, more recently, devices in which the dyes have been replaced by perovskites.[3] Such devices have attractive features such as high power conversion efficiency (above 10% for an organic-based device with a single layer[4] and above 15% with perovskite-based structures[5]), ease of processing (in particular by printing processes), deposition over large areas or flexibility. However, they currently suffer from drawbacks limiting their penetration in the market, in particular limited lifetime or toxicity (in lead-based perovskite devices). It is thus not clear

RSC Energy and Environment Series No. 16
Unconventional Thin Film Photovoltaics
Edited by Enrico Da Como, Filippo De Angelis, Henry Snaith and Alison Walker

Published by the Royal Society of Chemistry, www.rsc.org

at present which technologies, if any, will make it to the market in technological niches making them competitive compared to silicon-based solar cells. In all cases, development of these new technologies requires a fundamental understanding of the operation of the devices at both the macroscopic and microscopic scales, in order to define new device architectures or new materials leading to improved performance. In this context, we will concentrate here on organic-based photovoltaic cells (OPVs) and will describe with the help of theoretical modeling what is known in terms of the key opto-electronic processes governing the device efficiency at the molecular- and meso-scale. Most importantly, we will also highlight what has not been covered so far and where efforts should be dedicated to boost the current figures of merit of OPVs.

The physics of organic-based devices has major differences compared to traditional silicon solar cells. The most important one for photovoltaic application is the nature of the primary excitations. When a molecule (or polymer chain) is excited by promoting one electron from an occupied electronic level to an unoccupied electronic level, the photogenerated electron–hole pair is strongly bound by Coulomb interactions; the binding energy is typically of the order of 0.4 eV for polymers and larger for smaller molecules (due to the stronger spatial confinement of the charge carriers) as a result of: (i) the low dielectric constant of organic materials, typically in between 3–4 (for comparison, the static dielectric constant in silicon is around 11.5, leading to electron–hole pairs bound only by a few meV so that they can be readily separated after photoexcitation); (ii) the weaker van der Waals interactions between the individual chains instead of the covalent bonding structure in silicon, the weak contact between molecular units favors the localization of the excitations after an initial relaxation step (*vide infra*). An excitation can only be dissociated in a given organic material in the presence of a high electric field ($>10^6$ V cm^{-1},[6] typically much larger than the field created in operating devices). This is why organic solar cells are typically made by mixing an electron donor material (with a low ionization potential) and an electron accepting material (with a high electron affinity) to convert the excitations into electrical charges at their interface. Prototypical cells use polymer chains or small molecules as donors and fullerene derivatives as acceptors.

In the simplest picture, the conversion of light into electricity in organic-based solar cells relies on the succession of three simple events: (i) light absorption by the material; (ii) exciton dissociation at the interface; and (iii) collection of the charges at the electrodes. However, this view masks the complexity of the processes allowing for the generation of free charge carriers.[7] This will be discussed here by introducing some key concepts underlying the physics of the devices, see Figure 11.1. Another degree of complexity originates from the fact that the actual working mechanism appears to be affected by a large number of experimental parameters (nature of the donor and acceptor units, morphology of the blends, compositional ratio of the blends, processing conditions…), contributing to the many

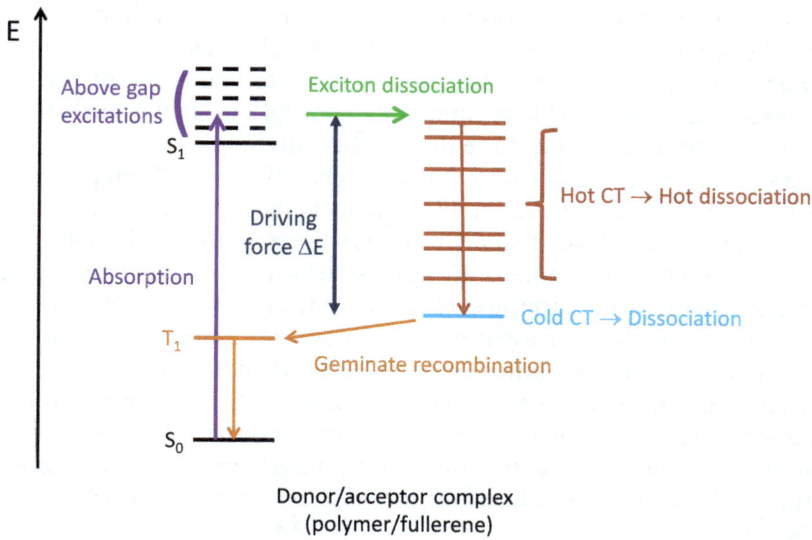

Figure 11.1 Illustration of the key electronic processes occurring at a polymer/fullerene interface, with the polymer acting as the absorbing material.

conflicting results reported in the literature. The different concepts described here are thus not all operative in a single device, and a current challenge in the field is to design the right materials and morphologies to favor the mechanisms leading to the highest charge generation efficiency.

In organic thin films, it is well accepted that the excitations generated at early times (in the fs timescale) are typically delocalized over several molecule/polymer segments. Indeed, if adjacent chains have similar transition energies, the electronic interactions between them will tend to promote a delocalization of the excitation. The actual degree of delocalization, at least at an early stage, results from a trade-off between the amplitude of the electronic interactions between the molecules and the amplitude of the energetic disorder among the transition energies; strongly disordered structures with a large dispersion of transition energies will favor the formation of excitations localized over a limited number of chains. Intramolecular and intermolecular vibrations then come into play in the picosecond timescale to relax the excitation, *via* nuclear geometric relaxation within the chains or loss of intermolecular coherence by dephasing effects, respectively. As a result, the spatial extent of the excitation shrinks progressively down to a single molecule (or a limited number of chains in ordered structures) after photoexcitation. When an exciton is fully relaxed spatially, it can only migrate by hopping from site to site within the thin film; since the lifetime of an excitation is typically on the order of nanoseconds, this translates into a rather small exciton diffusion range (typically in between 10–50 nm) depending on the degree of order.[8] In most cases, this precludes the use of a bilayer structure in which a donor layer is deposited on top of an acceptor layer (or *vice versa*, with film thicknesses of the order of 100 nm) since many excitations generated by the incident light will never reach the interface to be dissociated. This is why most of the cells built nowadays feature a homogeneous mixing of donor and acceptor regions at the nanoscale (the bulk heterojunction concept) to make sure that all excitations can meet an interface during their lifetime.

The spatial coherence of the excitation at early times has been put forward as a necessary property to rationalize the fact that free charge carriers can be generated below instrumental response times (<100 fs) in solar cells.[9] Indeed, such a fast timescale is not consistent with diffusion by hopping of excitations towards the interface. However, one can alternatively claim that the carriers generated at early times originate from excitations created next to the interface, thus alleviating the need for a diffusion process. A rise of the generated current with time is then associated with the conversion of the excitations generated further away from the interface.[10] This picture is supported by the fact that organic/organic interfaces are not, in general, sharply defined and typically exhibit broad mixed regions in between pure phases.[11]

It is also widely accepted that the second step of current generation after light absorption is the conversion of the intramolecular excitation into a charge-transfer exciton at the interface, with the hole residing over the donor and the electron over the acceptor.[12] It is not clear, however, whether the excitation has to reach the molecules lying at the forefront of the interface to be dissociated. Indeed, simulations of the morphologies of crystalline polythiophene/fullerene (P3HT/PCBM) interfaces at the atomistic scale *via* force-field calculations show that the polymer chains are more distorted at the interface due to the presence of a larger free volume;[13,14] these chains therefore have a larger transition energy compared to those lying in the bulk and might prevent the excitation reaching the interfacial border. In this situation, phenomenological theoretical models have suggested the possible occurrence of long-range exciton dissociation processes *via* tunneling across the interfacial region;[15] this process would generate charges that are already spatially separated, lowering their Coulomb attraction and facilitating their separation. It should also be noted that other models consider that the initially generated state is described by a coherent superposition of excitonic and charge-transfer excitations, thus allowing for the fast formation of charge separated pairs upon electronic relaxation.[16]

The complexity of the mechanism leading to free carrier generation originates from the fact that many different types of charge-transfer excitons can be created at the interface. Consider for example a charge-transfer (CT) exciton with the hole localized over a single donor molecule and an electron over a single acceptor molecule. A manifold of CT states can be generated by changing the nature of the molecular orbital supporting the hole and the nature of the level bearing the electron; the lowest energy CT state corresponds to the situation where the hole is located in the HOMO of the donor and the electron in the LUMO of the acceptor.[17] Moreover, a given excited CT state can be vibronically excited by populating high-energy vibrational levels.

Since the lowest CT state must be lower in energy than the lowest absorption band of the absorbing material (mostly the donor in polymer/fullerene solar cells) in order to provide a driving force for exciton dissociation,[18] the generated CT states have initially an excess energy to be dissipated upon relaxation. The initial models introduced to explain the efficiency of free carrier generation suggested that the thermalization of these initially "hot" excited states leads to a distribution of relaxed "cold" CT states, varying by the spatial separation between the electron and the hole.[6] The probability of CT dissociation is then given by a classical Onsager model:[19]

$$\phi(r_0) = \exp\left(-\frac{r_c}{r_0}\right) \tag{11.1}$$

with r_0 the actual separation between the electron and the hole and r_c the Coulomb capture radius corresponding to the distance at which the Coulomb attraction is equal to the thermal energy k_BT (r_c is equal to 16 nm when considering a dielectric constant of 3.5 for the organic medium). In this picture, a CT state localized on molecules lying at the interface cannot dissociate since the Coulomb attraction is typically of the order of a few tenths of an eV and it will therefore recombine geminately. Kinetic Monte Carlo simulations performed in this framework within a hopping regime further demonstrate that the rate of free carrier generation is enhanced when the mobility ratio between the electron in the acceptor phase and the hole in the donor phase is increased; this essentially reflects the fact that the distance between the two charges can be easily increased (and hence the Coulomb attraction reduced) when one charge is fixed while the other is mobile. Similar simulations indicate that the charge separation is also assisted by energetic disorder at the interface (in the extreme case by trapping one charge at the interface)[20] and by increasing the delocalization of the hole over the polymer backbone.[21] The efficiency of charge separation provided by such simulations is of course highly dependent on the value chosen for the geminate charge recombination rate.[22]

All of the previous simulations neglect the role of another key ingredient: solid-state electronic polarization effects. Indeed, in a condensed medium, a positive (negative) charge is stabilized by the formation of induced dipoles in the neighboring molecules, making the ionization potential (electron affinity) in the solid state much lower (higher) than in the gas phase. When two charges of opposite sign are involved, the degree of stabilization afforded by solid-state polarization effects clearly depends on the actual separation between the charges and is strongly reduced when the charges are next to one another at the interface. This indicates that electronic polarization effects act as a driving force for the separation of a cold CT state at the interface.[23] Moreover, quantum-chemical calculations have also shown that the inhomogeneity of the electrostatic potential at donor/acceptor (D/A) interfaces can provide a bending in the energy profile favoring the displacement of the charges from the interfaces towards the bulk.[24,25] We note that very recent calculations further demonstrate that the actual shape of the energy profile is strongly governed by long-range interactions, giving different trends when the interface has a 2D (slab) *versus* 3D (spherical) geometry.[26] Altogether, electrostatic and dynamic polarization effects can make the dissociation of cold CT exciton barrierless, as will be discussed in more detail later in this chapter.[14,23] Two other ingredients are also often neglected in the models discussed so far. The first is the role of the interface dipoles created in the ground state at the donor/acceptor interface following partial charge transfer from the acceptor to the donor or due to polarization effects.[27,28] Interestingly, quantum-chemical calculations point to the appearance of an induced dipole in fullerene molecules lying at interfaces due to an abrupt discontinuity in the electrostatic potential, though no charge transfer is observed between the donor units and the fullerene derivatives.[29] A simple electrostatic model has shown that a hole delocalized over a polymer chain lying parallel to a donor/acceptor interface could be repelled to the bulk by the repulsive potential created by the interface dipoles.[30] In the case of the pentacene/C_{60} interface, the contribution of such interface dipoles to charge separation was found to be much lower compared to the dynamic and electrostatic effects described above.[23] Another overlooked parameter is the entropy, which should favor charge separation by increasing the number of available sites for the charge carriers when they leave the interface; however, the actual contribution of entropic effects is not easy to evaluate when working in non-equilibrium conditions and when accounting for the energetic disorder in the thin films.[31]

Instead of focusing on cold CT states, another view is that the excess energy initially provided to the (vibronically or electronically) hot CT exciton can be exploited to overcome the Coulomb barrier. The charge carrier generation efficiency therefore results from a kinetic competition between the charge separation in the hot state and the thermal relaxation down to the cold CT state (on a picosecond timescale), as further supported by mixed classical/quantum dynamic simulations.[32] Experimental measurements have clearly demonstrated that the charge carrier efficiency increases with the amplitude of the driving force (energy difference between the intramolecular exciton and lowest CT exciton) in blends where the ionization potential of the donor or the electron affinity of the acceptor is systematically varied;[33,34] the excess energy can be further modulated by exciting the molecules above their optical bandgap. However, this does not appear to be an universal property since other blends yield the same amount of photogenerated charge carriers when exciting the polymer chains in their lowest absorption band or when directly exciting the cold CT states.[35–37] Additional studies suggest that another central effect controlling the efficiency of charge generation is the access to delocalized states featuring a larger separation between the electron and the hole.[38] The latter can be promoted by the motion of the hole along a single polymer chain[39] or by the delocalization of the electron over several fullerene molecules, highlighting the additional role of the cluster size.[10,40,41] Such a hot dissociation mechanism is further facilitated by the increase in the density of CT states with energy.[42] Quantum-chemical calculations have provided clear evidence that the separation between the electron and the hole increases with the energy of the CT states in polymer–fullerene complexes;[43,44] the role of excess energy and delocalization has also been illustrated by quantum dynamics simulations.[45]

The efficiency of charge carrier generation depends on the kinetics and energetics of the charge recombination processes. A very efficient mechanism for geminate recombination (implying that the electron and hole originating from the same exciton recombine) is the conversion of the lowest singlet CT state initially generated into a triplet CT state, followed by an electron transfer to generate an intramolecular triplet excited state (on the polymer in the case of polymer/fullerene blends).[18] This recombination pathway can be efficiently suppressed by choosing the blend components in such a way that the CT state lies below the intramolecular triplet states.[43] On the other hand, non-geminate recombination processes (implying a recombination after initial charge separation processes) can be limited by ensuring good charge transport properties in the donor and acceptor phases (in particular *via* a high degree of ordering). A recent experimental study has further highlighted that the recombination of triplet CT states into intramolecular triplet excitons can also be limited by charge delocalization effects.[46]

All the previous considerations demonstrate that a detailed understanding of the many electronic processes (exciton migration, exciton dissociation, charge separation, and charge transport) at the molecular scale is highly desirable in order to design new materials or morphologies aimed towards improved performance in organic solar cells. Here, theoretical modeling can prove very useful by shedding light on parameters that cannot be accessed easily experimentally, for instance the strength of the electronic couplings driving exciton and charge transport, the actual morphology at the donor/acceptor interface, and the electronic couplings between the donor and acceptor states. Since electronic couplings are known to be extremely sensitive to the relative position of the interacting molecules,[47] this also requires modeling tools to depict in a realistic manner the packing of the molecules/polymer in the bulk or at the interfaces. This is reflected by the growth of multiscale theoretical methods combining different techniques to extend the length- and timescales prevailing for a given approach. For the sake of illustration, classical force-field calculations are typically used to model the supramolecular packing of organic semiconductors (*i.e.* over a few tens of nanometers); quantum-chemical calculations can be performed next on small fragments extracted from such morphologies to evaluate relevant electronic parameters; in a final stage, these parameters can be injected into mesoscopic models, such as Kinetic Monte Carlo simulations to propagate charges or excitons over large dimensions and long timescales[48] or micro-electrostatic models used to depict the electrostatic landscape over large areas.[14] Some of these multiscale approaches will be described further in the remainder of this Chapter and used to illustrate some of the important concepts in organic solar cells discussed above.

11.2 Light Absorption and Exciton Dissociation: Review of Recent Modeling Work

The Internal Quantum Efficiency (IQE) of an OPV cell is the ratio between the number of charge carriers collected by the solar cell to the number of photons of a given energy that are incident on the solar cell from outside *and* are absorbed by the cell. Thus, the IQE embodies two important physical processes, the harvesting of solar light on the one hand and the conversion of the absorbed photons into free charge carriers on the other hand. The need to ensure good overlap with the solar AM1.5 radiation spectrum has led a large community of synthetic chemists to design new molecular strategies to harvest a larger fraction of the solar spectrum. In the field of conjugated polymers, this has led to the development of a new class of donor–acceptor conjugated copolymers with frontier molecular orbitals that can be tuned, both in energy and in shape, as a function of the strength of the electron-donating/withdrawing moiety. In addition to providing a simple tool for tailoring the width of the optical bandgap and therefore the absorption wavelength of the polymer materials, such a molecular engineering approach entails a strong degree of control of the energy level alignment at organic–organic but also organic–metal interfaces in OPVs, thought to be critical for exciton dissociation and charge extraction, respectively.

If, in most cases, a single photon promotes, at best, the formation of a single electron–hole pair, there is much hope from recent experimental and theoretical investigations that the resulting limit in quantum efficiency, also referred to as the Schockley–Queisser detailed balance limit, can be overcome by combining low-bandgap molecules or polymers with high-energy photon absorbing molecules susceptible to undergo singlet fission (SF). SF is a spin-conserving process by which the initially photogenerated singlet exciton generates a pair of spin-coupled triplets, each of which can subsequently experience dissociation at a donor–acceptor interface. Organic solar cells that exploit singlet exciton fission in pentacene to generate more than one electron per incident photon (with an IQE of $\sim$160%) have recently been demonstrated.[49]

Organic synthesis offers access to a virtually unlimited set of tailored molecular and polymeric photo-responsive materials, which have been largely exploited in the last twenty years or so, resulting in a dramatic improvement in the light harvesting ability of these materials. Thus, the real question that has been around for some time is not so much related to the collection of the incoming photons, but rather to their conversion into free charges, and in particular to understand how positive and negative charges photogenerated at D/A interfaces can escape from their mutual Coulomb attraction to produce free charge carriers. Understanding the factors driving the efficiency of free carrier generation requires a realistic description of the local morphology of the interfaces and of the ensuing static energetic disorder of the electron–hole (e–h) pairs. Moreover, the dynamical disorder due to thermally-induced structural fluctuations can significantly affect the charge separation energetics and kinetics, and should also be included in any computational scheme.[50] As mentioned above, these requirements can be achieved by resorting to multiscale modeling schemes that combine atomistic molecular dynamics (MD) to quantum mechanical (QM) and classical microelectrostatic (ME) calculations. Microelectrostatic approaches use classical expressions to describe the interaction of a local charge with the electric multipoles of the adjacent molecules and were proved successful in the evaluation of site energies and dielectric tensors in molecular crystals and thin films.[51,52] The results reviewed in this section were obtained assuming a fully localized picture considering the interactions between an excess electron/hole localized on a given molecular fragment and both the permanent atomic charges and induced dipoles of the surrounding molecules. The latter are estimated self-consistently, in order to account for mutual polarization effects, *via* consideration of a polarizability tensor. In practice, all quantities are broken down artificially into subgroup contributions (here by assigning a site for each non-hydrogen atom of the target molecules) in order to account for the molecular topology. Atomic charges of neutral and charged molecules are determined independently using QM calculations carried out at the DFT level using polarized basis sets, while atomic polarizability tensors are assigned by distributing uniformly on heavy atoms the molecular polarizability calculated at the same level. The energy of the system is thus expressed as:

$$U = \frac{1}{2}\sum_k \left(q_k V_k^0 - \boldsymbol{\mu}_k \cdot \boldsymbol{F}_k^0\right) \tag{11.2}$$

where $\boldsymbol{\mu}_k$ is the induced dipole at atomic site k, V_k^0 and $\boldsymbol{F}_k^0$ the electrostatic potential and field due to permanent charges, and the sum runs over sites bearing a charge and/or polarizability. The polarization energy of a specific charge configuration is then computed as the energy difference between the charged and the neutral system. For instance, the polarization energy, P_{ij}, associated to an electron–hole pair located at molecules i and j reads:

$$P_{ij}^{\pm} = U_{ij}^{\pm} - U^0 \tag{11.3}$$

This quantity represents the intermolecular energetic contribution to the creation of the e–h pair, and includes (i) the e–h interaction, (ii) the electrostatic interactions between the excess charges and the charge distributions of the surrounding neutral molecules (referred to as *static polarization energy*, $S_{ij}^{\pm}$), as well as (iii) the dynamic polarization effects originating from the electronic polarization of the surrounding molecules in presence of the charge (referred to as *dynamic polarization energy*, $D_{ij}^{\pm}$). Accounting for the additional intramolecular contributions, the total energy to create an e–h pair at sites i–j from the neutral system reads:

$$E_{ij}^{\pm} = \mathrm{IP}_i^{\mathrm{gas}} - \mathrm{EA}_j^{\mathrm{gas}} + P_{ij}^{\pm} \tag{11.4}$$

where IP_i^{gas} (EA_i^{gas}) is the gas-phase ionization potential (electron affinity) of molecule *i*. The energy necessary to split a geminate e–h pair bound at the interface into free charge carriers in the bulk materials can then be calculated as the difference between the corresponding energies:

$$\Delta E^{\pm} = E^{\pm}_{\text{free}} - E^{\pm}_{\text{interface}} = \Delta \text{IP}^{\text{gas}} - \Delta \text{EA}^{\text{gas}} + \Delta P^{\pm} \tag{11.5}$$

where non zero values of ΔIP^{gas} and ΔEA^{gas} arise from conformational fluctuations in disordered systems. Although ME models cannot grasp subtle electronic effects such as the hybridization of the molecular orbitals of adjacent molecules, these approaches have the advantage of handling highly extended systems. Therefore, ME calculations performed on large-scale snapshots extracted at regular time steps from the simulated morphologies provide a statistical time sampling of the energetics landscape spanned by local charges moving in the vicinity of the interface.

Such a framework was used recently to investigate the energy profile of e–h pairs at the prototypical P3HT/PCBM interface.[14] The interface morphology was obtained by means of MD simulations using periodic boundary conditions and a force field adapted from the AMBER united atom parameterization. The sample considered, shown in Figure 11.2a, has the typical size of P3HT crystallites and amorphous PCBM clusters, as measured in bulk heterojunctions after thermal annealing.[53] Simulated structures feature a certain amount of structural disorder that translates into the spreading by several hundreds of meV of the polarization energies for e–h pairs ($P^{\pm}$, blue dots, Figure 11.2b), calculated at the ME level for all possible combinations of electron and hole positions within the inner region of P3HT/PCBM clusters. This broadening of the energy levels is mostly due to the random orientation of the PCBM polar groups, as evidenced by the much thinner $P^{\pm}$ profile calculated for an artificial interface built by replacing PCBM by C_{60} molecules. Decomposing the total polarization energies into their static ($S^{\pm}$) and dynamic ($D^{\pm}$) contributions shows that the $S^{\pm}$ profile increases upon e–h pair separation and reaches a plateau at about 4–5 nm. Accounting for electrostatic interactions only (*i.e.*, excluding all polarization effects), the energy required to unbind geminate e–h pairs is estimated in this way as ~0.7 eV, which is much lower than the 1.2 eV value calculated considering the unscreened electrostatic interaction between the two charged molecules. In addition, $D^{\pm}$ decreases by ~0.4 eV upon e–h pair separation, also favoring the splitting of geminate pairs. This latter effect originates from the fact that two separate charges (in their respective bulks) are more stabilized by the polarizable environment than an e–h pair at the interface. Therefore, both the intermolecular electrostatic interactions and the dynamic polarization effects act synergistically to overcome the e–h attractive interaction.

Intramolecular contributions to the energy of e–h pairs, which arise from the dependence of IP^{gas} and EA^{gas} on the molecular geometry (see eqn (11.4)), were evaluated at the semi-empirical AM1 level using

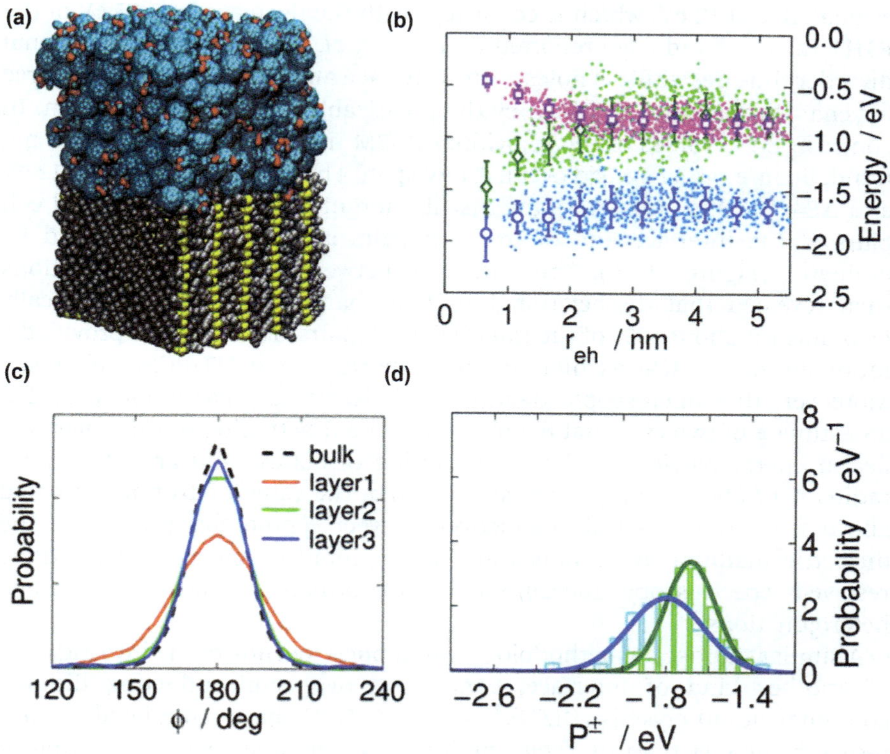

Figure 11.2 (a) Snapshot of the P3HT/PCBM interface, consisting in 400 molecules of amorphous PCBM on top of a crystalline slab of regioregular P3HT. The P3HT phase is composed by 5 lamellae of 16 π-stacked 20-*mers*, bonded to their periodic replica across the simulation box to simulate an infinite polymer chain. (b) ME polarization energy of e–h pairs as a function of the distance between electron and hole centers of charge: total ($P_{ij}^{\pm}$, blue dots), static ($S_{ij}^{\pm}$, green dots) and dynamic ($D_{ij}^{\pm}$, magenta dots) e–h polarization energy. Circles, diamonds and squares show mean values and standard deviations (error bars) of $P_{ij}^{\pm}$, $S_{ij}^{\pm}$ and $D_{ij}^{\pm}$, respectively, calculated over intervals of 0.5 nm. (c) Probability distribution of thiophene–thiophene dihedral angles of P3HT at PCBM interface and in the bulk, obtained from MD trajectories. (d) Histograms and probability distributions of the ME polarization energy of e–h pairs at the interface ($r_{eh} < 1.5$ nm, cyan and blue line) and free e–h pairs ($r_{eh} > 5.0$ nm, green and dark green line).
Adapted with permission from G. D'Avino *et al.*, *J. Phys. Chem. C*, **117**, 12981. Copyright (2013) American Chemical Society.[14]

molecular structures extracted from MD simulations. The distributions of thiophene–thiophene dihedral angles for polymer chains taken at the interface and in inner layers evidence a high degree of torsional disorder of the P3HT chains at the interface (Figure 11.2c), which translates into a decrease of IP^{gas} of the individual P3HT chains upon moving away from the interface. The average difference between interfacial and bulk chains was evaluated as 0.12 eV, which is consistent with the decrease by 0.15 eV of the P3HT valence band edge reported by Troisi *et al.*[13] The interfacial torsional disorder thus destabilizes holes at the interface and adds to the driving force for charge separation provided by the polarizable surrounding medium. In contrast, EA^{gas} values of the individual PCBM molecules do not show a net trend, but are rather spread over a 0.1 eV span. Thus, using $\Delta IP^{gas} = -0.12$ eV and $\Delta EA^{gas} = 0$ in eqn (11.5), the distribution of the energies of bound e–h pairs at the interface and of free e–h pairs at large distances could be evaluated (Figure 11.2d). The difference between these two distributions then revealed that dissociation into free charge carriers is energetically favorable for about 50% of the interfacial e–h pairs. These results provided a rationale for the IQE as high as 80% reported for P3HT/PCBM blends.[54] Moreover, this microscopic treatment of electronic polarization provided an estimate of two essential parameters to the functioning of this solar cell, *i.e.* an energy barrier for charge separation of 0.2 eV, and an e–h capture radius r_c of about 3 nm, much smaller than the value of 16 nm estimated above from the classical interaction between point charges within the dielectric medium. Both values are in quantitative agreement with time-resolved spectroscopy measurements performed on polymer–fullerene heterojunctions.[10]

A similar multiscale methodology was applied recently to the investigation of another kind of interface, connecting two semiconducting discotic columnar liquid crystals (DLCs).[48] Although DLCs have stimulated intense research interest towards their application in organic solar cells owing to their relatively long exciton diffusion lengths (about 50–100 nm) and high charge carrier mobilities (up to 1 $cm^2\ V^{-1}\ s^{-1}$), DLC-based heterojunctions have so far led to disappointing outcomes in terms of efficiency. Motivated by the recent synthesis of suitable discotic acceptors,[55,56] the interface between a perylene (PE) donor and a benzoperylene diimide (BP) acceptor was considered (see Figure 11.3a), with the double intent of investigating the morphology of the interface and understanding the relationships between the molecular organization and the energetics and kinetics of the charge separation processes. To mimic a typical experimental setup, the interface was built by stacking BP and PE columnar phases oriented homeotropically on top of a silica slab (Figure 11.3b). MD simulations conducted using a modified AMBER force field revealed that, owing to the incommensurability of the hexagonal lattices of the two DLCs, various degrees of mismatch can exist between contiguous D/A columns. To address whether having well-matched D and A columns at the interface is beneficial or not for the performance of a planar heterojunction organic solar cell, the relevant electronic processes were compared in two limiting situations: in the first one, the interface between BP and PE consists of molecules with largely superimposed aromatic cores (hereafter termed "*good matching*"); in the second type, the BP and PE cores display a large horizontal displacement ("*bad matching*"), each one facing the alkyl chains of the opposite semi-column.

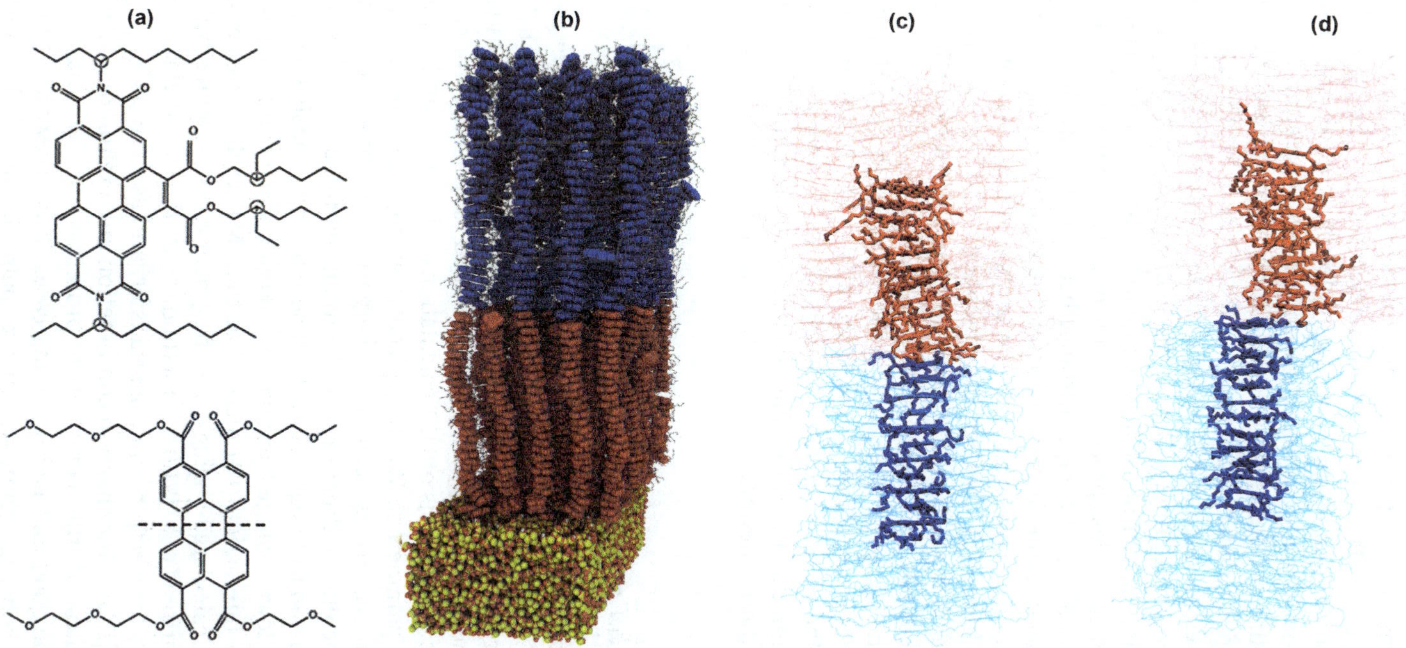

Figure 11.3 (a) Molecular structure of the benzoperylene diimido diester acceptor (BP, top, circles indicate the chiral centers) and of the perylene donor (PE, bottom, the dashed line indicates the symmetry mirror plane). (b) Snapshot of the simulated SiO_2/PE/BP stack with highlighted aromatic cores (PE in red, BP in blue). (c, d) Good/bad matching cylindrical clusters employed for electronic structure calculations.
Reprinted with permission from J. Idé *et al.*, *J. Am. Chem. Soc.*, 2014, **136**, 2911. Copyright (2014) American Chemical Society.[48]

Applying the same computational scheme as for the P3HT/PCBM heterojunction discussed above, the intra- and intermolecular energetic contributions to the energy of e–h pairs generated in the vicinity of the interface were evaluated at the AM1 and ME levels, respectively. In the case of the BP/PE interface, it was found that the intramolecular contribution to the transport levels does not have a net effect on the e–h separation, but represents only a source of energetic disorder. Intermolecular electron–hole polarization energies calculated along the MD trajectory for the good and bad matching cases revealed qualitatively different energy profiles for e–h separation: for good matching columns, the e–h separation requires overcoming an activation barrier of $\sim$0.8 eV, while in the bad matching case the e–h separation is a barrierless process for 31% of the e–h pairs, owing to the weaker e–h binding due to the larger distance between the interfacial D and A cores. Similar beneficial effects of reducing the Coulomb e–h binding by increasing the distance between the D and A units at the interface were also demonstrated at interfaces between polythiophene and various types of acceptor, where the introduction of bulky side groups induced the increase in the OSC photocurrent.[57]

In addition to the energetic landscape of the e–h pairs, the kinetics of the competitive electronic processes taking place at the PE/PB interface were also addressed. The rates of exciton dissociation (ED), hole and electron transfer away from the interface, and charge recombination (CR) were evaluated within the framework of the Marcus–Levich–Jortner theory from the expression:

$$k=\frac{2\pi}{\hbar}t^2\sqrt{\frac{1}{4\pi\lambda_{\text{ext}}k_{\text{B}}T}}\sum_{n=0}^{\infty}\exp\left(-S_{\text{eff}}^{n}\right)\frac{S_{\text{eff}}^{n}}{n!}\exp\left(-\frac{\left(\Delta G+\lambda_{\text{ext}}+n\hbar\omega_{\text{eff}}\right)^2}{4\lambda_{\text{ext}}k_{\text{B}}T}\right) \quad (11.6)$$

where $S_{\text{eff}}=\lambda_{\text{int}}/\hbar\omega_{\text{eff}}$ is the Huang–Rhys factor measuring the strength of electron–phonon coupling, which is directly related to the internal reorganization energy (λ_{int}) and to the frequency of an effective vibrational mode assisting the process (ω_{eff}). λ_{int} was estimated for the different charge transfer processes on the basis of DFT calculations, while $\hbar\omega_{\text{eff}}$ and λ_{ext} were set as adjustable parameters. When neglecting entropy contributions, the free energy difference ΔG associated to charge transfer processes can be identified with the energy difference ΔE between the initial and final electronic states evaluated from eqn (11.4). To ensure internal consistency in the calculations, the lowest singlet excitation energies were computed using the configuration interaction (CI) method at the AM1 level. The electronic couplings were calculated using the fragment-orbital semi-empirical Valence Bond/Hartree-Fock (VB/HF) method,[58] as $t^-=\langle A^-A|\hat{H}|AA^-\rangle$ and $t^+=\langle D^+D|\hat{H}|DD^+\rangle$ for electron and hole transfer, and as $t^{\text{CR}}=\langle D^+A^-|\hat{H}|DA\rangle$ and $t^{\text{ED}}=\langle D^*A|\hat{H}|D^+A^-\rangle$ for charge recombination and exciton dissociation. The resulting rates of the competitive processes are summarized in Figure 11.4 for the good and bad matching cases. It is noteworthy that the weaker binding between geminate charge pairs in the bad matching

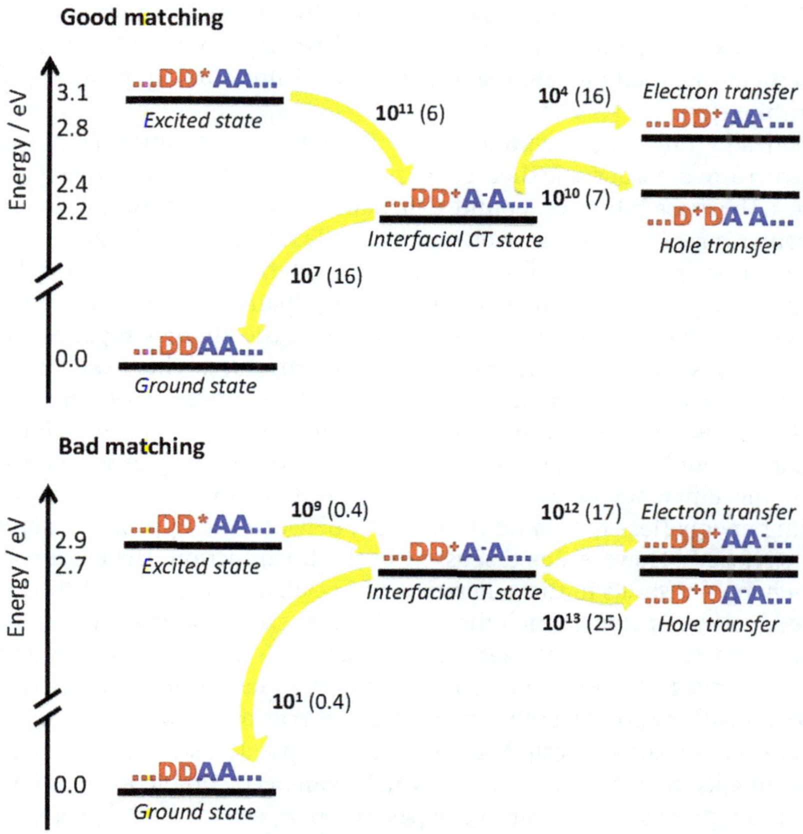

Figure 11.4 Schematic representation of the charge transfer processes occurring at the BP/PE interface for the case of good (top) and bad (bottom) matching columns, with mean energies of the initial and final states, mean absolute values of the relevant electronic couplings (between brackets, in meV) and rates (bold, order of magnitude in s^{-1}). Reproduced with permission from ref. 50 with permission from the PCCP Owner Societies.

columns results in high rates for hole and electron transfer away from the interface, even for the first steps of charge separation, which instead constitute the bottleneck for e–h separation in the good matching case. For the latter geometry, we found that the ground-state recombination of interfacial e–h pairs competes with electron transport, providing a channel for energy loss. Conversely, in bad matching columns, the smaller electronic couplings, but also the less stable D^+A^- state, makes CR many orders of magnitude slower than e–h separation. The weak *D*/*A* couplings in the case of bad matching columns lead to ED rates two orders of magnitude smaller than in the good matching case, although still acceptable in terms of absolute values ($\sim 10^9$ s^{-1}). These results suggest therefore that a certain amount of interfacial mismatch between the π-conjugated cores of discotic mesogens should be beneficial for the performance of DLC-based OSCs, as it favors the formation of weakly bound CT states that can efficiently mediate charge separation without being prone to fast recombination.

Besides highlighting the importance of the interfacial morphology and of the structural disorder, the theoretical results obtained for these two systems suggest the possibility for bound e–h pairs to separate through an intermolecular hopping process involving thermally relaxed "*cold*" states only. In essence, this is made possible due to a cancellation of the large Coulomb binding energy of the photogenerated charge pair at the interface by the increase in solid-state polarization effects when the charges separate, together with possible favorable band bending processes. This result is consistent with the similar quantum yields of carrier photogeneration measured in P3HT/PCBM interfaces after optical irradiation above *versus* below the band gap,[36] and with time-resolved microwave conductivity experiments, in which the efficiency of charge generation was attributed to a very small binding energy (less than a tenth of an eV) between the opposite charges at the interface.[59] However, as discussed in the Introduction, the exact role of high-energy CT states in the generation of free charge carriers is still elusive. Indeed, pathways along which the local Frenkel excitation first decays into a higher-lying CT state could lead to competitive charge generation mechanisms, either in helping to reach the lowest CT state from which geminate e–h pairs will eventually split, or through a direct dissociation of the "*hot*" CT state *via* a resonant coupling with charge-separated states. Femtosecond transient spectroscopy experiments, which evidenced sub-picosecond charge separation processes for various types of solar cells,[32,38,60,61] were interpreted as indications of the existence of such high-energy channels to dissociation, possibly because of the larger e–h separation in the hot CT states.[39,45] Quantum chemical (QC) calculations also attributed the origin of the efficient charge separation in fullerene-based heterojunctions to the availability of many high-energy states in PCBM crystallites.[41] However, the ultrafast relaxation of the high-energy CT states within the CT band evidenced by time-delayed collection field experiments contradicts the picture of charge generation through hot channels.[62] Instead, the high yields of free charge carrier generation observed in polymer:fullerene devices were attributed to the weakly bound character of the lowest-energy CT states. From QC calculations carried out on idealized D/A interfaces, it was shown that such highly delocalized cold CT states, with energy close to the energy of free holes and electrons, could be kinetically accessible from the exciton states due to the coupling between the orbitals of the donor and acceptor molecules.[63,64] A deep understanding of the competitive charge generation mechanisms in OPV cells thus calls for the development of modeling schemes that accurately describe the manifold of relevant local excited states in D/A interfaces, with a full account of electrostatic and polarization effects induced by the molecular environment, beyond the single-molecule picture for charge localization.

11.3 The Challenges Ahead: Towards the Prediction of CT Energies and Decay Dynamics

Without going into details, it suffices here to stress that the relevant overall quantum efficiency of a solar cell is a *power* efficiency, namely the product of a (short-circuit) current and a (open-circuit) voltage. Though some of the molecular details are still obscure, *vide supra*, it is now very clear that photoconversion in organic solar cells can be quasi-quantitative, namely all absorbed photons can be transformed into free charge carriers in properly designed device architectures and material combinations. For instance, Liu *et al.* have achieved high-performance (efficiencies up to 10.8%, fill factors up to 77%) thick-film polymer solar cells *via* the formation of a near-ideal polymer:fullerene morphology that contains highly crystalline and reasonably small polymer domains.[65] This morphology yields an external quantum yield above ~85% across a broad spectral range (~400–700 nm) and translates into short circuit currents close to 20 mA cm^{-2}. These currents are comparable to those measured in methylammonium lead iodide perovskites solar cells, yet the latter produce power conversion efficiencies (PCE) in excess of 15%. The main culprit for the limited power conversion efficiencies (PCE) in organics is their low open-circuit voltage, V_{oc}, around 0.77 V for the cells in ref. 65, against $V_{oc} > 1$ V in perovskite-based cells.

In a seminal contribution, Vandewal and colleagues have developed a microscopic model relating the open-circuit voltage to interface molecular properties of donor–acceptor bulk heterojunction solar cells (ref. 66 and refs therein). Based on the premise that the photogenerated current exactly compensates for the recombination current at open circuit and using a Marcus-like shape for the absorption cross section to the lowest CT state (see detailed derivation in ref. 66), they derived the following expression for V_{oc}:

$$V_{oc} \approx \frac{E_{CT}}{q} + \frac{kT}{q}\ln\left(\frac{J_{ph}h^3c^2}{fq2\pi E_{CT}}\right) + \frac{kT}{q}\ln(\mathrm{EQE_{EL}}), \tag{11.7}$$

where E_{CT} denotes the vertical CT transition energy, q is the elementary charge, f the CT optical absorption strength, J_{ph} the photogenerated current, and $\mathrm{EQE_{EL}}$ the electroluminescence external quantum efficiency. From eqn (11.7), the open-circuit voltage thus includes three contributions, with the first one being directly proportional to the energy of the CT state and the second and third temperature-dependent terms accounting for radiative and non-radiative losses, respectively. Very interestingly, independent measurements of V_{oc} and E_{CT} across a wide range of donor–acceptor interfaces show a nice linear correlation with an offset of ~0.6 eV attributed to radiative (~0.25 eV) and non-radiative (~0.35 eV) decay loss mechanisms of the CT pairs.[66] Clearly, the next challenge for computational scientists will be to develop and apply methods for a quantitative assessment of the CT energetics and dynamics, including the very demanding task of predicting rates for radiationless transitions. With the development of fast and robust quantum dynamics simulation approaches such as MultiConfigurational Time Dependent Hartree (MCTDH) methods (so far applied only to describe charge separation at model interfaces),[67] we are convinced that multiple endeavors in that direction will be proposed in the near future.

As alluded to in the previous section, getting the first term in eqn (11.7) right is already challenging. We have already emphasized the need to properly account for both static and dynamic polarization effects as well as delocalization effects in the solid state. Ideally, this should be done from first principles if the overall formalism is to be predictive. Here, we would like to review briefly our first attempt to predict the relative energetic positions of charge-transfer *vs.* local excitations in simple P3HT/PCBM complexes from *ab initio* calculations. As described in ref. 43, out of the lowest 20 electronic excitations computed at the TDDFT/B3LYP/6–31G(d,p) level for a complex between a 10-*mer* polymer chain and a PCBM acceptor, the lowest electronic excited state with a charge-transfer character ($^1CT_1$) was calculated at about 1.1 eV. It involves mainly the transition from the HOMO (lying on the polymer chain) to the LUMO (on PCBM) which, together with the $^1CT_2$ (HOMO→LUMO+1) and $^1CT_3$ (HOMO→LUMO+2), yield a set of three 'cold' charge-transfer excited states (Figure 11.5). Above these states lie multiple 'hot' (CT) charge-transfer states, originating from deeper and more delocalized occupied molecular orbitals, mostly HOMO–1 ($^1CT_4$–$^1CT_6$) and HOMO–2 ($^1CT_7$–$^1CT_9$). They extend in energy up to ~2 eV, just below the on-chain HOMO→LUMO electronic transition to the S_1 state of P3HT, calculated at 2.05 eV, in excellent agreement with the experimental value of 1.91 eV.[68] In order to complete the energy state diagram, the transition

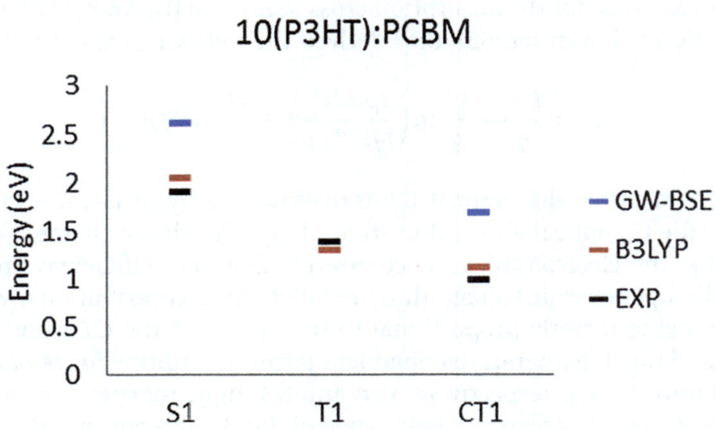

Figure 11.5 State diagram of the energy alignment in the P3HT/PCBM complex, as predicted by the GW/BSE and B3LYP methods, and as measured. S_1 and T_1 are the lowest-energy polymer singlet and triplet excitons, respectively. The $^1CT_1$ energy is the energy of the relaxed, Coulombically bound electron–hole pair across the heterojunction.

energy to the polymer triplet (T_1) has been calculated at 1.31 eV, which is ~0.1 eV lower than the corresponding measured value.[69] Compared to the experimental data, the DFT/B3LYP results for the P3HT:PCBM interface reproduce surprisingly well the relative position of the E_{S1}, E_{T1} and E_{CT1} on the energy state diagram (Figure 11.5); in line with spectroscopic data, the P3HT:PCBM heterojunction is a clear example of a blend showing photo-induced electron transfer (PET) without subsequent charge recombination to the triplet.

Complementing our theoretical results with more reliable self-consistent many-body GW/BSE calculations reveals a striking qualitative agreement between both levels of theory and the experimental observations.[43] In particular, both theoretical methods match well in terms of the energy ordering of the most distinctive excited states (S_1/CT_1 alignment and the 'hot' CT states lying below the S_1 state), the topologies of the wavefunctions representing the hole and electron, and the extent of delocalization in the 'hot' CT excited states (*e.g.* $^1CT_6$) with respect to the lowest-lying CT state (*e.g.* $^1CT_1$). Although the excitation energies in the case of GW/BSE are hypsochromically shifted by ~0.5 eV with respect to the TDDFT/B3LYP and the experimental values, there is excellent agreement in the driving force for PET (ΔG_{CT}), both fitting the experimental estimation of *ca.* −0.9 eV. Also the TDDFT/B3LYP electron–hole separation, as assessed from the state dipoles (Figure 11.6a), shows the same trend with the excitation energy as the GW/BSE calculated electron–hole distances (Figure 11.6b). Both approaches suggest a rather modest increase of the electron–hole separation by ~1–2 Å (according to GW/BSE) when going from the lower ($^1CT_1$–$^1CT_3$) to the higher ($^1CT_4$–$^1CT_6$) CT states. We conclude from this preliminary study that the lower-lying excited states are somehow delocalized in P3HT/PCBM complexes and might promote long-range electron–hole separations, while there is a limited increase in delocalization upon IR absorption from 'cold' to 'hot' CT states. We note that: (i) this picture might be affected by the degree of intermolecular order in the P3HT- and/or PCBM-rich regions, which should generally lead to extended delocalization of the electronic states; and (ii) localized excitations on the PCBM are calculated to lie in between the singlet charge transfer $^1CT_1$ and polymer exciton S_1 states at both TDDFT and GW/BSE levels. The high PET driving force and the position of the 'hot' states just below the singlet S_1 state in P3HT:PCBM can promote an effective pathway for charge dissociation *via* 'hot' delocalized CT states.

We conclude with a short discussion on the comparison between the results provided by the different theoretical approaches adopted here. It is well documented that, mostly owing to self-interaction error and the too short-sighted exchange–correlation potential, TDDFT/B3LYP underestimates the energy of charge transfer excitations. However, medium dielectric polarization effects, not included here, also largely renormalize the electronic bandgap upon going from isolated molecules to the corresponding solid-state materials; typical electronic polarization energies in organic conjugated crystals are in the ~1 eV range, so that a free charge-transfer pair is

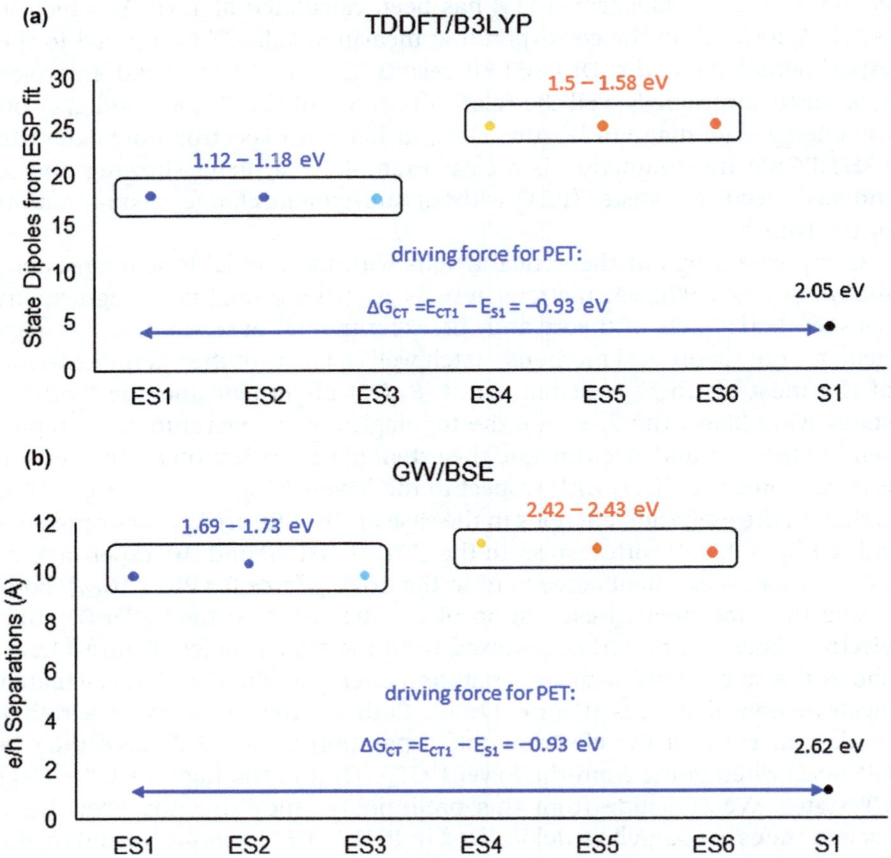

Figure 11.6 (a) TDDFT/B3LYP state dipoles and (b) GW/BSE electron–hole separation for relevant excited states in the 10P3HT:PCBM complex. The blue arrows represent the driving force calculated with both methods. Both approaches suggest a modest (~2 Å, according to GW/BSE) increase in the electron–hole separation when going from the lower to the higher-lying CT states.
Adapted from ref. 43 with permission from John Wiley and Sons © 2014 WILEY-VCH Verlag GmbH & Co. KGaA, Weinheim.

expected to be stabilized by up to 2 eV when embedded in the polarizable environment of the surrounding conjugated chromophores, as demonstrated in the previous section. From the study of the model systems investigated here, it seems these two main sources of errors (local character of the kernel and lack of polarization effects) almost perfectly cancel out, providing an overall energetic scheme for the local *vs.* charge-transfer excitations in gas-phase model complexes, in excellent agreement with the measured spectroscopic data on the corresponding bulk heterojunctions. While this is good news because it validates the relatively inexpensive gas-phase B3LYP approach to perform fast screening calculations of potentially interesting donor–acceptor interfaces, one must keep in mind that the quantitative agreement with the experiment obtained here for B3LYP results is borne out by a lucky cancellation of errors. Rather, there is a need to develop embedding techniques, such as the family of continuous polarizable models (PCM) or discrete microelectrostatic models, to be adapted to the Bethe–Salpeter formalism, which should allow to treat on an equal footing all electronic excitations in complex multi-chromophoric systems, from the most covalent to the most ionic.

11.4 Concluding Remarks

From this brief overview of some of our own recent modeling work, it seems that despite the fact organic photovoltaic cells have been around for a while and have now reached a stage of maturity close to commercialization, there is a need for further development or refinement of the existing theoretical models, in particular:

i. to account for the existence of multiple phases (pure donors, pure acceptors and mixed domains) in bulk heterojunction cells, while including molecular details on the microstructure of the interfacial regions;
ii. to set up kinetic models that rely on a fully atomistic description of the energy landscape for charge carriers across interfaces and in the bulk and where electronic polarization, coupling to vibrations, disorder and delocalization effects can be easily implemented;
iii. and most importantly, to identify which are the (non)radiative decay mechanisms that rule the CT state population and how one can tune these through molecular design to solve the open-circuit voltage bottleneck of all actual OPV devices.

Acknowledgements

The authors acknowledge fruitful collaborations with P. Aurel, I. Avilov, A. A. Bakulin, X. Blase, H. Bock, T. Branquinho de Queiroz, J. L. Brédas, D. Cheyns, G. D'Avino, L. Ducasse, I. Duchemin, R. H. Friend, V. Geskin, M. Guillaume, P. Heremans, J. Idé, S. Kümmel, C. Lennartz, M. Linares, A. Mityashin, R. Méreau, S. Mothy, L. Muccioli, D. Niedzalek, Y. Olivier, S. Osella, V. G. Pavelyev, M. S. Pshenichnikov, A. Rao, C. Rolin, O. M. Roscioni, P. H. M. van Loosdrecht, S. Verlaak, L. Wang and C. Zannoni. The results reported in this Chapter were obtained thanks to financial support of the European Community (project MINOTOR, FP7-NMP-228424). F. C. thanks the FNRS for his visiting researcher position in Mons. The work in Mons has also been supported by the Interuniversity Attraction Pole programs of the Belgian Federal Science Policy Office (PAI 7/05) and FNRS-FRFC. D. B. and J. C. are research directors of FNRS.

References

1. C. J. Brabec, N. S. Sariciftci and J. C. Hummelen, *Adv. Funct. Mater.*, 2001, **11**, 15.
2. B. O'Regan and M. Grätzel, *Nature*, 1991, **353**, 737.
3. M. M. Lee, J. Teuscher, T. Miyasaka, T. N. Murakamai and H. J. Snaith, *Science*, 2012, **2**, 643.
4. M. C. Scharber and N. S. Sariciftci, *Prog. Polym. Sci.*, 2013, **38**, 1929.
5. N. J. Jeon, J. H. Noh, Y. C. Kim, W. S. Yang, S. Ryu and S. I. Seok, *Nat. Mater.*, 2014, **13**, 897.
6. P. Peumans and S. R. Forrest, *Chem. Phys. Lett.*, 2004, **398**, 27.
7. S. Few, J. M. Frost and J. Nelson, *Phys. Chem. Chem. Phys.*, 2015, **17**, 2311.
8. H. Najafov, B. Lee, Q. Zhou, L. C. Feldman and V. Podzorov, *Nat. Mater.*, 2010, **9**, 938.
9. L. G. Kaake, D. Moses and A. J. Heeger, *J. Phys. Chem. Lett.*, 2013, **4**, 2264.
10. S. Gélinas, A. Rao, A. Kumar, S. L. Smith, A. W. Chin, J. Clark, T. S. van der Poll, G. C. Bazan and R. H. Friend, *Science*, 2014, **343**, 512.
11. D. C. Coffey and D. S. Ginger, *Nat. Mater.*, 2006, **5**, 735.
12. D. Amarasinghe Vithanage, A. Devizis, V. Abramavicius, Y. Infahsaeng, D. Abramavicius, R. C. I. MacKenzie, P. E. Keivanidis, A. Yartsev, D. Hertel, J. Nelson, V. Sundström and V. Gulbinas, *Nat. Commun.*, 2013, **4**, 2334.
13. D. P. McMahon, D. L. Cheung and A. Troisi, *J. Phys. Chem. Lett.*, 2011, **2**, 2737.
14. G. D'Avino, S. Mothy, L. Muccioli, C. Zannoni, L. Wang, J. Cornil, D. Beljonne and F. Castet, *J. Phys. Chem. C*, 2013, **117**, 12981.
15. D. Caruso and A. Troisi, *Proc. Natl. Acad. Sci. U. S. A.*, 2012, **109**, 13498.
16. E. R. Bittner and C. Silva, *Nat. Commun.*, 2014, **5**, 3119.
17. J. L. Brédas, J. S. Norton, J. Cornil and V. Coropceanu, *Acc. Chem. Res.*, 2009, **42**, 1691.
18. D. Veldman, S. C. J. Meskers and R. A. J. Janssen, *Adv. Funct. Mater.*, 2009, **19**, 1939.
19. P. K. Nayak, K. L. Narasimhan and D. Cahen, *J. Phys. Chem. Lett.*, 2013, **4**, 1707.
20. T. Offermans, S. C. J. Meskers and R. A. J. Janssen, *J. Chem. Phys.*, 2003, **119**, 10924.
21. C. Deibel, T. Strobel and V. Dyakonov, *Phys. Rev. Lett.*, 2009, **103**, 036402.
22. V. Mihailetchi, L. Koster, J. C. Hummelen and P. Blom, *Phys. Rev. Lett.*, 2004, **93**, 19.
23. S. Verlaak, D. Beljonne, D. Cheyns, C. Rolin, M. Linares, F. Castet, J. Cornil and P. Heremans, *Adv. Funct. Mater.*, 2009, **19**, 3809.
24. S. Mothy, M. Guillaume, J. Idé, F. Castet, L. Ducasse, J. Cornil and D. Beljonne, *J. Phys. Chem. Lett.*, 2012, **3**, 2374.
25. S. R. Yost and T. Van Voorhis, *J. Phys. Chem. C*, 2013, **117**, 5617.
26. C. Poelking, M. Tietze, C. Elschner, S. Olthof, D. Hertel, B. Baumeier, F. Würthner, K. Meerholz, K. Leo and D. Andrienko, *Nat. Mater.*, 2015, **14**, 434.

27. H. Ishii, K. Sugiyama, E. Ito and K. Seki, *Adv. Mater.*, 1999, **11**, 605.
28. I. Avilov, V. Geskin and J. Cornil, *Adv. Funct. Mater.*, 2009, **19**, 624.
29. M. Linares, D. Beljonne, J. Cornil, K. Lancaster, J. L. Brédas, S. Verlaak, A. Mityashin, P. Heremans, A. Fuchs, C. Lennartz, J. Idé, R. Méreau, P. Aurel, L. Ducasse and F. Castet, *J. Phys. Chem. C*, 2010, **114**, 3215.
30. V. Arkhipov, P. Heremans and H. Bässler, *Appl. Phys. Lett.*, 2003, **82**, 4605.
31. B. Gregg, *J. Phys. Chem. Lett.*, 2011, **2**, 3013.
32. A. E. Jailaubekov, A. P. Willard, J. R. Tritsch, W. L. Chan, N. Sai, R. Gearba, L. G. Kaake, K. J. Williams, K. Leung, P. Rossky and X. Y. Z. Zhu, *Nat. Mater.*, 2013, **12**, 66.
33. S. Shoaee, T. M. Clarke, C. Huang, S. Barlow, S. R. Marder, M. Heeney, I. McCulloch and J. Durrant, *J. Am. Chem. Soc.*, 2010, **132**, 12919.
34. S. D. Dimitrov, A. A. Balukin, C. B. Nielsen, B. C. Schroeder, J. Du, H. Bronstein, I. McCulloch, R. H. Friend and J. Durrant, *J. Am. Chem. Soc.*, 2012, **134**, 18189.
35. A. A. Balukin, D. Martyanov, D. Y. Paraschuk, P. H. M. van Loosdrecht and M. S. Pshenichnikov, *Chem. Phys. Lett.*, 2009, **99**, 482.
36. J. Lee, K. Vandewal, S. R. Yost, M. E. Bahlke, L. Goris, M. A. Baldo, J. V. Manca and T. Van Voorhis, *J. Am. Chem. Soc.*, 2010, **132**, 11878.
37. T. G. J. van der Hofstad, D. Di Nuzzo, M. van den Berg, R. A. J. Janssen and S. C. J. Meskers, *Adv. Funct. Mater.*, 2012, **2**, 1095.
38. G. Grancini, M. Maiuri, D. Fazzi, A. Petrozza, H. J. Egelhaaf, D. Brida, G. Cerullo and G. Lanzani, *Nat. Mater.*, 2013, **12**, 29.
39. A. A. Bakulin, A. Rao, V. G. Pavelyev, P. H. M. van Loosdrecht, M. S. Pshenichnikov, D. Niedzalek, J. Cornil, D. Beljonne and R. H. Friend, *Science*, 2012, **335**, 1340.
40. B. Bernardo, D. Cheyns, B. Verreet, R. D. Schaller, B. P. Rand and N. C. Giebink, *Nat. Commun.*, 2014, **5**, 3245.
41. B. M. Savoie, A. Rao, A. A. Bakulin, S. Gelinas, B. Movaghar, R. H. Friend, T. J. Marks and M. A. Ratner, *J. Am. Chem. Soc.*, 2014, **136**, 2876.
42. X. Y. Zhu, Q. Yang and M. Muntviler, *Acc. Chem. Res.*, 2009, **42**, 1779.
43. D. Niedzialek, I. Duchemin, T. Branquinho de Queiroz, S. Osella, A. Rao, R. H. Friend, X. Blase, S. Kümmel and D. Beljonne, *Adv. Funct. Mater.*, 2015, **25**, 1972, DOI: 10.1002/adfm.201402682.
44. S. Few, J. M. Frost, J. Kirkpatrick and J. Nelson, *J. Phys. Chem. C*, 2014, **16**, 8253.
45. H. Tamura and I. Burghardt, *J. Am. Chem. Soc.*, 2013, **135**, 16364.
46. A. Rao, P. C. Y. Chow, S. Gélinas, C. W. Schlenker, C. Z. Li, H. L. Yip, A. K. Y. Jen, D. S. Ginger and R. H. Friend, *Nature*, 2013, **500**, 435.
47. J. L. Brédas, J. P. Calbert, D. A. da Silva Filho and J. Cornil, *Proc. Natl. Acad. Sci. U. S. A.*, 2009, **99**, 5804.
48. J. Idé, R. Méreau, L. Ducasse, F. Castet, H. Bock, Y. Olivier, J. Cornil, D. Beljonne, G. D'Avino, O. M. Roscioni, L. Muccioli and C. Zannoni, *J. Am. Chem. Soc.*, 2014, **136**, 2911.

49. D. N. Congreve, J. Lee, N. J. Thompson, E. Hontz, S. R. Yost, P. D. Reusswig, M. E. Bahlke, S. Reineke, T. Van Voorhis and M. A. Baldo, *Science*, 2013, **19**, 334.
50. F. Castet, G. d'Avino, L. Muccioli, J. Cornil and D. Beljonne, *Phys. Chem. Chem. Phys.*, 2014, **16**, 20279.
51. S. Verlaak and P. Heremans, *Phys. Rev. B: Condens. Matter Mater. Phys.*, 2007, **75**, 115127.
52. G. D'Avino, L. Muccioli, C. Zannoni, D. Beljonne and Z. G. Soos, *J. Theor. Comput. Chem.*, 2014, **10**, 4959.
53. M. Y. Chiu, U.-S. Jeng, C. H. Su, K. S. Liang and K.-H. Wei, *Adv. Mater.*, 2008, **20**, 2573.
54. J. M. Guo, H. Ohkita, H. Benten and S. Ito, *J. Am. Chem. Soc.*, 2010, **132**, 6154.
55. O. Thiebaut, H. Bock and E. Grelet, *J. Am. Chem. Soc.*, 2010, **132**, 6886.
56. J. Eccher, G. C. Faria, H. Bock, H. von Seggern and I. H. Bechtold, *ACS Appl. Mater. Interfaces*, 2013, **5**, 11935.
57. T. W. Holcombe, J. E. Norton, J. Rivnay, C. H. Woo, L. Goris, C. Piliego, G. Griffini, A. Sellinger, J.-L. Brédas, A. Salleo and J. M. Fréchet, *J. Am. Chem. Soc.*, 2011, **133**, 12106.
58. F. Castet, P. Aurel, A. Fritsch, L. Ducasse, D. Liotard, M. Linares, J. Cornil and D. Beljonne, *Phys. Rev. B: Condens. Matter Mater. Phys.*, 2008, **77**, 115210.
59. D. H. K. Murthy, M. Gao, M. J. W. Vermeulen, L. D. A. Siebbeles and T. J. Savenije, *J. Phys. Chem. C*, 2012, **116**, 9214.
60. H. Ohkita, S. Cook, Y. Astuti, W. Duffy, S. Tierney, W. Zhang, M. Heeney, I. McCulloch, J. Nelson, D. D. C. Bradley and J. R. Durrant, *J. Am. Chem. Soc.*, 2008, **130**, 3030.
61. D. Veldman, O. Ipek, S. C. J. Meskers, J. Sweelssen, M. M. Koetse, S. C. Veenstra, J. M. Kroon, S. S. van Bavel, J. Loos and R. A. J. Janssen, *J. Am. Chem. Soc.*, 2008, **130**, 7721.
62. K. Vandewal, S. Albrecht, E. T. Hoke, K. R. Graham, J. Widmer, J. D. Douglas, M. Schubert, W. R. Mateker, J. T. Bloking, G. F. Burkhard, A. Sellinger, J. M. J. Fréchet, A. Amassian, M. K. Riede, M. D. McGehee, D. Neher and A. Salleo, *Nat. Mater.*, 2014, **13**, 6368.
63. A. Troisi, *Faraday Discuss.*, 2013, **163**, 377392.
64. G. Raos, M. Casalegno and J. Idé, *J. Chem. Theory Comput.*, 2013, **10**, 364.
65. Y. Liu, J. Zhao, Z. Li, C. Mu, W. Ma, H. Hu, K. Jiang, H. Lin, H. Ade and H. Yan, *Nat. Commun.*, 2014, **5**, 5293.
66. K. Vandewal, K. Tvingstedt, A. Gadisa, O. Inganäs and J. V. Manca, *Phys. Rev. B: Condens. Matter Mater. Phys.*, 2010, **12**, 125204.
67. H. Tamura and I. Burghardt, *J. Am. Chem. Soc.*, 2013, **135**, 16364.
68. G. Li, V. Shrotriya, J. Huang, Y. Yao, T. Moriarty, K. Emery and Y. Yang, *Nat. Mater.*, 2005, **4**, 864.
69. A. Rao, P. C. Y. Chow, S. Gélinas, C. W. Schlenker, C.-Z. Li, H.-L. Yip, A. K.-Y. Jen, D. S. Ginger and R. H. Friend, *Nature*, 2014, **500**, 435.

CHAPTER 12

Molecular Dynamics Simulations and their Application to Thin-film Devices

OTELLO MARIA ROSCIONI AND CLAUDIO ZANNONI*

Dipartimento di Chimica Industriale "Toso Montanari",
Università di Bologna, viale Risorgimento 4, 40136 Bologna, Italy
*Email: claudio.zannoni@unibo.it

12.1 Introduction

Photovoltaic technology based on organic semiconductors (OSCs) has the potential to be processed using simple and cost-efficient techniques, yielding solar cells which are both lightweight and mechanically flexible.[1,2] The active materials in organic solar cells are typically present as thin films of π-conjugated organic molecules, playing the role of electron donors and acceptors. Absorption of a photon leads to the generation of an exciton, formed by a tightly coupled pair of charges of opposite sign, which diffuses into the organic layer. Some of the excitons hitting the donor–acceptor interface can dissociate into free charge carriers, which can separately migrate to the electrodes, generating a photo-current which is then harvested.

The performance of devices based on organic semiconductors depends critically on the morphology of the donor–acceptor interface and, therefore, a considerable effort has been devoted to the understanding of the factors

RSC Energy and Environment Series No. 16
Unconventional Thin Film Photovoltaics
Edited by Enrico Da Como, Filippo De Angelis, Henry Snaith and Alison Walker

Published by the Royal Society of Chemistry, www.rsc.org

governing the charge separation phenomena. Indeed, quantities such as Fermi levels and charge transport depend not only on the chemical nature of the materials considered but also on their molecular organisation.[3] To complicate the matter, OSCs often display polymorphic phases due to interaction with a substrate during the fabrication process.[4–6] The presence of defects, their nature and distribution, also affects the electronic transport properties of OSC devices.

A typical transport mechanism for OSC consists of a chain of hopping events, whose kinetics depends on transfer integrals between two molecules as described, for instance, by Marcus theory.[3,7,8] Since the hopping time scale is significantly faster than that of molecular motions, transport can be calculated on a sequence of successive molecular configurations. In principle, it is then possible to fully assess the hopping transport of an OSC by employing quantum mechanics tools to calculate the transfer integrals between all molecules in a system. To do so, a set of coordinates describing the position of the atoms of every molecule of the system at a certain time (*i.e.* a configuration) and its time evolution (*i.e.* a trajectory) is obtained using the classical molecular dynamics briefly described in this chapter.

The task of obtaining reasonable morphologies and validating them is not an easy one. First and foremost, a good theoretical framework is needed to describe as accurately as possible the system under investigation. Secondly, in order to achieve good statistical sampling, the simulated sample should contain at least a few thousands (or more) molecules, while the time span has to be long enough to cover all the relevant processes of equilibration first and of evolution later for the system at hand (from tens to hundreds of ns).

Fortunately, we can normally consider systems where all the constituent parts, atoms and molecules, are subject to physical processes not involving the formation and breaking of bonds. Within the assumption that the quantum properties of the atoms and molecules and their interaction potential do not change significantly during the time window considered in a given simulation, the time evolution of the system is determined by solving Newton's equations of motion of classical mechanics. This methodology, as already alluded to, is know as Molecular Dynamics (MD), and in this chapter we will outline its applications to the field of organic electronics, starting with a brief overview of the basic concepts involved. We refer to specialised books for a more comprehensive treatment.[9–13]

A classic MD simulation is concerned with the determination of the time-dependent behaviour of a molecular system composed of a given number N of interacting particles, each described by a set of coordinates $\mathbf{r} \equiv \{\mathbf{r}_i\}$ and momenta $\mathbf{p} \equiv \{\mathbf{p}_i\}$ in a volume V. The internal energy of the system is described by a Hamiltonian $\mathscr{H}$, which is the sum of the potential $U(\mathbf{r})$ and kinetic $K(\mathbf{p})$ energy contributions:

$$\mathscr{H}(\mathbf{r},\mathbf{p}) = U(\mathbf{r}) + K(\mathbf{p}), \tag{12.1}$$

while its time evolution is defined by Hamilton's equations:[14,15]

$$\frac{\mathrm{d}\mathbf{r}}{\mathrm{d}t} = \frac{\partial \mathcal{H}}{\partial \mathbf{p}}, \quad \frac{\mathrm{d}\mathbf{p}}{\mathrm{d}t} = -\frac{\partial \mathcal{H}}{\partial \mathbf{r}} = -\frac{\partial U}{\partial \mathbf{r}}. \tag{12.2}$$

In the absence of dissipative contributions, the force acting on a particle i corresponds to the gradient of the potential energy U acting on that particle by effect of all others:

$$\mathbf{f}_i = -\frac{\partial U(\mathbf{r})}{\partial \mathbf{r}_i} = -\nabla_i U. \tag{12.3}$$

In classical MD simulations, the set of functions and parameters used to express the potential energy U is called the Force Field (FF). The proper determination of FFs is therefore of pivotal importance to achieve a good level of accuracy: the critical elements are the mathematical expressions used to approximate the internal energy and their associated parameters, which are usually based on a combination of accurate quantum mechanical calculations and on empirical parameter tuned by comparison with some selected experimental quantities.

Given an initial set of of coordinates $\mathbf{r}_0$ and momenta $\mathbf{p}_0$ defining the initial state of a system, and a FF to express its internal energy, it is possible to compute the evolution time step after time step by *finite difference* integration of the coupled Newton equations of motion (eqn (12.2)). The resulting *trajectory* is recorded as a series of snapshots at discrete time steps and contains the positions and velocities of all particles as they vary with time.[16] The numerical integration is performed with time increments Δt (typically of the order of 1 fs) using one of the several algorithms available. To be more specific, let us suppose we know the positions $\mathbf{r}_i(t)$ and velocities $\dot{\mathbf{r}}_i(t)$ of every particle at a time t. A Taylor expansion gives a simple approximation for the derivatives $\dot{\mathbf{r}}$ and $\ddot{\mathbf{r}}$ at time $(t+\Delta t)$, which is, to the first term:

$$\dot{\mathbf{r}}_i(t+\Delta t) = \dot{\mathbf{r}}_i(t) + \frac{1}{m}\mathbf{f}_i(\mathbf{r}(t))\Delta t + \mathcal{O}(\Delta t^2), \tag{12.4}$$

where the terms neglected are of the order of Δt^2, and $\mathbf{f}_i(\mathbf{r}(t))$ is the force acting on the i-th particle at time t. The new positions are then:

$$\mathbf{r}_i(t+\Delta t) = \mathbf{r}_i(t) + \dot{\mathbf{r}}_i(t)\Delta t + \mathcal{O}(\Delta t^2). \tag{12.5}$$

The accuracy of any finite difference method with respect to the true solution is affected by truncation errors intrinsic to the algorithm. Although seemingly negligible, these errors build up over the millions (or more) time steps in a typical MD simulations, producing results which diverge with time from the true solution. A better approximation is given by the popular

Verlet algorithm,[17] which is based on the Taylor expansion (eqn (12.5)) to the third order for positive and negative time increments:

$$\mathbf{r}_i(t+\Delta t)=\mathbf{r}_i(t)+\dot{\mathbf{r}}_i(t)\Delta t+\frac{1}{2}\ddot{\mathbf{r}}_i(t)\Delta t^2+\frac{1}{6}\dddot{\mathbf{r}}_i(t)\Delta t^3+\mathcal{O}(\Delta t^4), \tag{12.6}$$

$$\mathbf{r}_i(t-\Delta t)=\mathbf{r}_i(t)-\dot{\mathbf{r}}_i(t)\Delta t+\frac{1}{2}\ddot{\mathbf{r}}_i(t)\Delta t^2-\frac{1}{6}\dddot{\mathbf{r}}_i(t)\Delta t^3+\mathcal{O}(\Delta t^4). \tag{12.7}$$

By adding eqn (12.6) and (12.7) together, the new positions at time $t+\Delta t$ are now expressed with a third order predictor:

$$\mathbf{r}_i(t+\Delta t)=2\mathbf{r}_i(t)-\mathbf{r}_i(t-\Delta t)+\frac{\Delta t^2}{m}\mathbf{f}_i(t)+\mathcal{O}(\Delta t^4), \tag{12.8}$$

where $\mathbf{f}_i(t)\equiv\mathbf{f}_i(\mathbf{r}(t))$. Thus, the truncation error is now $\mathcal{O}(\Delta t^4)$ instead of $\mathcal{O}(\Delta t^2)$. The forces in this formula are computed from eqn (12.3) as the gradient of the internal energy, which is itself expressed through a FF. The immediate consequence is that every MD simulation is as good as the force field employed. It should be noted that the velocity does not appear in eqn (12.8). It can be obtained afterwards as the central difference:

$$\dot{\mathbf{r}}_i(t)=\frac{1}{2\Delta t}[\mathbf{r}_i(t+\Delta t)-\mathbf{r}_i(t-\Delta t)]+\mathcal{O}(\Delta t^2), \tag{12.9}$$

which has however an error of the order of $\mathcal{O}(\Delta t^2)$ rather than $\mathcal{O}(\Delta t^4)$. To overcome this problem, variant versions of the Verlet algorithm have been developed. The most commonly used time-integration algorithms are the velocity-corrected Verlet integrator,[18] the leap-frog algorithm,[19,20] the RESPA multi-timescale integrator[21] and the family of predictor–corrector integrators.[22]

A necessary condition for a good integration algorithm is that the energy should be conserved as long as the system is isolated. An additional condition is that the computed trajectory should also be time reversible,[23,24] as for Verlet's equation (eqn (12.8)).

12.2 Force Fields

The central approximation of molecular mechanics is to avoid an explicit usage of the electronic structure of the atoms and to express the total potential energy U of a molecular system as a sum of effective multi-body interaction terms:[25,26]

$$U=\sum_i U_i+\sum_{i<j}U_{ij}+\sum_{i<j<k}U_{ijk}+\sum_{i<j<k<l}U_{ijkl}+\ \dots, \tag{12.10}$$

where the first term represents the self-energy of the particles (*e.g.* atoms), the second the pairwise interaction between two particles and so on. The set of mathematical functions and parameters used to express the terms of eqn (12.10) is referred to as *force field* (FF). This approach has been successfully applied in the fields of physics, chemistry and materials science, in particular for modelling inorganic solids, and extended to cover covalent and molecular solids, as well as organic and biological systems. As a consequence, several formulae and parametrisation schemes have been proposed, which can be organised in three broad classes of FFs:[27]

Class I This class is based on harmonic interaction terms which have a direct physical meaning, for example the total energy of the system is divided into bonded (bonds, angles and torsions) and non-bonded (dispersion and electrostatic) interactions. This class of FFs aims at reproducing molecular morphologies and thermochemical properties such as density, transition temperatures and phase organisation.

Class II This class includes anharmonic terms and cross-terms to describe the coupling between internal coordinates (*e.g.* the stretching of two adjacent bonds coupled to their bending motion). This class of FFs aims at reproducing not only the molecular geometry but also vibrational frequencies.

Class III This class includes quantum-chemical effects *via* multi-body potentials, taking into account features such as electronegativity, polarisability and hyperconjugation.

Several formulae and parametrisation schemes have been proposed, giving rise to a variety of force fields (see ref. 28 for a brief summary). Among the most used FFs for the simulation of organic molecules, we shall cite those belonging to class I: AMBER-GAFF,[29,30] OPLS,[31] and CHARMM,[32,33] all relying on the division of the internal energy into bonded (bonds, angles and torsions) and non-bonded (dispersion and electrostatic) interactions:

$$U_{\text{total}} = U_{\text{bonded}} + U_{\text{non-bonded}}. \tag{12.11}$$

A pictorial view of the bonded interactions in eqn (12.11) is given in Figure 12.1, while their general expression is:

$$\begin{aligned} U_{\text{bonded}} = {} & \sum_{\text{bonds}} K_r(r - r_{\text{eq}})^2 + \sum_{\text{angles}} K_\omega(\omega - \omega_{\text{eq}})^2 \\ & + \sum_{\text{dihedrals}} \sum_n V_n[1 + \cos(n\phi - \gamma_n)] \\ & + \sum_{\text{impropers}} K_\psi(\psi - \psi_{\text{eq}})^2. \end{aligned} \tag{12.12}$$

The force constants are usually fitted to reproduce spectroscopic observables, or derived from quantum-chemical calculations. The terms describing the bond stretching, the angle bending and the improper torsional (*i.e.* out-of-plane) bending are simple harmonic potentials around an equilibrium

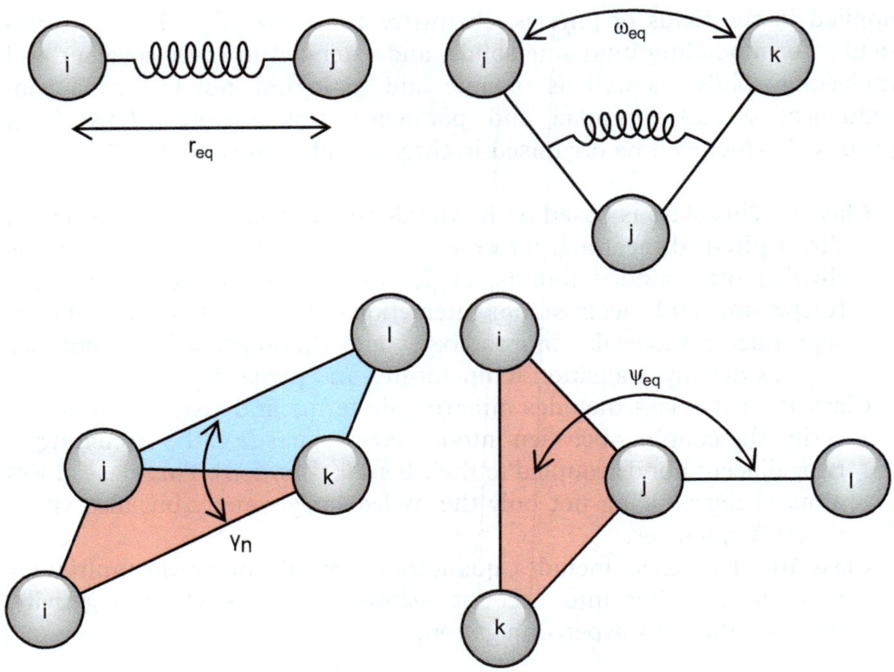

Figure 12.1 Pictorial view of the bonded terms listed in eqn (12.12).

value, while the terms describing the torsion along a dihedral angle is a cosine expansion with coefficients V_n and phases γ_n.

An example of torsional potential for the two different dihedral angles of sexithiophene (T6) is given in Figure 12.2. The plot shows several minima with a main energy barrier which can be easily overcome at room temperature, giving the T6 molecules a certain degree of flexibility along the backbone chain.[34]

The second part of eqn (12.11) deals with non-bonded interactions, which are assumed to be pairwise and additive. These interactions are computed between different molecules, or within atoms of the same molecule which are not directly bonded. They include a 12-6 Lennard-Jones term,[35] itself composed of a steric repulsion term A_{ij}/r_{ij}^{12} and a dispersive van der Waals attraction term B_{ij}/r_{ij}^{6}, and an unscreened electrostatic Coulomb term between charges q_i, q_j at distance r_{ij}:

$$U_{\text{non-bonded}} = \sum_{i<j}\left[\frac{A_{ij}}{r_{ij}^{12}} - \frac{B_{ij}}{r_{ij}^{6}} + \frac{q_i q_j}{r_{ij}}\right]. \tag{12.13}$$

The atomic charges reflect local differences in the electronegativity and chemical environment. In practice, partial charges are obtained, often together with the geometrical parameters in eqn (12.12), from preliminary quantum chemistry calculations on isolated molecules. Non-bonded interactions are normally excluded for directly bonded atoms (also labelled as

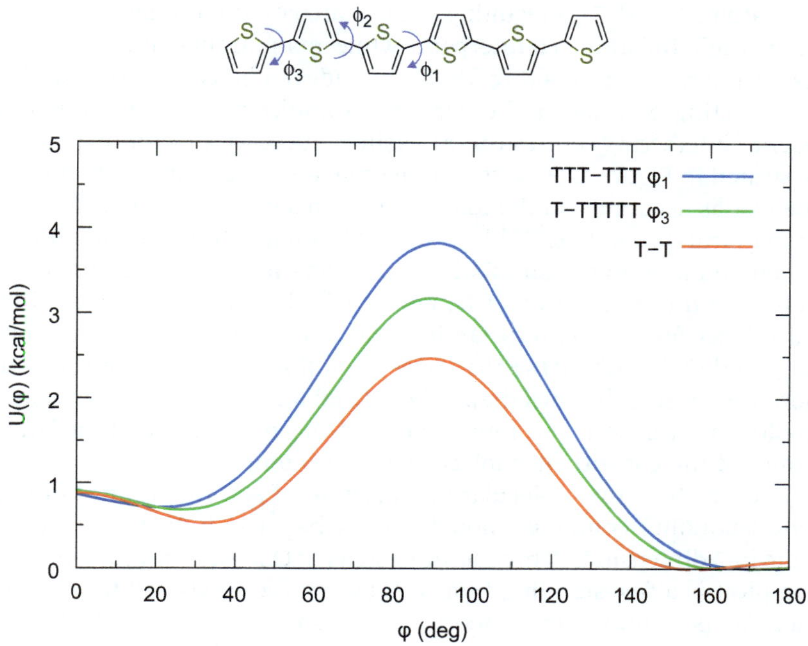

Figure 12.2 Torsional potentials for the internal angle of bithiophene (red) and the ϕ_1 (blue), ϕ_3 (green) angles of sexithiophene (top panel) calculated at the B3LYP//cc-pVDZ level of theory.
Adapted from ref. 34 with permission from the Royal Society of Chemistry.

1–2 interactions) and for atoms sharing a common bonded atom (1–3), while for atoms which are 1–4 connected they are reduced with respect to 1–5 or intermolecular ones by a factor which depends on the force field adopted.

Class I FFs offer a reasonable level of description for the simulation of soft-matter systems such as OSCs and, when optimised for a specific family of molecules, even achieve high accuracy in predicting their thermodynamical properties at equilibrium.[36] A general rule is to avoid mixing different FFs together, since each set of parameters is usually optimised to reproduce a certain kind of interaction, *e.g.* between water and biological molecules.[37] As mentioned already, a viable strategy to improve the accuracy of a FF is to follow a customisation scheme, which includes the computation of atomic charges,[30,38,39] the tuning of equilibrium values for bond lengths and angles, and the fitting of torsional potentials to *ab initio* calculation results.[40,41] A good parametrisation of torsional terms is crucial in order to obtain the correct molecular shape, in particular when the target system is composed of several interconnected aromatic units, for instance, in semiconducting oligomers or polymers.[34,42,43]

A different, slightly coarse grained, variety of FFs is built by combining hydrogen atoms with the heavier atoms they are bonded to (in particular, carbon), to form a single interacting body.[44–46] This approximation is called "united-atoms" (UA). The resulting UA-FFs are computationally less expensive that their full-atom counterparts while maintaining much of the essential physics, often making them the ideal choice for the study of semiconducting polymers and organic macromolecules. As an example, the number of interacting centres can be easily reduced by a factor of 2 or more, as illustrated in Figure 12.3. At the same time, it has been shown that UA-FFs can be just as accurate as full-atom FFs in terms of phase organisation and static physical properties.[36,44,47,48] An additional advantage of removing hydrogen atoms from the simulation system is that it allows an increase of the integration timestep from 1 fs, typical of full-atom FFs, to 2 fs. This in turn produces faster system dynamics, accelerating them by about an order of magnitude,[44,49] enhancing the space sampling and reducing the equilibration time. While this speed up is beneficial for structural determination, it should however be taken into account if the rotational or translational dynamics of the constituent molecules are of interest.

In order to study the molecular organisation of soft matter in thin films, the corresponding computer model usually has a slab geometry, *i.e.* it is periodic in 2 dimensions. Three basic designs of OSC thin films are possible: the simplest is a free-standing film with two surfaces exposed to vacuum;[50] the second is a thin film supported on a solid substrate, which leaves only the opposite interface exposed to vacuum,[51] and the third is a thin film in a confined geometry, *i.e.* bounded by solid walls. When the molecular model includes a solid support, the surface can be described with an explicit atomistic model,[52–59] with an effective potential wall[60–64] or through a combination of the two, where only the first few layers are treated explicitly and the underlying ones with an effective potential.[53] FFs with atomistic

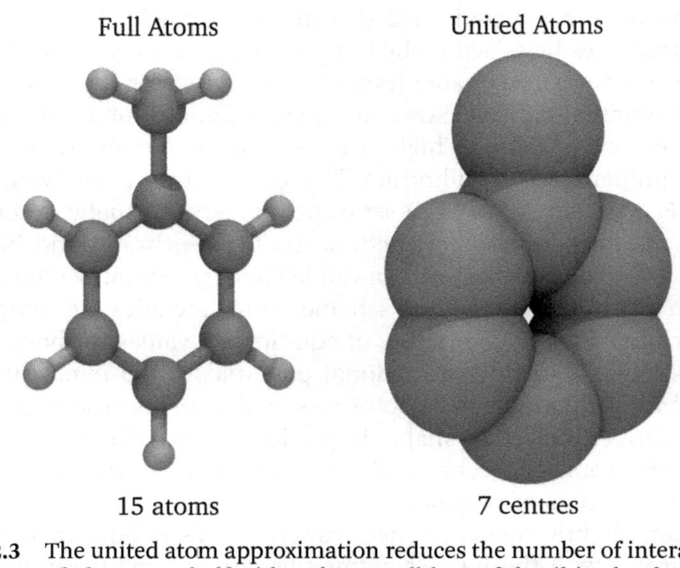

Figure 12.3 The united atom approximation reduces the number of interacting sites of about one half with only a small loss of detail in the description of molecular shape. The molecule of toluene is shown here as an example.

resolution should ensure a good transferability of the potential parameters and provide flexibility of all atoms of the support, thereby allowing for the exchange of energy and momentum between all components of the system.[65] Conversely, when the surface is represented by a single layer of interaction centres or by an effective potential (*e.g.* a Steele surface potential[66]), specific chemical interactions are hidden, and the general principles and trends governing the molecular alignment can be assessed. In addition, a significant saving in computer time is achieved, compared to explicit atomistic models.[61]

Computational materials science has a long history of modelling and simulations of the structure and properties of a variety of solid substrates. In particular, a great deal of attention has been devoted to the simulation of ionic materials, which are one of the key constituents of many technologies, *e.g.* high-*k* insulators in field-effect transistors (FET) and ceramic substrates in heterogeneous catalysis.[67] Typically, electrostatic interactions provide the dominant contribution to the total energy in ionic systems. Therefore, most ionic materials are successfully described by simple FFs based entirely on pairwise interactions. In this case, the structure of the material is determined by the position and shape of the energy minimum, which in turn depends on the steric repulsive terms. Extensive libraries of potentials have been developed to describe a high number of ionic materials.[68,69]

One of the most challenging problems in materials science is, however, to derive a FF capable of reliably describing interfaces between ionic and non-ionic systems. Indeed, a FF based solely on pairwise interactions has been designed to describe the structure and behaviour of water, hydroxyl surface species and intercalates in clay, hydroxide and oxyhydroxide phases.[65]

An accurate treatment of long-range Coulomb interactions for systems with charged particles is crucial to achieve a physically correct microscopic model, and this is particularly important for interfaces in systems with slab geometry. The calculation of electrostatic interactions is complicated by the fact that the Coulomb summation over all the particles interacting with a given one from various distances is non-convergent and cannot be simply handled including only particles within a cut-off separation. The problem has been the object of much attention for a long time and the methods developed for dealing with it for bulk systems with periodic boundary conditions in three dimensions (3D), such as the Ewald summation technique[70–72] and the derived particle-mesh approximations,[73,74] are the most used and accepted methods for the calculation of electrostatic interactions. For a slab geometry, the most accurate treatment of electrostatics is given by a two-dimensional modification of the Ewald summation, originally developed by Parry.[75,76] Although formally exact, the Parry method has the disadvantage that it is computationally very expensive. A widely used approach is thus to apply the conventional 3D Ewald summation to a simulation box obtained adding a sufficiently thick vacuum layer above the film, so as to decouple the interaction between periodic images of the slab.[51,59] Properties such as hydration shell structures, atom density profiles and pair correlation functions do not depend strongly on the long-range electrostatic interactions and, as a consequence, they are in qualitative agreement with the values obtained using the computationally more demanding Parry method.[77] However, a simple correction term to the standard Ewald summation method has been proposed for systems with slab geometry,[78] yielding accurate forces and a quantitative agreement with the Parry method.

12.3 Observables and Characterisation

The ultimate goal of any MD simulation is to describe the evolution of a given molecular system at a very fine level of detail, which includes the positions and velocities of all particles (*i.e.* atoms) at successive time steps within a certain time window. The rough information contained in MD trajectories is the starting point for the computation of any arbitrary property A. The average value at equilibrium of a given physical property A can be defined as the time integral over the trajectory in the phase space:

$$\langle A \rangle = \lim_{\tau \to \infty} \frac{1}{\tau} \int_0^{\tau} dt A(t) \tag{12.14}$$

$$\approx \frac{1}{N_t} \sum_{n=1}^{N_t} A(n\Delta t), \tag{12.15}$$

where $A(t) \equiv A[\mathbf{r}(t),\mathbf{p}(t)]$ is the value of the physical property at time t and τ is the length of the production time, *i.e.* the simulated time interval after the system has reached thermal equilibrium. In practical terms, the property A is computed with a discrete summation over a number N_t of configurations. These are not necessarily contiguous, as it is important that the analysed configurations provide samplings of the phase space of the system that are, as far as possible, uncorrelated, to ensure the statistical significance of the average value $\langle A \rangle$ (eqn (12.15)). It should be noted that the observables computed from MD simulations are not ensemble averages as, *e.g.*, in Monte Carlo methods,[79] even if, as long as the system can fully explore the phase space (*i.e.* when ergodic conditions apply), the equivalence between time and ensemble averages holds.[80]

The structural changes of organic materials in thin film phases are a direct consequence of the balance between the inter-molecular interactions within the film and the molecule–surface ones. Although a wealth of experimental techniques can be used to investigate the structure of thin films, revealing the detailed molecular organisation remains a challenge. In this regard, the structural observables which can be computed from MD trajectories are important both to validate the MD simulations and to establish the phase organisation obtained, as well as to provide the molecular origin of quantities, like the type of molecular alignment and the anchoring energy of molecules at the surface, that are normally treated as empirical parameters.

As a concrete example of the determination of observables relevant to thin films, we consider nano-thick (10–20 nm) films of a nematic liquid crystal,

4-*n*-pentyl-4′-cyano biphenyl (5CB), supported on different solid substrates that have been recently investigated by means of MD simulations. These studies revealed the alignment of the liquid crystal across the films and the relationship between the morphology of the surface and the molecular organisation of the 5CB at the interface,[51,59] but the methodology can be applied to films of other functional organic materials. A first insight into the molecular organisation of the 5CB films can be gained from the side views of the equilibrated samples (Figure 12.4), where the molecules are colour coded according to their orientation. Four substrates are considered. Two of them expose an atomically flat crystal surface, one being the hydrogen-capped Si(001) surface and the other the cristobalite (001) surface (cristobalite is a crystal form of SiO_2). The other substrates are composed of amorphous SiO_2 (*silica*) with different surface roughness: one, here called *smooth*, has a RMS

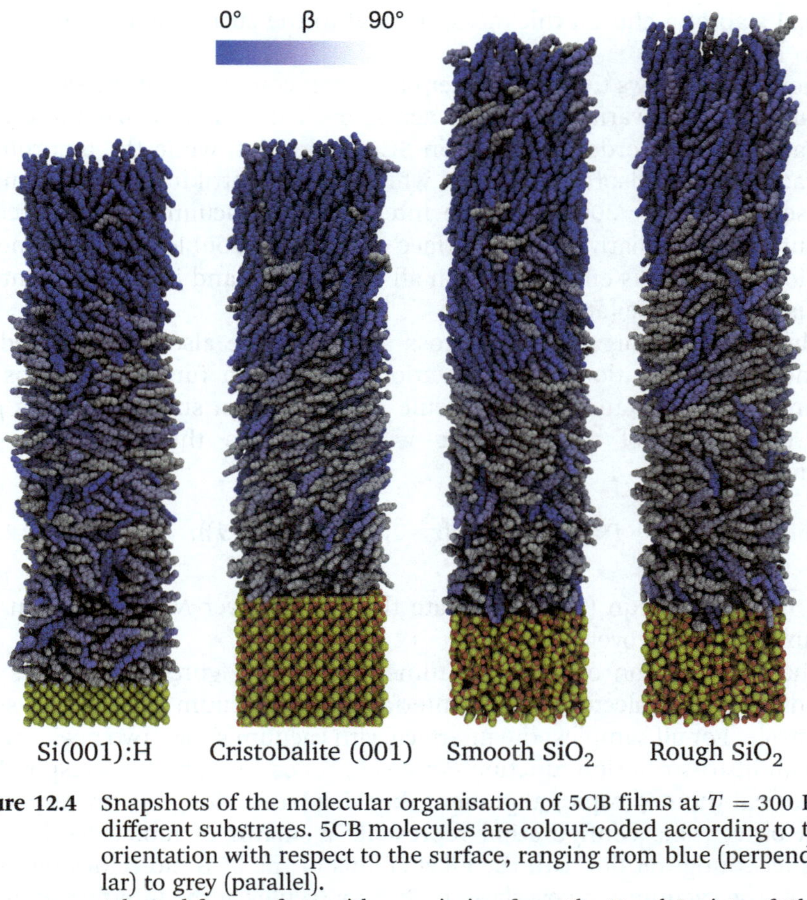

Figure 12.4 Snapshots of the molecular organisation of 5CB films at $T = 300$ K on different substrates. 5CB molecules are colour-coded according to their orientation with respect to the surface, ranging from blue (perpendicular) to grey (parallel).
Adapted from ref. 51 with permission from the Royal Society of Chemistry and ref. 59 with permission from O. M. Roscioni *et al.*, *Langmuir*, 2013, **29**, 8950–8958. Copyright (2013) American Chemical Society.

roughness of 1.5 Å and the other, named *rough*, has a RMS roughness of 3.2 Å.

A fundamental physical observable to explore molecular organisations is the scalar order parameter $\langle P_2 \rangle$, which is used to characterise the average degree of alignment of a molecule with respect to the local preferred direction (or director) $\mathbf{n}(\mathbf{r})$. For thin films, it is convenient to define the scalar order $\langle P_2 \rangle$ and the preferred direction $\mathbf{n}$ as a function of the distance z from the surface, as the largest eigenvalue and the relative eigenvector of the so-called ordering matrix.[81] Here, $\mathbf{Q}(z)$ is defined as

$$\mathbf{Q}(z) = \left\langle \sum_{i=1}^{N(z,t)} [3\mathbf{u}_i(t) \otimes \mathbf{u}_i(t) - \mathbf{I}]/2N(z,t)) \right\rangle_t, \tag{12.16}$$

where $\mathbf{u}_i(t)$ is a unit vector giving the orientation of molecule i (*e.g.* the principal axis of the inertia tensor corresponding to the lowest eigenvalue), $\mathbf{I}$ is the identity matrix, and the sum runs over all the $N(z, t)$ molecules of the virtual slab of a chosen thickness, parallel to the surface and at distance z from it.[36,51,81]

Figure 12.5 shows the scalar order parameter computed for the samples of Figure 12.4. The variation of $\langle P_2 \rangle$ across each film shows that the crystal surfaces have an ordering effect on 5CB molecules, while the amorphous surfaces have a disordering effect which is more striking in the nematic phase (here at $T = 300$ K). At the interface with vacuum, 5CB molecules orient perpendicularly to the interface and form a double layer, enhancing the local order: this effect is seen in all the samples and is more evident for the isotropic phase (at $T = 320$ K).

The molecular organisation across a film can be also characterised by a one-particle position and orientation distribution function $P(z, \cos\beta)$, where z is the distance of a molecule from a chosen surface and $\cos\beta$ is the orientation of the molecule with regard to the normal to the reference plane:

$$P(z, \cos\beta) = \langle \delta(z - z_i)\delta(\cos\beta - \hat{\mathbf{u}}_i \cdot \hat{\mathbf{z}}) \rangle. \tag{12.17}$$

The brackets in eqn (12.17) indicate the average over N molecules in the sample as well as over time.

The cross section of four 5CB films, shown in Figure 12.6, reveals the orientation of molecules at the interfaces with vacuum and with a solid substrate. For all samples, the interface with vacuum shows two peaks of the probability distribution function $P(z, \cos\beta)$ at $\cos\beta \pm 1$, both corresponding to vertical (homeotropic) alignment, but displaced along z with respect to each other. This configuration represents a bilayer of molecules in antiparallel configuration, with the alkyl chains of the 5CB molecules pointing toward the vacuum. The overlayer of 5CB molecules at the interface with the Si(001):H and cristobalite (001) surfaces displays a strong planar alignment characterised by an intense peak at $\cos\beta = 0$. Further layers of flat-lying

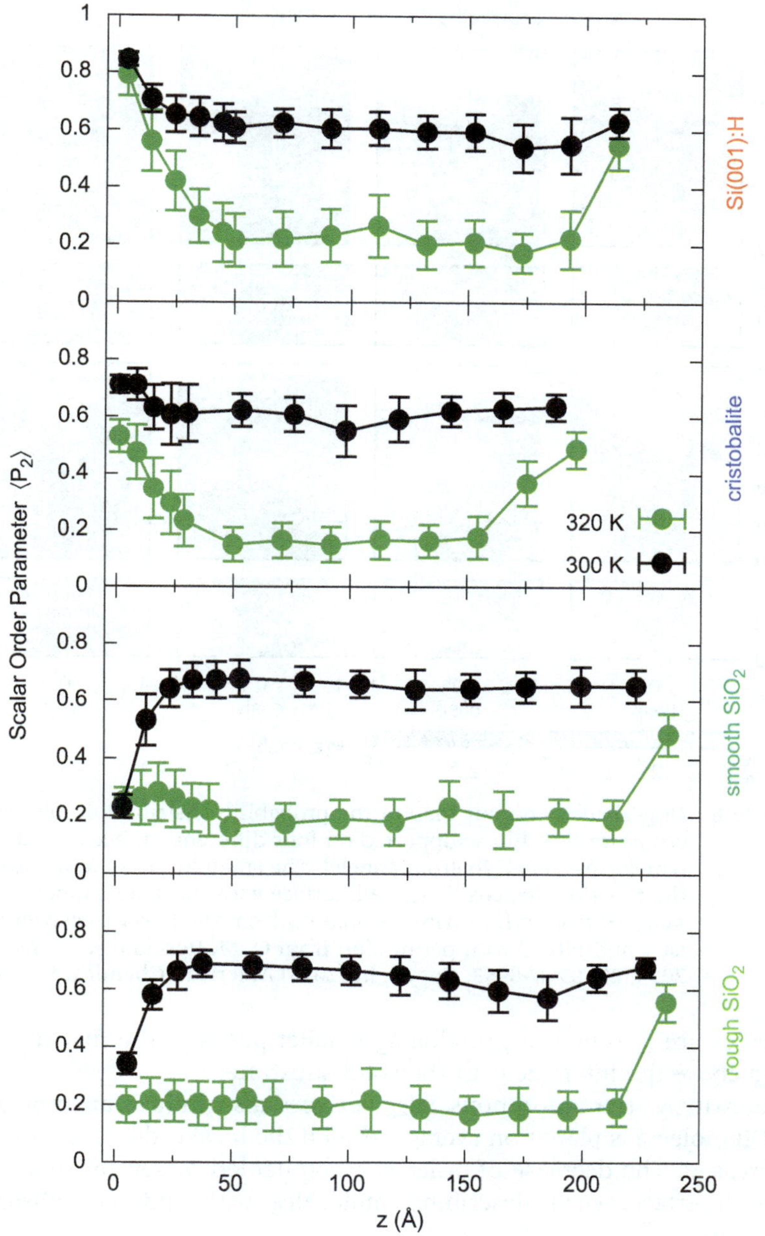

Figure 12.5 Local order parameter of 5CB films as a function of the distance from four different surfaces at $T=300$ K (black) and $T=320$ K (green).
Adapted from ref. 51 with permission from the Royal Society of Chemistry and ref. 59 with permission from O. M. Roscioni *et al.*, *Langmuir*, 2013, **29**, 8950–8958. Copyright (2013) American Chemical Society.

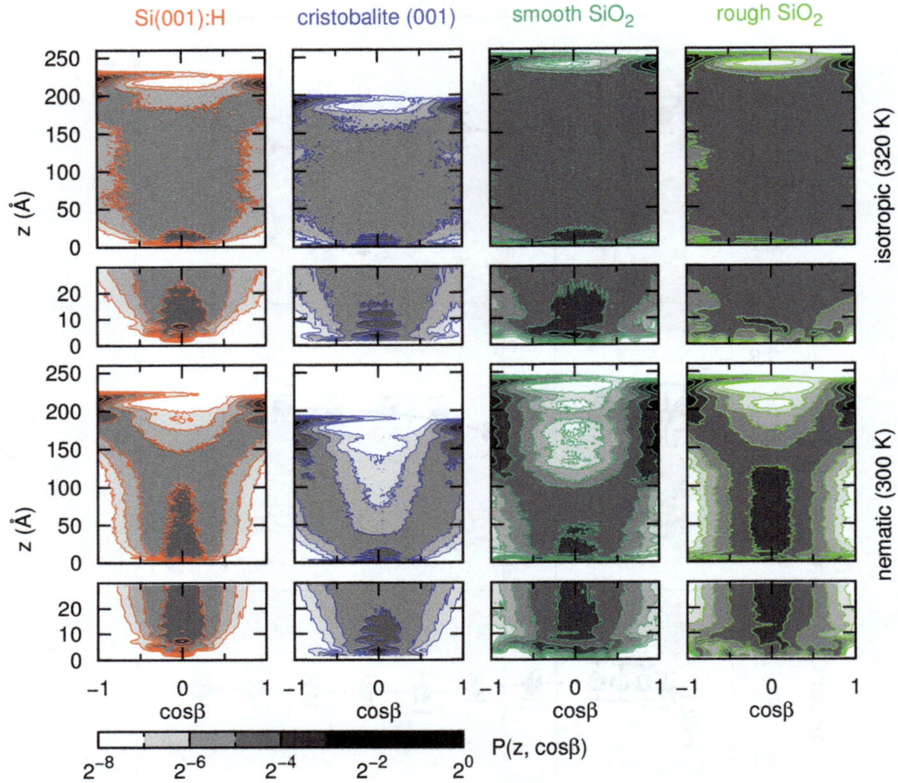

Figure 12.6 Gray-shaded contour map of the probability distribution function $P(z, \cos\beta)$ for 5CB films supported on four different surfaces at 320 K (top panels) and 300 K (bottom panels). The small insets show a close-up of the function $P(z, \cos\beta)$ at the interface with the solid support. Adapted from ref. 51 with permission from the Royal Society of Chemistry and ref. 59 with permission from O. M. Roscioni *et al.*, *Langmuir*, 2013, **29**, 8950–8958. Copyright (2013) American Chemical Society.

molecules are also present, producing smaller peaks in the first few nanometres above the interface with the solid substrate.

Interestingly, the amorphous SiO_2 surfaces also induce an ordering of the 5CB molecules planar on average, even if the local order at the interface is decreased. The decrease of order is accompanied by several small peaks at the interface, each describing molecules with different absorption geometries.

Complementary information of the molecular organisation across the film can be obtained by expanding $P(z, \cos\beta)$ in eqn (12.17) in terms of Legendre polynomials:[59]

$$P(z, \cos\beta) = \frac{\rho(z)}{\langle\rho\rangle}\left[\frac{1}{2} + \frac{3}{2}\langle P_1(\cos\beta)\rangle_z P_1(\cos\beta) + \frac{5}{2}\langle P_2(\cos\beta)\rangle_z P_2(\cos\beta) + \ldots\right], \tag{12.18}$$

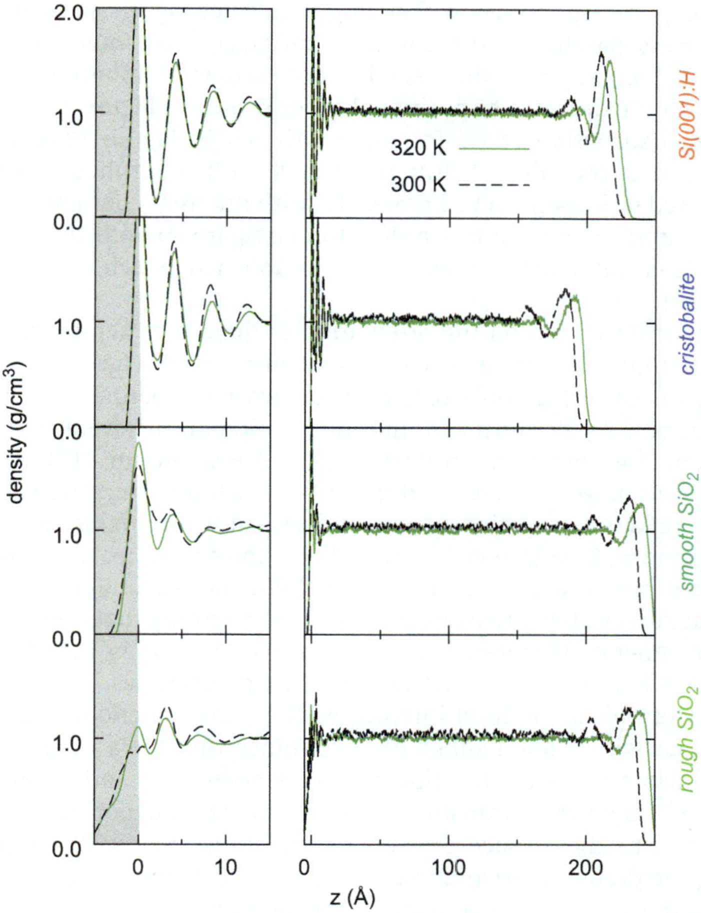

Figure 12.7 Density $\rho(z)$ of different 5CB films at $T = 300$ K and 320 K across the film (left panel) and close to the solid substrate (right panel), computed with respect to the normal at the surface, z. The 5CB/support interface is located at the grey-white border.
Adapted from ref. 51 with permission from the Royal Society of Chemistry and ref. 59 with permission from O. M. Roscioni *et al.*, *Langmuir*, 2013, **29**, 8950–8958. Copyright (2013) American Chemical Society.

where $\rho(z)$ is the density, $\cos\beta \equiv \mathbf{u}\cdot\mathbf{z}$, $\langle P_1(\cos\beta)\rangle_z \equiv \langle P_1(z)\rangle$ is the polar order parameter, $\langle P_2(\cos\beta)\rangle_z \equiv \langle P_2(z)\rangle$ is the quadrupolar order parameter and so on. Each term is calculated at a distance z from the solid surface.

The density of 5CB thin films shown in Figure 12.7 reveals significant fluctuations close to the two boundaries. The interface of the thin-film of 5CB with vacuum is characterised by an oscillation of $\rho(z)$ corresponding to two double layers of 5CB molecules in anti-parallel arrangement. This feature is common to all films, irrespective of the solid surface. On the other side of the film, the interface with the crystal surfaces shows strong oscillations, corresponding to flat-lying molecules tightly packed in contiguous layers parallel to the surface. Conversely, the interface with the amorphous SiO_2 surfaces show smaller oscillations. Combining this information with the decrease of positional order shown in Figure 12.5 and with the broad distribution of conformations shown in Figure 12.6, it can be concluded that the film of 5CB in contact with amorphous surfaces is characterised by loosely packed molecules with different configurations. The resulting interface layer is less ordered in comparison with that on crystal surfaces. However, a planar alignment is established, which propagates well into the film.

Detailed information on the orientation of molecules across the films is obtained from the polar order parameter $\langle P_1(z)\rangle$ (green curves in Figure 12.8). Once the molecular director has been defined, the value of $\langle P_1(z)\rangle$ can be directly translated into information on the average molecular orientation. For instance, positive values correspond to 5CB molecules pointing their alkyl chains toward the solid substrates, negative values correspond to alkyl chains pointing toward the vacuum, and values around zero correspond to 5CB molecules lying parallel to the substrate. Non-zero values of the third rank order parameter $\langle P_3(z)\rangle$ (blue dashed line) determine the non-linear susceptibility tensor element χ^2_{zzz}, confirming that optical second-harmonic generation measurements can detect anchoring transitions, *e.g.* from planar to tilted, at the interface with solid substrates.[82,83]

The anisotropic strength of interaction of organic functional materials at the interface with a solid substrate determines the film's properties such as its thickness[84] and its thickness-dependent[85,86] and temperature-dependent[87] anchoring transitions. Physically, the anchoring strength is proportional to the surface free energy per unit area $W(\beta)$ required for changing the preferred orientation (*i.e.* the director) of molecules from their equilibrium value, β_{eq}. The anisotropic anchoring strength w_2^A is commonly described by a phenomenological law proposed originally by Rapini and Papoular (RP):[88–90]

$$W(\beta) = w_0^A - \frac{1}{2} w_2^A \sin^2(\beta - \beta_{eq}). \tag{12.19}$$

The RP expression is useful to summarise the behaviour of molecules close to a surface. The energy that anchors each molecule to the surface, and its surface-induced variations, can be measured directly from atomistic MD simulations[51] by means of the Boltzmann inversion of eqn (12.17):

$$W(z_i, \cos\beta) = -k_B T \ln P(z_i, \cos\beta) N(z_i)/A, \tag{12.20}$$

where the subscript i indicates that z_i is a discrete variable, and $P(z_i, \cos\beta)$ is normalised to 1 for each layer z_i. Hence, the anchoring coefficient is computed (and varies) as a function of the distance from the solid substrate.

The calculated w_2^A coefficients for the samples discussed here are given in Figure 12.9. These values measure the strength of the orienting mean field

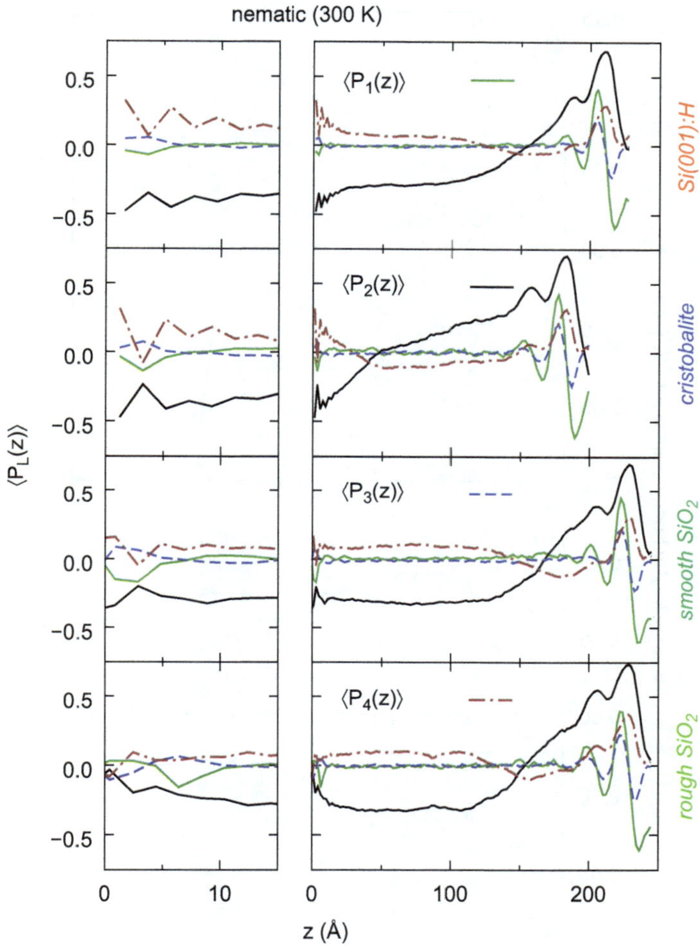

Figure 12.8 Order parameters $\langle P_1(z)\rangle$ (green line), $\langle P_2(z)\rangle$ (black line), $\langle P_3(z)\rangle$ (blue dashed line), $\langle P_4(z)\rangle$ (red dot-dashed line) of different 5CB films at $T = 300$ K across the film (left panel) and close to the silica (right panel), computed with respect to the normal at the surface, z.
Adapted from ref. 51 with permission from the Royal Society of Chemistry and ref. 59 with permission from O. M. Roscioni *et al.*, *Langmuir*, 2013, **29**, 8950–8958. Copyright (2013) American Chemical Society.

exerted by the combined effect of surface and the surrounding molecules on a 5CB molecule in each layer. The highest anchoring coefficients are found at the interface with vacuum and with the crystal surfaces. The anchoring strength at the interface with the amorphous SiO_2 surfaces is small, but their range of effect is longer and clearly correlated to the extent of the region of planar alignment. The anchoring strengths computed from MD trajectories at the very surface are rather different from the values reported in the literature for thick (micron) films of cyanobiphenyls *via* electro-optical measurements.[84,91–93] However, it should be noted that the experimental

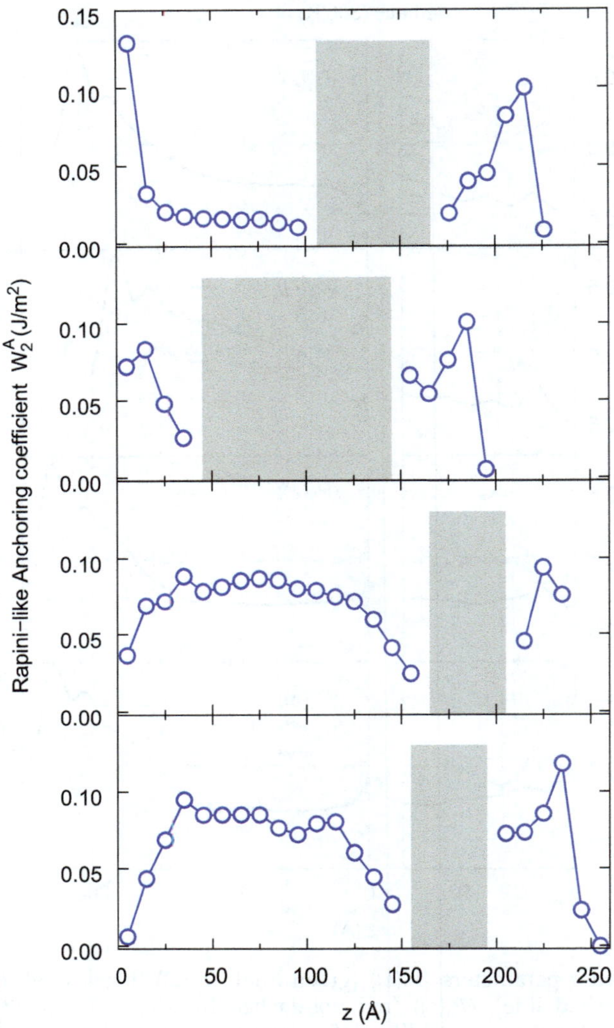

Figure 12.9 Rapini-like anchoring coefficient W_2^A for 5CB as a function of the distance from the cristobalite, smooth and rough silica surfaces at 300 K. The coefficients were calculated for sections of the sample with a thickness of 10 Å. Anchoring coefficients were not calculated in the inner regions corresponding to a switch over between the two antagonistic surfaces (grey areas).
Adapted from ref. 51 with permission from the Royal Society of Chemistry and ref. 59 with permission from O. M. Roscioni *et al.*, *Langmuir*, 2013, **29**, 8950–8958. Copyright (2013) American Chemical Society.

techniques used until recently do not probe the immediate neighbourhood of the substrate surface, and the computed w_2^A coefficients vary dramatically in a few nanometres from the surface (see Figure 12.9). Moreover, the simulations clearly show that tiny morphological differences at the nanometre scale, and in particular in the preparation of the surface and in its roughness, deeply change the anchoring characteristics of the surfaces with a given chemical composition.[59] Although it is difficult to assess experimentally the relationship between the morphology of a substrate and the properties of an overlying OSC film, a decrease in performance has been reported[94] for organic field-effect transistors (OFETs) based on pentacene deposited on photolithographically patterned polyimide surfaces, due to a small increase in the surface roughness from 2.9 to 8.2 Å.

Another class of physical observables relevant to organic photovoltaics are the electronic couplings governing charge carrier and exciton diffusion,[3,95–97] discussed in other chapters of this book. The charge transport in organic solar cells begins with the absorption of light and the creation of an exciton, an electronically excited state of an OSC. The energy change associated with the diffusion of excitons plays a central role both in the operation and degradation of organic devices. For instance, in organic blends and hybrid photovoltaic devices, excitons must diffuse to an interface between the donor and acceptor materials in order to dissociate into separate charges. The efficiency of this process is strictly related to the exciton diffusion length, *i.e.* the average distance over which an exciton moves before it decays.

Exciton transfer rates can be directly evaluated *via* quantum-chemical calculations, for instance using realistic models of organic materials and blends obtained from atomistic MD simulations. In the weak coupling regime, excitons are localised on single molecular sites and exciton transfer takes place *via* Förster type resonant energy transfer from a donor molecule D in the excited state to an acceptor molecule A in its ground state. The rate of hopping is

$$k_{\mathrm{DA}} = \frac{2\pi}{h} |V_{\mathrm{DA}}|^2 J_{\mathrm{DA}} \tag{12.21}$$

with V_{DA} being the excitonic coupling and J_{DA} the spectral overlap between the donor emission $F_{\mathrm{D}}(\omega)$ and the acceptor absorption $A_{\mathrm{A}}(\omega)$, given by

$$J_{\mathrm{DA}} = \int_0^{x} F_{\mathrm{D}}(\omega) A_{\mathrm{A}}(\omega) \mathrm{d}\omega. \tag{12.22}$$

V_{DA} can be calculated using the distributed monopole approximation[98,99]

$$V_{\mathrm{DA}} = \frac{1}{4\pi\varepsilon_0} \sum_{i\in \mathrm{D}} \sum_{j\in \mathrm{A}} \frac{\rho_{\mathrm{D}}(\mathbf{r}_{\mathrm{D}}) \rho_{\mathrm{A}}(\mathbf{r}_{\mathrm{A}})}{|\mathbf{r}_{\mathrm{D}} - \mathbf{r}_{\mathrm{A}}|} \tag{12.23}$$

The exciton diffusion length L_{d} can be obtained modelling the motion of excitons as a stochastic Markovian process, corresponding to hopping from one site to another. It is useful to recall that the exciton diffusion length is related to the exciton lifetime τ and the exciton diffusion coefficient D_{e} by the equation:

$$L_{\mathrm{d}} = \sqrt{\tau D_{\mathrm{e}}} \tag{12.24}$$

A kinetic Monte Carlo (KMC) procedure can be used to model the exciton diffusion as a random walk, from which L_d is determined.[100] This approach was used in ref. 48 to study the exciton mobility of indenofluorene trimer (IF3), an OSC exhibiting an ordered smectic phase at room temperature. The study of transport properties as a function of the temperature revealed that the increased orientational disorder at elevated temperatures barely affects the ability of excitons to be transported over large distances, though disorder influences the directionality of the energy diffusion process. Detailed quantum-chemical calculations have shown that this result arises from a combination of reduced excitonic couplings and increased spectral overlap as the temperature increases, which would have been difficult to predict without using realistic configurations obtained from MD simulations.

12.4 Thin Films and Interfaces

OSCs are molecular solids held together by weak intermolecular interactions such as van der Waals dispersion forces, electrostatic interactions and steric repulsion, all contributing to determine the extent of the long-range order in the condensed phase. Most crucially, they determine the molecular organisation at the interface with a metallic electrode, a dielectric substrate or another OSC (as for instance in bulk heterojunctions). Due to the weak nature of these interactions, the resulting morphology is characterised by a rich display of molecular arrangements with similar energy.[101] The relative orientation of the molecules is, in turn, responsible for the formation of conduction pathways and directly determines the transport properties of an OSC.[102] A multi-scale approach able to assess the complexity of the free energy landscape of OSCs, *i.e.* a theoretical model describing the morphology of OSCs under certain thermodynamic conditions, requires the use of computer models including hundreds of thousand of atoms, mimicking as close as possible the experimental sample, on both size[103] and disposition of the constituent elements.[97]

Unfortunately, this task is unavoidable because OSC technology relies on the precise control of the interfaces, the morphology of the film, and the structure of the active layers[104] to determine the performance of the resulting OSC device.[105] The fabrication of the OSC active layer relies on high vacuum vapour phase deposition or simpler techniques such as spin coating. The growth mechanism of OSC thin films is a complex and often non-equilibrium process which involves competing phenomena such as secondary nucleation, growth transitions, dewetting, and recrystallisation.[105] Although it is difficult to fully characterise the growth mechanism of OSCs, it is possible to get precious insight from MD simulations, which are better suited to follow dynamical and out of equilibrium processes than, for instance, MC simulations.

One such example is the growth of a p-type semiconductor, pentacene, on top of an n-type semiconductor, C_{60} fullerene, by a vapour-phase deposition process, which has been recently investigated by means of atomistic MD simulations.[57] The vapour deposition of pentacene was mimicked by inserting one molecule every 250 ps into a simulation box already containing the C_{60} (001) slab. Each new molecule was inserted with the same initial geometry but with random position and velocity. This scheme was repeated until two pentacene layers were formed. The first molecules deposited lie flat and disordered on C_{60} and diffuse quite rapidly, until the nucleation process begins. At 0.5 units of monolayer formation (ML), a single film is formed covering the entire surface. Crucially, at 0.6 ML, a collective reorientation takes place, with pentacene molecules self-assembling into a herringbone lattice and changing their orientation from an almost planar to a roughly perpendicular alignment.[57] A similar mechanism is observed for the formation of the second layer, although some molecules diffuse on the first monolayer until they find an empty lattice site to fill. The morphology of the resulting two layers is similar (Figure 12.10), with an average tilt angle of about 35°. The interface between the two OSCs is found not to be favourable for charge separation. Therefore, it can be concluded that, as long as the model is realistic enough, vapour deposition of pentacene on C_{60} (001) is not ideal for the production of organic solar cells.

More recently, a theoretical investigation of the molecular organisation of sexithiophene (T6) on a C_{60} (001) surface revealed that two different film deposition processes produce dramatic effects on the resulting morphology of the interface.[106] As in the previous example, the vapour deposition of T6 molecules was mimicked by means of atomistic MD simulations in which one T6 molecule at a time is deposited on a C_{60} (001) surface. The observed growth mechanism proceeds in a layer-by-layer fashion and is characterised by a collective reorientation of T6 molecules, initially adsorbed flat on the

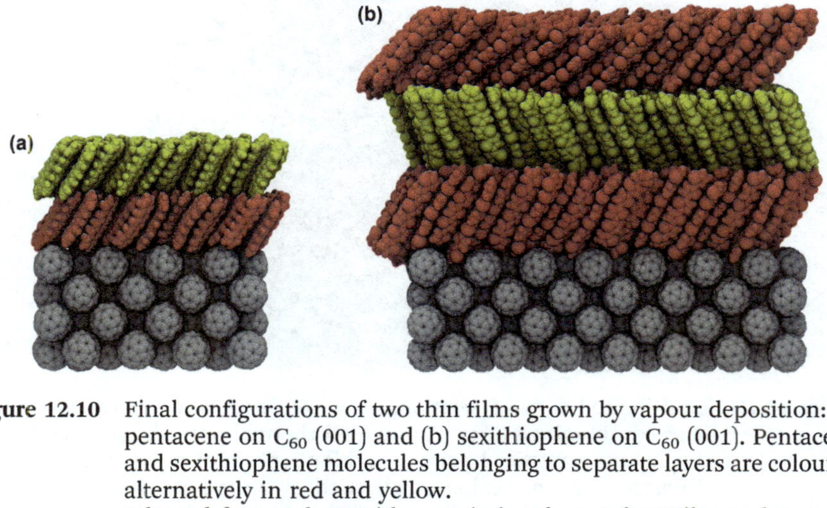

Figure 12.10 Final configurations of two thin films grown by vapour deposition: (a) pentacene on C_{60} (001) and (b) sexithiophene on C_{60} (001). Pentacene and sexithiophene molecules belonging to separate layers are coloured alternatively in red and yellow.
Adapted from ref. 57 with permission from John Wiley and Sons © 2011 WILEY-VCH Verlag GmbH & Co. KGaA, Weinheim and ref. 106 with permission from John Wiley and Sons © 2014 WILEY-VCH Verlag GmbH & Co. KGaA, Weinheim.

surface. Each layer displays a crystalline herringbone pattern with an average tilt angle of about 30°, in agreement with experimental determinations.[106] For T6, the simulated deposition process yields layers with uncorrelated molecular orientations, as shown in Figure 12.10(b). Furthermore, epitaxial grow is observed in the first monolayer, which does not propagate to the subsequent layers. The growth of the second and third layers is characterised by the spontaneous formation of supramolecular aggregates ("islands") with chiral edges and a core of molecules in a herringbone configuration. A different deposition process consists in placing an amorphous T6 film between two flat C_{60} (001) surfaces, followed by temperature annealing at 600 K. In this case the MD simulations showed that, contrary to the vapour deposited structure, T6 molecules at the interface have a planar alignment, a situation which may occur locally in bulk heterojunctions. As already clear from the snapshot in Figure 12.11, T6 forms smectic layers extending across the whole film, perpendicular to the C_{60} surface, and aligned along the [110] direction. The planar T6 interface obtained is expected to favour charge separation and efficient two-dimensional charge transport in organic solar cells, suggesting a viable way of optimising this organic interface for specific applications.

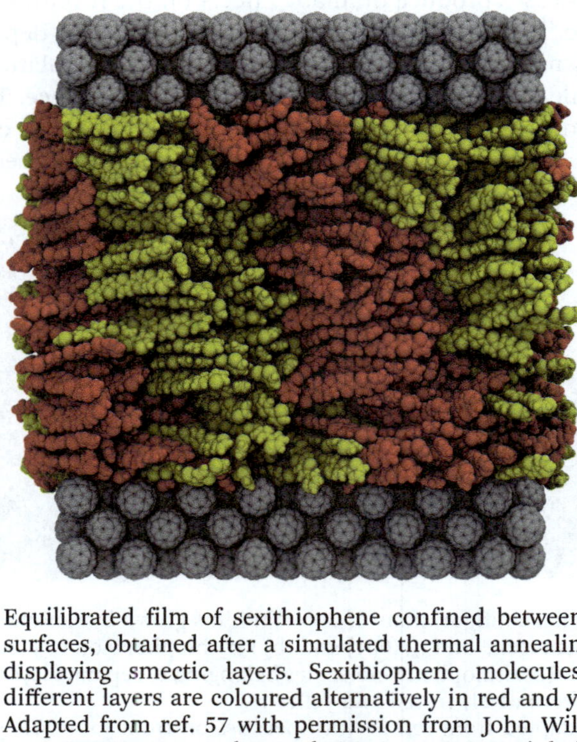

Figure 12.11 Equilibrated film of sexithiophene confined between two C_{60} (001) surfaces, obtained after a simulated thermal annealing at 600 K and displaying smectic layers. Sexithiophene molecules belonging to different layers are coloured alternatively in red and yellow.
Adapted from ref. 57 with permission from John Wiley and Sons © 2011 WILEY-VCH Verlag GmbH & Co. KGaA, Weinheim and ref. 106 with permission from John Wiley and Sons © 2014 WILEY-VCH Verlag GmbH & Co. KGaA, Weinheim.

The interaction between OSC materials and a solid substrate critically determines the electronic properties of the resulting device. As established experimentally for OFETs, the charge carrier accumulation occurs at the interface between the OSC and the dielectric support, and therefore the charge transport layer is expected to be localised within the first few semiconductor monolayers above the insulator surface.[107,108] In order to capture this effect theoretically, an explicit model of the surface should be included in MD simulations.[109] Furthermore, the examples discussed previously show that the morphology of OSC thin films is determined not only by the interaction with a surface, but also by their growth mechanism.[110,111] Atomistic models including explicitly the various components of a system, as well as mimicking the experimental preparation protocols, are necessary in order to obtain realistic MD simulations. The resulting models provide an accurate description of the molecular organisation of OSC thin films, which can be used to pinpoint the factors influencing their charge transport properties. For instance, the packing of pentacene deposited over two polymer dielectric substrates, poly(methyl methacrylate) (PMMA) and polystyrene (PS), was studied by means of classical MD simulations. The structure of the resulting pentacene–insulator interface was then used to carry out quantum-chemical calculations in order to evaluate the energetic disorder in the organic semiconducting layer. The whole picture reveals that the polar groups on the polymeric chains increase the energetic disorder and decrease significantly the hole mobility.[112]

The efficiency of organic solar cells can be improved by either designing new device geometries or by optimising the electrical contact at the different interfaces of the OSCs. Regarding the latter strategy, a promising possibility is to mix donor (D) and acceptor (A) materials in a blend, thus maximising the D/A interface of the resulting bulk heterojunctions. The energetics of the exciton dissociation at the D/A interface can be assessed at various levels of theory,[101] once a reasonable atomistic model of the interface is obtained. A combination of MD simulations with quantum-chemical and classical microelectrostatic calculations has proved successful in explaining the reported high quantum yields of P3HT/PCBM blends, which originates from different factors promoting electron–hole separation and pushing electrons and holes from the interface to the bulk.[113] The same approach was used to model a realistic planar heterojunction between two columnar discotic liquid crystals, one being a perylene donor and the other a benzoperylene diimide acceptor, deposited on an amorphous silica surface. The super-cell used to model the system was big enough to achieve the best commensurism between the hexagonal columnar phases of the donor and acceptor liquid crystals. As a consequence, the molecules at the interface were found to display different degrees of alignment. Two cases were then studied: one where the aromatic cores were largely superimposed, representing about a quarter of the interface, and one with a large horizontal displacement of the cores. The energetics of exciton dissociation in these two limiting configurations showed that column mismatch has a favourable effect on the electron–hole separation process, mainly due to a low rate of charge recombination.[97] These examples show that energetic disorder at the D/A interface governs the performance of bulk heterojunctions and highlights the relevance of theoretical tools to probe the energy landscapes in molecular assemblies.[114]

Conjugated polymers are a class of OSC materials that are more challenging to model compared to molecular solids, since their theoretical description requires a multi-scale approach. The semiconducting nature of conjugated polymers depends on molecular-scale details, such as the degree of inter-ring twisting, while bulk transport efficiencies depend on polymer folding, aggregation, and the morphology of grain boundaries.[115] Force fields with molecular details provide some of the required information, while coarse-grained (CG) models are necessary to assess the self-assembly of structures spanning tens or even hundreds of nanometres.[116–121] Several coarse graining schemes have been proposed to assess large systems over long timescales. We refer to a recent review for a systematic comparison of different CG models developed for amorphous polystyrene.[122]

Given the availability of very advanced computer codes,[28] it is expected that molecular growth by self-assembly and in non-equilibrium conditions will become a common topic for complex MD simulations, even though we should never forget that the timescale of MD simulations is much shorter than that of real experiments. Deposition by spin coating, despite being relatively straightforward to execute experimentally, is a formidable testing ground for MD simulations, and it is not surprising that, to our knowledge, no studies have been produced yet as they would require a multi-component model taking into account the evaporation of a liquid phase and the crystallisation of the organic phase in a confined environment.[123]

In conclusion, molecular dynamics has already proved to be an essential tool for understating and predicting the molecular organisation of organic functional materials in the bulk and at interfaces, while its role in studying the kinetic growth of films, although very important, is still in the early stages of development.

Acknowledgements

We would like to acknowledge the European projects MINOTOR (FP7-NMP-228424) and One-P (FP7-NMP-212311) for financial support. We also thank D. Beljonne, J. Cornil, G. D'Avino, J. Idé, A. Mityashin, L. Muccioli, Y. Olivier, M. F. Palermo, A. Pizzirusso, M. Ricci, A. Walker for stimulating discussions.

References

1. N. Kaur, M. Singh, D. Pathak, T. Wagner and J. M. Nunzi, *Synth. Met.*, 2014, **190**, 20–26.
2. A. M. Haruk and J. M. Mativetsky, *Int. J. Mol. Sci.*, 2015, **16**, 13381.

3. *Topics in Current Chemistry. Multiscale Modelling of Organic and Hybrid Photovoltaics*, ed. D. Beljonne and J. Cornil, Springer, Berlin, 2014, vol. 352.
4. R. G. Della Valle, A. Brillante, E. Venuti, L. Farina, A. Girlando and M. Masino, *Org. Electron.*, 2004, **5**, 1–6.
5. H.-L. Cheng, Y.-S. Mai, W.-Y. Chou, L.-R. Chang and X.-W. Liang, *Adv. Funct. Mater.*, 2007, **17**, 3639–3649.
6. C. Lercher, C. Röthel, O. M. Roscioni, Y. H. Geerts, Q. Shen, C. Teichert, R. Fischer, G. Leising, M. Sferrazza, G. Gbabode and R. Resel, *Chem. Phys. Lett.*, 2015, **630**, 12–17.
7. R. A. Marcus, *Rev. Mod. Phys.*, 1993, **65**, 599–610.
8. N. Martinelli, Olivier, L. Muccioli, A. Minoia, P. Brocorens, M.-C. Ruiz Delgado, C. Zannoni, D. Beljonne, R. Lazzaroni, J.-L. Bredas and J. Cornil, *Functional Supramolecular Architectures: For Organic Electronics and Nanotechnology*, ed. P. Samori and F. Cacialli, Wiley - VCH, Weinheim, 2011, vol. 1, pp. 3–37.
9. M. P. Allen and D. J. Tildesley, *Computer Simulation of Liquids*, Oxford University Press, Oxford, 1989.
10. J. M. Haile, *Molecular Dynamics Simulation*, Wiley, New York, 1992.
11. D. Rapaport, *The Art of Molecular Dynamics Simulation*, Cambridge University Press, Cambridge, 2004.
12. H. J. C. Berendsen, *Simulating the Physical World. Hierarchical Modeling from Quantum Mechanics to Fluid Dynamics*, Cambridge University Press, Cambridge, 2007.
13. M. E. Tuckerman, *Statistical Mechanics: Theory and Molecular Simulation*, Oxford University Press, Oxford, 2010.
14. L. D. Landau and E. M. Lifshitz, *Mechanics*, Butterworth-Heinemann, Oxford, 1993.
15. H. Goldstein, *Classical Mechanics*, Addison-Wesley, Reading, MA, 1980.
16. D. Frenkel and B. Smit, *Understanding Molecular Simulations*, 2nd edn, Academic Press, San Diego, 2001.
17. L. Verlet, *Phys. Rev.*, 1967, **165**, 201.
18. H. C. Andersen, *J. Comput. Phys.*, 1983, **52**, 24.
19. J. A. Snyman, *Appl. Math. Modeling*, 1983, **7**, 216–218.
20. J. A. Snyman, *Comput. Math. Appl.*, 2000, **40**, 1085–1096.
21. M. Tuckerman, B. J. Berne and G. J. Martyna, *J. Chem. Phys.*, 1992, **97**, 1990–2001.
22. C. V. Gear, *Numerical Initial Value Problems in Ordinary Differential Equation*, Prentice-Hall, Englewood Cliffs, NJ, 1971.
23. S. Toxvaerd, *J. Comput. Phys.*, 1982, **47**, 444–451.
24. S. Toxvaerd, *J. Comput. Phys.*, 1983, **52**, 214–216.
25. A. J. Stone, *The Theory of Intermolecular Forces*, Oxford University Press, Oxford, 1996, vol. 32.
26. J. D. Gale, *Philos. Mag. B*, 1996, **73**, 3–19.
27. A. R. Leach, *Molecular Modelling Principle and Applications*, Longmans, Harlow, 1996, p. 481.

28. L. Muccioli, G. D'Avino, R. Berardi, S. Orlandi, A. Pizzirusso, M. Ricci, O. M. Roscioni and C. Zannoni, *Topics in Current Chemistry. Multiscale Modelling of Organic and Hybrid Photovoltaics*, ed. D. Beljonne and J. Cornil, Springer, Berlin, 2014, vol. 352, pp. 39–102.
29. W. D. Cornell, P. Cieplak, C. I. Bayly, I. R. Gould, K. M. Merz, D. M. Ferguson, D. C. Spellmeyer, T. Fox, J. W. Caldwell and P. A. Kollman, *J. Am. Chem. Soc.*, 1995, **117**, 5179–5197.
30. J. Wang, R. M. Wolf, J. W. Caldwell, P. A. Kollman and D. A. Case, *J. Comput. Chem.*, 2004, **25**, 1157–1174.
31. W. L. Jorgensen, D. S. Maxwell and J. Tirado-Rives, *J. Am. Chem. Soc.*, 1996, **118**, 11225–11236.
32. A. D. MacKerrel, J. Wirkeiwicz-Kuczera and M. Karplus, *J. Am. Chem. Soc.*, 1995, **117**, 11946.
33. A. D. Mackerell, D. Bashford, Bellott, R. L. Dunbrack, J. D. Evanseck, M. J. Field, S. Fischer, J. Gao, H. Guo, S. Ha, D. Joseph-McCarthy, L. Kuchnir, K. Kuczera, F. T. K. Lau, C. Mattos, S. Michnick, T. Ngo, D. T. Nguyen, B. Prodhom, W. E. Reiher, B. Roux, M. Schlenkrich, J. C. Smith, R. Stote, J. Straub, M. Watanabe, J. Wiorkiewicz-Kuczera, D. Yin and M. Karplus, *J. Phys. Chem. B*, 1998, **102**, 3586–3616.
34. A. Pizzirusso, M. Savini, L. Muccioli and C. Zannoni, *J. Mater. Chem.*, 2011, **21**, 125–133.
35. J. E. Lennard-Jones, *Proc. R. Soc. London, Ser. A*, 1924, **106**, 463–477.
36. G. Tiberio, L. Muccioli, R. Berardi and C. Zannoni, *ChemPhysChem*, 2009, **10**, 125–136.
37. J. Wang, P. Cieplak and P. A. Kollman, *J. Comput. Chem.*, 2000, **21**, 1049.
38. C. Kramer, P. Gedeck and M. Meuwly, *J. Comput. Chem.*, 2012, **33**, 1673–1688.
39. B. Wang and D. G. Truhlar, *J. Chem. Theory Comput.*, 2012, **8**, 1989–1998.
40. J. Sancho-García and A. Karpfen, *Chem. Phys. Lett.*, 2009, **473**, 49–56.
41. R. S. Bhatta, Y. Y. Yimer, M. Tsige and D. S. Perry, *Comput. Theor. Chem.*, 2012, **995**, 36–42.
42. V. Marcon, N. van der Vegt, G. Wegner and G. Raos, *J. Phys. Chem. B*, 2006, **110**, 5253–5261.
43. D. L. Cheung, D. P. McMahon and A. Troisi, *J. Phys. Chem. B*, 2009, **113**, 9393–9401.
44. L. Yang, C.-h. Tan, M.-J. Hsieh, J. Wang, Y. Duan, P. Cieplak, J. Caldwell, P. A. Kollman and R. Luo, *J. Phys. Chem. B*, 2006, **110**, 13166–13176.
45. O. A. von Lilienfeld and D. Andrienko, *J. Chem. Phys.*, 2006, **124**, 054307.
46. W. R. P. Scott, P. H. Hünenberger, I. G. Tironi, A. E. Mark, S. R. Billeter, J. Fennen, A. E. Torda, T. Huber, P. Krüger and W. F. van Gunsteren, *J. Phys. Chem. A*, 1999, **103**, 3596–3607.
47. Y. Olivier, L. Muccioli, V. Lemaur, Y. H. Geerts, C. Zannoni and J. Cornil, *J. Phys. Chem. B*, 2009, **113**, 14102–14111.

48. T. A. Papadopoulos, L. Muccioli, S. Athanasopoulos, A. B. Walker, C. Zannoni and D. Beljonne, *Chem. Sci.*, 2011, **2**, 1025–1032.
49. M. F. Palermo, A. Pizzirusso, L. Muccioli and C. Zannoni, *J. Chem. Phys.*, 2013, **138**, 204901.
50. M. F. Palermo, L. Muccioli and C. Zannoni, *Phys. Chem. Chem. Phys.*, 2015, **17**, 26149–26159.
51. A. Pizzirusso, R. Berardi, L. Muccioli, M. Ricci and C. Zannoni, *Chem. Sci.*, 2012, **3**, 573–579.
52. A. R. van Buuren, S.-J. Marrink and H. J. Berendsen, *Colloids Surf., A*, 1995, **102**, 143–157.
53. V. Palermo, F. Biscarini and C. Zannoni, *Phys. Rev. E: Stat. Phys., Plasmas, Fluids, Relat. Interdiscip. Top.*, 1998, **57**, R2519–R2522.
54. N. H. de Leeuw and S. C. Parker, *Mol. Simul.*, 2000, **24**, 71–86.
55. S. Chen and J. Ma, *J. Chem. Phys.*, 2012, **137**, 074708.
56. B. Meredig, A. Salleo and R. Gee, *ACS Nano*, 2009, **3**, 2881–2886.
57. L. Muccioli, G. D'Avino and C. Zannoni, *Adv. Mater.*, 2011, **23**, 4532–4536.
58. R. A. Cantrell, C. James and P. Clancy, *Langmuir*, 2011, **27**, 9944–9954.
59. O. M. Roscioni, L. Muccioli, R. G. Della Valle, A. Pizzirusso, M. Ricci and C. Zannoni, *Langmuir*, 2013, **29**, 8950–8958.
60. J. Stelzer, P. Galatola, G. Barbero and L. Longa, *Phys. Rev. E: Stat. Phys., Plasmas, Fluids, Relat. Interdiscip. Top.*, 1997, **55**, 477–480.
61. D. R. Binger and S. Hanna, *Liq. Cryst.*, 2001, **28**, 1215–1234.
62. I. M. Sivebaek, V. N. Samoilov and B. N. J. Persson, *Eur. Phys. J. E: Soft Matter Biol. Phys.*, 2008, **27**, 37–46.
63. M. Yoneya, M. Kawasaki and M. Ando, *J. Mater. Chem.*, 2010, **20**, 10397–10402.
64. I. M. Sivebaek, V. N. Samoilov and B. N. J. Persson, *Phys. Rev. Lett.*, 2012, **108**, 036102.
65. R. T. Cygan, J.-J. Liang and A. G. Kalinichev, *J. Phys. Chem. B*, 2004, **108**, 1255–1266.
66. W. A. Steele, *Surf. Sci.*, 1973, **36**, 317–352.
67. S. Yip, *Handbook of Materials Modeling*, Springer, Dordrecht, 2007.
68. T. S. Bush, J. D. Gale, C. R. A. Catlow and P. D. Battle, *J. Mater. Chem.*, 1994, **4**, 831–837.
69. S. M. Woodley, P. D. Battle, J. D. Gale, C. Richard and A. Catlow, *Phys. Chem. Chem. Phys.*, 1999, **1**, 2535–2542.
70. P. Ewald, *Ann. Phys.*, 1921, **64**, 253.
71. S. W. de Leeuw, J. W. Perram and E. R. Smith, *Proc. R. Soc. London, Ser. A*, 1980, **373**, 27.
72. S. W. de Leeuw, J. W. Perram and E. R. Smith, *Proc. R. Soc. London, Ser. A*, 1983, **388**, 177.
73. T. Darden, D. York and L. Pedersen, *J. Chem. Phys.*, 1993, **98**, 10089–10092.
74. U. Essmann, L. Perera, M. L. Berkowitz, T. A. Darden, H. Lee and L. G. Pedersen, *J. Chem. Phys.*, 1995, **103**, 8577–8593.

75. D. E. Parry, *Surf. Sci.*, 1975, **49**, 433–440.
76. D. E. Parry, *Surf. Sci.*, 1976, **54**, 195.
77. E. Spohr, *J. Chem. Phys.*, 1997, **107**, 6342–6348.
78. I.-C. Yeh and M. L. Berkowitz, *J. Chem. Phys.*, 1999, **111**, 3155–3162.
79. D. Frenkel and B. Smit, *Understanding Molecular Simulations. From Algorithms to Applications*, Academic Press, New York, 2002.
80. R. Balescu, *Equilibrium and Non Equilibrium Statistical Mechanics*, Wiley, New York, 1975.
81. C. Zannoni, *The Molecular Dynamics of Liquid Crystals*, ed. G. R. Luckhurst and C. A. Veracini, Kluwer, Dordrecht, 1994, vol. 431, pp. 11–36.
82. B. Jérôme and Y. R. Shen, *Phys. Rev. E: Stat. Phys., Plasmas, Fluids, Relat. Interdiscip. Top.*, 1993, **48**, 4556–4574.
83. X. Zhuang, L. Marrucci, D. Johannsmann and Y. R. Shen, *Mol. Cryst. Liq. Cryst. A*, 1995, **262**, 1323–1331.
84. G. R. Luckhurst, T. Miyamoto, A. Sugimura, T. Takashiro and B. A. Timimi, *J. Chem. Phys.*, 2001, **114**, 10493–10503.
85. A. M. Cazabat, U. Delabre, C. Richard and Y. Y. C. Sang, *Adv. Colloid Interface Sci.*, 2011, **168**, 29–39.
86. M. Ruths and B. Zappone, *Langmuir*, 2012, **28**, 8371–8383.
87. E. Castellon, M. Zayat and D. Levy, *Phys. Chem. Chem. Phys.*, 2009, **11**, 6234–6241.
88. A. Rapini and M. Papoular, *J. Phys., Colloq.*, 1969, **30**, C4-54–C4-56.
89. P. de Gennes, *The Physics of Liquid Crystals*, Oxford University Press, 1974.
90. J. Proust and L. Ter-Minassian-Saraga, *J. Phys., Colloq.*, 1979, **40**, C3-490–C3-496.
91. L. M. Blinov, A. Y. Kabayenkov and A. A. Sonin, *Liq. Cryst.*, 1989, **5**, 645–661.
92. M. M. Wittebrood, D. H. Luijendijk, S. Stallinga, T. Rasing and I. Muševič, *Phys. Rev. E: Stat. Phys., Plasmas, Fluids, Relat. Interdiscip. Top.*, 1996, **54**, 5232–5234.
93. S. Całus, D. Rau, P. Huber and A. V. Kityk, *Phys. Rev. E: Stat., Nonlinear, Soft Matter Phys.*, 2012, **86**, 021701.
94. S. Pyo, H. Son, K.-Y. Choi, M. H. Yi and S. K. Hong, *Appl. Phys. Lett.*, 2005, **86**, 133508.
95. J. Kirkpatrick, V. Marcon, K. Kremer, J. Nelson and D. Andrienko, *J. Chem. Phys.*, 2008, **129**, 094506.
96. C. Bacchiocchi, E. Hennebicq, S. Orlandi, L. Muccioli, D. Beljonne and C. Zannoni, *J. Phys. Chem. B*, 2008, **112**, 1752–1760.
97. J. Idé, R. Mereau, L. Ducasse, F. Castet, H. Bock, Y. Olivier, J. Cornil, D. Beljonne, G. D'Avino, O. M. Roscioni, L. Muccioli and C. Zannoni, *J. Am. Chem. Soc.*, 2014, **136**, 2911–2920.
98. S. Marguet, D. Markovitsi, P. Millie, H. Sigal and S. Kumar, *J. Phys. Chem. B*, 1998, **102**, 4697–4710.
99. D. Beljonne, J. Cornil, R. Silbey, P. Millie and J. Bredas, *J. Chem. Phys.*, 2000, **112**, 4749–4758.
100. A. B. Walker, *Topics in Current Chemistry. Multiscale Modelling of Organic and Hybrid Photovoltaics*, ed. D. Beljonne and J. Cornil, Springer, Berlin, 2014, vol. 352, pp. 237–256.

101. D. Beljonne, J. Cornil, L. Muccioli, C. Zannoni, J.-L. Brédas and F. Castet, *Chem. Mater.*, 2011, **23**, 591–609.
102. M. Mas-Torrent and C. Rovira, *Chem. Rev.*, 2011, **111**, 4833–4856.
103. M. Ricci, M. Mazzeo, R. Berardi, P. Pasini and C. Zannoni, *Faraday Discuss.*, 2010, **144**, 171–185.
104. D. K. Hwang, C. Fuentes-Hernandez, J. B. Kim, W. J. Potscavage Jr. and B. Kippelen, *Org. Electron.*, 2011, **12**, 1108–1113.
105. S. Casalini, T. Cramer, F. Leonardi, M. Cavallini and F. Biscarini, *Low-Dimensionality Effects in Organic Field Effect Transistors*, John Wiley & Sons, Inc., 2013, ch. 18, pp. 397–419.
106. G. D'Avino, L. Muccioli and C. Zannoni, *Adv. Funct. Mater.*, 2015, **25**, 1985–1995.
107. P. Ostoja, S. Guerri, S. Rossini, M. Servidori, C. Taliani and R. Zamboni, *Synth. Met.*, 1993, **54**, 447–452.
108. A. Shehu, S. D. Quiroga, P. D'Angelo, C. Albonetti, F. Borgatti, M. Murgia, A. Scorzoni, P. Stoliar and F. Biscarini, *Phys. Rev. Lett.*, 2010, **104**, 246602.
109. P. Clancy, *Chem. Mater.*, 2011, **23**, 522–543.
110. J. E. Goose and P. Clancy, *J. Chem. Phys. C*, 2007, **111**, 15653–15659.
111. I. Lyubimov, L. Antony, D. M. Walters, D. Rodney, M. D. Ediger and J. J. de Pablo, *J. Chem. Phys.*, 2015, **143**, 094502.
112. N. G. Martinelli, M. Savini, L. Muccioli, Y. Olivier, F. Castet, C. Zannoni, D. Beljonne and J. Cornil, *Adv. Funct. Mater.*, 2009, **19**, 3254–3261.
113. G. D'Avino, S. Mothy, L. Muccioli, C. Zannoni, L. Wang, J. Cornil, D. Beljonne and F. Castet, *J. Phys. Chem. C*, 2013, **117**, 12981–12990.
114. J. Cornil, S. Verlaak, N. Martinelli, A. Mityashin, Y. Olivier, T. Van Regemorter, G. D'Avino, L. Muccioli, C. Zannoni, F. Castet, D. Beljonne and P. Heremans, *Acc. Chem. Res.*, 2013, **46**, 434–443.
115. K. H. DuBay, M. L. Hall, T. F. Hughes, C. Wu, D. R. Reichman and R. A. Friesner, *J. Chem. Theory Comput.*, 2012, **8**, 4556–4569.
116. G. Rossi, L. Monticelli, S. R. Puisto, I. Vattulainen and T. Ala-Nissila, *Soft Matter*, 2011, **7**, 698–708.
117. N. di Pasquale, D. Marchisio and P. Carbone, *J. Chem. Phys.*, 2012, **137**, 164111.
118. C. K. Lee, C. C. Hua and S. A. Chen, *J. Chem. Phys.*, 2012, **136**, 084901.
119. A. Makke, O. Lame, M. Perez and J.-L. Barrat, *Macromolecules*, 2012, **45**, 8445–8452.
120. H. S. Marsh and A. Jayaraman, *J. Polym. Sci., Part B: Polym. Phys.*, 2013, **51**, 64–77.
121. P.-H. Lin, I. Lyubimov, L. Yu, M. D. Ediger and J. J. de Pablo, *J. Chem. Phys.*, 2014, **140**, 204504.
122. H. A. Karimi-Varzaneh, N. F. A. van der Vegt, F. Müller-Plathe and P. Carbone, *ChemPhysChem*, 2012, **13**, 3428–3439.
123. M. Pichumani, P. Bagheri, K. M. Poduska, W. Gonzalez-Vinas and A. Yethiraj, *Soft Matter*, 2013, **9**, 3220–3229.

CHAPTER 13

3D Simulations of Organic Solar Cells

DAVIDE BARTESAGHI,[a,b] NIELS VAN DER KAAP[a] AND L. JAN ANTON KOSTER*[a]

[a] Zernike Institute for Advanced materials, Nijenborgh 4, NL-9747AG Groningen, The Netherlands; [b] Dutch Polymer Institute, P. O. Box 902, 5600AX Eindhoven, The Netherlands
*Email: l.j.a.koster@rug.nl

13.1 Introduction

The performance of organic solar cells has increased rapidly in the past two decades.[1–4] Developments are driven not only by design and engineering of new materials,[2–7] but also by improving fabrication conditions and device structure.[3,8,9] Stretching the limits of these cells even further requires physical insight into the mechanisms involved in the device operation. The net effect of all these processes makes it difficult to distinguish directly between the separate contributions. Numerical simulations allow us to overcome these issues because the influences of the different mechanisms can be studied independently. The ultimate goal of simulations is to become predictive, *i.e.* predicting the performance of certain materials or configurations, without having to synthesise the materials and fabricate the devices in a large number of configurations.

Simulations of charge transport in organic semiconductors originate from 1D drift-diffusion calculations that were first introduced in 1964 for describing charge transport in inorganic semiconductors.[10] Solving systems of coupled non-linear equations has become a fast and well-understood

RSC Energy and Environment Series No. 16
Unconventional Thin Film Photovoltaics
Edited by Enrico Da Como, Filippo De Angelis, Henry Snaith and Alison Walker

Published by the Royal Society of Chemistry, www.rsc.org

method to obtain insight into the macroscopic physics of semiconductor devices. The ease with which this type of simulation can be implemented has also made it attractive for modelling the transport in organic (light emitting) diodes[11] and solar cells.[12] Many features that are specific to organic semiconductors have been added. These features include mobility models that provide an improved description of charge transport,[13] exciton (strongly bound electron–hole pair) transport[14] and charge separation mechanisms.[12] For organic bulk heterojunctions, 1D drift-diffusion simulations use an effective medium approach, where the effective medium is made up by the electron transporting level of the acceptor and the hole transporting level of the donor. This approach permits fast calculations, but neglects the influence of the morphology. In contrast, 2D and 3D simulations contain a separate donor and acceptor phase, so calculations to determine the influence of the morphology are possible. Although drift-diffusion simulations are capable of providing a good macroscopic description of device operation, some phenomena require a more fundamental approach. For instance, a description in terms of charge carrier densities is unable to include particle–particle interactions between separate charges adequately, nor can it treat exciton separation on a microscopic scale. As a solution, Kinetic Monte Carlo (KMC) simulations allow 3D treatment at the molecular level.

The goal of this chapter is to give a description of the operation and possibilities of 2D and 3D drift-diffusion and Kinetic Monte Carlo simulations. First, the fundamental processes involved in the operation of organic solar cells are introduced briefly. Next, an overview is given of both the drift-diffusion and Kinetic Monte Carlo techniques, and some of the technical challenges are discussed. The treatment of KMC is more elaborate, because the implementation of drift-diffusion simulations for organic semiconductors is largely based on well-known drift-diffusion simulations of inorganic semiconductors, while this is not the case for KMC simulations. Furthermore, KMC simulations are computationally more demanding, which makes the implementation details more important. Finally, some results of drift-diffusion simulations and Kinetic Monte Carlo simulations are compared.

13.2 Fundamental Processes in Organic Solar Cells

Organic bulk heterojunction (BHJ) solar cells consist of an interpenetrating network of one electron donor and one electron acceptor material.[15,16] The absorption of light excites electrons into the conduction band of the absorber material, creating excitons. These excitons move to the donor–acceptor interface, where they may dissociate into a free electron and hole. The free charge carriers are transported to the contacts and extracted from the device, thus generating an electric current. The understanding of the basic physical phenomena that take place in the device is crucial for making meaningful simulations.

13.2.1 Generation of Free Charges

The dissociation of excitons and the generation of free carriers at the donor–acceptor interface is a crucial process in BHJ solar cells. A geminate pair is defined as dissociated once the distance between the charges exceeds the capture radius: the mutual Coulomb attraction between the charges is then negligible compared to the energetic disorder. The overall generation rate of free charges, G, depends on the efficiency of the optical absorption and on the efficiency of the dissociation of the photogenerated excitons into free charges. The absorption can be modelled by taking into account the interference between the incoming light and the radiation reflected by the opaque electrode.[17] For example, several optical models involve the transfer matrix formalism. Several mechanisms have been proposed for the dissociation of excitons. A model for charge generation at organic donor–acceptor interfaces was developed by Braun[18] by adapting the Onsager's analytical model for ionic separation in weak electrolytes.[19] The subject is still highly debated, as the mechanism that dominates the generation of charges following photoexcitation has not been determined yet.[20] After the dissociation of excitons, an additional electron (and/or hole) movement away from the dissociation site produces further charge separation that results in the generation of long-lived charge-separated states.[21] If the electron and hole generated from the same exciton do not overcome their Coulombic attraction, they recombine. This recombination process is referred to as geminate recombination, and it has been observed for both small molecule and polymer semiconductors.[22,23]

13.2.2 Charge Transport

Energetic and spatial disorder in organic semiconductors causes charge transport to occur *via* hopping between localised states (Figure 13.1). This charge transport mechanism is characterised by lower electron and hole mobilities compared to the values found for inorganic semiconductors.

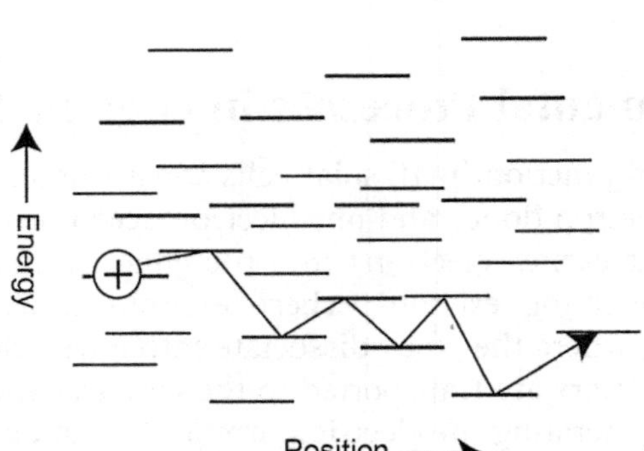

Figure 13.1 Hopping through a disordered density of states.

Hopping rate expressions describe the rate at which a particle at site *i* will hop to another site, *j*, and depend on the spatial separation and the total site energy difference between the initial and final sites. The most commonly used rate expression was derived by Miller and Abrahams:[24]

$$\upsilon_{ij} = \upsilon_0 \exp\left(-2\alpha r_{ij} - \frac{\Delta E_{ij} + |\Delta E_{ij}|}{2kT}\right), \tag{13.1}$$

where ΔE_{ij} contains the energy difference between two sites, k is Bolzmann's constant, T is the temperature, ν_0 is the attempt to jump frequency, r_{ij} is the intersite distance and α is the inverse delocalisation length. The semi-classical expression is based on single-phonon assisted hopping, and it was derived for describing charge hopping through shallow trap states in crystalline semiconductors at low temperatures. Charge carrier transport calculations based on this expression explain the dependence of charge carrier mobility on the temperature,[25] electric field and carrier density.[13]

Another important semi-classical approach was developed by Marcus:[26]

$$\upsilon_{ij} = \upsilon_0 \sqrt{\frac{\pi}{4\lambda kT}} \exp\left(-2\alpha r_{ij} - \frac{(\Delta E_{ij} + \lambda)^2}{4\lambda kT}\right). \tag{13.2}$$

This expression assumes that the initial and final sites are located in a potential well, separated by an energy barrier. The size of the energy barrier is provided by the parameter λ. The barrier mimics the polaronic nature of charge carriers in the organic semiconductors: charges polarise their close surroundings and are therefore located in a potential well. Charges hop from well to well, and have to overcome the barriers in between.

Semi-classical approaches break down when temperature approaches 0 K: both Miller–Abrahams and Marcus hopping predict that the charge carrier mobility decreases as a function of the electric field and that the charge carrier mobility vanishes at temperatures close to 0 K.[27] Alternatively, quantum mechanical descriptions are able to deal with these situations. An example of a quantum mechanical hopping rate expression is nuclear tunnelling mediated hopping:[28]

$$k_{ij} = \frac{H_{ij}^2}{h\omega_c}\left(\frac{h\omega_c}{2\pi kT}\right)^{1-2\alpha_k} \frac{\left|\Gamma\left(\alpha_k + i\left(\frac{\Delta E_{ij}}{2\pi kT}\right)\right)\right|^2}{\Gamma(2\alpha_k)} \exp\left(\frac{\Delta E_{ij}}{2kT}\right) \exp\left(\frac{|\Delta E_{ij}|}{h\omega_c}\right). \tag{13.3}$$

Here, H_{ij} is the overlap integral between the initial and final site, ω_c represents the ground state frequency of the material under investigation, and α_k describes the coupling of the system to the heat bath. The hopping model assumes that the system always possesses a ground state energy that drives the hopping process, even at 0 K. This results in a non-vanishing mobility for temperatures down to 0 K. For increased temperatures, the expression reduces to the semi-classical Marcus expression.

For some donor–acceptor blends, the transport of either electrons or holes can be limited by charge trapping. A charge trap is a defect site within the forbidden energy gap of the donor–acceptor blend.[29] Charge trapping is often a limitation for the transport of electrons in conjugated polymers.[30] Analysing the trap-limited electron transport in different conjugated polymers, Nicolai *et al.* found that all exhibit a common trap distribution, located at an energy of ~3.6 eV below the vacuum level,[31] indicating that the trap states have a common origin. Although the origin of the electron-trapping states is not known, it is likely that they are due to chemical defects related to water or oxygen.[32,33] Trapping of holes has been shown to be less significant, although recent studies have provided experimental evidence for hole trapping in organic materials.[34,35]

The first Kinetic Monte Carlo algorithm investigating hopping transport was reported in 1966 and described the migration of defects in binary ordered alloys.[36] Monte Carlo simulations were first applied to disordered systems in 1971 by Silver *et al.*,[37] who simulated the transient photocurrent after an incident pulse of light. They attributed the dispersion of the photocurrent to carrier trapping in both the surface and bulk of the material. During the 1970s, the theory on charge transport was further refined,[38] although there was still debate whether the dispersion was caused by an energetic distribution of trap states or an energetic distribution of hop sites.

In 1981, Schönherr *et al.* performed KMC simulations using Miller–Abrahams hopping rates in a disordered material.[39] They demonstrated that hopping through a Gaussian density of states could also explain the dispersive behaviour of the photocurrent. Moreover, the results could explain the dependence of charge carrier mobility in time of flight measurements on the temperature and electric field.[40] Research on KMC methods for time of flight measurements continued during the 1980s, providing relations for the scaling of charge carrier mobility with the electric field, disorder and charge carrier density.[41]

During the 1990s, KMC simulations remained focused on time of flight calculations: device physics was still treated using classic transport equations. Simulations were extended with models describing exciton transport and dissociation.[42] In 1999, Wolf *et al.* were the first to work on KMC simulations of charge injection from metallic contacts into a Gaussian density of states.[43] Although no full device was modelled, they discarded the presence of long-range (Fowler–Nordheim) tunnelling injection and showed that the injection occurs in a way similar to thermionic emission.

13.2.3 Charge Recombination

An important loss mechanism in organic solar cells is the non-geminate recombination of free charges. The process of bimolecular recombination was first described by Langevin for ions in a gas;[44] the rate of bimolecular recombination depends quadratically on the charge density and is determined by the time required for electrons and holes to diffuse towards each other.[45]

According to Langevin's theory, the rate of bimolecular recombination is given by

$$R_{\text{bimolecular}} = k_{\text{R}}(np - n_{\text{i}}^2), \tag{13.4}$$

where n and p are the density of electrons and holes respectively, n_{i} is the intrinsic carrier concentration of electrons and holes and k_{R} is the bimolecular recombination coefficient,

$$k_{\text{R}} = \frac{q}{\varepsilon}(\mu_{\text{n}} + \mu_{\text{p}}). \tag{13.5}$$

Here, q is the elementary charge, ε is the dielectric constant of the donor–acceptor blend, and $\mu_{\text{n(p)}}$ the electron (hole) mobility. However, in some organic solar cells, a bimolecular recombination rate which it is significantly lower than the one following from Langevin's theory has been reported.[46–48] Bobbert and Greenham[49] used KMC calculations to show that disorder introduces a weak dependence on charge density in k_{R}.

If charge traps are present, an additional recombination channel is active, the description of which is given by the Shockley–Read–Hall (SRH) equation:[50,51]

$$R_{\text{SRH}} = \frac{C_{\text{n}}C_{\text{p}}N_{\text{t}}}{[C_{\text{n}}(n + n_i) + C_{\text{p}}(p + p_i)]}, \tag{13.6}$$

where R_{SRH} is the rate of trap-assisted recombination, C_{n} and C_{p} are the capture coefficients for electrons and holes, respectively, and N_{t} is the density of charge traps. Furthermore, recombination losses due to surface recombination may have a significant impact on the device performance.[52–55] Surface recombination is governed by the presence of minority carriers at the contacts. Electrons (holes) diffusing to the anode (cathode) recombine with injected holes (electrons). Thus, the current at the electrodes is related to the density of minority carriers:[53,56]

$$J_{\text{n(p)}} = qS_{\text{n(p)}}[n(p) - n(p)_{\text{eq}}], \tag{13.7}$$

where $J_{\text{n(p)}}$ is the electron (hole) current at the anode (cathode), $S_{\text{n(p)}}$ is the surface recombination velocity, and $n(p)_{\text{eq}}$ is the equilibrium carrier density at the contact. Typically, the surface recombination velocity is (implicitly) assumed to be infinite, meaning that the densities of carriers at the contacts are equal to their equilibrium values.[57,58]

13.3 Operation of Drift-diffusion Simulations

The band diagram of a donor–acceptor bulk heterojunction solar cell under short-circuit conditions is shown in Figure 13.2. The excitons are dissociated into free charges thanks to the energy offset between the lowest unoccupied molecular orbitals (LUMO) or between the highest occupied molecular orbitals (HOMO) of the two materials. Once free electrons and holes are

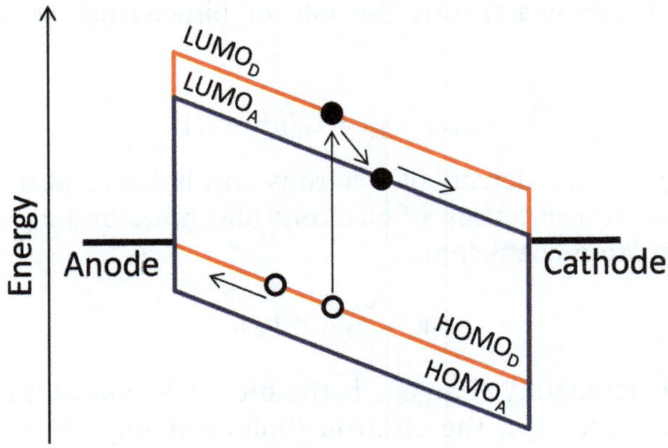

Figure 13.2 Band diagram of a bulk heterojunction solar cell at short-circuit conditions (for simplicity, any band bending is ignored). The notation D and A refers to donor and acceptor material, respectively.

present in the device, they move driven by the built-in and the external electric field (drift) and by their concentration gradient (diffusion).

A widely used approach for the simulation of organic semiconductor devices consists in considering the effects of both drift and diffusion on the transport of charges (drift-diffusion simulations).

The electric field is obtained by solving the Poisson equation:[59]

$$\frac{\partial^2}{\partial x^2}\psi(x) = \frac{q}{\varepsilon}[n(x) - p(x)], \tag{13.8}$$

where ψ is the electric potential. In our notation, x denotes the distance to the cathode.

The continuity equations for the electrons and holes in the steady state are derived from the conservation of charges:[59]

$$\frac{\partial J_{\mathrm{n}}(x)}{\partial x} = q(G - R), \tag{13.9}$$

$$\frac{\partial J_{\mathrm{p}}(x)}{\partial x} = -q(G - R). \tag{13.10}$$

Here, $J_{\mathrm{n(p)}}$ is the electron (hole) current density, G and R are the rates of generation and recombination of free charges, respectively. By incorporating the drift and diffusion terms, the expressions for the electron and hole current densities are[59]

$$J_{\mathrm{n}} = -qn\mu_{\mathrm{n}}\frac{\partial}{\partial x}\psi(x) + qD_{\mathrm{n}}\frac{\partial}{\partial x}n(x), \tag{13.11}$$

$$J_{\mathrm{p}} = -qp\mu_{\mathrm{p}}\frac{\partial}{\partial x}\psi(x) - qD_{\mathrm{p}}\frac{\partial}{\partial x}p(x), \tag{13.12}$$

where $D_{n,p}$ are the carrier diffusion coefficients, which are related to the mobilities of the two species through the classical or the generalised Einstein relation.[60–62]

The generation and recombination rates of free charges are calculated as described above. Although the absorption of light, and hence the generation rate G, is generally not uniform across the active layer, it has been shown that including an optical profile in the modelling of the device does not significantly change the results for a thickness of the active layer smaller than 300 nm.[63] Thus, a constant generation profile through the active layer can be used if the profile is not available.

Eqn (13.8)–(13.12) have to be solved to calculate the total current extracted from the device. To do so, it is necessary to introduce boundary conditions that specify the carrier densities and the potential at both contacts. The boundary condition on the potential is given by

$$(V_L - V_0 + V_{\text{applied}}) = W_{\text{anode}} - W_{\text{cathode}} \tag{13.13}$$

where V_L and V_0 are the potentials at the two electrodes, V_{applied} is the applied voltage, W_{anode} and W_{cathode} are the work functions of the anode ($x = L$) and the cathode ($x = 0$), respectively. The boundary conditions on the electron and hole densities at the contacts depends on the type of contacts and on the surface recombination velocity. For ohmic contacts (no energy barrier for the injection of charges) and infinite surface recombination velocity, the boundary conditions are given by

$$n(0) = p(L) = N_{\text{cv}}, \tag{13.14}$$

$$n(L) = N_{\text{cv}} \exp\left(-\frac{\varphi_{\text{A}}}{kT}\right), \tag{13.15}$$

$$p(0) = N_{\text{cv}} \exp\left(-\frac{\varphi_{\text{D}}}{k_{\text{B}}T}\right), \tag{13.16}$$

where N_{cv} is the effective density of states and $\varphi_{\text{A(D)}}$ is the energy barrier between the LUMO (HOMO) of the acceptor (donor) and the anode (cathode).

Eqn (13.8)–(13.16) are written in one dimension (1D); by solving them, it is implied that the charges move only in the direction x, normal to the surface of the device. Moreover, the 1D approach treats the active layer as an effective blend, in which every node can be at the same time donor and acceptor, thus neglecting the effect of the morphology on the device performance. However, it is well known that the morphology of the donor–acceptor blend plays a significant role.[64–66] The phase separation between the donor and acceptor materials provides interfaces at which exciton dissociation occurs. The diffusion length of excitons in organic materials is around 10 nm.[67,68] Therefore, if the phase separation is too coarse, with domains of pure acceptor and donor larger than the exciton diffusion length, the excitons generated too far from the interface will not reach it within their lifetime; on the other hand, too fine a phase separation may hinder the transport of charges, which requires the presence of continuous percolation pathways to the electrodes. The importance of the morphology is treated in more detail in Chapter 12 of this volume.

In order to include the effects of the blend microstructure in drift-diffusion simulations, one has to leave the 1D approach and add one or two more dimensions. The morphology of the blend is simulated by means of a 2D or 3D grid, in which each node represents either the donor, the acceptor phase, or a mixed phase of the two materials.

Eqn (13.8)–(13.16) are rewritten in 2D or 3D; furthermore, the number of excitons that yield free charges at the interface is calculated by solving the exciton diffusion equation

$$0 = -\frac{X(\boldsymbol{r})}{\tau} + D_x \nabla^2 X(\boldsymbol{r}) + g, \tag{13.17}$$

where $X(\boldsymbol{r})$ is the exciton density in the position defined by the vector $\boldsymbol{r}$, τ is the exciton lifetime, D_x is the exciton diffusion constant and g is the generation rate of excitons. We note that the exciton diffusion equation can be also solved for the 1D case, if a bilayer device is considered.[69]

The numerical solution of the 2D or 3D equations involves the discretisation of the equations over 2D or 3D grids and requires a significantly larger computational effort. At the price of longer simulation times than in the 1D case, the implementation of 2D and 3D models can provide insight into the effect of the morphology on the device performance.

13.4 Operation of KMC Simulations

In contrast to the large delocalisation lengths in crystalline semiconductors, charge carriers in disordered materials are only delocalised over a distance between one or two nanometres. This causes charge transport to be driven by phonon-activated hopping of carriers between delocalised sites. Because the discrete hop events cannot be translated into a simple set of mathematical transport equations, an alternative method is required to model the charge transport. Kinetic Monte Carlo simulations predict the charge transport in disordered systems by simulating the individual behaviour of its particles.

The disordered material is represented by a spatial grid, where each node is a hopping site for localised electrons, holes and excitons. A static energy level is assigned to each node that represents the local LUMO or HOMO level of the materials. Most simulations use a cubic grid and Gaussian distributed site energies. These configurations have been able to predict many of the charge transport phenomena in organic semiconductors.[13,41,70,71] Alternatively, material-specific values can be used for the locations and site energies by performing molecular dynamics and quantum chemistry calculations, respectively.[72,73]

The type of boundary conditions determines the applications of the KMC simulation. Two types of boundary conditions can be distinguished. The first type has periodic boundary conditions in all directions: charges that leave the device are reintroduced on the opposite side of the volume. This type of boundary conditions is used for investigating bulk effects like recombination[49] and charge transport.[41] The second type contains metallic contacts from which carriers can be injected or extracted.[74] This type allows full devices to be simulated such as, for example, light emitting diodes[75] and solar cells.[76]

Although KMC simulations are used for many purposes, they can all be seen as a collection of a countable number of states. In our case, each state corresponds to a particular distribution of particles over all available hopping sites. Hopping events can be regarded as transitions that take the system from one state to another. The hopping rates in eqn (13.1)–(13.3) give the occurrence rate of each transition as a function of the difference in potential energy before and after the transition. Kinetic Monte Carlo simulations determine the time evolution of the system by starting in one state, and repeatedly performing transitions that each take a certain amount of time.[77]

First, the system is set up in an initial state. This might be done in different ways: for transient photo absorption techniques, excitons are randomly generated throughout the layer; for time of flight studies, charge carriers are initialised in a slab of the device; for charge carrier transport calculations, charge carriers are generated in their equilibrium position. Next, all transition rates are calculated. The rate s of each possible transition is given by a hopping expression, and the sum of all transition rates from a single state gives the overall transition rate S of that state. Because the time of the previous event is independent of the time of the next event, hopping can be seen as a Poisson process. Therefore, the lifetime of the current state can be found by drawing an exponentially distributed random number with the overall rate S. Given a uniformly distributed random number r between 0 and 1, this becomes:

$$\tau_{\text{next}} = \frac{\ln(r)}{S}. \tag{13.18}$$

Choosing which transition occurs is done by defining an interval between 0 and S, where each transition is assigned a subinterval with a width of s. Next, a uniformly distributed random number between 0 and S hits one of the intervals and determines that the corresponding transition will occur.

In general, the process of selecting and performing transitions is repeated many times during one simulation instance. Current densities and charge carrier mobilities are obtained by counting the number of charges that move in the applied field direction within a given time frame. The stopping criteria depend on the application of the simulation. For time of flight and transient photo absorption, the simulation may continue until all charges have reached an electrode or have recombined. For charge carrier mobility or full device calculations, the process may continue until the mobility or current density has stabilised or has reached a steady-state value.

13.5 Implementation

The solution to the basic equations of the drift-diffusion model is obtained numerically with an iterative method;[59] this approach requires the discretisation of the equations.[10] An example of the numerical solution for the 1D case is given in a paper from Koster *et al.*, in which the continuity and Poisson equations were solved with a modified generation rate for free charge carriers, to include geminate recombination.[12]

The numerical solution of the 2D or 3D equations involves the discretisation of the equations over 2D or 3D grids and requires a significantly larger computational effort. At the price of longer simulation times than in the 1D case, the implementation of 2D and 3D models can provide insight on the effect of the morphology on the device performance.

Performing KMC simulations is generally time consuming: performing the millions of transitions that are required for a single run is computationally very demanding. Moreover, for concentrations below 1% and for high degrees of disorder (for low temperatures or a wide density of states), the simulation volume needs to be increased (or multiple simulations need to be executed) in order to obtain valid results.[78] Therefore, a fast implementation is crucial in order to obtain useful results. Different strategies exist for the reduction of the computational effort. The first reaction method calculates the individual waiting time for each possible transition and stores these in a queue.[79] The transition with the shortest waiting time is executed first. After performing a transition, all new transitions are added to the queue. All other rates are reused. This approach reduces the effort because only a few hopping rates need to be recalculated after each move. A disadvantage is that changes in the interaction potential due to nearby charge hops are not taken into account, although this does not seem to be an important process at the low charge concentrations in OPV.[80]

Alternatively, simulations can be parallelised.[81] The KMC algorithm treats all possible transitions as a Poisson process: transitions are independent and the interval between their occurrence is exponentially distributed. Performing multiple transitions simultaneously will therefore invalidate the simulation results. Still, parallel implementations can be obtained by dividing the simulation volume in different segments and treating each segment as a separate simulation. The validity of the simulation is guaranteed by making sure that only one transition is performed in each segment simultaneously. An important detail is that the overall transition rate of all segments must be equal, otherwise the simulation time will be different for the different segments. Particles that leave a certain segment are introduced in an adjacent segment. The effectiveness of parallelising the simulation depends on the number of charges and their distribution within the volume: the parallel efficiency for many homogenously spread charges is better than for a few charges that are localised.[82]

One of the most critical parts in KMC simulations in OPV is the implementation of Coulomb interactions between the charge carriers. The low dielectric constant causes the interaction potential between charges to be relatively large compared to most other semiconductors. This influences both the charge transport[83] and the recombination rates.[84] A proper description of the operation of a solar cell therefore requires the implementation of Coulomb interactions. Including these interactions is computationally expensive: the number of interaction calculations scales quadratically with the number of charges in the system. Moreover, the transition rates of nearby particles need to be updated when a charge moves.

The number of interactions is typically reduced by introducing a cut-off radius for the interactions: only the Coulomb interactions of nearby particles are taken into account. Different studies have been carried out to determine the required cut-off radius. Casalegno *et al.* showed that using a cut-off, even for values as large as 16 nm, overestimates the carrier densities and underestimates the device performance of solar cells.[85] A study on charge carrier transport in volumes with full periodic boundary conditions has shown that the charge carrier mobility is only affected at high carrier concentrations and low electric fields.[86]

Another approach for accelerating the calculations is to change the algorithm. Instead of fully recalculating all interactions after a nearby charge move, the interaction potentials of other charges can be corrected by adding a dipole contribution associated with the move.[82] Only the interaction potentials of the moving particle need to be recalculated completely. This approach requires the dipole addition to be commutative: undoing a charge move should result in the same interaction potential.

13.6 Impact of Morphology on OPV Performance

The performance of organic bulk heterojunctions is strongly dependent on the morphology of the active layer. Defining an ideal morphology would be of great benefit for the optimisation of the devices, and device modelling can guide the experiments towards the achievement of the optimal morphology. The easiest method of looking into these morphological effects is by using 2D finite element simulations.[87,88] This reduces greatly the computational effort and reproduces some of the characteristics. However, comparison with real devices requires 3D morphologies. Therefore, this section focuses on 3D KMC simulations.

During the rapid developments in organic bulk heterojunction solar cells, it was clear that a more detailed device description required information about the blend morphology.[1] In 2003, Nelson introduced the first KMC simulation for the description of charge recombination in a polymer–fullerene blend of MDMO and PCBM.[54] The simulation modelled the photocurrent after an incident pulse of laser light, and reproduced the recombination kinetics for different intensities, temperatures and level of background radiation. The dependence of the separation yield on the blend morphology was first investigated for planar heterojunctions, showing an improved yield with respect to the yield for bulk separation.[89] In 2005, Watkins *et al.* developed a KMC simulation including extracting electrodes that could also deal with the blend morphology.[76] The results indicated that the internal quantum efficiency (IQE) depends strongly on the length scale of the phase separation. This behaviour is also shown in Figure 13.3: the optimum internal quantum efficiency was found for an interfacial area of $0.5 \times 10^6\,\text{nm}^2$. This optimum arises at the point where the product of the exciton dissociation efficiency and the charge collection efficiency is maximum. Moreover, an increase in IQE of a factor of 1.5 was found between a disordered and a checkered morphology.

Because the optimal checkered morphology is hard to fabricate in real devices, more knowledge is required about the disordered morphologies that are encountered in most experiments. A common method for the characterisation of the morphology is by determining the size of the features in the morphology. This measure is defined as three times the ratio between the simulation volume and the total interfacial area between the donor and acceptor phases.[90] Different studies have been performed to determine the optimal feature size for use in PV devices. All the values obtained fall around 10 nm. Although the size of the features and the spacing between the donor and acceptor phases are a good way of characterising the system, the performance is also determined by the type of interface between the two phases: Lyons *et al.* investigated this for different types of interfaces and domain purities.[91] Figure 13.4 shows the short circuit current, the extraction efficiency and the recombination efficiency as a function of the domain size. The blue squares correspond to a pure domain with a sharp interface, the red triangles to a pure domain with an interface that is not sharp, and the black circles to an impure domain without sharp interface. The morphology itself was kept constant for the three types of interfaces. The results show that pure domains result in good performances because charges are extracted more easily. Also, the effect of sharpening diffuse interfaces is more pronounced than coarsening the interface. Moreover, an optimum feature size of around 7 nm is observed for all types of configurations.

Efficient carrier extraction requires good connectivity of both phases with the corresponding electrodes, as for the rod-type morphology discussed above. Isolated islands and dead-ends prevent the extraction of charges and therefore limit the device performance. A special type of morphology is the gyroid morphology: two interpenetrating, continuous and identical phases.[92] The absence of dead-ends and isolated regions in the phase would be expected to result in an increase of the overall performance. Moreover, experimental systems that could reproduce this morphology are known.

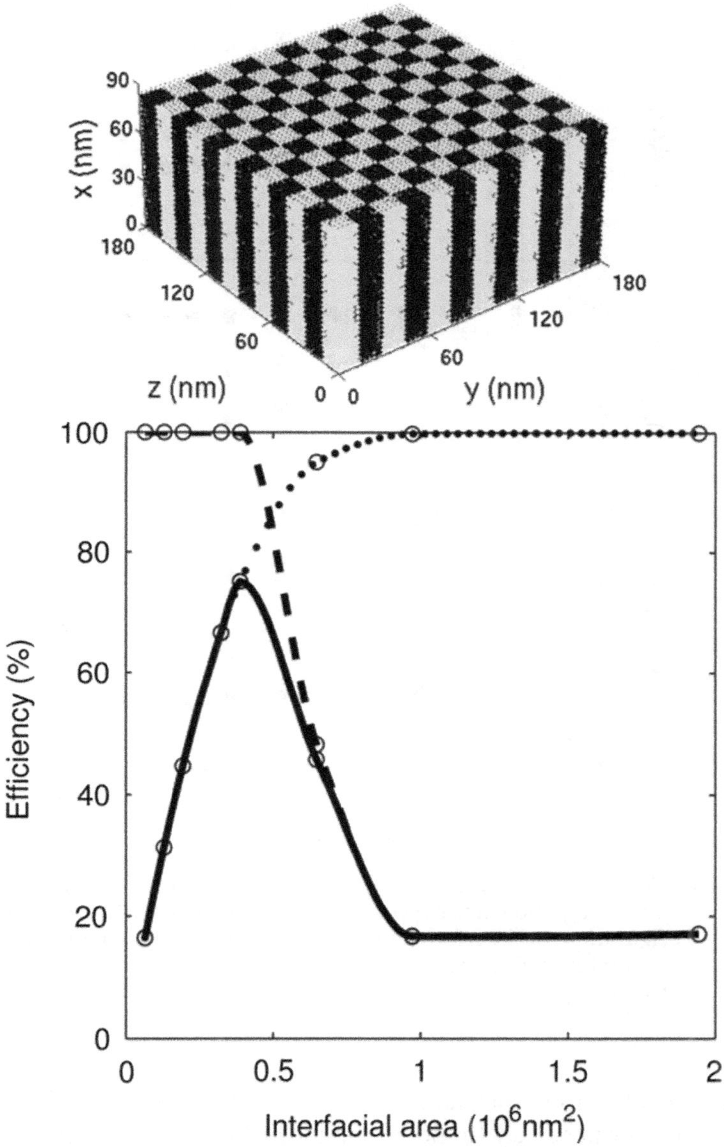

Figure 13.3 The top figure shows a checkered morphology. The bottom plot shows the exciton dissociation efficiency (dotted line), charge collection efficiency (dashed line), and internal quantum efficiency (solid line) *vs.* interfacial area in units of 106 nm^2. Simulation data points, to which the lines are fitted, are shown as circles.
Reprinted with permission from P. K. Watkins *et al.*, *Nano Lett.*, 2005, **5**, 1814–1818. Copyright © 2012 American Chemical Society.

However, KMC calculations on this material showed that recombination losses would prevent an increase in performance. This is a good example of the predictive power of simulations in the field of OPV.

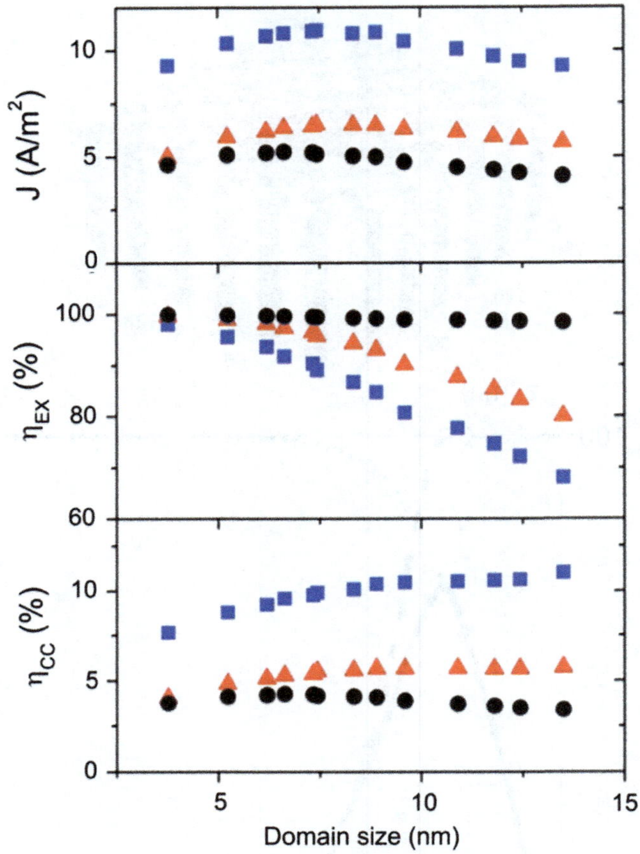

Figure 13.4 Current density at an electric field of 1×10^7 V m^{-1}, charge extraction efficiency, and carrier recombination efficiency for three different configurations of interface and domain, for varying domain size. Reproduced from ref. 91 with permission from the Royal Society of Chemistry.

Except for calculating the characteristics of full devices, KMC simulations have been used extensively to model the individual charge dissociation and extraction mechanisms for solar cells. One example are the calculations on charge transport in disordered materials that have been discussed above.[41] Other examples are simulations on the impact of using heterojunctions on the carrier generation and recombination, for instance in the studies by Peumans *et al.*[89] and Nelson[54] that were already mentioned above. The generation of free charges in organic semiconductors is a complex process: the high exciton binding energy requires an energetic offset of the donor and acceptor phases, but the disordered nature and strong Coulomb interaction between the dissociated carriers make the separation process complicated. A straightforward approach to modelling the separation process leads to lower rates than those observed experimentally. Much research has therefore been done on the origin of the relatively high separation yields that are encountered in OPV.

The influence of the energetic disorder on charge dissociation was first investigated by Albrecht and Bässler in 1995.[93] It was found that if charge carriers are generated at a high position in the density of states, the energetic difference with adjacent sites can be enough to overcome the mutual attraction between a pair of geminate charge carriers. This effect is more pronounced in the case of correlated disorder or with the presence of energetic cascades that lead the charges away from the interface.[94,95]

Another example of KMC studies on the dissociation of charges is the work of Groves *et al.*, who showed that the dissociation yield increases rapidly once the geminate pair has separated a few nanometers.[90] This effect is shown in Figure 13.5, which shows the geminate separation efficiency as a function of the initial spatial separation. For relatively small electric fields, this distance decreases below 10 nm. This is much smaller than the thermal capture radius, and a large initial separation could therefore explain the relative low separation yield. This mechanism is further supported by recent studies that show that the initial charge separation distance is larger than previously assumed.[20,96,97]

In state-of-the-art OPV devices, the internal quantum efficiency approaches 100 percent. This indicates that charge separation is no longer important for the device operation. Therefore, the main loss mechanism in these devices is non-geminate recombination of electrons and holes that originate from different excitons. The first KMC studies on recombination focused on the generation of light in organic light emitting diodes.[93] The recombination of electrons and holes in the single material was described as a function of the binding energy between the opposite carriers.

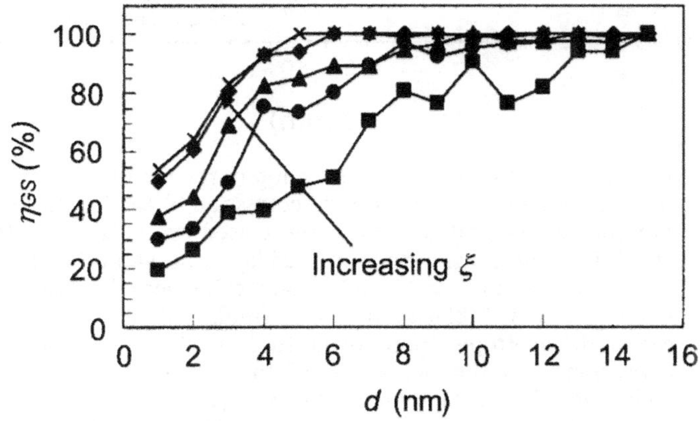

Figure 13.5 Geminate pair separation efficiency for increasing electric fields as a function of the initial separation.
Reprinted with permission from C. Groves *et al.*, *J. Chem. Phys.*, 2008, **129**, 114903. Copyright © 2008 AIP Publishing LLC.

For large binding energies, recombination according to the Langevin theory was found. For decreasing energies, the recombination efficiency was found to fall rapidly. In another study, the impact of recombination in 3D bulk heterojunctions was investigated.[70] This paper shows why 3D KMC simulations are required. Figure 13.6 contains calculations of the recombination constant as a function of the number of charge transport layers considered. For a layer thickness of 1, transport occurs in only 2 dimensions. For low densities, the recombination constant for low dimensionality is much lower than the value obtained from Langevin theory. Moreover, these calculations show that the recombination rate of 3D bulk heterojunction solar cells deviates by about one order of magnitude from the values obtained by Langevin theory, as shown in Figure 13.7.

The first to use the drift-diffusion model in order to include information about the morphology in a device model were Buxton and Clarke in 2006;

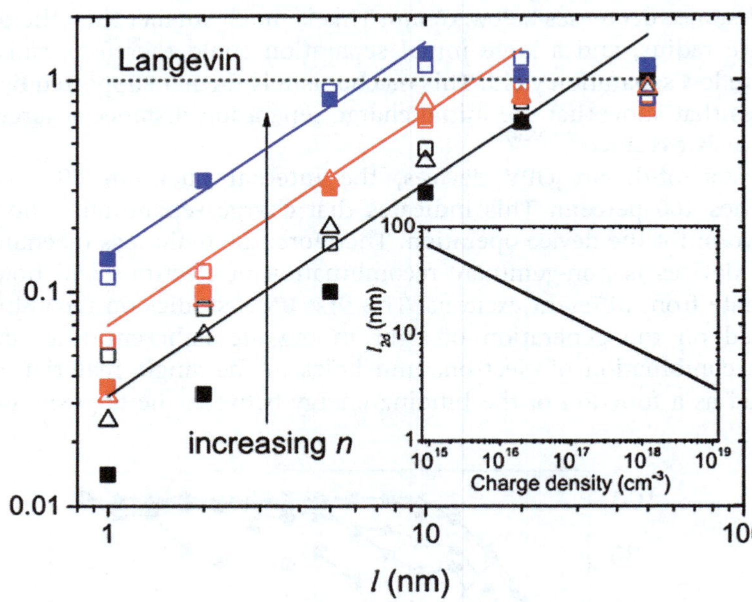

Figure 13.6 Recombination constant as a function of the simulation volume extent in one dimension, l. l = 1 nm, corresponds to the 2D limit. Open symbols correspond to $F = 0$ while closed symbols to $F = 107$ V m^{-1} which is in the plane of the confined layer. We show data for sigma = 0 meV (squares) and $\sigma = 75$ meV (triangles), where the charge density is $n = p = 10^{16}$ cm^{-3} (black), 10^{17} cm^{-3} (red), and 10^{18} cm^{-3} (blue). In all cases, the isotropic mobility is 10^{-4} cm^2 V^{-1} s^{-1}. The dotted line corresponds to the prediction of the Langevin relation, while the solid lines show the predictions of eqn (13.4) for the charge densities examined. The inset shows the volume thickness below, in which two-dimensional transport is expected from eqn (5) of ref. 70.
Reprinted figure with permission from C. Groves and N. C. Greenham, *Phys. Rev. B: Condens. Matter Mater. Phys.*, 2008, **78**, 155205. Copyright © 2008 by the American Physical Society.

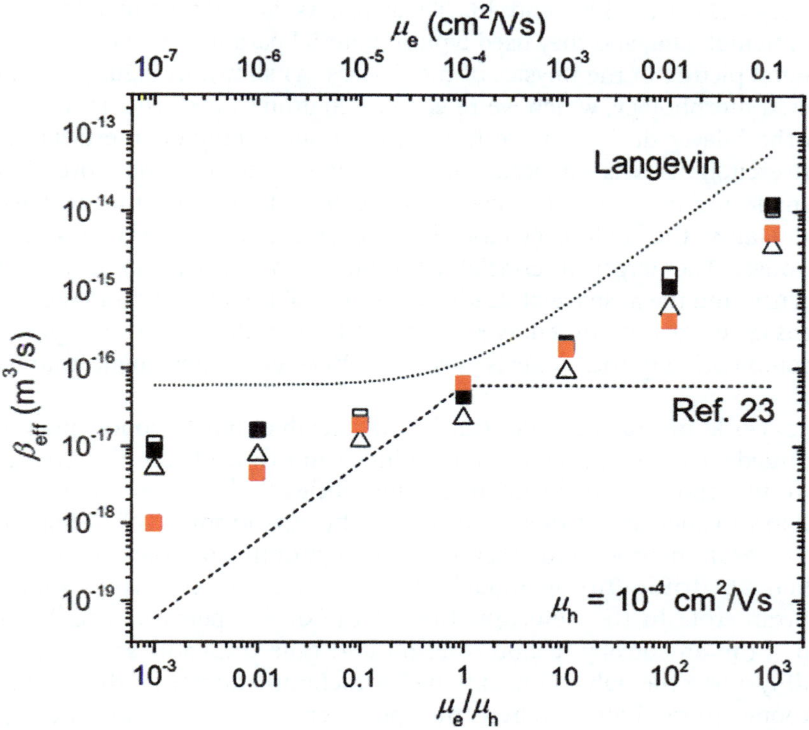

Figure 13.7 Measured efficiency for a 1 : 1 blend of electron and hole-accepting polymers with an average feature size of $d = 4$ (black) and 35 nm (red). Squares correspond to $\sigma = 0$ and triangles correspond to $\sigma = 75$ meV. Open symbols represent $F = 0$ and closed symbols represent $F = 107\ \mathrm{V\,m^{-1}}$. In all cases the isotropic hole mobility is $10^{-4}\ \mathrm{cm^2\,V^{-1}\,s^{-1}}$. The dotted line shows the prediction of the Langevin equation and the dashed line shows the prediction of the Langevin equation when using only the smallest mobility, as suggested by Koster *et al.* ("ref. 23" in the figure).[98]
Reprinted figure with permission from C. Groves and N. C. Greenham, *Phys. Rev. B: Condens. Matter Mater. Phys.*, 2008, **78**, 155205.

they investigated the effect of complex diblock copolymer morphologies on the device performance[99] by solving the equations for the electrostatic potential and for the electron, hole and exciton density in two dimensions as explained before. Semiconducting diblock copolymers have an intrinsic length scale of phase separation that makes them ideal materials for organic bulk heterojunctions. The phase separation in diblock copolymers leads to a bicontinuous network of donor and acceptor domains, which can be oriented through surface-induced ordering or by applying an electric field.[100,101] Buxton and Clarke employed the 2D drift-diffusion approach to analyse the photovoltaic performance of devices with a bicontinuous morphology that was either disordered or oriented perpendicularly to the electrodes (Figure 13.8a and b). Their results are not compared to any experimental data, and they used typical material parameters in order to have a general picture of the physics of the devices. As shown in Figure 13.8c, the oriented morphology, which we refer to as columnar, performs much better than the bilayer device. In the former, the donor–acceptor interfacial area, where exciton dissociation occurs, is larger than in the latter. Moreover, in the columnar morphology there are no dead-ends that force the charges to move against the built-in electric field for part of their travel towards the electrodes. The larger interfacial area increases the exciton dissociation efficiency, and the absence of dead-ends in the columnar morphology ensures an easier extraction of charges compared to a disordered morphology; the combination of these effects yields an efficiency enhancement from 0.6% to 2.2%.

The work of Buxton and Clarke on flexible block copolymers was continued by Shah and Ganesan to include semi-rigid or rod-like structures and to incorporate anisotropy in the mobilities.[102,103] Exciton and charge transport in such structures is expected to be anisotropic.[104,105] First, Shah and Ganesan incorporated anisotropy in the mobility of holes and excitons in their 2D drift-diffusion model, showing that this anisotropy plays an important role in the determination of the device performance.[102] They defined the anisotropy in both exciton and hole transport as the ratio of mobility along the polymer backbone (intrachain) and perpendicular to the backbone (interchain). When the polymer chains are oriented perpendicularly to the donor–acceptor interface, there is an optimum value of anisotropy, resulting from a balance between the transport of excitons to the interface and the transport of holes to the electrodes. If, instead, the chains are oriented in parallel, the effect of such anisotropy is less strong.[102] In a second publication,[103] these authors used the model to analyse realistic systems based on donor–acceptor materials of rod–coil block copolymers, though they noted that their model is generally valid for BHJ systems with a continuous spatial variation of the morphology and in which the donor phases are characterised by anisotropy in hole and exciton mobilities. The morphology of rod–coil copolymers can be represented by a lamellar structure, with layers of rod (donor) and coil (acceptor) blocks alternated. The lamellae can be oriented by tuning the surface interaction and/or applying an electric field (Figure 13.9a and b). Shah and Ganesan analysed the effect of the orientation of the lamellae and showed that a perpendicular orientation with respect to the surface of the device is preferable. Beside the lamellar structure, they investigated the performance of "non-layered" morphologies, such as the "puck morphology" that is formed for high volume fractions of coil blocks (Figure 13.9c). In this case, the advantage offered by the high interfacial area between rod and coil blocks is counterbalanced by the lack of continuous pathways for holes to reach the anode, and the performance of devices with such morphology is intermediate between the parallel and perpendicular lamellar structures (Figure 13.9d).

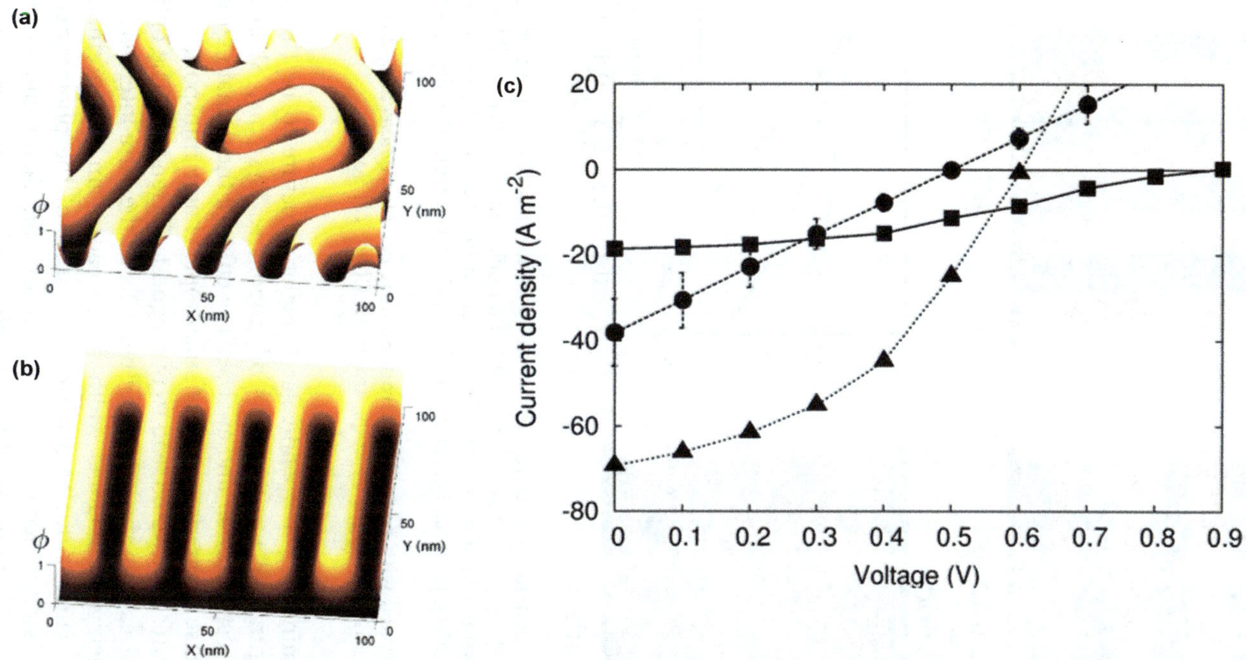

Figure 13.8 (a, b) Diblock copolymer thin film morphologies showing disordered (a) and columnar (b) structure. (c) Simulated current–voltage (*J–V*) characteristics for systems with planar (squares), disordered bulk heterojunction (circles) and columnar (triangles) morphology.
Reprinted figure with permission from G. A. Buxton and N. Clarke, *Phys. Rev. B: Condens. Matter Mater. Phys.*, 2006, **74**, 085207. Copyright © 2006 by the American Physical Society.

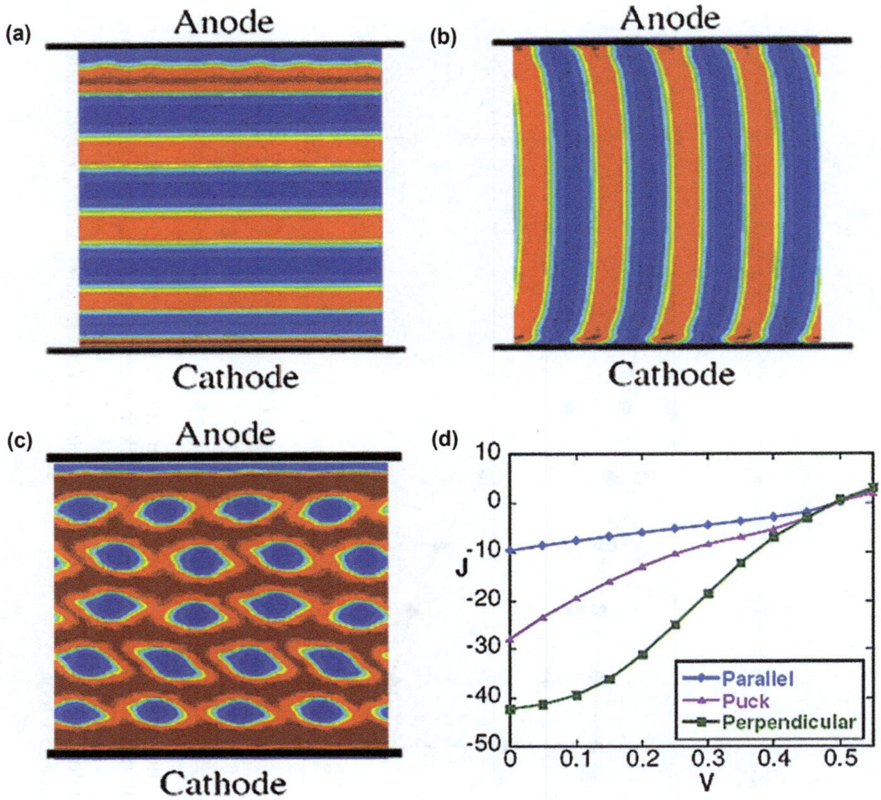

Figure 13.9 (a–c) Density profiles for donor and acceptor phases for parallel lamellae (a), perpendicular lamellae (b) and puck phases (c). The coil block is shown in red, and the rod block is shown in blue. (d) *J–V* curves for the preceding phases. The units are in A m^{-2} for current density and V for voltage.
Reprinted with permission from Shah *et al.*, *Macromolecules*, 2010, **43**, 543–552. Copyright © 2009 American Chemical Society.

Having determined the perpendicular lamellar structure to be the most desirable configuration for the rod–coil copolymer systems, Shah and Ganesan moved a step further and analysed the effect of the domain spacing, the degree of phase separation and the anisotropic charge transport, finding the optimal features that would lead to the highest efficiency.[103] The optimal column width arises from the balance of two effects: the extra interfacial area created by reducing the channel width which causes an enhancement of the short circuit current due to a better harvesting of the photogenerated excitons, and a simultaneously decrease in the fill factor due to extra recombination at the interface. Their results indicated that the optimal column width is in the order of the exciton diffusion length. The same was obtained *via* KMC methods by Lei *et al.*,[106] who showed that the optimal domain width increases upon increasing the exciton lifetime.

Shah and Ganesan also analysed the effect of chemical incompatibility: when this is increased, the phase separation is enhanced and the interface width is decreased. This reduces the recombination of charges and improves the short-circuit current.[103] In agreement with the work of Shah and Ganesan, KMC simulations in a recent publication[91] showed that the length scale of the phase separation has an influence on the separation yield, and that sharply defined interfaces give improved results over blurred ones.

An ordered bulk heterojunction was also suggested to be ideal by Martin *et al.*[107] They represented the ordered heterojunction as an interpenetrating network of columns, each of which is composed of either pure acceptor or pure donor material. They proposed a 2D representation of this morphology in which the key parameters are the width and height of the channels (C_W and C_L, respectively) and the thickness of the electron (d_n) and hole (d_p) blocking layers (Figure 13.10). They investigated the effect of varying the channel height and width. Increasing the channel height has two effects. On one hand, it increases the interface area, thus enhancing the exciton dissociation efficiency; on the other hand, longer channels involve a longer average distance that the charges must travel in order to reach the electrodes, which increases recombination. The optimal channel height is therefore achieved by a trade-off between these two effects, and it is in the order of 100 nm, depending on the recombination strength (weaker recombination increases the optimal thickness of the device). About the

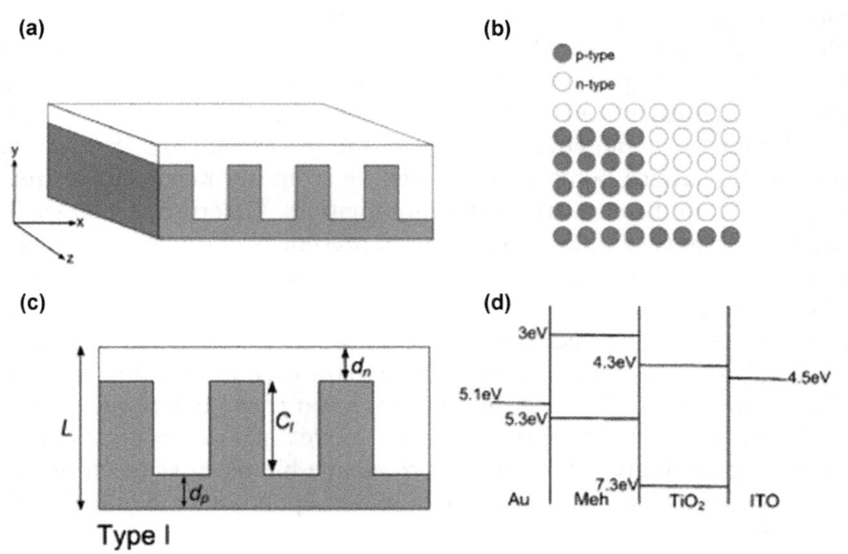

Figure 13.10 (a) Schematic diagram of the devices modelled by Martin *et al.* (b) Representation of the devices on a 2D lattice. (c) Device parameters. (d) Energy diagram for an ITO/TiO_2/MEH-PPV/Au device. Reprinted with permission from C. M. Martin *et al.*, *J. Appl. Phys.*, 2007, **102**, 104506. Copyright © 2007 AIP Publishing LLC.

optimal channel width, they agreed with other works in the field pointing out that column sizes of the order of the exciton diffusion length result in the best performances.

Other publications focused on determining the characteristics of the optimal morphology.[108–110] However, the fabrication of devices with ordered structure *via* morphology engineering[109,111,112] is challenging and requires expensive techniques. Moreover, Ray and Alam showed that the intrinsic random morphology obtained by inexpensive solution-processing does not necessarily translate into a lower performance compared to ordered heterojunctions.[113] Their work is also based on drift-diffusion modelling and sets a fundamental limit for the performance gain that can be achieved through morphology engineering. Although previous work showed a beneficial effect of morphology engineering (*e.g.* Watkins *et al.* found an increase in the internal quantum efficiency (IQE) of a factor of 1.5 between a disordered and a checkered morphology[76]), Ray and Alam demonstrated that an optimised disordered bulk heterojunction performs almost as well as an ordered heterojunction (in their example, the best PCE values for the disordered and ordered heterojunction were 5.5% and 6.1%, respectively).

The examples reported up to this point highlight the possibility of using drift-diffusion and KMC modelling as theoretical tools to study the effect of morphology on device efficiency. The results of the simulations in these papers have not been compared to experimental data. The first attempt to describe the effect of the morphology quantitatively by comparing simulations and experimental data is the work of Maturová *et al.*[114–116] They proposed a 2D drift-diffusion model in which an alternation of slabs of two different phases (*i.e.* mixed phase and pure acceptor phase) mimics the complex morphology of a polymer–fullerene blend (Figure 13.11a). The *J–V* characteristics calculated from their model agreed well with the experimental data;[114,115] moreover, they were able to explain the contrast and shifts of the surface potential measured by scanning Kelvin probe microscopy (SKPM) in the dark and under illumination.[116] Maturová *et al.* showed that, when extensive phase separation is present in the polymer–fullerene blends, the lateral transport of electrons towards the acceptor-rich phase[117] is the critical process for the operation of the device, overcoming the effect of exciton diffusion.[114] Thus, the optimal feature size, which they determined to be smaller than 50 nm,[115] is not related to the exciton diffusion length but rather to the lateral distance that electrons can travel before recombining with holes. Their results explain the *J–V* curves at low electric fields, for which the probability for the charges to escape bimolecular recombination strongly depends on the length scale of phase separation, and at high fields, for which the lateral transport of electrons is suppressed.

The work of Maturová also indicates how the electron and hole currents and recombination are distributed inside the device (Figure 13.11b and c). Similarly, a recent publication from Bartesaghi and Koster shows that, in the case of coarse phase separation in the form of blob-like aggregates of an acceptor material embedded into an homogeneous mixed phase, it is

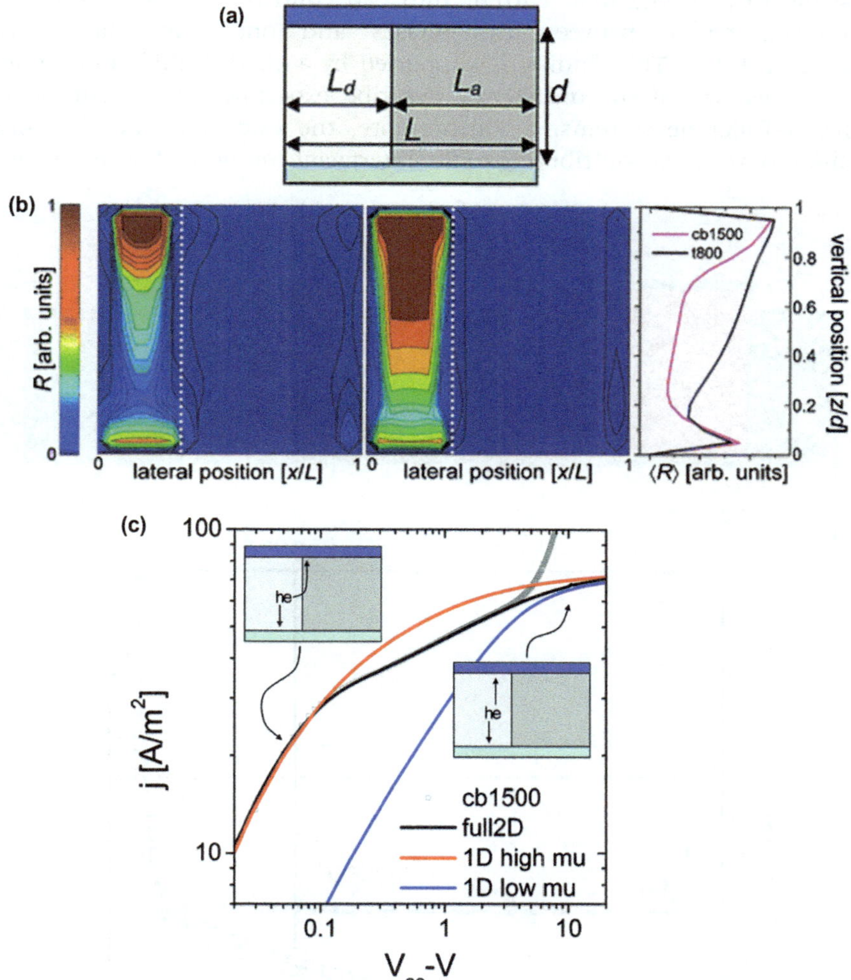

Figure 13.11 (a) Schematic representation of the sample layout used in the numerical model by Maturová *et al.* L_a and L_d are the width of the acceptor-rich and the donor-rich phase, respectively, d is the thickness of the active layer. (b) Calculated recombination rates under short-circuit conditions inside MDMO-PPV:PCBM devices spin-cast from chlorobenzene at 1500 rpm (left, $L = 40$ nm) and from toluene at 800 rpm (centre, $L = 800$ nm) against normalised vertical and lateral positions; dotted lines indicate the phase boundary. The right panel shows the average recombination rate of each device. (c) J–V characteristic of a device spin-cast from chlorobenzene at 1500 rpm. Symbols represent the experimental data, solid lines represent the results of numerical simulations. "high mu" and "low mu" in the legend refer to the electron mobility used in the 1D simulations.
Reprinted with permission from K. Maturová *et al.*, *Nano Lett.*, 2009, **9**, 3032–3037.

possible to split the total current into two contributions, one from the interfacial region between the phases and one from the matrix (Figure 13.12).[118] This finding is supported by a 3D drift-diffusion model, and it was proved to consistently describe experimental J–V curves for polymer–fullerene systems.[119] Furthermore, the extension and the composition of the area contributing to the interfacial component of the current

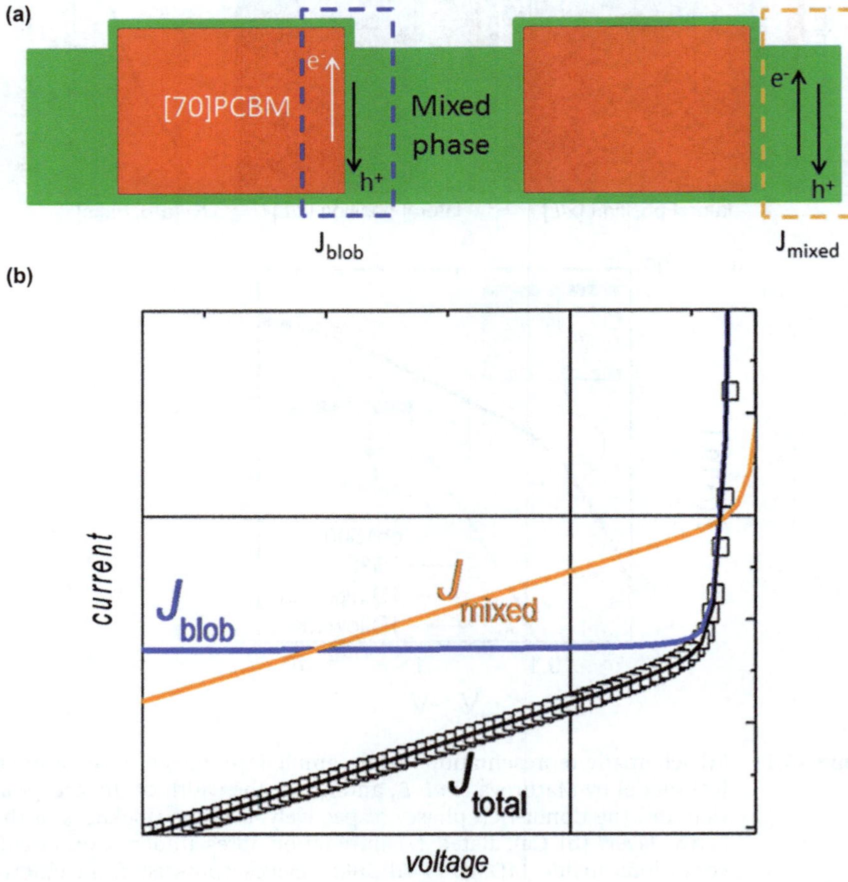

Figure 13.12 (a) Schematic representation of the device morphology of a PDPP5T:[70]PCBM solar cell with coarse phase separation. The regions giving rise to J_{blob} (blue) and J_{mixed} (orange) are highlighted. (b) Experimental (symbols) and simulated (solid lines) J–V curves for a PDPP5T:[70]PCBM solar cell with coarse phase separation. The two contributions J_{blob} and J_{mixed} are calculated with a 1D drift-diffusion model and then added to give J_{total}.
Reprinted from D. Bartesaghi *et al.*, Charge transport and recombination in PDPP5T:[70]PCBM organic solar cells: The influence of morphology, *Org. Electron.*, **15**, 3191–3202, Copyright 2014, with permission from Elsevier.

explain the changes in shape and magnitude of the *J–V* curve upon changing the extent of phase separation and the composition of the mixed phase.

One further step in the field was made by Koster *et al.* by using a 3D optoelectronic model to explain the role of morphology in hybrid P3HT/ZnO solar cells.[120] The high density of ZnO, and the lack of a mixed donor and acceptor phase in this system, allows detailed morphological information to be obtained using electron microscopy and tomography.[121,122] For the first time, this paper uses an experimentally imaged 3D morphology as direct input for optoelectronic simulations, showing that a quantitative link between morphology and device performance can be realised *via* drift-diffusion approaches. They showed that the IQE in P3HT/ZnO solar cells is a trade-off between charge transport and exciton harvesting. Furthermore, they analysed the impact of isolated clusters of acceptor and of dead-ends in the morphology on the overall device efficiency.

13.7 KMC for Other Types of Devices

KMC simulations are also used extensively for other types of devices. A study by van der Holst *et al.* used KMC simulations to study the current voltage characteristics of sandwiched-type single carrier devices, and compared these with drift-diffusion simulations.[123] Although good agreement between the two methods was found for high injection barriers, the current densities obtained by KMC simulations were lower for reduced barriers. In 2009, Kwiatkowski *et al.* studied the charge transport through C60 thin film transistors.[72] The morphology in this model was obtained by the simulated evaporation of a film of C60 molecules. A study by Mesta *et al.* in 2013 simulated a fully operational multilayer white organic light emitting diode.[75] Good agreement between experiments and theory was found for both charge and exciton transport.

13.8 Comparing Drift-diffusion and KMC Simulations

Drift-diffusion simulations and Kinetic Monte Carlo (KMC) simulations are both commonly used methods for the understanding of the physics related to OPV. Drift-diffusion simulations allow a description using macroscopic properties like mobility, charge carrier density and generation and recombination rates. The fact that this approach works well enough to describe the experimental data suggests that the molecular length scale is decoupled from the largest length scale of phase separation (>10 nm).[115] KMC simulations describe the system from a microscopic point of view, considering the dynamics of individual charges.

Historically, drift-diffusion simulations have been used for the description of space charge limited devices, while KMC simulations were more common for describing processes involving low carrier concentrations. The main reason for this difference are the computational requirements of both methods. A single current voltage sweep using drift-diffusion takes at most a few minutes, while KMC may need several days to perform the same job. This makes drift-diffusion more suitable for fitting experimental data, while KMC gives more information about microscopic behaviour. Often, detailed mechanisms from KMC are included in the drift-diffusion simulations to mimic the microscopic behaviour in the macroscopic simulation.[13,123]

Despite the large computational resources that are required for KMC simulations involving high charge densities, advances in algorithm design and computationally available power have made these simulations possible. One example compares drift-diffusion simulations of single carrier diodes including injection barriers with KMC simulations.[123] Although the drift-diffusion simulation gives similar results to the KMC simulation for large injection barriers, a deviation is found when ohmic contacts are used. The difference is attributed to the inclusion of short range Coulomb interactions between charge carriers in the KMC simulation, a feature that cannot be implemented in a drift-diffusion simulation. A similar comparison was made by Koster *et al.* between the impact of the morphological feature size on the bimolecular recombination loss in a ZnO BHJ for KMC and drift-diffusion simulations.[120] Here, the KMC data were taken from a previous publication.[92] Although the comparison is only qualitative, both methods predict that the bimolecular recombination losses will decrease with the increasing feature size. From the previous comparisons, it can be concluded that drift-diffusion simulations and KMC simulations can give similar results, as long as the drift-diffusion simulation is supplied with enough details on the mechanisms under consideration. However, some mechanisms, for example short range Coulomb interactions, cannot be implemented due to the macroscopic description using charge densities.

13.9 Outlook

Drift-diffusion and KMC simulations have provided insight into many physical properties of charge transport in organic semiconductors and into the operation mechanisms of electronic devices. KMC simulations have mainly been used to investigate the fundamental processes involved in the operation of devices, while drift-diffusion simulations have been used for explaining device characteristics. With research shifting towards smaller length scales, drift-diffusion and KMC simulations are increasingly coupled to other methods. The first works that included details on the chemical structure of individual molecules dates from 2006.[124,125] Molecular dynamics calculations are used to obtain a more realistic particle grid, while quantum chemistry calculations provide the energy levels of each hopping site. In 2007, Kirkpatrick *et al.* used this multiscale approach to model time of flight experiments for different derivatives of hexabenzocoronene.[126] The results of the KMC simulations were compared to experimental values, and agreement was found for both the magnitude and trend of the mobilities. Another example is the work by Rühle *et al.*, where charge transport through conjugated oligomers was modelled on different lengthscales.[127] Multiscale methods allow nanoscale properties to be coupled to macroscopic phenomena like charge carrier mobilities, current densities, or even the energy conversion efficiency of a solar cell. These methods not only offer insight into the physics behind different processes in disordered materials, but they also open up opportunities to predict the performance of newly designed materials, even before they have been synthesised.

Acknowledgements

D. B. is supported by the Dutch Polymer Institute (DPI), project number 734. The work by N. J. v. d. K. is part of the research program of the Foundation for Fundamental Research on Matter (FOM), which is part of the Netherlands Organization for Scientific Research (NWO).

References

1. S. E. Shaheen, C. J. Brabec, N. S. Sariciftci, F. Padinger, T. Fromherz and J. C. Hummelen, *Appl. Phys. Lett.*, 2001, **78**, 841.
2. G. Li, V. Shrotriya, J. Huang, Y. Yao, T. Moriarty, K. Emery and Y. Yang, *Nat. Mater.*, 2005, **4**, 864.
3. Y. Liang, Z. Xu, J. Xia, S.-T. Tsai, Y. Wu, G. Li, C. Ray and L. Yu, *Adv. Mater.*, 2010, **22**, E135.
4. Y. Liu, J. Zhao, Z. Li, C. Mu, W. Ma, H. Hu, K. Jiang, H. Lin, H. Ade and H. Yan, *Nat. Commun.*, 2014, **5**, 5293.
5. C. B. Nielsen, R. S. Ashraf, N. D. Treat, B. C. Schroeder, J. E. Donaghey, A. J. P. White, N. Stingelin and I. McCulloch, *Adv. Mater.*, 2015, **27**, 948.
6. Y. Liu, Y. M. Yang, C. C. Chen, Q. Chen, L. Dou, Z. Hong, G. Li and Y. Yang, *Adv. Funct. Mater.*, 2013, **25**, 4657.
7. V. S. Gevaerts, A. Furlan, M. M. Wienk, M. Turbiez and R. A. J. Janssen, *Adv. Mater.*, 2012, **24**, 2130.
8. Q. Wei, T. Nishizawa, K. Tajima and K. Hashimoto, *Adv. Mater.*, 2008, **20**, 2211.
9. J. J. Intemann, K. Yao, Y.-X. Li, Y.-X. Xu, P.-W. Liang, C.-C. Chueh, F.-Z. Ding, X. Yiang, X. Li, Y. Chen and A. K.-Y. Jen, *Adv. Funct. Mater.*, 2014, **24**, 1465.
10. H. K. Gummel, *IEEE Trans. Electron Devices*, 1964, **11**, 455.
11. H. T. Nicolai, G. A. H. Wetzelaer, M. Kuik, A. J. Kronemeijer, B. de Boer and P. W. M. Blom, *Appl. Phys. Lett.*, 2010, **96**, 172107.
12. L. J. A. Koster, E. C. P. Smits, V. D. Mihailetchi and P. W. M. Blom, *Phys. Rev. B: Condens. Matter Mater. Phys.*, 2005, **72**, 085205.
13. W. F. Pasveer, J. Cottaar, C. Tanase, R. Coehoorn, P. A. Bobbert, P. W. M. Blom, D. M. de Leeuw and M. A. J. Michels, *Phys. Rev. Lett.*, 2005, **94**, 206601.
14. M. Kuik, L. J. A. Koster, A. G. Dijkstra, G. A. H. Wetzelaer and P. W. M. Blom, *Org. Electron.*, 2012, **13**, 969.

15. G. Yu, J. Gao, J. C. Hummelen, F. Wudl and A. J. Heeger, *Science*, 1995, **270**, 1789.
16. J. J. M. Halls, C. A. Walsh, N. C. Greenham, E. A. Marseglia, R. H. Friend, S. C. Moratti and A. B. Holmes, *Nature*, 1995, **376**, 498.
17. G. Burkhard, E. T. Hoke and M. D. McGehee, *Adv. Mater.*, 2010, **22**, 3293.
18. C. L. Braun, *J. Chem. Phys.*, 1984, **80**, 4157.
19. L. Onsager, *J. Chem. Phys.*, 1934, **2**, 599.
20. S. Few, J. M. Frost and J. Nelson, *Phys. Chem. Chem. Phys.*, 2015, **17**, 2311.
21. T. M. Clarke and J. R. Durrant, *Chem. Rev.*, 2010, **110**, 6737.
22. B. Schweitzer, V. I. Arkipov, U. Scherf and H. Bässler, *Chem. Phys. Lett.*, 1999, **313**, 57.
23. D. Credgington, F. C. Jamieson, B. Walker, T.-Q. Nguyen and J. R. Durrant, *Adv. Mater.*, 2012, **24**, 2135.
24. A. Miller and E. Abrahams, *Phys. Rev.*, 1960, **120**, 745.
25. S. D. Baranovskii, I. P. Zvyagin, H. Cordes, S. Yamasaki and P. Thomas, *Phys. Status Solidi B*, 2002, **230**, 281.
26. R. A. Marcus and N. Sutin, *Biochim. Biophys. Acya, Bioenerg.*, 1985, **811**, 265.
27. J. D. Yuen, R. Menon, N. E. Coates, E. B. Namdas, S. Cho, S. T. Hannahs, D. Moses and A. J. Heeger, *Nat. Mater.*, 2009, **8**, 572.
28. K. Asadi, A. J. Kronemeijer, T. Cramer, L. J. A. Koster, P. W. M. Blom and D. M. de Leeuw, *Nat. Commun.*, 2013, **4**, 1710.
29. M. A. Lampert, *Phys. Rev.*, 1956, **103**, 1648.
30. M. M. Mandoc, B. de Boer, G. Paasch and P. W. M. Blom, *Phys. Rev. B: Condens. Matter Mater. Phys.*, 2007, **75**, 193202.
31. H. T. Nicolai, M. Kuik, G. A. H. Wetzelaer, B. de Boer, C. Campbell, C. Risko, J. L. Bredas and P. W. M. Blom, *Nat. Mater.*, 2012, **11**, 882.
32. H. E. Tseng, K. Y. Peng and S. A. Chen, *Appl. Phys. Lett.*, 2003, **82**, 4086.
33. V. Kažukauskas, *Semicond. Sci. Technol.*, 2004, **201**, 1236.
34. J. Schafferhans, A. Baumann, C. Deibel and V. Dyakonov, *Appl. Phys. Lett.*, 2008, **93**, 093303.
35. H. S. Woo, R. Czerw, S. Webster, D. L. Carroll, J. Ballato, A. E. Strevens, D. O'Brien and W. J. Blau, *Appl. Phys. Lett.*, 2000, **77**, 1393.
36. W. M. Young and E. W. Elcock, *Proc. Phys. Soc., London*, 1966, **89**, 735.
37. M. Silver, K. S. Dy and I. L. Huang, *Phys. Rev. Lett.*, 1971, **27**, 21.
38. G. Pfister and H. Scher, *Adv. Phys.*, 1978, **27**, 747.
39. G. Schönherr, H. Bässler and M. Silver, *Philos. Mag. B*, 1981, **44**, 369.
40. H. Bässler, G. Schönherr, M. Abkowitz and D. M. Pai, *Phys. Rev. B: Condens. Matter Mater. Phys.*, 1982, **26**, 3105.
41. H. Bässler, *Phys. Status Solidi B*, 1993, **175**, 15.
42. M. Scheidler, U. Lemmer, R. Kersting, S. Karg, W. Riess, B. Cleve, R. F. Mahrt, H. Kurz, H. Bässler, E. O. Göbel and P. Thomas, *Phys. Rev. B: Condens. Matter Mater. Phys.*, 1996, **54**, 5536.

43. U. Wolf, V. I. Arkhipov and H. Bässler, *Phys. Rev. B: Condens. Matter Mater. Phys.*, 1999, **59**, 7507.
44. P. Langevin, *Ann. Chim. Phys.*, 1903, **28**, 433.
45. G. J. Adriaenssens and V. I. Arkhipov, *Sol. State Commun.*, 1997, **103**, 541.
46. G. Juška, K. Arlauskas, J. Stuchlik and R. Österbacka, *J. Non-Cryst. Solids*, 2006, **352**, 1167.
47. A. Facchetti, *Mater. Today*, 2013, **16**, 123.
48. P. Zalar, M. Kuik, N. A. Ran, J. A. Love and T.-Q. Nguyen, *Adv. Energy Mater.*, 2014, **4**, 1400438.
49. N. C. Greenham and P. A. Bobbert, *Phys. Rev. B: Condens. Matter Mater. Phys.*, 2003, **68**, 245301.
50. W. Shockley and W. T. Read, *Phys. Rev.*, 1952, **87**, 835.
51. R. N. Hall, *Phys. Rev.*, 1952, **87**, 387.
52. A. Wagenpfahl, C. Deibel and V. Dyakonov, *IEEE J. Sel. Top. Quantum Electron.*, 2010, **16**, 1759.
53. T. Kirchartz, B. E. Pieters, K. Taretto and U. Rau, *Phys. Rev. B: Condens. Matter Mater. Phys.*, 2009, **80**, 035334.
54. J. Nelson, *Phys. Rev. B: Condens. Matter Mater. Phys.*, 2003, **67**, 155209.
55. J. C. Scott and G. G. Malliaras, *Chem. Phys. Lett.*, 1999, **299**, 115.
56. J. Kniepert, I. Lange, N. J. van der Kaap, L. J. A. Koster and D. Neher, *Adv. Energy Mater.*, 2014, **4**, 1301401.
57. C. Deibel, A. Wagenpfahl and V. Dyakonov, *Phys. Status Solidi*, 2008, **2**, 175.
58. S. Lacic and O. Inganas, *J. Appl. Phys.*, 2005, **97**, 124901.
59. S. Selberherr, *Analysis and Simulation of Semiconductor Devices*, Springer Verlag, Wien, 1984.
60. A. Einstein, *Ann. Phys.*, 1905, **322**, 549.
61. Y. Roichman and N. Tessler, *Appl. Phys. Lett.*, 2002, **80**, 1948.
62. G. A. H. Wetzelaer, L. J. A. Koster and P. W. M. Blom, *Phys. Rev. Lett.*, 2011, **107**, 066605.
63. J. D. Kotlarski, P. W. M. Blom, L. J. A. Koster, M. Lenes and L. H. Slooff, *J. Appl. Phys.*, 2008, **103**, 084502.
64. S. van Bavel, S. Veenstra and J. Loos, *Macromol. Rapid Commun.*, 2010, **31**, 1835.
65. M. A. Ruderer and P. Müller-Buschbaum, *Soft Matter*, 2011, **7**, 5482.
66. F. Liu, Y. Gu, J. W. Jung, W. H. Jo and T. P. Russel, *J. Polym. Sci., Part B: Polym. Phys.*, 2012, **50**, 1018.
67. O. V. Mikhnenko, H. Azimi, M. Scharber, M. Morana, P. W. M. Blom and M. A. Loi, *Energy Environ. Sci.*, 2012, **5**, 6960.
68. S. D. Dimitrov, C. B. Nielsen, S. Shoaee, P. S. Tuladhar, J. Du, I. McCulloch and R. Durrant, *J. Phys. Chem. Lett.*, 2012, **3**, 140.
69. V. S. Gevaerts, L. J. A. Koster, M. M. Wienk and R. A. J. Janssen, *Appl. Mater. Interfaces*, 2011, **3**, 3252.
70. C. Groves and N. C. Greenham, *Phys. Rev. B: Condens. Matter Mater. Phys.*, 2008, **78**, 155205.

71. J. O. Oelerich, D. Huemmer and S. D. Baranovskii, *Phys. Rev. Lett.*, 2012, **108**, 226403.
72. J. J. Kwiatkowski, J. M. Frost and J. Nelson, *Nano Lett.*, 2009, **9**, 1085.
73. V. Rühle, A. Lukyanov, F. May, M. Schrader, T. Vehoff, J. Kirkpatrick, B. Baumeier and D. Andrienko, *J. Chem. Theory Comput.*, 2011, **7**, 3335.
74. J. J. M. van der Holst, M. A. Uijttewaal, B. Ramachandhran, R. Coehoorn, P. A. Bobbert, G. A. de Wijs and R. A. de Groot, *Phys. Rev. B: Condens. Matter Mater. Phys.*, 2009, **79**, 085203.
75. M. Mesta, M. Carvelli, R. J. de Vries, H. van Eersel, J. J. M. van der Holst, M. Schober, M. Furno, B. Lüssem, K. Leo, P. Loebl, R. Coehoorn and P. A. Bobbert, *Nat. Mater.*, 2013, **12**, 652.
76. P. K. Watkins, A. B. Walker and G. L. B. Verschoor, *Nano Lett.*, 2005, **5**, 1814.
77. D. T. Gillespie, *J. Comput. Phys.*, 1976, **22**, 403.
78. S. Kim, J. Yeo, C. Im and D. Kim, *J. Korean Phys. Soc.*, 2012, **60**, 1897.
79. M. A. Gibson and J. Bruck, *J. Phys. Chem. A*, 2000, **104**, 1876.
80. C. Groves, R. G. E. Kimber and A. B. Walker, *J. Chem. Phys.*, 2010, **133**, 144110.
81. Y. Shim and J. G. Amar, *Phys. Rev. B: Condens. Matter Mater. Phys.*, 2005, **71**, 125432.
82. N. J. van der Kaap and L. J. A. Koster, *J. Comput. Phys.*, 2016, **307**, 321.
83. J. Zhou, Y. C. Zhou, J. M. Zhao, C. Q. Wu, X. M. Ding and X. Y. Hou, *Phys. Rev. B: Condens. Matter Mater. Phys.*, 2007, **75**, 153201.
84. L. J. A. Koster, S. E. Shaheen and J. C. Hummelen, *Adv. Energy Mater.*, 2012, **2**, 1246.
85. M. Casalegno, G. Raos and R. Po, *J. Chem. Phys.*, 2010, **132**, 094705.
86. J. J. M. van der Holst, *Three-dimensional Modeling of Charge Transport, Injection and Recombination in Organic Light-emitting Diodes*, Technische Universiteit Eindhoven, Eindhoven, 2010.
87. A. G. Gagorik, J. W. Mohin, T. Kowalewski and G. R. Hutchison, *J. Phys. Chem. Lett.*, 2013, **4**, 36.
88. J. Williams and A. B. Walker, *Nanotechnology*, 2008, **19**, 424011.
89. P. Peumans and S. R. Forrest, *Chem. Phys. Lett.*, 2004, **398**, 27.
90. C. Groves, R. A. Marsh and N. C. Greenham, *J. Phys. Chem.*, 2008, **129**, 114903.
91. B. P. Lyons, N. Clarke and C. Groves, *Energy Environ. Sci.*, 2012, **5**, 7657.
92. R. G. E. Kimber, A. B. Walker, G. E. Schröder-Turk and D. J. Cleaver, *Phys. Chem. Chem. Phys.*, 2010, **12**, 844.
93. U. Albrecht and H. Bässler, *Chem. Phys.*, 1995, **199**, 207.
94. H. van Eersel, R. A. J. Janssen and M. Kemerink, *Adv. Funct. Mater.*, 2012, **22**, 2700.
95. T. M. Burke and M. D. McGehee, *Adv. Mater.*, 2013, **26**, 1923.
96. D. Amarasinghe Vithanage, A. Devižis, V. Abramavičius, Y. Infahsaeng, D. Abramavičius, R. C. I. MacKenzie, P. E. Keivanidis, A. Yartsev,

D. Hertel, J. Nelson, V. Sundström and V. Gulbinas, *Nat. Commun.*, 2013, **4**, 2334.
97. S. Gélinas, A. Rao, A. Kumar, S. L. Smith, A. W. Chin, J. Clark, T. S. van der Poll, G. C. Bazan and R. H. Friend, *Science*, 2014, **343**, 512.
98. L. J. A. Koster, V. D. Mihailetchi and P. W. M. Blom, *Appl. Phys. Lett.*, 2006, **88**, 052104.
99. G. A. Buxton and N. Clarke, *Phys. Rev. B: Condens. Matter Mater. Phys.*, 2006, **74**, 085207.
100. T. Xu, C. J. Hawker and T. P. Russell, *Marcomolecule*, 2005, **38**, 2802.
101. T. L. Morkved, M. Lu, A. M. Urbas, E. E. Ehrichs, H. M. Jaeger, P. Mansky and T. P. Russell, *Science*, 1996, **273**, 931.
102. M. Shah, V. Pryamitsyn and V. Ganesan, *Appl. Phys. Lett.*, 2009, **95**, 194101.
103. M. Shah and V. Ganesan, *Macromolecules*, 2010, **43**, 543.
104. K. Coakley and M. McGehee, *Chem. Mater.*, 2004, **16**, 4533.
105. K. Coakley, B. Srinivasan, J. Ziebarth, C. Goh, Y. Lu and M. McGehee, *Adv. Funct. Mater.*, 2005, **15**, 1927.
106. B. Lei, Y. Yao, A. Kumar, Y. Yang and V. Ozolins, *J. Appl. Phys*, 2008, **104**, 024504.
107. C. M. Martin, V. M. Burlakov, H. E. Assender and D. A. R. Barkhouse, *J. Appl. Phys.*, 2007, **102**, 104506.
108. Y. Min Nam, J. Huh and W. H. Jo, *Sol. Energy Mater. Sol. Cells*, 2010, **94**, 1118.
109. A. A. Gorodetsky, C. Y. Chiu, T. Schiros, M. Palma, M. Cox, Z. Jia, W. Sattler, I. Kymissis, M. Steigerwald and C. Nuckolls, *Angew. Chem., Int. Ed.*, 2010, **49**, 7909.
110. E. Krali and R. J. Curry, *ACS Nano*, 2011, **5**, 3069.
111. Z. Fan, H. Razavi, J. won Do, A. Moriwaki, O. Ergen, Y. L. Chueh, P. W. Leu, J. C. Ho, T. Takahashi, L. A. Reichertz, S. Neale, K. Yu, M. Wu, J. W. Ager and A. Javey, *Nat. Mater.*, 2009, **8**, 648.
112. J. Kong, I.-W. Hwang and K. Lee, *Adv. Mater.*, 2014, **26**, 6275.
113. B. Ray and M. A. Alam, *Sol. Energy Mater. Sol. Cells*, 2012, **99**, 204.
114. K. Maturová, S. S. van Bavel, M. M. Wienk, R. A. J. Janssen and M. Kemerink, *Nano Lett.*, 2009, **9**, 3032.
115. K. Maturová, S. S. van Bavel, M. M. Wienk, R. A. J. Janssen and M. Kemerink, *Adv. Funct. Mater.*, 2011, **21**, 261.
116. K. Maturová, M. Kemerink, M. M. Wienk, D. S. H. Charrier and R. A. J. Janssen, *Adv. Funct. Mater.*, 2009, **19**, 1379.
117. W. Geens, T. Martens, J. Poortmans, T. Aernouts, J. Manca, L. Lutsen, P. Heremans, S. Borghs, R. Mertens and D. Vanderzande, *Thin Solid Films*, 2004, **498**, 451.
118. D. Bartesaghi, M. Turbiez and L. J. A. Koster, *Org. Electron.*, 2014, **15**, 3192.
119. D. Bartesaghi and L. J. A. Koster, *Adv. Funct. Mater.*, 2015, **25**, 2013.

120. L. J. A. Koster, O. Stenzel, S. D. Oosterhout, M. M. Wienk, V. Schmidt and R. A. J. Janssen, *Adv. Energy Mater.*, 2013, **3**, 615.
121. S. D. Oosterhout, M. M. Wienk, S. S. van Bavel, R. Thiedmann, L. J. A. Koster, J. Gilot, J. Loos, V. Schmidt and R. A. J. Janssen, *Nat. Mater.*, 2009, **8**, 818.
122. S. S. van Bavel, E. Sourty, G. de With and J. Loos, *Nano Lett.*, 2009, **9**, 507.
123. J. J. M. van der Holst, F. W. A. van Oost, R. Coehoorn and P. A. Bobbert, *Phys. Rev. B: Condens. Matter Mater. Phys.*, 2011, **83**, 085206.
124. J. M. Frost, F. Cheynis, S. M. Tuladhar and J. Nelson, *Nano Lett.*, 2006, **6**, 1674.
125. S. Athanasopoulos, J. Kirkpatrick, D. Martínez, J. M. Frost, C. M. Foden, A. B. Walker and J. Nelson, *Nano Lett.*, 2007, **7**, 1785.
126. J. Kirkpatrick, V. Marcon, J. Nelson, K. Kremer and D. Andrienko, *Phys. Rev. Lett.*, 2007, **98**, 227402.
127. V. Rühle, J. Kirkpatrick and D. Andrienko, *J. Chem. Phys.*, 2010, **132**, 134103.

CHAPTER 14

Continuum-limit Modelling of Structure Evolution in Active Blends for Organic Solar Cells

JASPER J. MICHELS*[a] AND CHARLEY SCHAEFER[b,c]

[a] Max Planck Institute for Polymer Research, Ackermannweg 10, 55128 Mainz, Germany; [b] Eindhoven University of Technology, P.O. Box 513, 5600 MB Eindhoven, The Netherlands; [c] Dutch Polymer Institute, P.O. Box 902, 5600 AX Eindhoven, The Netherlands

*Email: michels@mpip-mainz.mpg.de

14.1 Introduction

In the past two decades, the prospect of facile processing, device flexibility and transparency have stimulated the active research and development of organic photovoltaics (OPV). The active layer of solution-processed organic solar cells (OSCs) consists of a blend of an electron donor and an electron acceptor, which form a bulk heterojunction upon phase separation during solution-processing. The donor and acceptor species can either be polymeric or monomeric, the most often encountered combination being a polymeric donor and a fullerene-based acceptor (usually phenyl-C_{61}-butyric acid methyl ester ($PC_{61}BM$) or $PC_{71}BM$). It has been widely recognised that, for efficient charge separation, the average domain size of the donor:acceptor morphology should not significantly exceed the exciton diffusion length (typically ~10 nm). Conversely, mixing at a molecular level limits the device performance due to reduced charge transport rates.

RSC Energy and Environment Series No. 16
Unconventional Thin Film Photovoltaics
Edited by Enrico Da Como, Filippo De Angelis, Henry Snaith and Alison Walker

Published by the Royal Society of Chemistry, www.rsc.org

As the donor:acceptor phase morphology has a decisive influence on device performance, many experimental and theoretical studies target aspects concerning a priori prediction of the active layer morphology as a function of the material's properties and processing conditions. Experimentally, this brings about elaborate trial and error efforts leading to optimised procedures often unique to one specific blend system. On the theoretical side, dynamic models that allow accurate description of the length scales encountered in optimal morphologies, such as (coarse grained) molecular dynamics and kinetic Monte Carlo methods, do not access time and length scales compatible with real-time solution-casting.

Continuum-limit modelling, on the other hand, does allow for the description of macroscopic time and length scales but sacrifices too much molecular detail to accurately address dynamics at length scales comparable to the exciton diffusion length. Nevertheless, one mode of dynamic structure formation in (solvent-borne) blends that can be described very well in the continuum limit, *e.g.*, using square gradient theory, is liquid-phase demixing. As we shall see, this mode is frequently observed during casting of polymer:fullerene blends, and yields domains of "macroscopic size", typically in the range 0.1–10 μm, depending on the stage in which the morphology is kinetically trapped by divergence of the viscosity. As liquid-phase demixing is dominated by interfacial tension, the resulting morphological features are usually rounded or droplike, either bicontinuous or dispersed. Depending on the sign of the curvature in the mixing free energy of the blend, demixing occurs spontaneously *via* spinodal decomposition or activated *via* nucleation and growth.

Donor:acceptor morphologies resulting from liquid phase demixing usually yield domain sizes that significantly exceed the exciton diffusion length. For this reason, it is of paramount importance to understand: (i) under what conditions liquid phase demixing dominates the dynamic behaviour of the blend, and (ii) how it is influenced by physical parameters such as mixing free energy, diffusivity, and evaporation rate. Only *via* prior understanding of these aspects can one sensibly predict strategies towards suppressing liquid-phase demixing in order to obtain more intimately phase-separated morphologies.

Section 14.2 gives several examples of liquid-phase demixing encountered in solution-processed polymer:fullerene blends. In Section 14.3, square gradient theory is summarised, together with transport dynamics and its application to modelling morphology formation in polymer:PCBM blends. Special focus is placed on: (i) the effect of evaporation on the emerging length scale during spinodal decomposition, and (ii) the relation between blend phase behaviour and OPV cell operation.

14.2 Experimental Observation of Liquid-phase Demixing in Polymer:PCBM Blends

Prior to deposition, the donor:acceptor blend solution represents a stable single phase, typically containing between 1 and 5 wt% solids. Optimal donor : acceptor ratios depend on the molecular structure and directly relate to the phase behaviour of the blend. Poly(*p*-phenylene-vinylene) (PPV) : $PC_{61}BM$-based systems require a 1 : 4 (w/w) ratio for optimal device performance.[1–4] More recent systems usually require a higher relative amount of polymer with optimal ratios of 1 : 0.8 or 1 : 1 (w/w) for P3HT : $PC_{61}BM$[5–7] and 1 : 2 (w/w) for low bandgap polymers, such as diketopyrrolopyrrole (DPP)-based polymers.[8] During film casting (*e.g.*, by spin-coating) the solvent evaporates at a rate that depends on the material properties and processing conditions. Generally, the increase in solid content due to solvent evaporation drives the system into some coexistence region, either leading to liquid–liquid or liquid–solid demixing.

Despite the fact that it was not until recently that the OPV community acknowledged that liquid-phase demixing is indeed frequently observed in solution-processed polymer : PCBM blends,[9–11] a substantial bulk of much earlier results already showed morphologies that seem to have resulted from binodal or spinodal decomposition during casting. Over the past twenty years, morphologies characterised by rounded, droplet-like PCBM domains embedded in a polymer-rich matrix have been reported for a wide range of blends, including systems based on traditional semiconducting polymers, such as PPV's,[3,4,12] poly(9,9-dioctylfluorenyl-2,7-diyl) (F8),[9,12,13] and poly-[(9,9-dioctylfluorenyl-2,7-diyl)-*alt-co*-(*N*,*N'*-diphenyl-*N*,*N'* di(*p*-butylphenyl)-1,4-diaminobenzene)] (PFB),[9,13] as well as mixtures containing low bandgap polymers, such as diketopyrrolopyrrole,[8,14] thienopyrroledione,[15] thienothiophene,[16–18] iso-indigo,[19,20] benzodifuran,[21] Si-dithienyl,[22] and thiophene-quinoxaline.[23]

Figure 14.1 shows a representative selection of polymer : PCBM morphologies formed *via* liquid-phase demixing. Usually, the PCBM domains account for significant height variation, whereby the polymer-dominated phase seems to have collapsed around the domains, sometimes exhibiting wetting layers at the substrate and vapor interfaces (Figure 14.1d,e). Occasionally a deviation from the "ideal" droplet-like shape (see *e.g.* Figure 14.1c) is observed, probably due to the occurrence of viscoelastic forces during demixing which are especially prominent if there is strong dynamic asymmetry between the components in different phases,[24] a feature not inconceivable for polymer:PCBM blends.

The fact that liquid-phase demixing is consistently observed when the blend is processed from a single solvent having good compatibility with the polymer suggests that it occurs at the expense of competing processes such as aggregation or crystallisation.[10] It is further noted that the domain size increases with the decreasing polymer : PCBM solids ratio (see Figure 14.2, left).[9] Besides lateral structure formation, demixing in polymer : PCBM blends has also shown to be surface-directed, as observed for the fluorene-dithienyl-benzothiadiazole copolymer APFO-3, yielding vertically resolved morphologies (see Figure 14.2, right).[25]

Increasing the spin-coating speed while maintaining the solids concentration results in a reduction of the size of the PCBM domains (Figure 14.3).[10] For slow spinning (*i.e.* when the fluid film is thicker), the size

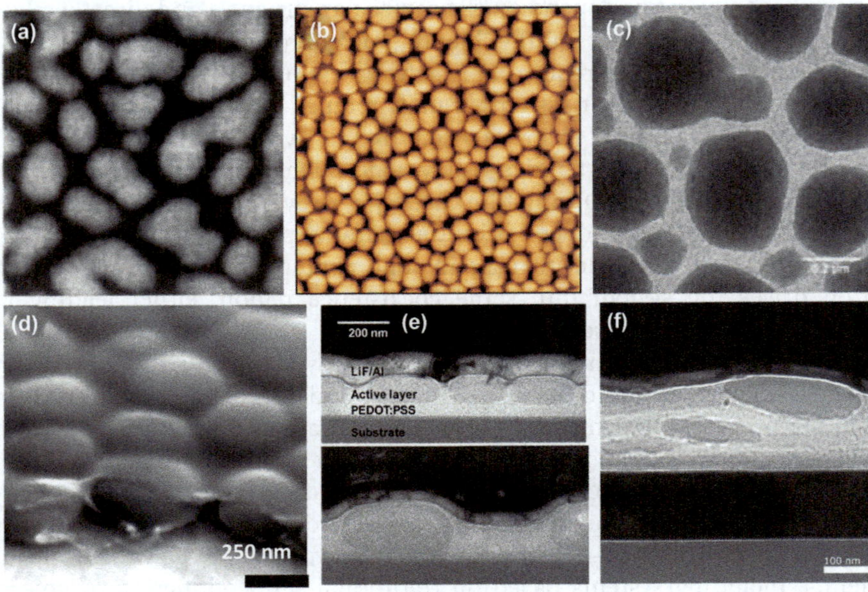

Figure 14.1 Representative selection of liquid-phase demixed polymer:PCBM morphologies; top row: top-view images, bottom row: cross-sectional images; (a) F8:$PC_{61}BM$ (1×1 μm AFM),[13] (b) TQ1:$PC_{71}BM$ (5×5 μm AFM),[23] (c) PDPPTPT:$PC_{71}BM$ (TEM),[14] (d) PTB7:$PC_{71}BM$ (SEM),[17] (e) PDPP-5T:$PC_{71}BM$ (TEM),[10] (f) Si-PDTBT:$PC_{71}BM$ (TEM).[22]
Figure 1(a) adapted from C. M. Bjorstrom *et al.*, Control of phase separation in blends of polyfluorene (co)polymers and the C60-derivative PCBM, *Synth. Met.*, **152**. Copyright (2005) with permission from Elsevier.[13] Figure 1(b) adapted from ref. 23 with permission from the Royal Society of Chemistry.[23] Figure 1(c) adapted from ref. 14 with permission from John Wiley and Sons © 2010 WILEY-VCH Verlag GmbH & Co. KGaA, Weinhiem.[14] Figure 1(d) adapted from ref. 17 with permission from NPG.[17] Figure 1(e) Adapted with permission from S. Kouijzer *et al.*, *J. Am. Chem. Soc.*, 2013, **135**, 12057. Copyright (2013) American Chemical Society.[10] Figure 1(f) J. S. Moon *et al.*, *Nano Lett.*, 2010, **10**, 4005. Copyright (2010) American Chemical Society.[22]

of the domains increases with an associated decrease in their number density. This observation strongly suggests that the blend components reside in a mobile state for a longer duration in case of a thick fluid film, allowing coarsening to proceed.

Several authors have reported that the formation of large domains can be avoided by use of a few percent of a high boiling additive, such as *o*-dichlorobenzene (*o*-DCB),[10] diiodooctane,[14,17,18] or chloronaphthalene.[22,23,26] The fact that phase separation becomes more intimate causes a rise in cell performance. The mechanistic details are usually not provided. Recently, this hiatus has been lifted by the work of van Franeker *et al.*, who used a combination of *in situ* light-interference, -scattering and -absorption to track the structure formation during drying of chloroform/*o*-DCB-based blends of

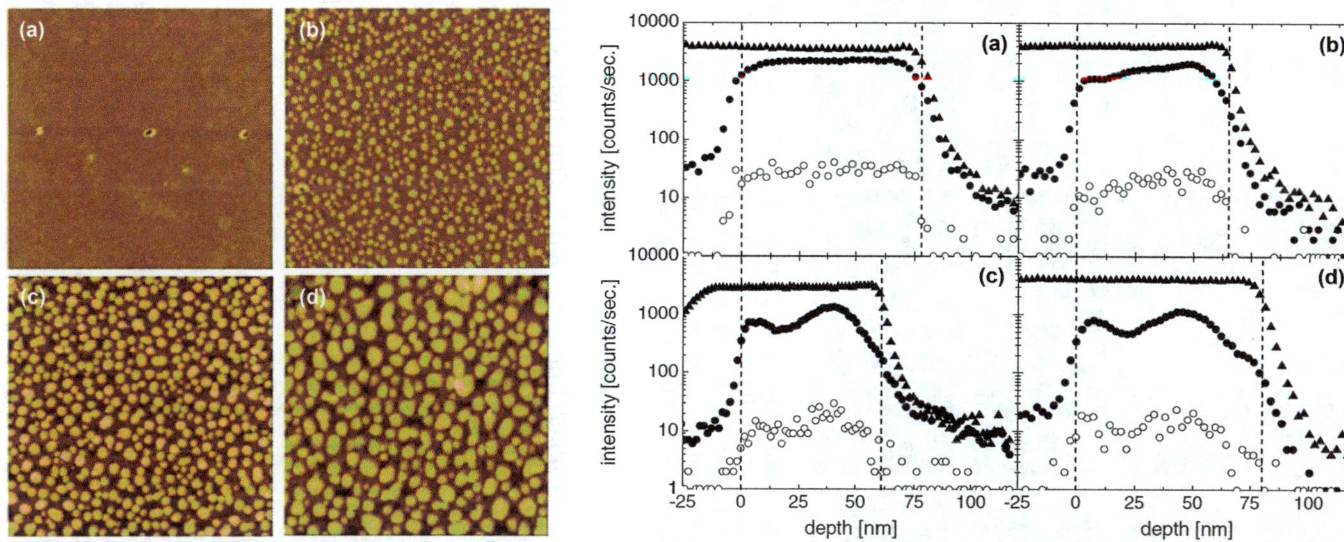

Figure 14.2 Left: AFM height images (5 μm×5 μm) of spin-casted PFB:$PC_{61}BM$ blends from xylene at blend ratios 1 : 1, 1 : 2, 1 : 3, 1 : 4 (a–d), adapted with permission from S. Nilsson, *et al.*, *Macromolecules*, **40**, 8291. Copyright (2007) American Chemical Society.[9] Right: SIMS traces of APFO-3:$PC_{61}BM$ films processed form CHCl3 at the same blend ratio's, adapted from ref. 25. © IOP Publishing. Reproduced with permission.

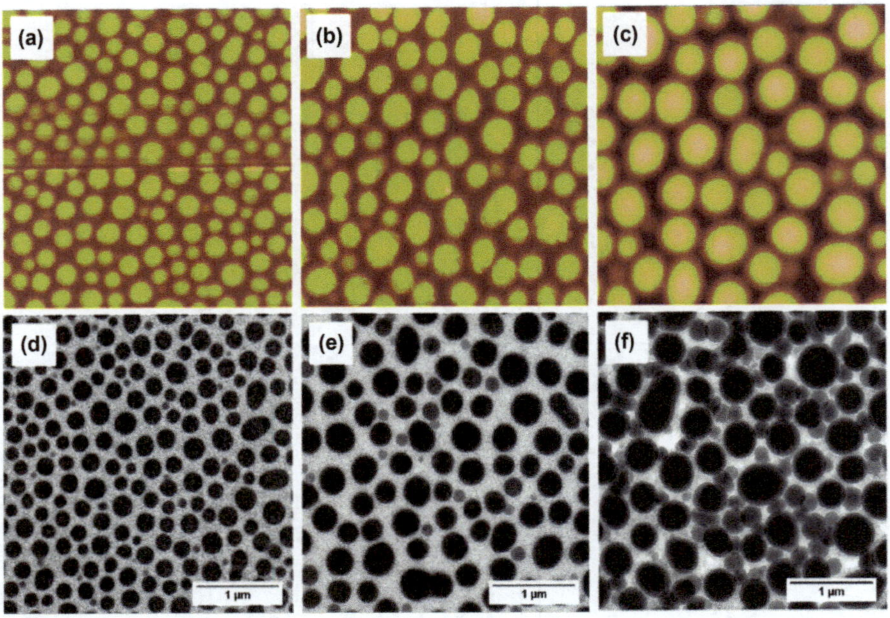

Figure 14.3 AFM (a, b, c) and TEM (d, e, f) images (3.5×3.5 µm) of 64 nm (a, d), 104 nm (b, e), and 180 nm (c, f) thick layers of PDPP-5T:$PC_{71}BM$ spin-coated from chloroform; height scale is 80 nm in panel (a) and 100 nm in panels (b) and (c), adapted with permission from S. Kouijzer *et al.*, *J. Am. Chem. Soc.*, **135**, 12057. Copyright (2013) American Chemical Society.[10]

a DPP-based polymer (PDPP-5T) and $PC_{71}BM$.[27] The study shows that, as *o*-DCB is a non-solvent for the polymer, evaporation of chloroform significantly suppresses solvent quality. As a result, polymer aggregation competes with liquid-phase demixing, whereby a strong tendency towards aggregation effectively frustrates demixing, probably due to gelation (see Figure 14.4). The authors show that the minimum amount of *o*-DCB required to fully prevent liquid-phase demixing coincides with saturation of the dependence of the light-to-power conversion efficiency on the initial additive fraction.

14.3 Continuum Modelling of Morphology Formation by Liquid-phase Demixing

One of the most popular theories employed for modelling liquid-phase demixing is square gradient theory. Square gradient theory has been known for several decades and has therefore been extensively reviewed. Hence, for reasons of space, this chapter provides a concise description of the theory and rather focusses on aspects and extensions that have so far shown to be of specific relevance to modelling liquid-phase demixing in solvent-borne OPV donor:acceptor blends.

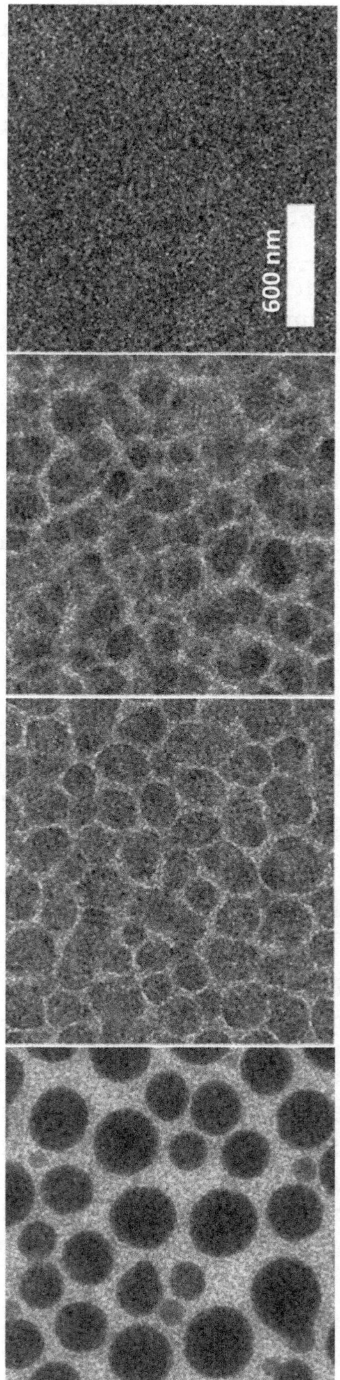

Figure 14.4 TEM mages of phase separated morphologies of PDPP5T:PC71BM processed from chloroform with an increasing amount of *o*-dichlofobenzene (from left to right: 0%, 1%, 2% and 3%), adapted by permission from Macmillan Publishers Ltd: *Nat. Commun.*, ref. 27. Copyright (2015).

14.3.1 Free Energy

Phase separation of a blend solution in a d-dimensional volume is accomplished by letting chemical potential gradients drive diffusive transport. The chemical potential μ_i is the functional derivative $\delta F/\delta\phi_i$ of the dimensionless free energy $\mathscr{F}[\phi_i,\phi_j,\phi_k,\ldots]$ with respect to the volume fraction ϕ. In units of thermal energy, the free energy is the volume integral:

$$\mathscr{F} = a^{-d}\int \mathbf{dr}(f_{\text{loc}} + f_{\text{nonloc}}) \tag{14.1}$$

The integrand of eqn (14.1) represents a sum of the local and nonlocal free energy per segment volume a^d. Analogous to the phenomenological treatment of the liquid–vapor interface by van der Waals,[28] Cahn and Hilliard[29] showed, for an incompressible blend of small molecules, that this sum of two contributions follows from a multivariate Taylor expansion of the local free energy in terms of the local composition and its spatial derivatives (typically up to second order).

If estimates for molecular size and interaction strength are available, it is convenient to invoke a dimensionless Flory–Huggins free energy density for the local contribution to $\mathscr{F}$, which for an N-component mixture reads:

$$f_{\text{loc}} = \sum_{i=1}^{N}\frac{\varphi_i}{L_i}\ln\phi_i + \sum_{i=1}^{N-1}\sum_{j>i}^{N}\chi_{ij}\phi_i\phi_j, \tag{14.2}$$

with L_i the degree of polymerisation of component $i = 1,\ldots,N$ in number of lattice segments and χ_{ij} the Flory parameter for binary interaction. The nonlocal term in eqn (14.1) represents a free energy contribution due to the formation of gradients in the composition:

$$f_{\text{nonloc}} = \frac{1}{2}\sum_{i,j=1}^{N-1}\kappa_{ij}\nabla\phi_i\nabla\phi_j. \tag{14.3}$$

Here, κ_{ij} are "stiffness" coefficients. As the mixture is assumed to be incompressible, the summation in eqn (14.3) runs over the $N-1$ independent components.

It is important to realise that the approach by Cahn and Hilliard assumes a binary blend of *monomeric* components (*e.g.*, a metal alloy). In case the blend contains (a) polymeric constituent(s), the square gradient approach only holds if composition gradients are weak on the scale of the size of the chains. Assuming unperturbed Gaussian chain statistics, the stiffness of an incompressible binary polymeric blend can be defined as:[30]

$$\kappa(\varphi) = \frac{a^2}{36\phi(1-\phi)}, \tag{14.4}$$

in which the prefactor of 1/2 has been absorbed. The composition-dependence of κ in eqn (14.4) is a result of the reduction in chain conformational entropy near the interface. Central to the derivation of eqn (14.4) is a Taylor expansion of the structure factor of the blend about $q=0$. As a result, eqn (14.4) is still only valid if concentration gradients are shallow on length scales comparable to the polymer chain size. If gradients are steep on the scale of the chain size, the influence of chain statistics becomes much more prominent.[31] It is noted that computational stability demands κ to be sufficiently large to avoid gradients becoming steep on the scale of the grid spacing of the discretising mesh assisting in the numerical integration of the dynamic equations that minimise the free energy. As long as this numerical constraint is satisfied, composition-invariant stiffness coefficents suffice to model structure evolution on a qualitative level.

14.3.2 Transport Dynamics

Liquid-phase demixing in polymer:PCBM blends has so far mostly been addressed by involving purely diffusive material transport, the effect of hydrodynamics and viscoelasticity in these particular systems being a possible subject for further study. Structure evolution during demixing is modelled by tracking local changes in the composition as a function of time by implementing the set of continuity equations:

$$\frac{\partial \phi_i}{\partial t} = -\nabla \mathbf{J}_i + \eta_i. \tag{14.5}$$

Here, $\mathbf{J}_i$ represents the current of the ith component and η_i is a noise term accounting for thermal fluctuations.[32] The Gaussian stochastic variable η_i is δ-correlated in time and subject to the following constraints:

$$\langle \eta_i(\mathbf{r},t)\rangle = 0,\ \langle \eta_i(\mathbf{r}',t')\eta_i(\mathbf{r},t)\rangle = -2\nabla_{\mathbf{r}}\cdot \Lambda_{ij}\nabla_{\mathbf{r}}\delta(\mathbf{r}'-\mathbf{r})\delta(t'-t), \tag{14.6}$$

where Λ_{ij} are mobility coefficients, as defined below. The noise term is spatially correlated inasmuch as mass conservation is guaranteed.

In order to describe diffusion in blends composed of species of different size and structure, it is customary to assume the mixture to be close to equilibrium locally, so that the current $\mathbf{J}_i$ can be linearly related to the thermodynamic driving force *via* the Onsager relation,

$$\mathbf{J}_i = -\lambda_i \nabla \mu_i, \tag{14.7}$$

wherein μ_i denotes the chemical potential and $\lambda_i = \phi_i L_i D_i$, the diagonal bare mobility coefficient, with D_i being the self-diffusivity. Usually, the cross mobilities that relate the mass current of component i to gradients in the chemical potential of component j are ignored. However, cross dependencies do come into play by assuming incompressibility, $\sum_{i=1}^{N} \mathbf{J}_i = 0$.

The incompressibility approximation leads to $N-1$ transport equations of the form:

$$\mathbf{J}_i = -\sum_{k=1}^{N-1} \Lambda_{ik} \nabla \mu_{kN}, \tag{14.8}$$

with $\mu_{kN} \equiv \mu_k - \mu_N$ being the so-called exchange chemical potential, obtained by taking the functional derivative $\mu_{kN} = \dfrac{\delta a^d \mathscr{F}}{\delta \phi_k}$, and Λ_{ij} being the effective mobility coefficients that are constituted by the bare Onsager coefficients. The chemical potentials can, according to the Gibbs–Duhem equation $\sum_{i=1}^{N} \phi_i \nabla \mu_i = 0$, be expressed as a function of these exchange chemical potentials giving

$$\nabla \mu_i = \nabla \mu_{iN} + \nabla \mu_N = \nabla \mu_{iN} - \sum_{j=1}^{N-1} \phi_j \nabla \mu_{jN}. \tag{14.9}$$

Two different forms of Λ_{ij} are obtained by modifying the Onsager equations such that they obey incompressibility.[33] In 'slow-mode' theory, movement of a constituent is counteracted by movement of another species in the opposite direction, *i.e.*, by exchange. On the other hand, according to 'fast-mode' theory, an advective current field is introduced and defined to be a result of instantaneously equilibrating vacant sites on the molecular lattice. The presence of vacancies suggests that compressibility is taken into account. However, the theory treats the "vacancies" as incompressible, so that in fact the current merely serves to counteract mass accumulation in an otherwise incompressible mixture.

First focusing on slow-mode theory, the Onsager equations are modified with an 'external' potential, U, that is experienced by all components, $\mathbf{J}_i = -\lambda_i \nabla(\mu_i + U)$.[34] As follows from the incompressibility constraint, this potential should obey $\nabla U = -\sum_{i=1}^{N} \lambda_i \nabla \mu_i / . \sum_{i=1}^{N} \lambda_i$. Inserting this into the modified Onsager equation, and subsequent substitution of the chemical potentials for exchange chemical potentials using eqn (14.9) gives $N-1$ mass currents as a function of the exchange chemical potentals. After some algebra, these currents are cast in the form of eqn (14.8), where Λ_{ii} and Λ_{ij} represent the diagonal and off-diagonal components of the mobility matrix, respectively, with

$$\Lambda_{ii} = \frac{\lambda_i \sum_{j=1, j\neq i}^{N} \lambda_j}{\sum_{k=1}^{N} \lambda_k}, \quad \text{and} \quad \Lambda_{ij} = \Lambda_{ji} = \frac{\lambda_i \lambda_j}{\sum_{k=1}^{N} \lambda_k}. \tag{14.10}$$

Hence, if the volume fraction of one of the components tends to zero, this results in $N-2$ transport equations of a mixture of $N-1$ components, as it should, so that in this limit the dynamics is determined by the other components. If, on the other hand, component N has a finite concentration and a mobility that by far exceeds the mobility of the other constituents, so that $\lambda_N \gg \lambda_{i,i\neq N}$, eqn (14.10) is reduced to:

$$\Lambda_{ii} = \lambda_i, \quad \Lambda_{ij} = \Lambda_{ji} = 0. \tag{14.11}$$

Fast-mode theory introduces an advective field $\mathbf{J}_\mathrm{v}$ that modifies the Onsager equations in eqn (14.7) into $\mathbf{J}_i = -\lambda_i \nabla \mu_i + \phi_i \mathbf{J}_\mathrm{v}$.[31] The way this field counteracts the accumulation of material becomes apparent by imposing the incompressibility constraint:

$$\mathbf{J}_\mathrm{v} = \sum_{i=1}^{N} \lambda_i \nabla \mu_i. \tag{14.12}$$

Substitution of this result in the modified Onsager equation yields:

$$\mathbf{J}_i = -\lambda_i \nabla \mu_i + \phi_i \sum_{i=1}^{N} \lambda_i \nabla \mu_i. \tag{14.13}$$

Finally, using eqn (14.9) to express the chemical potentials in terms of exchange chemical potentials and rearranging into the form of eqn (14.8) gives the fast-mode mobility coefficients:

$$\Lambda_{ik} = \phi_i \phi_k \sum_{j=1}^{N} \lambda_j + (\delta_{ik} - \phi_i)\lambda_k - \phi_k \lambda_i, \tag{14.14}$$

where δ_{ik} is the Kronecker delta.

The difference between slow- and fast-mode theory arises from the fact that, in the former, it is assumed that interdiffusion results from exchange of two species of which the kinetics is dictated by the slowest component, whereas in fast-mode theory the slow species are "dragged along" by the fast ones through advective motion. Not counterintuitively, slow-mode theory seems to hold for densely packed polymer blends close to the glass transition temperature (T_g), whereas fast-mode theory applies rather to highly mobile mixtures at temperatures far above T_g.[35] Hence, one may expect that during processing of solvent-borne polymer blends, during which the mixture approaches T_g from an initially highly mobile situation,[36] the kinetics exhibits a crossover from fast- to slow-mode dynamics. At present, this does not seem to have been verified by experiment or simulation.

14.3.3 Substrate Interaction

Due to the large surface-to-volume ratio of a thin liquid blend film on a solid support, the dynamics are usually significantly affected by phenomena arising from differences in substrate- and vapour-phase interactions between the mixture's constituents. Hence a complete model requires the bulk free energy to be supplemented with a surface contribution. Wetting and wetting transitions of polymeric fluids near a substrate have been reviewed extensively.[37–39]

A straightforward and often quoted formulation is by Schmidt and Binder who, in a manner consistent with the approach by Cahn and Hilliard, expand the surface free energy of a binary blend in terms of the local composition and compositon gradients at the substrate, arriving at a half-space free energy expressed as a sum of bulk ("b") and surface ("s") contributions:

$$\mathscr{F} = \mathscr{F}_{\mathrm{b}} + \mathscr{F}_{\mathrm{s}} = a^{-d}\int_{-\infty}^{\infty}\mathrm{d}x\int_{-\infty}^{\infty}\mathrm{d}y\int_{-\infty}^{\infty}\mathrm{d}z(f_{\mathrm{loc}} + f_{\mathrm{nonloc}}) + a^{1-d}\int_{-\infty}^{\infty}\mathrm{d}x\int_{-\infty}^{\infty}\mathrm{d}y(f_{\mathrm{loc}}^{\mathrm{s}} + f_{\mathrm{nonloc}}^{\mathrm{s}}), \tag{14.15}$$

If only nearest-neighbor interactions are considered, the surface concentrations rapidly relax to equilibrium so that nonlocal contributions are ignored.[40] What remains is a bare surface-free energy density ($f_{\mathrm{s}}^{(\mathrm{b})}(\phi_1)$) that quadratically depends on the local surface composition $\phi_1 \equiv \phi(z=0)$:[41]

$$f_{\mathrm{s}}^{(\mathrm{b})}(\phi_1) = -\mu_1\phi_1 - \frac{1}{2}g\phi_1^2. \tag{14.16}$$

The expansion coefficients μ_1 and g respectively represent a surface potential and an interaction parameter different from the bulk value due to the presence of the substrate.[40] By crude argumentation,[37] Jones has shown that μ_1 and g can be written in terms of the surface energy difference $\Delta\gamma = \gamma_{\mathrm{A}} - \gamma_{\mathrm{B}}$ between the two components and the bulk Flory parameter χ:

$$\mu_1 = \frac{b^3\Delta\gamma}{kT} + \frac{z'}{2z}b\chi, \quad g = -\frac{z'}{z}b\chi \tag{14.17}$$

Here, b is the length of an interacting lattice site and z and z' are the bulk and surface coordination numbers, respectively.

The surface energy difference $\Delta\gamma$ should be properly defined as the interfacial tension between fluid and substrate. Hence, it should be based not only on the chemistry of the blend components, but also on the functional groups at the substrate. In this respect, it seems convenient to define interaction parameters for the nearest neighbor interactions between monomeric units in the liquid and at the substrate. The same procedure might be followed at the vapour interface. A combination of this approach with Jones' "bond cutting arguments"[37] allows for the definition of substrate- ("S") and vapour ("v")-interaction parameters for a multi-component mixture:[11]

$$\chi_{i\mathrm{S}} = \frac{zb^2}{z'kT}\left[\gamma_i + \gamma_{\mathrm{S}} - 2\left(\sqrt{\gamma_i^{\mathrm{d}}\gamma_{\mathrm{S}}^{\mathrm{d}}} + \sqrt{\gamma_i^{\mathrm{p}}\gamma_{\mathrm{S}}^{\mathrm{p}}}\right)\right] \tag{14.18}$$

$$\chi_{i\mathrm{v}} = \frac{zb^2\gamma_i}{z'kT} \tag{14.19}$$

Following the approaches of Owens and Wendt[42] and Kaelbe,[43] superscripts d and p denote the disperse and polar contributions to the total (liquid–vapor, or solid–vapor) surface tension. Advantageously, this approach provides a practical and direct link between measured surface energy data and physical input data for the square gradient model.

14.3.4 Solvent Evaporation

Solvent evaporation strongly influences the morphology formation in solution-processed OPV blends, and is therefore an essential element to consider when modelling phase separation dynamics. A straightforward way to implement evaporation in 3D or 2D-cross-sectional geometries is to let the upper boundary of the computational domain ($z = h$) coincide with the liquid–vapour interface, and to apply a boundary condition that relates the evaporative flux to the decrease in height due to the loss of liquid material:

$$J_i^{\mathrm{v}}\big|_z = \Lambda_i \frac{\partial \mu_i}{\partial z} = \frac{\mathrm{d}h}{\mathrm{d}t}, \tag{14.20}$$

with Λ_i a collective diffusivity. The change in height is related to the evaporation rate *via*:

$$\frac{\mathrm{d}h}{\mathrm{d}t} = -k_{\mathrm{e}}\bar{a}_{\iota}, \tag{14.21}$$

with $\bar{a}_{\iota}$ the average solvent activity at the top of the film ($z = h$). Mass transfer coefficient k_{e} absorbs all circumstantial and geometrical contributions and can be controlled experimentally, *e.g.*, *via* the spin-coating speed.[44–46] Examples of the implementation of this treatment, or similar, can be found in ref. 47–50. Numerically, the reduction in vertical dimension can be accounted for by downscaling the grid spacing at each time step.[47,48] Alternatively, the need for mesh rescaling can be removed by solving the equations in a reference frame in which the vertical coordinate is scaled with respect to the height $h(t)$ of the film.[49]

Despite the fact that this treatment is severely limited in the sense that it does not allow for: (i) lateral gradients in the evaporation rate, and (ii) topology development; it provides general insight into how demixing is influenced by the ratio of the diffusive and evaporative time scales,[50] expressed by the dimensionless Biot number: $\mathrm{Bi} = k_{\mathrm{e}}h(t=0)/\Lambda_i$. In effect, if evaporation is fast compared to diffusion ($\mathrm{Bi} \gg 1$), vertical structure formation is dominant due to solute accumulation at the top of the liquid film. *Vice versa*, if $\mathrm{Bi} \ll 1$, diffusion easily counteracts evaporation-induced vertical composition gradients.

The above mentioned restrictions can, perhaps at the expense of computational tractability, be lifted by treating the vapour (gas) phase explicitly.[51] The liquid–vapour then becomes diffuse and is rendered susceptible to deformation under the influence of interfacial forces. Recently,[52] an approach has been proposed that combines both methods by rescaling the mesh elements in a columnar fashion, allowing for local vertical extension or contraction of the film upon lateral material transport. This treatment offers the possibility to integrate fully isotropic diffusive transport with lubrication-based hydrodynamics. Although it is outside the scope of this chapter to give a detailed account of free interface simulation, we do stress its importance as prediction of the surface topology is of interest to any technology relying on thin film solution-processing.

14.3.5 The Effect of Solvent Evaporation on Structure Size Evolution

As the exciton splitting efficiency in OPV cells strongly depends on the typical length scales of the donor:acceptor morphology, it is of particular interest to address the effect of evaporation on the characteristic feature size that emerges during phase separation.[53] In the limit $\mathrm{Bi} \ll 1$ and under the assumption of uniform solvent evaporation and compatibility, the free energy functional eqn (14.1) for a solvent:polymer:polymer ternary is effectively reduced to one for a binary mixture in a two-dimensional lateral space:[29,54]

$$F_\rho[\phi] = a^{-2}\int \mathbf{dr}\left(f_\rho + \frac{1}{2}\kappa_\rho|\nabla\phi|^2\right) \tag{14.22}$$

where ρ is the density of solid components and $\phi = \phi_A/(\phi_A + \phi_B)$ the fraction of component A in the dry blend. Upon evaporation, ρ increases until the blend is dry ($\rho = 1$). At the same time, concentration gradients are increasingly strongly penalised according to $\kappa_\rho = \rho^2\kappa$. The local free energy is given by:

$$f_\rho(\phi) = \rho[\phi\ln\phi/L_A + (1-\phi)\ln(1-\phi)/L_B + \phi(1-\phi)\rho\chi] \tag{14.23}$$

Here, $\rho\chi$ is the effective Flory interaction parameter, which also increases upon evaporation, so that the mixture becomes unstable and ultimately phase-separated. The fact that this free energy is similar to that of the simpler case of a single polymer in an evaporating solvent allows mapping of the early-stage solution of the latter model onto the former by choosing $\rho = 1$, so that the local free energy is reduced to eqn (14.2).

From the free energy functional eqn (14.1) and the continuity eqn (14.5), the following dynamic equation is obtained:

$$\frac{\partial\phi}{\partial t} = \nabla\cdot\lambda\nabla(\partial_\phi f - \kappa\nabla^2\phi) + \alpha + \eta \tag{14.24}$$

Here, ϕ now represents the polymer volume fraction and the slow-mode mobility coefficient $\lambda \approx \phi D_P$ assuming the mobility of the solvent to be much higher than that of the polymer. The evaporation rate α is taken constant. As an initial condition, the concentration field homogeneously adopts the value ϕ_0 at the low-concentration branch of the spinodal ($\partial_{\phi\phi}f(\phi_0) = 0$), for which it is assumed that the metastable region is traversed owing to a much shorter time scale of evaporation than that of nucleation. Hence, the mean concentration increases as $\phi = \phi_0 + \alpha t$, such that $\partial_{\phi\phi} f$ becomes negative, leading to spinodal decomposition.

Figure 14.5 shows representative morphologies obtained by solving eqn (14.24) numerically. During structure development, three kinetic regimes can be identified: (i) early-stage spinodal demixing, (ii) late-stage coarsening, and (iii) redissolution. Strikingly, and in stark contrast to spinodal decomposition in the absence of evaporation, as the mixture destabilises, the dominant wave number initially *increases* (solid blue symbols in Figure 14.5b). After a well-defined "lag time",[55] τ_L, rapid bifurcation takes place so that the concentrations (green symbols) approach the binodal values and the coarsening regime is entered (open blue symbols). The third kinetic regime, where all structural features vanish after the high-concentration branch of the binodal is reached, is irrelevant to binary blends in a solvent as the polymers remain immiscible at high concentrations.

In contrast, the cross-over between the first two regimes is universal and does carry over to phase separation of binary blends in a solvent. The fact that, for $t < \tau_L$, the concentration variations remain small allows for linearisation of eqn (14.24) to find analytical expressions to directly predict the emerging dominant length scale as a function of the evaporation rate, interaction parameter, degree of polymerisation and diffusivity. This linearisation,[28,53] followed by a Fourier transform, yields:

$$\ln \widehat{\delta\phi}(q,t)/\widehat{\delta\phi}(q,0) = -R(q,t)t, \tag{14.25}$$

wherein $\widehat{\delta\phi}(q,t)$ is the amplitude of concentration variations of wave number q at time t, and $R(q,t) = (q^2 D/a^2 t)\int_0^t dt'[\phi(t')\partial_{\phi\phi} f_{\text{loc}}(\phi(t')) + q^2\phi(t')\kappa]$ is the relaxation rate of these variations. By setting $\partial R/\partial q = 0$, one obtains an expression for the time-dependent predominant wave number:

$$q_*^2(t) = -\int_0^t dt'\phi(t')\partial_{\phi\phi} f_{\text{loc}}(\phi(t'))/2\kappa\int_0^t dt''\phi(t''), \tag{14.26}$$

where $\phi = \phi_0 + \alpha t$, with ϕ_0 the concentration at which the spinodal region is entered. As shown by the blue line in Figure 14.5b, eqn (14.26) shows an excellent agreement with the numerical results for $t < \tau_L$ but, expectedly, fails for $t > \tau_L$.

The lag time, τ_L, is obtained by inserting $q = q_*(t)$ and $t = \tau_L$ into $R(q,t)$ and eqn (14.25), and solving

$$\ln r_L = a^{-2}\kappa\lambda(\phi_0 + \alpha\tau_L/2)q_*^4(\tau_L)\tau_L \tag{14.27}$$

The fact that τ_L is a weak function of the amplitude of the concentration fluctuation at the lag time relative to that at time zero, $r_L \equiv \widehat{\delta\phi}(q_*(\tau_L),\tau_L)/\widehat{\delta\phi}(q_*(\tau_L),0)$ allows quantification of τ_L by using r_L as a free parameter to fit eqn (14.27) to the numerical results. Inserting the lag time so obtained into eqn (14.26) yields the predominant wave number at the lag time. The excellent agreement between the analytical estimates and the

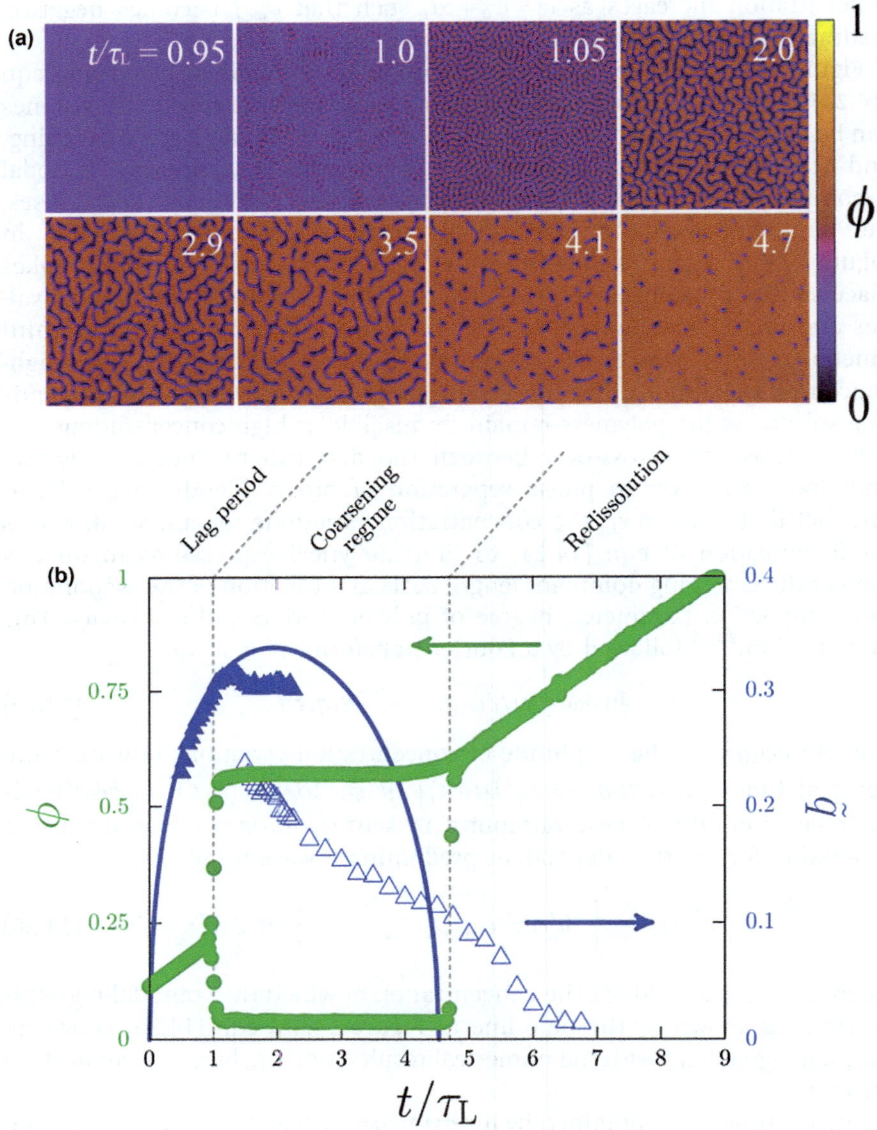

Figure 14.5 Modelled development of the morphology of a polymer:solvent mixture under evaporative conditions with parameters $\chi = 1$, $N = 10$, $\alpha = 5.5\times10^{-4}$; at $t = 0$ the spinodal is entered. (a) Concentration field of the morphology at different time intervals, and (b) volume fraction and dimensionless wave number plotted as a function of the time relative to the time at bifurcation; symbols represent numerical results, the drawn line represents eqn (14.27).
Reprinted figure with permission from C. Schaefer, P. van der Schoot and J. J. Michels, *Phys. Rev. E: Stat., Nonlinear Soft Matter Phys.*, 2015, **91**, 022602. Copyright (2015) by the American Physical Society.[53]

numerical results for lag time and the emerging morphological length scale is shown in Figure 14.6.

In Figure 14.6, wave number, lag time and evaporation rate are scaled appropriately to produce dimensionless quantities. The wave number is rendered dimensionless by multiplication with the square root of the stiffness, $\tilde{q}_* \equiv \sqrt{\kappa} q_*$. The lag time and evaporation rate are scaled by two different natural time scales obtained by linearising eqn (14.27) to yield $\tau_L^3 = 16(\kappa a^2/\phi_0\lambda)(\partial_{\phi\phi\phi} f_{\text{loc}}(\phi_0)\alpha)^2 \ln r_L$, which indeed shows that τ_L is a weak function of r_L. In this equation, $\tau_e \equiv 1/|\partial_{\phi\phi\phi} f_{\text{loc}}(\phi_0)\alpha|$ is a natural time scale measuring the rate at which the polymer solution destabilises due to evaporation, and $\tau_d \equiv \kappa a^2/\phi_0 \lambda$ is a time scale for self-diffusion. This result allows to define scaled parameters $\tilde{\alpha} \equiv \tau_d/\tau_e$ and $\tilde{\tau}_L \equiv \tau_L/\tau_d$ as well as universal scaling relations for the lag time

$$\tau_L \approx \tau_e^{2/3} \tau_d^{1/3} \propto \alpha^{-2/3} \tag{14.28}$$

and predominant wave number at the lag time

$$q_*(\tau_L) \approx \kappa^{-1/2} (\tau_d/\tau_e)^{1/6} \propto \alpha^{1/6}, \tag{14.29}$$

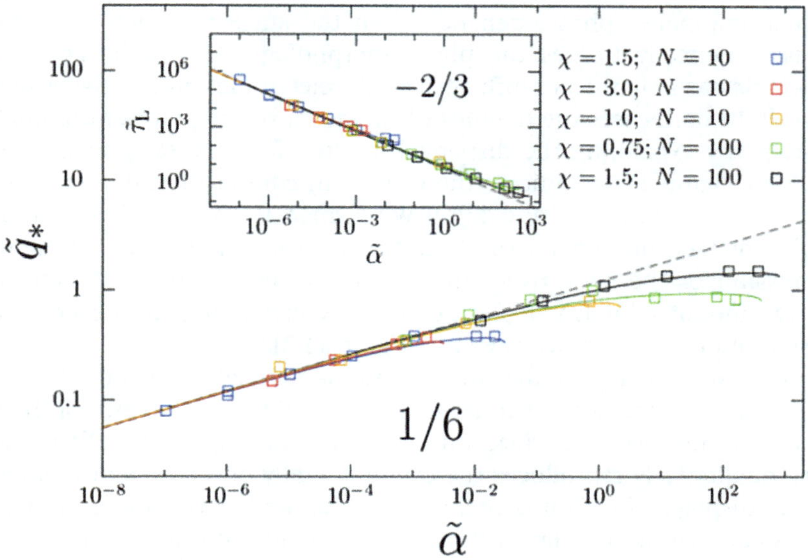

Figure 14.6 Dimensionless predominant wave number and lag time (inset) calculated as a function of the dimensionless evaporation rate for various values of the interaction parameter and the degree of polymerisation; symbols represent numerical results and solid lines result from linearised theory; a limiting power law is approached for slow evaporation (dashed lines).
Reprinted figure with permission from C. Schaefer, P. van der Schoot and J. J. Michels, *Phys. Rev. E: Stat., Nonlinear Soft Matter Phys.*, 2015, **91**, 022602. Copyright (2015) by the American Physical Society.[53]

where in the first approximate equalities a prefactor of order unity has been ignored. These scaling relations not only show how lag time and emerging morphology universally depend on the evaporation rate, but also on the two fundamental time scales of the problem, one set by diffusion and the other by the skewness of the free energy near the spinodal.

14.3.6 Polymer:PCBM Morphology Formation and Relation to OPV Device Performance

In order to model liquid-phase demixing in polymer:PCBM:solvent blends, input data is required for effective molecular size and interaction. Table 14.1 lists these parameters for the blends APFO-3:$PC_{61}BM$:$CHCl_3$ and PDPP-5T:$PC_{71}BM$:$CHCl_3$. The effective degrees of polymerisation are obtained by scaling molecular volumes to that of chloroform. Interaction parameters have been determined *via* surface energy measurements.[9]

Figure 14.7 shows the ternary phase diagrams for the blends mentioned above. The interaction between polymer and fullerene is more repulsive for DPP-5T:$PC_{71}BM$:$CHCl_3$ than for APFO-3:$PC_{61}BM$:$CHCl_3$, resulting in a smaller single phase region. The important consequence of the asymmetry in the binodal is the presence of a considerable fraction of PCBM in the polymer-dominated phase, depending on the stage at which the drying process kinetically freezes the phase morphology. As will be addressed below, this has a significant influence on the electronic characteristics of the OPV cell. In both cases the tie-lines tilt in favour of the polymer-dominated phase, partly explaining the dispersed nature of the PCBM domains. The arrows in Figure 14.7a indicate the change in composition due to evaporation for a 1 : 4, 1 : 3, 1 : 2, and 1 : 1 w/w polymer : PCBM ratio (left to right). Despite the fact that the length of the arrows is arbitrary, they indicate that suppression of the collective diffusivity is expected upon increasing the relative amount of polymer, *i.e.* qualitatively explaining the observed decrease in morphological feature size (Figure 14.2).

Numerically solving the dynamic equations for the two blends, either in 2D cross-section including surface-interactions (Figure 14.8 left), or 2D lateral in the absence of surface interactions assuming $\mathrm{Bi} \ll 1$ (Figure 14.8 right), qualitatively reproduces the experimentally observed morphological features displayed in Figures 14.2 and 14.3, as well as the above discussed aspects concerning binodal asymmetry and blend ratio. In the surface directed case, the pronunciation of the lamellar structure increases with the

Table 14.1 Flory–Huggins parameters for the blends APFO-3:$PC_{61}BM$:$CHCl_3$ and PDPP-5T:$PC_{71}BM$:$CHCl_3$.

	APFO-3	$PC_{61}BM$	$CHCl_3$		PDPP-5T	$PC_{71}BM$	$CHCl_3$
APFO-3	$L=66$	$\chi=0.4$	$\chi=0.5$	PDPP-5T	$L=89$	$\chi=0.9$	$\chi=0.4$
$PC_{61}BM$	$\chi=0.4$	$L=5$	$\chi=0.9$	$PC_{71}BM$	$\chi=0.9$	$L=7$	$\chi=0.9$
$CHCl_3$	$\chi=0.5$	$\chi=0.9$	$L=1$	$CHCl_3$	$\chi=0.4$	$\chi=0.9$	$L=1$

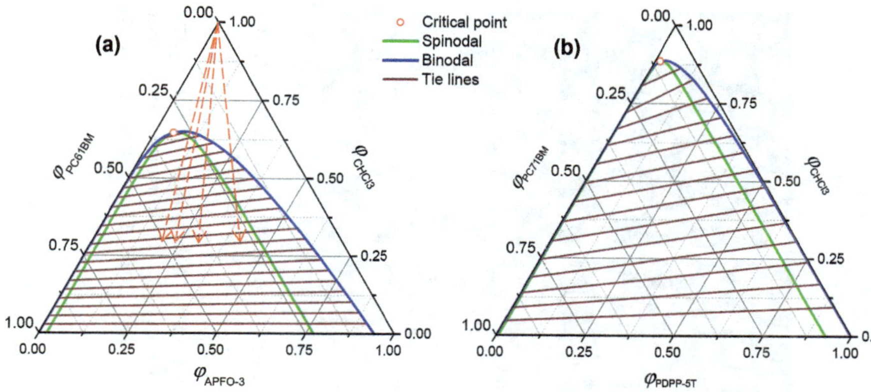

Figure 14.7 Ternary phase diagrams of the blends APFO-3:PC61BM:$CHCl_3$ (a) and PDPP-5T:PC71BM:$CHCl_3$ (b) calculated using Flory–Huggins theory. The arrows in (a) correspond to blends with a 1 : 4, 1 : 3, 1 : 2, and 1 : 1 (w/w) polymer : PC61BM blend ratio.
Figure 7(a) adapted with permission from J. J. Michels and E. Moons, *Macromolecules*, 2013, **46**, 8693. Copyright (2013) American Chemical Society.[11] Figure 7(b) adapted with permission from S. Kouijzer *et al.*, *J. Am. Chem. Soc.*, 2013, **135**, 12057. Copyright (2013) American Chemical Society.[10]

decreasing (relative) polymer content, whereas in the lateral situation, the PCBM concentration in the continuous phase decreases as long as demixing is allowed to proceed.

The modelled change in $PC_{71}BM$ concentration in the continuous PDPP-5T phase is reflected by the shape of the measured $J-V$ curves (Figure 14.9a). In thick PDPP-5T:$PC_{71}BM$ films, excitons predominantly dissociate at the interface between the two relatively pure phases (Figure 14.9b, left), resulting in the facile collection of charges giving a high fill factor. In thin blend layers, charges are also generated in the PDPP-5T-rich matrix because it contains an appreciable amount of $PC_{71}BM$ (Figure 14.9b, right). As a result, negative charges created in the continuous phase are more difficult to collect due to the absence of percolating pathways. These charges hence experience a strong electric field-dependent mobility, resulting in a photocurrent that increases with reverse bias.

Notably, besides for modelling liquid-phase demixing in polymer : PCBM blends, the square gradient method has also been used as a low-computational cost method to generate 'effective' morphologies that represent more or less optimised structures, *i.e.* containing feature sizes smaller than physically accessible with the square gradient method but also than typically expected for liquid-phase demixing. These studies do not consider solvent evaporation, but rather assume spinodal phase separation to take place during annealing of donor:acceptor blends that are symmetric in terms of molecular size. OPV cell performance is subsequently predicted using either Monte Carlo charge transport simulation[56,57] or drift-diffusion

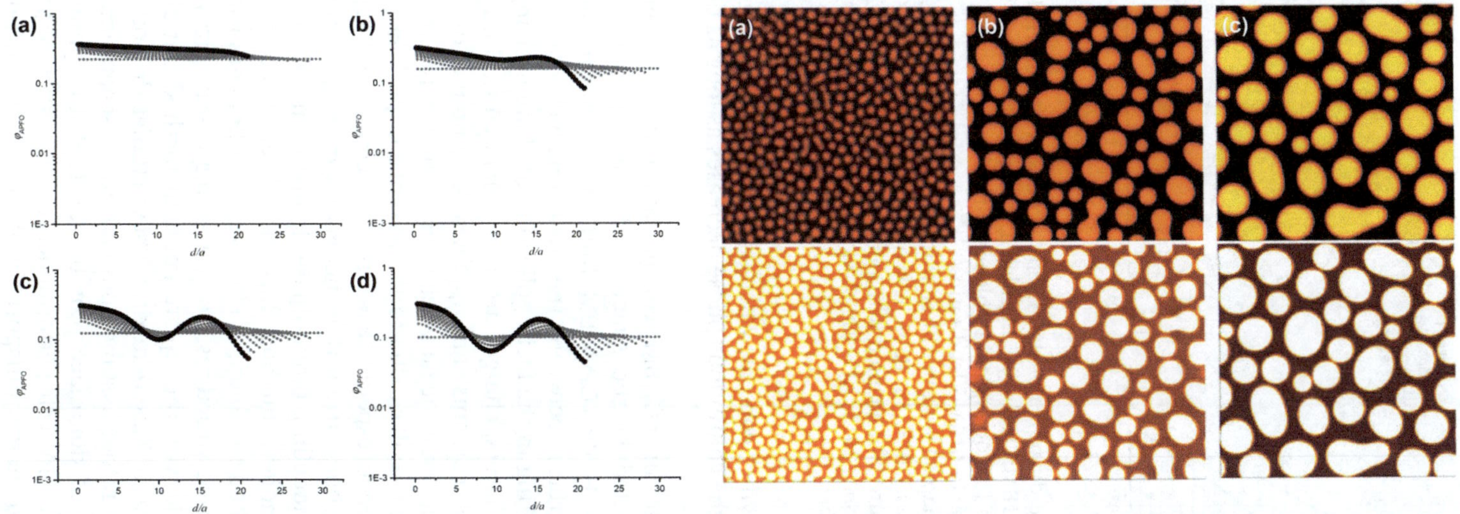

Figure 14.8 Left: cross-sectional 2D numerical simulation of the surface-directed spinodal decomposition of APFO-3 : $PC_{61}BM$: $CHCl_3$ with solids ratios 1 : 1, 1 : 2, 1 : 3, and 1 : 4 w/w (a–d), adapted with permission from J. J. Michels and E. Moons, *Macromolecules*, 2013, **46**, 8693. Copyright (2013) American Chemical Society.[11] Right: 2D topview numerical simulation of the spinodal decomposition of PDPP-5T:$PC_{71}BM$:$CHCl_3$, plotting the $PC_{71}BM$ volume fraction as a function of the spatial coordinate for three sequential time intervals (a–c); top: color scale interpolating $0 < \phi_{PCBM} < 1$, bottom: color scale interpolating $0 < \phi_{PCBM} < 0.1$, adapted with permission from S. Kouijzer *et al.*, *J. Am. Chem. Soc.*, 2013, **135**, 12057. Copyright (2013) American Chemical Society.[10]

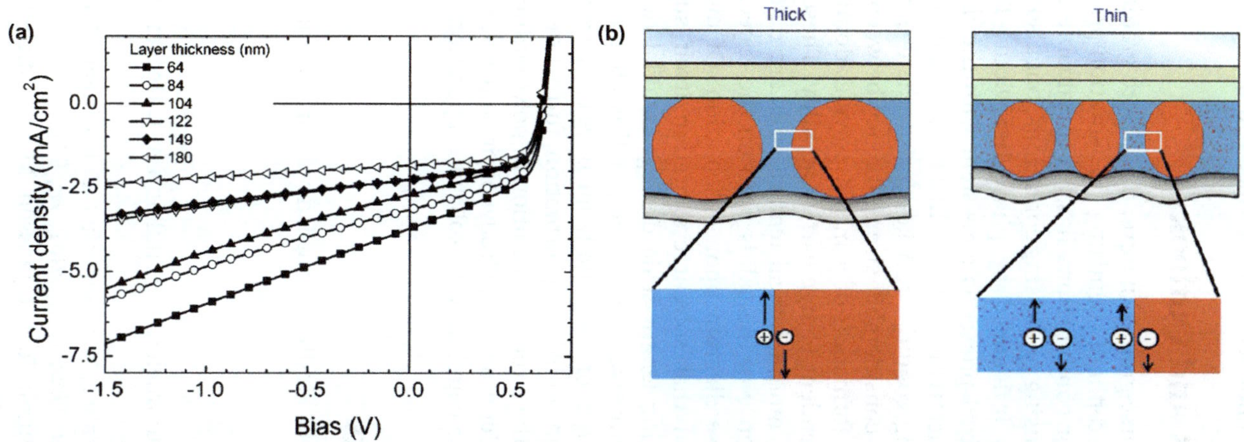

Figure 14.9 (a) $J-V$ characteristics as a function of the layer thickness for OPV cells based on PDPP-5T:PCBM blend films processed from chloroform; (b) charge generation and collection in thick and thin chloroform-processed blend layers of PDPP-5T (blue) and $PC_{71}BM$ (red).
Adapted with permission from S. Kouijzer *et al.*, *J. Am. Chem. Soc.*, 2013, **135**, 12057. Copyright (2013) American Chemical Society.[10]

modelling[58,59] in these computationally generated morphologies. Despite its somewhat unphysical nature, this methodology has allowed for the definition of general design rules related, for instance, to interface enrichment, phase composition, and donor:acceptor interface width.

14.4 Conclusions and Outlook

Despite its drastically coarse-grained nature, the square gradient approach towards modelling liquid-phase demixing in solution-processed polymer:PCBM donor:acceptor combinations has led to valuable new insights, not only relevant to optimising processing conditions, but also to explaining how device operation is influenced by phase behaviour. The modelled composition of the demixed phases is consistent with the dependence of the current–voltage characteristics on the thickness of the active layer, as determined by the casting conditions. A key feature here is that entropic arguments related to the considerable size difference between the polymer and fullerene favour the formation of a mixed polymer-dominated phase, giving rise to a field-dependent charge carrier mobility. These same calculations demonstrate that morphological feature sizes are determined to a large extent by the influence of the solids ratio on the trajectory followed through the ternary phase diagram upon evaporation. Of more profound nature is the observation that solvent evaporation itself suppresses the magnitude of the emerging length scale during spinodal decomposition. Scaling relations show universal dependence of this characteristic length scale and demixing kinetics on essentially all physical parameters in the system. Besides simulations of topology formation, further developments on continuum modelling of the structure evolution in polymer:fullerene blends could include the consideration of hydrodynamic and viscoelastic effects during demixing, with the coupling with solvent evaporation continuing to receive special attention.

Acknowledgements

Prof. P. van der Schoot is kindly acknowledged for critically reading the manuscript.

References

1. G. Yu, J. Gao, J. C. Hummelen, F. Wudl and A. J. Heeger, *Science*, 1995, **270**, 1789.
2. S. E. Shaheen, C. J. Brabec, N. S. Sariciftci, F. Padinger, T. Fromherz and J. C. Hummelen, *Appl. Phys. Lett.*, 2001, **78**, 841.
3. X. Yang, J. K. J. van Duren, R. A. J. Janssen, M. A. J. Michels and J. Loos, *Macromolecules*, 2004, **37**, 2151.
4. H. Hoppe, T. Glatzel, M. Niggeman, W. Schwinger, F. Schaeffler, A. Hinsch, M. C. Lux-Steiner and N. S. Sariciftci, *Thin Solid Films*, 2006, **511–512**, 587.

5. M. Al-Ibrahim, O. Ambacher, S. Sensfuss and G. Gobsch, *Appl. Phys. Lett.*, 2005, **86**, 201120.
6. M. Reyes-Reyes, K. Kim and D. L. Carroll, *Appl. Phys. Lett.*, 2005, **87**, 083506.
7. W. Ma, C. Yang, X. Gong, K. Lee and A. J. Heeger, *Adv. Funct. Mater.*, 2005, **15**, 1617.
8. M. M. Wienk, M. Turbiez, J. Gilot and R. A. J. Janssen, *Adv. Mater.*, 2008, **20**, 2556.
9. S. Nilsson, A. Bernasik, A. Budkowski and E. Moons, *Macromolecules*, 2007, **40**, 8291.
10. S. Kouijzer, J. J. Michels, M. van den Berg, V. S. Gevaerts, M. Turbiez, M. M. Wienk and R. A. J. Janssen, *J. Am. Chem. Soc.*, 2013, **135**, 12057.
11. J. J. Michels and E. Moons, *Macromolecules*, 2013, **46**, 8693.
12. D. T. Duong, B. Walker, J. Lin, C. Kim, J. Love, B. Purushothaman, J. E. Anthony and T.-Q. Nguyen, *J. Polym. Sci., Part B: Polym. Phys.*, 2012, **50**, 1405.
13. C. M. Björström, K. O. Magnusson and E. Moons, *Synth. Met.,*, 2005, **152**, 109.
14. J. C. Bijleveld, V. S. Gevaerts, D. Di Nuzzo, M. Turbiez, S. G. J. Mathijsen, D. M. de Leeuw, M. M. Wienk and R. A. J. Janssen, *Adv. Energy Mater.*, 2010, **22**, E242.
15. C. M. Amb, S. Chen, K. R. Graham, J. Subbiah, C. E. Small, F. So and J. R. Reynolds, *J. Am. Chem. Soc.*, 2011, **133**, 10062.
16. B. A. Collins, Z. Li, J. R. Tumbleston, E. Gann, C. R. McNeill and H. Ade, *Adv. Energy Mater.*, 2013, **3**, 65.
17. G. J. Hedley, A. J. Ward, A. Alekseev, C. T. Howells, E. R. Martins, L. A. Serrano, G. Cooke, A. Ruseckas and I. D. W. Samuel, *Nat. Commun.*, 2013, **4**, 2867.
18. D. Huang, Y. Li, Z. Xu, S. Zhao, L. Zhao and J. Zhao, *Phys. Chem. Chem. Phys.*, 2015, **17**, 8053.
19. E. G. Wang, Z. Ma, Z. Zhang, K. Vandewal, P. Hendriksson, O. Inganäs, F. Zhang and M. R. Andersson, *J. Am. Chem. Soc.*, 2011, **133**, 14244.
20. R. Stalder, C. Grand, J. Subbiah, F. So and J. R. Reynolds, *Polym. Chem.*, 2012, **3**, 89.
21. C. Hu, Y. Fu, S. Li, Z. Xie and Q. Zhang, *Polym. Chem.*, 2012, **3**, 2949.
22. J. S. Moon, C. J. Takacs, S. Cho, R. C. Coffin, H. Kim, G. Bazan and A. J. Heeger, *Nano Lett.*, 2010, **10**, 4005.
23. R. Hansson, L. K. E. Ericsson, N. P. Holmes, J. Rysz, A. Opitz, M. Campoy-Quiles, E. Wang, M. G. Barr, A. L. D. Kilcoyne, X. Zhou, P. Dastoor and E. Moons, *J. Mater. Chem. A*, 2015, **3**, 6970.
24. H. Tanaka, *Phys. Rev. Lett.*, 1996, **76**, 787.
25. C. M. Björström, A. Bernasik, J. Rysz, A. Budkowski, S. Nilsson, M. Svensson, M. R. Andersson, K. O. Magnusson and E. Moons, *J. Phys.: Condens. Matter*, 2005, **17**, L529.

26. J. A. Bartelt, J. D. Douglas, W. R. Mateker, A. El Labban, C. J. Tassone, M. F. Toney, J. M. J. Fréchet, P. M. Beaujuge and M. D. McGehee, *Adv. Energy Mater.*, 2014, 1301733.
27. J. J. van Franeker, M. Turbiez, W. Li, M. M. Wienk and R. A. J. Janssen, *Nat. Commun.*, 2015, **6**, 6229.
28. J. D. van der Waals, in *On the Continuity of the Gaseous and Liquid State*, ed. J. S. Rowlinson, Dover Publications, New York, 1988.
29. J. W. Cahn and J. E. Hilliard, *J. Chem. Phys.*, 1958, **28**, 258.
30. P.-G. de Gennes, *Scaling Concepts in Polymer Physics*, Cornell University Press, Ithaca, 1979.
31. P. van der Schoot, *Macromolecules*, 2000, **33**, 8497.
32. H. E. Cook, *Acta Metall.*, 1970, **18**, 297.
33. E. J. Kramer, P. Green and C. J. Palmstrøm, *Polymer*, 1984, **25**, 473.
34. F. Brochard, J. Jouffroy and P. Levinson, *Macromolecules*, 1983, **16**, 1639.
35. E. Jabbari and N. A. Peppas, *Rev. Macromol. Chem. Phys.*, 1994, **C34**, 205.
36. M. Tsige and G. S. Grest, *J. Phys.: Condens. Matter*, 2005, **17**, S4119.
37. R. A. L. Jones and R. W. Richards, *Polymers at Surfaces and Interfaces*, Cambrigde University Press, Cambridge, UK, 1999.
38. M. Geoghegan and G. Krausch, *Prog. Polym. Sci.*, 2003, **28**, 261.
39. K. Binder, S. Puri, S. K. Daqs and J. Horback, *J. Stat. Phys.*, 2010, **138**, 51.
40. S. Puri and K. Binder, *Phys. Rev. E: Stat. Phys., Plasmas, Fluids, Relat. Interdiscip. Top.*, 1994, **49**, 5359.
41. I. Schmidt and K. Binder, *J. Phys.*, 1985, **46**, 1631.
42. D. K. Owens and R. C. Wendt, *J. Appl. Polym. Sci.*, 1969, **13**, 1741.
43. D. H. Kaelbe, *J. Adhes.*, 1970, **2**, 66.
44. D. E. Bornside, W. W. Macosco and L. E. Scriven, *J. Electrochem. Soc.*, 1991, **138**, 317.
45. D. T. W. Toolan, E. ul Haq, A. Dunbar, S. Ebbens, N. Clarke, P. D. Topham and J. R. Howse, *J. Polym. Sci., Part B: Polym. Phys.*, 2013, **51**, 875.
46. J. J. van Franeker, D. Westhoff, M. Turbiez, M. M. Wienk, V. Schmidt and R. A. J. Janssen, *Adv. Funct. Mater.*, 2015, **25**, 855.
47. G. A. Buxton and N. Clarke, *EPL*, 2007, **78**, 56006.
48. C.–S. Kim, D. M. Saylor, M. K. McDermott, D. V. Patwardan and J. A. Warren, *J. Biomed. Mater. Res., Part B*, 2009, **90**, 688.
49. D. M. Saylor, J. E. Guyer, D. Wheeler and J. A. Warren, *Acta Biomater.*, 2011, **7**, 604.
50. O. Wodo and B. Ganapathysubramanian, *Comput. Mater. Sci.*, 2012, **55**, 113.
51. H. Morita, T. Kawakatsu and M. Doi, *Macromolecules*, 2001, **34**, 8777.
52. S. Coveney and N. Clarke, *Phys. Rev. Lett.*, 2014, **113**, 218301.
53. C. Schaefer, P. van der Schoot and J. J. Michels, *Phys. Rev. E: Stat., Nonlinear Soft Matter Phys.*, 2015, **91**, 022602.
54. J. D. Gunton, M. San Miguel and P. S. Sahni, in *Phase Transitions and Critical Phenomena*, ed. C. Domb and J. L. Lebowitz, Academic, New York, 1983, vol. 8.

55. Not to be confused with the lag time from nucleation theory, see: D. Kashchiev, *Nucleation: Basic Theory with Applications*, Butterworth-Heinemann, Oxford, 2000.
56. B. P. Lyons, N. Clarke and C. Groves, *J. Phys. Chem. C*, 2011, **115**, 22572.
57. B. P. Lyons, N. Clarke and C. Groves, *Energy Environ. Sci.*, 2012, **5**, 7657.
58. B. Ray, P. R. Nair and M. A. Alam, *Sol. Energy Mater. Sol. Cells*, 2011, **95**, 3287.
59. B. Ray and M. A. Alam, *Sol. Energy Mater. Sol. Cells*, 2012, **99**, 204.

Subject Index